NATIONAL GEOGRAPHIC

ALMANAC OF GEOGRAPHY

CONTRIBUTING GEOGRAPHERS

Chief Consultant: Roger M. Downs, Ph.D.
Professor and Head, Department of Geography,
Pennsylvania State University

Frederick A. Day, Ph.D.
Professor of Geography, Department of Geography,
Southwest Texas State University

Paul L. Knox, Ph.D.
University Distinguished Professor and Dean,
College of Architecture and Urban Studies,
Virginia Polytechnic Institute and State University

Peter Haynes Meserve, Ph.D.
Associate Professor of Geography and Geology,
Department of Natural Sciences and Mathematics;
Head, Environmental Studies,
Columbia College, Missouri

Barney Warf, Ph.D.
Professor and Chair,
Department of Geography, Florida State University

NATIONAL GEOGRAPHIC
ALMANAC OF GEOGRAPHY

NATIONAL GEOGRAPHIC

WASHINGTON, D.C.

CONTENTS

INTRODUCTION

GEOGRAPHY is one of the oldest sciences. Among the earliest geographers were Greeks who lived more than 2,000 years ago. In fact, the term "geography" comes from the Greek words *geo* and *graphia* meaning "to write about or describe Earth." Writing about or describing Earth has always been an important aspect of geography. When the National Geographic Society was founded in 1888 by educators, explorers, cartographers, geographers, and others, its mission was to increase and diffuse geographic knowledge. National Geographic's mission was built upon the goals of the first geographers—mapmakers, scientists, explorers who, in past centuries, charted unknown areas of the world. (See the comprehensive Geographers and Explorers chart on pages 12–17.) To both document their journeys and to record the dimensions and features of Earth, these early explorers made maps. Although the ancient Greeks created the science of mapmaking when they developed the first system of latitude and longitude, maps were used in many cultures: Polynesians wove networks of palm fibers showing wave patterns; Inuit fishermen carved pieces of driftwood to show coastal features; explorers in Columbus's time drew on parchment.

Mapping is a cornerstone of geography and a major element in the National Geographic *Almanac of Geography*. In these pages you will find almost 300 maps. They reflect the mission of geographers through many eras. The early geographer-explorers served rulers seeking to colonize new lands and establish trade routes. Their maps show early renderings of Earth and the continents. Present-day geographers render changes in Earth and the continents, too. But they also address other subjects: soils, migration, cultural identities and diffusion, specialization and trade in the world economy. Geographers help plan the layouts of new suburbs and the revitalization of cities. They help manage and conserve natural resources. They study voting patterns and the spread of disease. These topics are all covered in the text, photos, charts, and diagrams in this book, and reveal the importance of geographic knowledge to the preservation and enjoyment of life in the 21st century. The creation of this almanac is the National Geographic Society's way of providing insight for you, the reader, into this world of connections that is the world of modern geography.

—The Editors

To create this detailed image of Earth from space, scientists and visualizers used a collection of satellite-based observations to stitch together a true color mosaic of the planet's land surface, oceans, sea ice, and clouds.

GEOGRAPHY PAST & PRESENT

In this SeaWiFS satellite composite mosaic, high concentrations of chlorophyll from phytoplankton in the oceans range from green to yellow to red. On land, tan denotes little or no vegetation; dark green indicates dense vegetation.

WHAT IS GEOGRAPHY?

Geography gives us a way to look at Earth as a whole—both the physical world and its people. Geography is much more than the name of the world's highest mountain or longest river or the capital of Brazil. As a science, it seeks to understand where things are and how and why they got there by studying—with an emphasis on location—the connections and interactions among people, places, and environments. Geography looks at how people and their activities adapt to and change their surroundings. To do this, geography must draw upon and integrate the knowledge of many different disciplines.

Geographers divide this broad field of study into physical geography and human geography. Physical geography draws on the sciences of geology, climatology, biology, ecology, hydrology (the study of water), and other natural sciences. Human geography encompasses cultural anthropology, economics, political science, history, demography, and other social sciences. Geography links these disciplines to determine why things happen in a particular location. Cartography, the art and science of mapmaking, provides geographers with their most basic tool—a graphic representation of a geographic setting. Geographers also use other tools, including photographs, remotely sensed images, and computer generated graphics to understand the world in which we live.

More than ever, human activities—for example, clearing rain for-

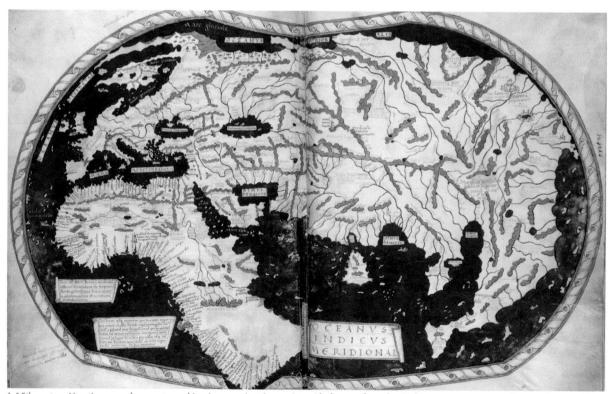

A 15th-century Venetian map shows geographic misconceptions increasing with distance from the Mediterranean. A century of successful European voyages of exploration saw dramatic improvements in map accuracy and a proliferation of globe- and mapmaking.

A GEOGRAPHICALLY INFORMED PERSON SEES AND UNDERSTANDS THE WORLD FROM FOUR PERSPECITVES:

- The **spatial perspective:** Geography is concerned with the essential issue of *"whereness."* The spatial perspective focuses on specific questions—such as Where is it? Why is it there?—in order to establish a context of spatial relationships where the human story unfolds.

- The **ecological perspective:** Human societies depend on diverse small and large ecosystems for food, water, and all other resources. The ecological perspective leads us to inquire about connections and relationships among life-forms, ecosystems, and human societies.

- The **historical perspective:** Questions—When? Why then? Why is the event significant?—help us gain a clearer understanding of past and contemporary events.

- The **economic perspective:** Diverse peoples are connected through trade in goods and services. Previously isolated economies are entering the global economy, and technological changes accelerate transportation and communication. The economic perspective helps us recognize the increasing interdependence among all societies in a world where decisions made in distant places regularly impact local economies.

est to make room for farmland—have direct consequences, both local and global, that affect the planet in ways that may not be immediately apparent. Geography gives us a framework for understanding those connections so that we can make informed decisions about them.

Geography also exposes us to worlds different from our own so we can see and appreciate their richness and beauty. This knowledge allows us to view Earth and its people through focused lenses, opening our minds to different points of view and ways of life. When we understand how humans and Earth's physical forces and places are intertwined and interdepend-ent—and constantly changing—the world becomes an even more fascinating place.

HISTORY OF GEOGRAPHY

The word "geography" comes from the Greek words *geo* and *graphia*, meaning "to write about or describe Earth." In ancient Greece, geography often meant tales of travel, adventure, and discovery such as Homer's *Iliad* and *Odyssey*, written in the ninth century B.C. Although Homer's epics survive, few accounts of travels by earlier peoples—the Phoenicians, Egyptians, Chinese, Polynesians, for instance—remain.

Beginning with ancient Greeks such as Aristotle, who attempted to explain the size and nature of Earth scientifically, curiosity about geography has been an important stimulus in the development of modern science. Later, the discoveries of European explorers in the 16th and 17th centuries reinforced the value of direct observation, accurate measurements, and the collection of specimens—a change in approach that countered long-standing and mystical beliefs about Earth and its people. Today, geographers and other scientists use a variety of sophisticated technologies not only to reveal Earth's secrets but also to help us better understand this planet that we call home.

GEOGRAPHERS AND EXPLORERS

Explorers, travelers, and trailblazers who ventured into new lands and encountered new peoples were, in a sense, all geographers. The reports of their journeys and the knowledge they gained helped mapmakers, geologists, biologists, and anthropologists—as well as the general public—better understand Earth. Today geographers and other scientists continue the search for knowledge and understanding of Earth in all its dimensions. In the timeline that follows, representatives of the world's well-known and some lesser-known geographers and explorers appear in chronological order.

EXPLORER	DATES OF DISCOVERY	WHERE THEY WENT AND WHY
Phoenicians	3000–800 B.C.	World's earliest sailors built trading posts throughout the Mediterranean. First to use the North Star for navigation. Explored as far as Britain and possibly the Azores and Africa's Cape of Good Hope.
Ancient Egyptians	2500–1493 B.C.	Maritime traders in the Mediterranean by 2500 B.C. Searched east coast of Africa for ivory, ebony, silver, and gold.
Ancient Polynesians	1500 B.C.– A.D. 1000	About 1500 B.C., traveled from southeast Asia to Australia; settled Fiji by 1300 B.C., most of Oceania between A.D. 300 and 800 (Hawaii, A.D. 500; New Zealand, A.D.800).
Alexander the Great (Greece)	334–323 B.C.	Conquests carried him eastward to Afghanistan, the Hindu Kush, and India and southward to Egypt. Took with him geographers, engineers, botanists, and historians.
Pytheas of Massalia (Greece)	ca. 325 B.C.	Sailed from the Mediterranean around Spain to Britain, and possibly to Iceland or Norway, in search of tin and amber.
Eratosthenes (Egypt)	190–126 B.C.	Plotted the known world from the British Isles to Sri Lanka and from the Caspian Sea to Ethhiopia on most detailed and accurate maps of the time. Calculated a remarkably accurate estimate of Earth's circumference.
Chang Ch'ien (China)	ca. 138–123 B.C.	Explorations of central and western Asia helped establish the ancient Silk Road, the largely overland trade route linking Asia and Europe.
Strabo (Greece-Rome)	ca. 64 B.C–A.D.23	Authored the 17-volume *Geographica* detailing his era's knowledge of geography, history, science, mathematics, and politics.
Claudius Ptolemy (Egypt)	ca. A.D. 90–168	Viewed geography as the location of places and things. Wrote an 8-volume *Geography* giving coordinates for 8,000 places. His book served as an early atlas, featuring a world map and 26 regional maps, as well as tables of latitudes and longitudes. Christopher Columbus used Ptolemy's map, which underestimated the circumference of Earth by 25 percent, in his attempt to reach Asia by sailing west from Europe.
Erik the Red (Norway)	A.D. 986	Explored Greenland, proposing its colonization by settlers from overcrowded Iceland. His son Leif Eriksson continued the search for new lands and is thought to have touched Canada at Baffin Island, Labrador, and Newfoundland.
Al-Idrisi (Morocco)	ca. 1100–1165	Traveled extensively throughout the Mediterranean. Compiled a geography and 79 sheets of detailed and fairly accurate regional maps.

EXPLORER	DATES OF DISCOVERY	WHERE THEY WENT AND WHY
Marco Polo (Italy)	1254–1324	Traveled through Central and East Asia. Returned to Europe after 24 years and wrote about the peoples, sights, and cultures of Asia, providing Europe new insights into China.
Ibn Battuta (Morocco)	1304–1369	Journeyed from Spain to China and from Timbuktu to the steppes of Russia over three decades. Wrote a narrative account of his travels.
Ibn Khaldun (Tunisia)	1332–1406	Wrote about his theories of how societies rise, decline, and fall. Referred often to the relationship of humankind and the physical environment.
Prince Henry (Portugal)	1394–1460	Supported advances in navigational science that raised Portugal to prominence as a center of cartography, astronomy, and nautical instrumentation. Encouraged exploration of Africa's west coast.
Cheng Ho (China)	1405–1433	Commanded fleets of as many as 317 ships on seven voyages to lands bordering the China Sea and Indian Ocean, including the east coast of Africa.
Bartholomeu Dias (Portugal)	1487–1488	First European to round the Cape of Good Hope and Cape Agulhas in southern Africa and enter the Indian Ocean, seeking an eastward route to Asia.
Sebastian Münster (Germany)	1488–1552	Made the first separate maps of the continents. Discussed many physical processes, such as the effects of rivers and floods on Earth's surface, in his book *Cosmography*.
Christopher Columbus (Italy/Spain)	1492–1504	Sailing for Spain in search of a westward route to Asia, landed in the Bahamas and the Caribbean islands and later reached Central and South America.
Vasco da Gama (Portugal)	1497–1499	Following Dias's route down the west coast of Africa, sailed around the Cape of Good Hope to Calicut, India, the first recorded voyage from Europe to India.
John Cabot (England)	1497	Reached coastal Canada believing it was Asia; established England's claims in North America.
Amerigo Vespucci (Spain and Portugal)	1499–1507	Sailed in Spanish and Portuguese explorations to South and Central America; recognized existence of a continent west of Europe and east of Asia, coining the term "New World."
Pedro Álvars Cabral (Portugal)	1500–1501	First European to reach Brazil.
Vasco Núñez de Balboa (Spain)	1502–1513	Founded Darien in northeast South America, the first stable European settlement on the American mainland; crossed the Isthmus of Panama and became the first European to see the Pacific Ocean.
Juan Ponce de Leon (Spain)	1513–1521	Discovered and explored Florida in his search for the Fountain of Youth.
Hernán Cortés (Spain)	1518–1521 1524–1526	Sailed from Cuba with 600 men and entered the interior of Mexico, conquering the Aztec empire.
Ferdinand Magellan (Spain)	1519–1521	Went in search of a westward route to Asia. Led the first expedition to circumnavigate the world. Threaded the Strait of Magellan, sailed across the Pacific, killed in the Philippines. The circumnavigation was completed by Juan Sebastián del Cano in 1522.
Francisco Pizarro (Spain)	1524–1525	Explored the Pacific coast of South America from Panama to Peru. Later conquered the Incas of Peru.

EXPLORER	DATES OF DISCOVERY	WHERE THEY WENT AND WHY
Jacques Cartier (France)	1535	Discovered passage past Newfoundland to the St. Lawrence River, later used by the French to settle in America.
Hernando de Soto (Spain)	1539–1542	Discovered the Mississippi River.
Francisco de Coronado (Spain)	1540–1542	Explored western Mexico and southwestern U.S., reaching Kansas; members of his party were the first Europeans to see the Grand Canyon.
Matteo Ricci (Italy)	1578–1610	Jesuit priest; traveled by sea to China to convert the Chinese. Produced a map of the world that persuaded Europeans that Cathay was actually China.
Yermak Timofeyevich (Russia)	1579–1585	Crossed the Ural Mountains and captured Siber, a Mongolian Empire outpost; launched Russian expansion across Siberia.
Samuel de Champlain (France)	1603–1615	Established Quebec City as the first French colony in the Americas and explored Lake Champlain, the Ottawa River, and the Great Lakes.
John Smith (England)	1607–1614	Explored and mapped Chesapeake Bay and the New England coast.
Henry Hudson (Holland, England)	1609–1611	Seeking a northern passage to Asia for Dutch interests, sailed up the Hudson River to Albany; on later English voyage discovered Hudson Bay.
Bernhard Varens [Varenius] (Germany)	1622–1650	Wrote about physical geography as well as mapmaking and navigation. Depended on travelers' direct observations of people and places for his writings.
Jean-Baptiste Tarvernier (France)	1638–1668	Traveled India as a gem dealer; wrote book about the caste system and other aspects of Indian culture.
Abel Janszoon Tasman (Holland)	1642–1644	Discovered New Zealand, Tasmania, and Fiji; circumnavigated Australia, proving it was an island continent.
Louis Jolliet and Jacques Marquette (France)	1673	Explored interior North America from Michigan down the Mississippi River, paving the way for France's claim to the Mississippi Valley.
Benjamin Franklin (United States)	1706–1790	Conducted scientific studies of the Gulf Stream. First to record the current's course, speed, temperature, and depth.
Immanuel Kant (Germany)	1724–1804	Defined geography's role as a discipline that could unify many disparate areas of knowledge into an understanding of the world as a whole.
Vitus Bering (Denmark)	1741	Explored Alaska's Pacific coast for Russia; first to reach North America from the west.
Thomas Jefferson (United States)	1743–1826	While President of the Untied States, sent Meriwether Lewis and William Clark on their western expedition (1804-1806) "to enlarge our knowledge of the geography of our continent."
Jedidiah Morse (United States)	1761–1826	First American to write about the geography of the New World. Became know as the father of American geography.
Samuel Wallis (England)	1766—1768	First European to land in Tahiti.
James Bruce (Scotland)	1768–1773	One of the first Europeans to explore Africa, mostly Ethiopia; located source of Blue Nile.
James Cook (England)	1768–1779	Made three Pacific voyages: to Tahiti, New Zealand, and Hawaii, where he was the first European to land. Mapped the west coast of Canada and was first to cross the Antarctic Circle (1773).
Alexander von Humboldt (Germany)	1769-1859	Often called the father of modern geography. Explored South America, collecting botanical and geological specimens and studying ocean currents and temperatures. Observed the rate at which temperature decreased with elevation.

EXPLORER	DATES OF DISCOVERY	WHERE THEY WENT AND WHY
Daniel Boone (United States)	1775	Blazed the Wilderness Road from Tennessee through the Cumberland Gap into Kentucky.
Carl Ritter (Germany)	1779–1859	Envisioned geography as a science that examines the relationship of all forms of nature to human beings, which exist in harmony and unity as evidence of God's plan. Prolific writer who relied largely on the observations of others.
Mary Somerville (England)	1780–1872	Published *Physical Geography* in 1848, a text widely used in schools and universities for fifty years. First woman elected to Royal Astronomical Society (1835).
Alexander Mackenzie (Scotland)	1789–1793	Explored Canada by boat from Great Slave Lake to the Beaufort Sea; later crossed the Rockies and descended the Bella Coola River to the Pacific. First to cross the continent north of Mexico.
Mungo Park (Scotland)	1795–1797; 1805–1806	Explored the basin of the Niger River in West Africa.
Jedediah Smith (United States)	1826–1829	Sought a link between the Great Salt Lake and the Pacific; first American to enter California from the east; explored the California coast.
René Caillé (France)	1827–1828	First European to report on Timbuktu in Africa; published a book of his travels across the Sahara.
Charles Darwin (Britain)	1831–1836	World-roaming voyage as an amateur naturalist on the HMS Beagle yielded observations that led to his theory of evolution.
Isabella Bird (England)	1831–1904	Published *An Englishwoman in America* (1856), detailing her travels in North America. Additional books recorded her travels to Australia, Hawaii, Japan, China, Indonesia, and the Middle East. First woman to become a fellow of the Royal Geographical Society.
Charles Wilkes (United States)	1838–1842	Explored Melanesia, including Fiji, and confirmed the existence of Antarctica.
David Livingstone (Britian)	1841–1873	First European to cross Africa, opening trade routes for Europe; sought the source of the Nile but died before reaching it.
John Charles Frémont (United States)	1842–1847	Mapped much of the U.S. west of the Mississippi and established U.S. claims in Oregon.
Annie Peck (United States)	1850–1935	Noted as mountain climber, author, and lecturer. Set records for highest Western Hemisphere peak (Mt. Huascaran, Peru) climbed by an American (1908).
Richard Francis Burton (Britain)	1857–1859	Found Lake Tanganyika; Speke discovered Lake Victoria and Ripon Falls, which proved to and John Hanning Speke be the source of the White Nile.
Samuel and Florence Baker (England)	1861–1865	Explored tributaries of the Nile in Ethiopia; discovered Lake Albert.
Mary Kingsley (England)	1862–1900	Traveled through Africa, observing the laws and customs of the people that she later described in her books. Identified the need for a society to study the interests of Africa (the Royal African Society was granted a charter in 1968).
Ellen Churchill Semple (United States)	1863–1932	Became first woman elected president of the Association of American Geographers (1921). Promoted the view that the physical environment determines human history and culture.
Marion Newbigin (Scotland)	1864–1931	Edited the *Scottish Geographical Magazine* (1902-1934) making it a leading geographical journal.

EXPLORER	DATES OF DISCOVERY	WHERE THEY WENT AND WHY
Gertrude Bell (England)	1868–1926	Only woman to participate in the 1921 conference to determine the future of Mesopotamia; played a key role in determining the borders of Iraq. Served as Oriental Secretary to the British High Commissioner in Baghdad.
John Wesley Powell (United States)	1869	Explored the Grand Canyon by boat, traveling some 1,000 miles in 98 days.
Nikolay Przhevalsky (Russia)	1870–1885	Crossed the Gobi and brought back scientific data and botanical collections from Central Asia.
Henry Morton Stanley (Britain)	1874–1877	Explored Lake Victoria, discovered Lake Edward, and traced the length of the Congo River.
Francis Younghusband (Britain)	1886–1904	Crossed China from Peking to India; explored the Karakoram Range and Pamirs; led a small army into Lhasa, Tibet, to establish trade and political relations for Britain.
Amelia Earhart (United States)	1897–1937	First woman to receive pilot's license. First woman to fly across the Atlantic and coast to coast ; first person to fly from the Red Sea to India. First woman to receive the Distinguished Flying Cross.
Roald Amundsen (Norway)	1903–1906; 1911–1912	First to sail through the Northwest Passage (1903–1906) and to reach the South Pole (1911).
Ernest Shackleton (England)	1908–1909; 1914–1916	Attempted to reach the South Pole; failed twice.
Robert E. Peary (United States)	1909	Believed to have reached the North Pole after four attempts.
Robert Falcon Scott (Britain)	1910–1912	Lost the race with Amundsen to the South Pole.
Evelyn Pruitt (United States)	1918–2000	Advanced knowledge of coastal environments. Encouraged use of satellites to collect geographic information. Coined the term "remote sensing." Served as chief administrator of the geography Branch of the Office of Naval Research (1959).
Marie Tharp (United States)	1920–	Created sea-floor maps that became the basis for theories of plate tectonics and continental drift.
Roy Chapman Andrews (United States)	1922–1930	Led five zoological expeditions to the Gobi; discovered important dinosaur fossils.
Richard E. Byrd (United States)	1926; 1929	First to fly over the North Pole (1926) and South Pole (1929).
C. William Beebe (United States)	1934	Dived in bathysphere off Bermuda to a record depth of 3,028 feet (923 m).
Sylvia Earle (United States)	1935–	First woman to serve as chief scientist at the National Oceanic and Atmospheric Administration (NOAA) (1990). One of first explorers to use SCUBA gear to study underwater plant and animals life.
Claudio Villas Boas (Brazil)	1943–1973	Explored the southern Amazon Basin, with his brothers Orlando and Leonardo, to study Indian cultures.
Jacques Piccard (Switzerland)	1960	Dived to deepest known point on Earth, the Mariana Trench, 35,800 feet (10,912 m) in the western Pacific Ocean.
Yuri Gagarin (U.S.S.R.)	1961	First man in space.
Alan B. Shepard, Jr. (United States)	1961	First American in space.

EXPLORER	DATES OF DISCOVERY	WHERE THEY WENT AND WHY
John Glenn (United States)	1962	First American to orbit the Earth.
Neil A. Armstrong (United States)	1969	First man to walk on the moon.
Ranulph Fiennes (Britain)	1979–1981	First to travel around the world via the Poles.
Sally Ride (United States)	1983	First American woman to orbit Earth. Head of the California Space Institute.
Kathryn Sullivan (United States)	1984	First American woman to walk in space. Member of the Space Shuttle Discovery crew that deployed the Hubble Space Telescope.

GEOGRAPHIC SOCIETIES

During the 1800s, societies began to emerge in Europe and North America to promote the study of geography. Today national and in many cases state or provincial geographic societies exist in most countries of the world. A selected listing follows:

GEOGRAPHICAL SOCIETIES	LOCATION AND DATE	PURPOSE
Societé de Géographie de Paris	France, 1821	Furthered the practical interests that geography served, in particular commercial and colonial ventures.
Royal Geographical Society	Great Britain, 1830	Sponsored many overseas explorations, including expeditions to the Arctic and Australia.
American Geographical Society	United States (New York City), 1851	Provided geographic and statistical information about less explored parts of the United States and supplied United States businessmen with similar information about foreign countries with which they traded.
Tokyo Geographical Society	Japan, 1879	Development and dissemination of the sciences of the earth and its human inhabitants.
Royal Scottish Geographical Society	Scotland, 1884	Advancement of the science of geography and creation of a greater understanding of the wider world.
Royal Geographical Society of South Australia	Australia (Adelaide), 1885	Promotion of an interest in the natural and cultural environment and the interpretation of the modified and natural landscape.
National Geographic Society	United States (Washington, D.C.), 1888	Promotion of the increase and diffusion of geographic knowledge.
Association of American Geographers	United States (Philadelphia), 1904	Advancement of professional studies in geography, encouraging the application of geographic research in education, government, and business.
Explorers Club	United States (New York City), 1904	Promotion of the scientific exploration of land, sea, air, and space by supporting research and education in the physical, natural, and biological sciences.
National Council for Geographic Education	United States, 1915	To enhance the status and quality of geography teaching and learning.
Society of Woman Geographers	United States (New York City), 1925	Provision of a forum for the exchange of experiences and ideas of women who have conducted research in geographic disciplines.
Royal Canadian Geographical Society	Canada, 1929	Making Canada better known to Canadians and to the world.

GEOGRAPHY TODAY

The desire to explore and understand the physical world and its peoples gave birth to geography as a field of study. Modern geography continues that tradition, with an increasing emphasis on understanding how humans are changing the physical Earth in ways both good and bad. A key role of geographers is to help explain, predict, and manage the global consequences of local and regional conditions and problems. Today geography goes beyond its original Greek definition, "to write about the Earth"; it is a science that can help us better understand and protect our home planet.

GEOGRAPHERS AT WORK

Geographers work in a wide variety of jobs in government, business, and education. The topics geographers pursue are as diverse as the world itself. Geographers study such phenomena as population growth and movements, the spread of diseases, and the depletion of resources. They work side by side with urban planners, developers, environmentalists, government officials, corporate managers, military leaders, politicians, economists, biologists, and practitioners and researchers in many other fields. Many geogra-

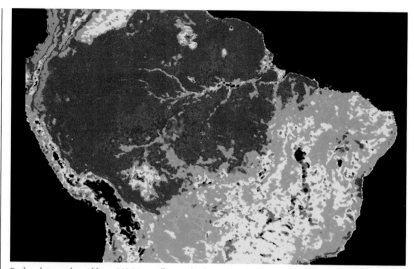

Radar data gathered by a NASA satellite and color enhanced by computers reveals vegetation patterns in northern South America. Rain forest appears in purple, while grasslands are yellow and green.

phers are teachers; others specialize in understanding a specific world region, such as Asia, Latin America, or Africa.

GEOGRAPHIC SKILLS AND TOOLS

Geographers rely on a variety of skills to collect and organize information (often in vast amounts), make sense of that information, and use it to generate sound conclusions. Data come from a variety of sources, including maps, field interviews, reference materials, and other statistical and published data.

Remote sensing, or the use of computers together with cameras or other instruments mounted in airplanes, space shuttles, or satellites to capture images of Earth features

from a distance, has revolutionized the ways geographers work with and analyze information to see Earth in ways not possible in the past.

Aerial photography has dramatically altered the visual aspects of geographic information and interpretation and changed the way we view Earth.

Computer technology is used to create, enhance, and interpret images taken from low Earth-orbit satellites giving geographers yet another tool for gathering information that was previously difficult or even impossible to obtain.

Another powerful tool used by geographers is the *geographic information system* (GIS). A GIS uses computers to store, revise, analyze, manipulate, model, and display geographic data. The data can be derived from maps, reports, statistics, satellite images, surveys, land records, and more. A GIS is a database with a difference: All its information is linked to a geographic, or spatial, reference. Products from a GIS include reports, statistical models, spatial analyses, and maps. One way to view a GIS is to imagine several layers of specialized, or thematic, maps resting on a basic reference map.

The most important application of a GIS is to help geographers identify and analyze spatial patterns and processes. To accomplish this, the systems pull together enormous amounts of information and produce complex models and analyses that would be time-consuming and difficult, if not impossible to prepare manually. As decision-making tools, GIS's are used in such diverse areas as environmental management, agriculture, utilities management, urban and land-use planning, marketing and demographic analysis, hazards management, and emergency and transportation planning. GIS's are revolutionizing the way geographers, as well as researchers and professionals in many other fields, seek answers and solutions to real world problems.

MAPS AND GLOBES

A globe is a scale model and the most accurate way to represent Earth in its true proportions. Globes, however, offer limited detail, are expensive, and are difficult to store or carry around. In addition, a person looking at a globe can see only half the world at a time, whereas a map can give the viewer a complete look at the Earth's surface.

Geography's traditional and most valuable tool is a map. Maps are graphic representations of

Maps Tell Many Different Stories

The state of California, for example, can be represented in many different ways: It can be shown as (**A**) a road map for travelers; as (**B**) a relief map demonstrating topography (the relative elevations and positions of such features as lakes, mountains, and canyons); and as (**C**) a demographic map showing the location and distribution of population; or as a satellite map, picturing Earth from space, to name just a few of the many kinds of thematic maps.

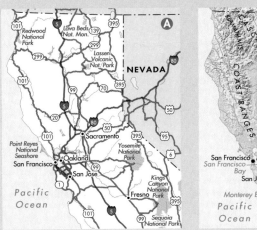

selected aspects of the surface of Earth. No map can cover every possible characteristic of a location—the result would contain so much information that the map would be unreadable. Geographers use a variety of maps to represent information about Earth.

MAPS OF THE PAST
Early Mapmakers

People probably created maps before they could write. Some of the oldest surviving maps—scratched on clay tablets and showing features such as canals, gates, walls, and houses—date back more than 4,000 years to early Mesopotamia. Early Egyptian maps illustrate similar features, along with tombs and shrines. Most of these maps had practical uses—to indicate boundaries or provide directions.

A web of palm-frond ribs and shells, depicting part of the Pacific Ocean, re-created the type of map used by Micronesian navigators perhaps as early as 1500 B.C. Usually 18 to 24 inches square, such maps showed wave patterns and swells, which early navigators interpreted to guide their Pacific Ocean migrations. Seashells or bits of coral marked the locations of islands.

Mapmaking in the Middle Ages

Although the science of mapmaking enjoyed little progress in Europe during the Middle Ages, scholars and mapmakers in China as well as in the Islamic world made significant contributions.

Rediscovery in the Renaissance

By the 1400s European scholars were rediscovering the works of Ptolemy. The introduction of the printing press in that century made his maps and manuscripts more

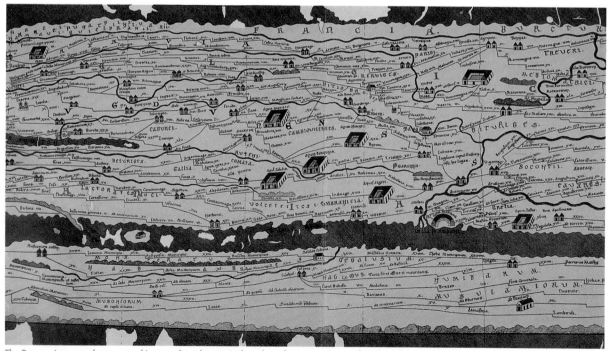

The Romans' approach to mapmaking was largely practical, to chart their extensive road system. This Roman map (above) is based on another from the first century A.D. that shows roads and towns.

available to scholars. At the same time, improvements in shipbuilding and navigation were heralding the beginning of the age of exploration. The information gathered by the great sailors and explorers of the day—Dias, Columbus, Cabot, Magellan, and others—significantly improved the accuracy of maps. Their voyages also convinced most educated Westerners that the world was indeed a sphere while China still clung to the concept of a flat Earth and round heaven.

Maps Since the 1700s

Mapmaking became more scientific with the invention of better tools and instruments, such as the telescope, the sextant, and an accurate timepiece known as the

The first map to contain the name "America" was published in 1507 by Martin Waldseemüller. The map supported Amerigo Vespucci's belief that North and South America were separate continents not connected to Asia.

Beginning in the 800s Islamic mapmakers began to build on Ptolemy's work, which had been translated into Arabic. The world map compiled by the Arab cartographer and geographer al-Idrisi for King Roger II of Sicily in 1154 surpassed any made elsewhere in Europe at the time.

chronometer. The demand for maps for both civil and military purposes increased in the 1700s and 1800s, and national survey organizations were established in several European countries to create territorial maps.

Today, almost every inch of Earth's land area has been mapped, some areas in greater detail than others; seafloors too have been extensively charted, though generally in less detail than the land. However, maps must be continuously revised to reflect changes in boundaries, place-names, human structures, and natural phenomena.

KINDS OF MAPS

Maps can be classified according to scale. Small-scale maps show large areas with little detail; large-scale show small areas in greater detail.

Maps can also be classified according to function, either general reference or thematic.

A general reference map shows location, either absolute (such as a place at 30° N latitude and 100° W longitude) or relative (the site of one place as it relates to the known position of another place: for example, a map that tells a tourist the direction and distance from Wall Street to Times Square in New York City). General reference maps present natural features such as rivers, coastlines, and mountains as well as man-made features such as highways, railways, and cities and other political subdivisions. General reference maps are sometimes referred to as base maps.

Thematic maps emphasize a specific theme or topic, perhaps illus-

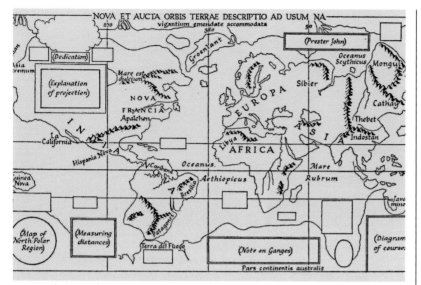

Cartographer Gerardus Mercator (1512–1594) of Flanders gave sailors a valuable navigational tool when he developed the Mercator projection in 1569. He adjusted the spacing of meridians and parallels in such a way that all lines of constant bearing—called rhumb lines—could be drawn as straight lines, even though the Earth is curved. Sailors using such a map could steer by a constant compass setting based on the bearing of the straight line that connected their home and destination ports, correcting for the variation of true from magnetic north.

HOW MAPS ARE MADE

Surveying

Cartographers create their varied maps from data derived from surveys, which today include data collected by remote sensing. Surveying is the science of determining the exact size, shape, and location of a given land or undersea area. Today the remote-sensing technology—including aerial photography, electronic distance measuring, and satellite imaging—has revolutionized the science of surveying.

Remote Sensing

Remote sensing is the gathering, storing, and extracting of geographic information from great distances without any physical contact with the target. The process usually covers large areas. Aerial photography was the first form of remote sensing, beginning in 1858 with photographs taken from a hot-air balloon near Paris.

Today remote sensing is most often performed by instruments mounted on high-altitude aircraft or satellites, which enable scientists and cartographers to capture information invisible to human eyes by recording electromagnetic waves. Once remote-sensing data have been gathered, scientists apply image-processing techniques to extract the specific information they need—for example, to create a dig-

trating the distribution of average annual rainfall for a region or the locations of various crops. (This information may also be represented on a general reference map, but on thematic maps, features such as towns and rivers are intended only as reference points.) Thematic maps can provide a graphic complement both to words and to statistics.

A *cartogram* is an abstract thematic map that presents geographic areas on a basis of statistical factors, such as population proportionate to land areas. For example, a cartogram comparing the populations of the United States and China, which are about the same size in land area, would show the United States as smaller because the U.S. population is much smaller.

A large-scale topographic map (such as of the Chugach Mountains of southern Alaska, on page 23) uses both color and contour lines to clearly identify land classes and elevations, respectively.

Not all maps are of Earth. NASA has sponsored projects to map the moon as well as other planets in the solar system. Today astronomers are even mapping the universe in hopes of uncovering clues to its origins.

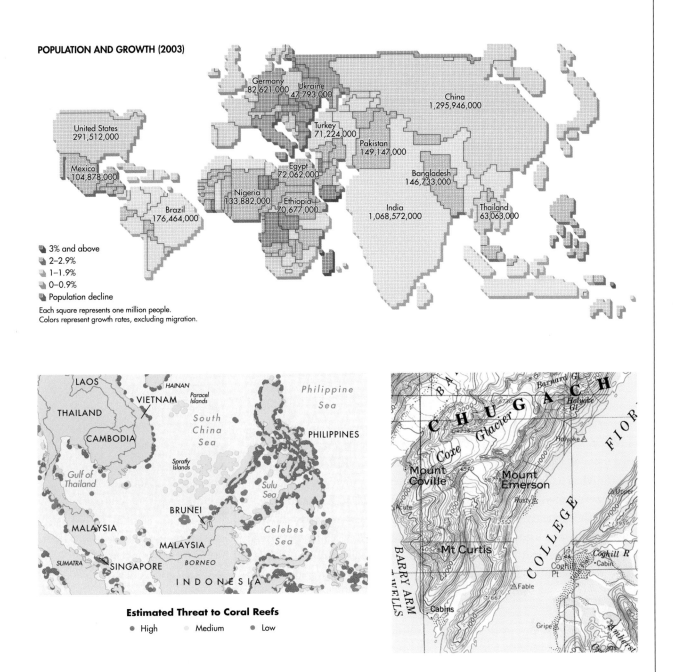

POPULATION AND GROWTH (2003)

Germany
82,621,000

Ukraine
47,793,000

China
1,295,946,000

United States
291,512,000

Turkey
71,224,000

Pakistan
149,147,000

Mexico
104,878,000

Egypt
72,062,000

Bangladesh
146,733,000

Nigeria
133,882,000

Ethiopia
70,677,000

India
1,068,572,000

Thailand
63,063,000

Brazil
176,464,000

3% and above
2–2.9%
1–1.9%
0–0.9%
Population decline

Each square represents one million people.
Colors represent growth rates, excluding migration.

LAOS
HAINAN
Philippine
Sea
THAILAND
VIETNAM
Paracel
Islands
CAMBODIA
South
China
Sea
PHILIPPINES
Gulf of
Thailand
Spratly
Islands
Sulu
Sea
BRUNEI
MALAYSIA
Celebes
Sea
SUMATRA
SINGAPORE
MALAYSIA
BORNEO
I N D O N E S I A

Estimated Threat to Coral Reefs
● High ● Medium ● Low

C H U G A C H
Barnard Gl
Halfyke
Gl
Core Glacier
FIORD
Mount
Coville
Mount
Emerson
Rusty's
Halyoke
Upper
Acute
BARRY ARM
Mt Curtis
Coghill R
Cabin
Coghill
Pt
COLLEGE
WELLS
Fable
Cabins
Gripe
Amherst

A cartogram designed to show world population by country uses color to depict rates of growth and proportional size to convey population sizes of various countries relative to one another. For example, the cartogram above (top) shows India and Nigeria, with large populations, as larger than their physical areas are. A color key makes clear the theme of a map of the South China Sea—red highlights threatened coral reefs (bottom left).

ital image showing temperature patterns in ocean water. They do this by telling the computer to process only certain spectral bands, a method known as theme extraction. The computer then converts these digital images into photographic images, which can be used to make maps, predict weather, locate areas of deforestation and *mineral* deposits, identify crops and their problems, track oil spills, find long-lost archaeological sites, assess flood damage, detect surface changes over time, and identify hundreds of other physical patterns and processes.

The Cartographer's Job

People who make maps are called cartographers. Before cartographers begin work on a map, they need to know who will use the map and for what purpose? Answers to these questions will influence decisions about area and features to display, scale and size, type of projection, level of detail, and appropriate symbols.

Fighting Disease from Space

Images from satellites such as Landsat are giving health officials worldwide a head start in fighting infectious diseases. The satellite data allow scientists to detect changes in the habitats of disease carrying organisms such as insects and rodents, which might trigger epidemics, and to alert local health officials to adopt defensive strategies. In NASA-sponsored tests, Landsat images and remote-sensing technology located areas of heavy mosquito breeding in rice fields in California, malaria-prone areas in Mexico, and in a New York State test, likely landscapes for Lyme disease.

After considering a map's purpose, cartographers must determine the level of detail to include so that the map does not become overcrowded with information and therefore hard to read. At times the cartographer must generalize or omit features because of limitations

Landsat satellites survey Earth in strips 115 miles (185 km) wide (top). More than 560 color images from satellite data were combined to make a portrait of the 48 contiguous states (bottom).

Cartographic Changes

In the three years between the first and second revisions to the sixth edition of the *National Geographic Atlas of the World*, Society cartographers made nearly 3,500 place-name changes on its maps. Between 1992 and 1995 they changed the spelling Kazakhstan to Kazakstan (an alteration made by its citizens to show independence from Russia and in 1997 changed back to Kazakhstan), added the route of the English Chunnel (the Channel Tunnel), revised the shrinking shoreline of the Aral Sea, and added the nations of Eritrea and Palau. They also adjusted the borders between Oman and Yemen and between Saudi Arabia and the United Arab Emirates.

A New Way to Pinpoint Location

Thanks to the 24 satellites of the U.S. Department of Defense's Global Positioning System (GPS), persons operating a car, boat, or airplane equipped with a GPS receiver will always know exactly where they are, often to within a few meters. Developed for military use in the 1970s, the GPS is also becoming widely available to the public, and the list of civilian applications is burgeoning. Oil companies employ GPS to pinpoint drilling sites in remote areas. Mappers take observations in the field with GPS and superimpose plots into GIS systems. To help

Time and location data from three satellites are needed to provide information on latitude and longitude to receivers on Earth; a fourth satellite is needed to provide information on altitude. Computers in the receiver, applying the geometric principle of triangulation, can then determine a receiver's location.

GPS Location:
N 66°22'39.5"
W 045°12'41.2"

forecast the weather, meteorologists can measure delays of GPS signals caused by changes in the atmosphere. Geophysicists using GPS and a technique called carrier tracking, monitor slight changes in Earth's crust that may help predict earthquakes. Other uses, such as systems designed to help blind persons move about safely, are also being developed.

Precise timekeeping is essential to GPS. Each satellite carries sophisticated atomic clocks, and signals broadcast by the satellites include their exact locations and the time sent.

of scale or space. For example, the coves, inlets, and bays of the Cape Cod shoreline would appear on a large-scale map of Massachusetts, but be smoothed out on a small-scale map of the United States.

Today's maps are increasingly created with the help of computers. By linking geographic information systems with graphics software and electronic systems, cartographers can create and revise maps much faster and cheaper than by hand.

Parts of a Map

To describe the essential components of a map, cartographers often use the acronym TODALSIGS, developed by educator Jeremy Anderson. (See chart, page 26.)

LEGEND

Cartographers depend on symbols and other typographic elements to present information that would be impossible to fit on a map in words. Knowing the meaning of a map's symbols is essential to gleaning all the information it contains.

SCALE

Three types of scale are used on maps.

The graphic or bar scale looks like a ruler and shows the length of a certain distance on the map. The verbal or word scale is a sentence that states the relationship of map distance to actual distance. The representative fraction indicates the ratio to which the map was drawn.

Graphic, or Bar, Scale

0   80 Miles

0   80 Kilometers

Verbal Scale

1 INCH represents 106 MILES
1 CENTIMETER represents 67 KILOMETERS

Representative Fraction

SCALE 1:6,700,000

A graphic or bar scale (top) looks like a ruler and shows distance on a map. A verbal or word scale (middle) states the relationship of map distance to actual distance. A representative fraction (bottom) indicates the ratio to which the map was drawn.

A GOOD MAP INCLUDES BASIC ELEMENTS THAT GUIDE THE MAP READER		
T	**Title**	Describes what the map shows and sometimes the time period
O	**Orientation**	Denotes the way the map is placed on the paper, traditionally but not always with north at the top, and with an arrow or compass rose
D	**Date**	Identifies when the map was made, an important indicator of reliability
A	**Author**	Cartographer, agency, or company that developed, researched, and drew the map
L	**Legend**	Information the user needs to understand the map's symbols
S	**Scale**	Distance measurement on the map proportionally related to the actual distance on Earth
I	**Index**	Alphabetical listing of the map's features and place-names, cross-referenced to their grid locations
G	**Grid**	Matrix of lettered rows and numbered columns that provide a frame of reference for locating points on the map
S	**Surrounding Places**	Places that border the mapped area

CONICAL
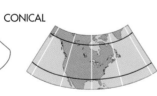

AZIMUTHAL

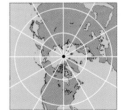

CYLINDRICAL
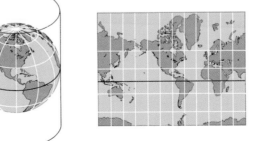

Cartographers through the centuries have used these three basic types of projections.

Latitude and Longitude

In order to create a flat map of the world, cartographers need lines of reference to pinpoint locations on Earth. The reference system, which originated with the ancient Greeks, uses a grid of lines known as latitude (parallels) and longitude (meridians) that is measured in degrees.

Latitude measures angular distance north and south of the Equator (0° latitude) and divides Earth into the Northern and Southern Hemispheres. All lines of latitude are parallel to the Equator. Longitude measures angular distance east and west of the *Prime Merid-ian* (0° longitude), which was located in Greenwich, England in 1884 by international agreement. All lines of longitude converge at the North and South Poles.

The distance covered by a degree of latitude or longitude varies depending on its distance from the Equator. At the Equator, a degree of latitude is 68.708 miles and a degree of longitude is 69.171 miles.

More at:

http://erg.usgs.gov/isb/pubs/Map-Projections/projections.html

MAP PROJECTIONS

Cartographers transfer Earth's curved surface onto a flat piece of paper by a process known as projection.

There are three basic types of *map projections*—cylindrical, conical, and azimuthal—but none is perfect. All map projections involve some degree of distortion in terms of shape, size, distance, or direction.

More than a hundred world map projections have been invented. Because all projections have advantages and disadvantages, the intended use of a map determines which projection is chosen.

More at:

http://erg.usgs.gov/isb/pubs/Map-Projections/projections.html

Longititude: Just a Matter of Time

Longitude is most easily calculated by using clocks—one that shows Greenwich time and one that shows local time. Here's how it works. Earth takes 24 hours to rotate a full 360°, traveling 15° of longitude each hour. If you know the difference in hours, minutes, and seconds between Greenwich Mean Time (also called Universal Time) and your local time, you can multiply that difference by 15 in order to determine the degrees longitude of your time zone. Be sure to take into account daylight saving programs.

For early mariners at sea, calculating longitude was not always so easy. Sailors lacked a seagoing clock that could withstand the stresses of swings of temperature and humidity and a ship's motion, and reliably tell the time in their home port or some other location of known longitude. During the 1730s English clockmaker John Harrison perfected a chronometer whose workings employed a mix of metals that achieved this reliability. Subsequently, the mariner, knowing the time at a fixed point and a local one (because he had a second clock or could calculate local time by sextant and astronomical tables), could pinpoint local longitude.

In 1884, 24 countries accepted the international time zone system. The 24 time zones are aligned with meridians at 15-degree intervals, each representing one hour. In some locations, the time-zone lines do not exactly follow the meridians, but adjust to accommodate local borders or other political features. The International Date Line, the point at which each new day begins and each old day ends, follows the 180th meridian—halfway around the world from Greenwich—for much of its length.

More at:

http://www.timeanddate.com/worldclock/

What's in a Name?

Cartographers must frequently determine the current and official name of a geographic place or feature. The official source for U.S. names is the U.S. Board on Geographic Names (BGN). Created in 1890 to ensure that the federal government uses geographic names consistently, it also fields questions about and resolves problems with geographic names and accepts proposals for new names from the general public. In cooperation with the U.S. Geological Survey (USGS), BGN maintains a database of nearly two million names of physical and cultural geographic features in the United States. This database gives users the federally recognized name of a feature, the *state* and county where it is located, latitude and longitude, elevation (where available), population (of incorporated cities and towns), and names that may have designated the feature in the past. Some 3.3 million foreign geographic features are maintained on the GEOnet names server by the National Geospatial-Intelligence Agency (NGA).

More at:
http://geonames.usgs.gov/

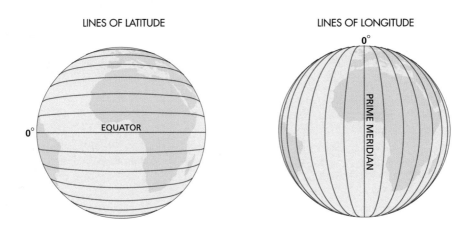

LINES OF LATITUDE

0°   EQUATOR

LINES OF LONGITUDE

0°   PRIME MERIDIAN

Latitude is the distance north or south of the Equator, and longitude is the distance east or west of the prime meridian.

The Shortest Distance Between

Great circle is the term used to describe the largest circle that can be drawn around a sphere such as a globe. Specifically, it is the circle made on the surface of the sphere by any plane that passes through the sphere's center, thus dividing the sphere into two halves. On any globe, there are an infinite number of great circles. On the Earth, the Equator is the only line of latitude that forms a great circle. Each meridian, or line of longitude, forms half of a great circle.

The phrase "great circle route" describes the shortest distance between two points on Earth's surface. When a string is stretched between any two points on a globe, it marks the great circle route. If a ruler connects the same two places on a flat map, the route is different and appears shorter. However, because of the distortion caused by projecting a round object onto a flat surface, the map route only looks shorter.

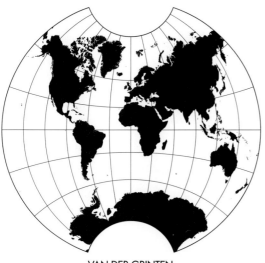

VAN DER GRINTEN

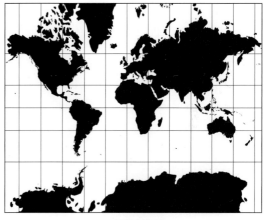

MERCATOR

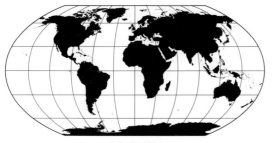

ROBINSON

MOLLWEIDE

WINKEL TRIPEL

The projection often chosen for general reference, world-coverage purposes is the Winkel Tripel, used since 1998 by the National Geographic Society. It replaced the Robinson projection, adopted in 1988, and the Van Der Grinten, which preceded the Robinson and was used by the Society for most of its political maps since 1922. All projections distort. The Mollweide projection compresses and shears shapes at high latitudes; the Mercator also distorts the relative sizes of landmasses in high latitudes, making them appear much larger.

SOURCES OF FURTHER INFORMATION

1. A full account of the history of geography may be found in *The Geographical Tradition*, by David N. Livingstone (Blackwell Publishers, 1992). The author tracks the evolution of geography from the times of the early explorers to contemporary applications that focus on social and environmental problems.

2. For a wealth of information on the dozens of different map projections and their histories, see John P. Snyder's book *Flattening the Earth: Two Thousand Years of Map Projections* (University of Chicago Press, 1993). This heavily illustrated volume, intended for a general audience but a little technical at times, traces map development chronologically from the time of Ptolemy to the present day.

3. To learn how maps reflect the beliefs and times of the cultures that developed them, read *Maps and Civilization: Cartography and Culture in Society*, by Norman J. W. Thrower (University of Chicago Press, 1996). The author, a professor of geography at the University of California, Los Angeles, traces the history of the map from prehistoric times to the present day and includes information on the modern-day mapmaking of governmental and institutional organizations.

4. For a fascinating account of how John Harrison solved the problem of determining longitude at sea, see Dava Sobel's book *Longitude: The True Story of a Lone Genius Who Solved the Greatest Scientific Problem of His Time* (Walker & Company, 1995).

5. Numerous U.S. government agencies provide map information. Among major sources, the Geography and Map Division of the Library of Congress has 4 million maps, 50,000 atlases, and 8,000 reference books—one of the most comprehensive collections in the world (phone 202-707-6277). The National Geospatial-Intelligence Agency (NGA), formerly the Defense Mapping Agency, makes available aeronautical, nautical, topographic, and hydrographic maps and charts (Internet site is http://www.nima.mil). Landsat images are available at NASA's Jet Propulsion Laboratory Internet site, http://www.jpl.nasa.gov. The U.S. Geological Survey offers educational pamphlets, brochures, and maps via their Internet site (http://www.usgs.gov) or by phone 1-800-USA-MAPS (1-800-872-6277).

6. The History of Cartography project at the University of Wisconsin, Madison, is publishing a multivolume series of illustrated books titled *The History of Cartography*. The University of Chicago Press has published three cartographic histories to date, and five more are planned. Each is devoted to a region and time period. The three currently available are *Cartography in Prehistoric, Ancient, and Medieval Europe and the Mediterranean* (1987); *Cartography in the Traditional Islamic and South Asian Societies* (1992); and *Cartography in the Traditional East and Southeast Asian Societies* (1994).

7. Popular books about maps and cartographers include *How to Lie with Maps* (University of Chicago Press, 1991), *Drawing the Line: Tales of Maps and Cartacontroversy* (Henry Holt, 1996), *Cartographies of Danger: Mapping Hazards in America* (University of Chicago Press, 1997), and Maps with the News (University of Chicago Press, 1989), all by Mark Monmonier; *The Map Catalog* (3rd. ed., Vintage/Tilden Press, 1992), which lists sources for maps of all kinds; *The Mapmakers*, by John Noble Wilford (Vintage Books, 1982); and *The Power of Maps*, by Denis Wood (The Guilford Press, 1992).

8. The Map Machine section of the National Geographic Society's Internet site (http://www.nationalgeographic.com) provides dozens of online links to libraries, government agencies, and organizations with information on cartography and related subjects.

PHYSICAL GEOGRAPHY

California's Yosemite National Park, located in the Sierra Nevada, is famous for its spectacular scenery. The Merced River passes through magnificent Yosemite Valley, lined with great precipices of granite that rise 4,000 feet from the valley floor.

PLANET EARTH

FORMATION AND COMPOSITION

Earth formed at the same time as the rest of our solar system, approximately 4.6 billion years ago. The foundation of understanding Earth's origins remains the solar nebular theory proposed by French astronomer Pierre-Simon Laplace (1749–1827) in 1796. Laplace held that the sun and the planets each coalesced from a rotating cloud of cooling gases and dust in one arm of the Milky Way galaxy. The sun and the individual planets gained mass as their gravitational pulls attracted debris from space, ranging in size from dust to asteroids.

Over the next billion years, Earth's surface evolved from a thin basalt crust to include granitic continental masses on shifting plates above a still molten interior. Gases from volcanic eruptions provided the beginnings of the first permanent *atmosphere*. Gradually precipitation accumulated to form Earth's first bodies of water. Then, about 3.8 billion years ago, life in the form of bacteria and algae first

The behavior of seismic waves during earthquakes has revealed that Earth's crust is less dense than the mantle, the outer core is liquid rock, and the inner core is **solid**.

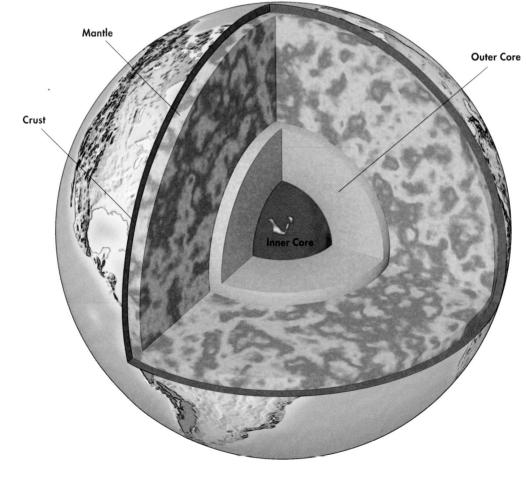

Mantle

Outer Core

Crust

Inner Core

Chicxulub Crater and the K-T Boundary

About 65 million years ago an asteroid more than 6 miles in diameter struck what is now the Yucatán Peninsula of Mexico. The impact created the Chicxulub Crater, more than 180 miles wide, and caused a gigantic dust cloud. The dust eventually settled as a layer of iridium-rich clay now found within sedimentary rock layers beneath the Atlantic and

Pacific Oceans and on most northern land masses. The dust cloud may have caused temperatures to drop around the world. This global cooling following the impact is linked to the extinction of dinosaurs and more than half of all plant and animal species.

Significant differences in the fossil record separate the Cretaceous (K) period of the

Mesozoic era from the Tertiary (T) of the Cenozoic era, marking the K-T boundary. The K stands for *Kreide*, German for Cretaceous. (See geologic time chart, page 177.) Other geologic events at this time, including lava eruptions in what is now India, caused atmospheric changes that may have contributed to the mass extinctions.

appeared within those bodies of water. Once airless, waterless, and lifeless, Earth became the blue planet. Oceans covered nearly three-fourths of its surface, making the entire planet appear blue from space. Earth continues to change today.

EARTH'S STRUCTURE

Earth's minerals and rocks, like all matter, are composed of elements, basic substances that cannot be broken down chemically. So far 112 elements have been discovered, including oxygen, silicon, aluminum, and iron. Though the relative percentages of elements making up the planet have remained constant since its formation, the internal structure of Earth itself is not homogeneous. During the first billion years, heat from three sources—meteorite impacts, gravity's compression of *magma* and other inner Earth material, and the

ABUNDANT ELEMENTS OF EARTH'S CRUST	
ELEMENT	PERCENTAGE IN WEIGHT
Oxygen	46.6%
Silicon	27.7%
Aluminum	8.1%
Iron	5.0%
Calcium	3.6%
Sodium	2.8%
Potassium	2.6%
Magnesium	2.1%
TOTAL	98.5%

radioactive decay of some elements—caused melting within the planet. The high temperatures caused the elements to separate into layers on the basis of their density. Heavy elements such as iron and nickel concentrated nearer Earth's center, and lighter elements such as oxygen and silicon combined with other elements to form most surface rocks and minerals.

DIMENSIONS AND SHAPE

The dimensions of the sun and planets reflect the total amount of space debris captured by each. Earth is now about 24,900 miles in circumference and 7,900 miles in diameter. It is the largest of the four planets closest to the sun.

Earth is an oblate ellipsoid rather than a perfect sphere, meaning that its circumference at the Equator is greater than at the Poles. Its roundish shape is caused by the planet's gravitational pull, which is equal in all directions and creates a sphere with all surface points an equal distance from the center. Because Earth rotates on its axis, however, the greatest centrifugal or outward force is at the Equator; this variation causes the relative difference between the equatorial diameter and the polar diameter. The diameter of Earth at the Equator is less than one percent greater than

Earth's Major Layers

Earth has four major layers: a crust, an underlying mantle, an outer core, and an inner core.

Earth's crust is the thinnest and least dense of the four layers and floats on the mantle. The oceanic crust forming the ocean floors is from three to seven miles thick and is composed of igneous rocks rich in magnesium and iron. Continental crust, ranging from 6 to 45 miles in thickness, makes up land-masses, and is thickest under mountain ranges. Rocks containing more feldspar and silica than oceanic crust make up the continental crust and are less dense than the rocks of the ocean floor.

Below the crust is the denser rock of the mantle, which extends about 1,790 miles toward the core. Temperature and pressure increase with depth in the mantle and result in some layering within the mantle in terms of composition and rigidity. Of particular significance for plate tectonics is that between depths of 60 and 200 miles, the increasing heat melts a small percentage of the mantle, allowing overlying tectonic plates to move around.

Beneath the mantle are the two layers of Earth's core: a liquid outer core 1,400 miles thick and a solid inner core with a 750-mile radius. Iron is the most common element in the outer and inner cores.

the diameter between the Poles. Visualizing Earth as a sphere, however, is sufficient for understanding the variations in amounts of solar *energy* reaching Earth—which affect the formation and patterns of climates and ecosystems—and the equal pull of gravity around the planet.

MOTIONS AND TILT

Earth's orbit around the sun is elliptical, or oval, and each complete orbit takes one terrestrial year, or about 365.25 days; each day equals one rotation on Earth's axis. Earth's distance from the sun varies from 91 million miles, the *perihelion* (from *peri*, Latin for "around," and *helios*, Greek for "sun"), in early January to 94 million miles, the *aphelion* (from *apo*, Greek for "away from"), in early July.

As Earth revolves around the sun, it also rotates on its axis. This rotation has immense consequences for Earth's environments and for human societies. First, our day-night, or diurnal-nocturnal, system is the direct result of each location on Earth rotating into the sun's radiation at dawn and out of the rays at dusk. Second, rotation produces the *Coriolis effect*. The effect is seen in the deflection of winds and ocean currents to the right of their original path in the Northern Hemisphere and to the left in the Southern Hemisphere. (See Global Winds and Ocean Currents, page 58.) Third, the axis of rotation forms our geographic North Pole and South Pole; the Poles and the Equator halfway between, constitute the basis for the coordinate grid system of latitude and longitude. (See Latitude and Longitude, page 27.) Finally, the differential rotation between Earth's inner and outer cores creates the magnetic field that deflects solar wind, thus protecting Earth's surface from harmful ionizing radiation; this rotation also produces the north magnetic pole and the south magnetic pole, used for navigational purposes.

Earth tilts on its axis 23.5 degrees away from perpendicular to the plane of its orbit around the sun. The angle of tilt is constant with reference to certain stars: In the Northern Hemisphere the geographic North Pole points toward Polaris, the North Star. The tilt, relative to the sun, changes during Earth's annual revolution, which causes seasonal variation in solar energy recieved on Earth.

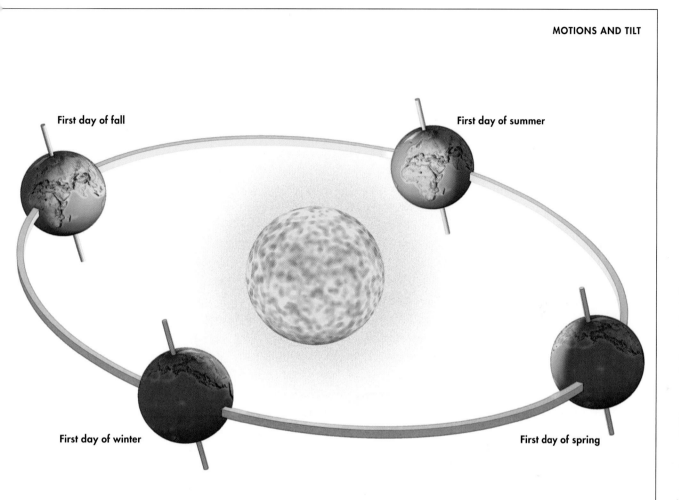

First day of fall

First day of summer

First day of winter

First day of spring

Seasons change when Earth reaches the solstice and equinox positions. Note that the North Pole is tilted away from the sun as winter begins in the Northern Hemisphere and tilted toward the sun as summer begins in the Northern Hemisphere.

Moving Magnetic Poles

Geographers have long been aware that Earth's north and south magnetic poles shift locations. While Earth rotates on its axis, rotating convection currents in the outer core may generate the magnetic field that surrounds Earth and may cause the solid inner core to spin slightly faster than the rest of the planet. The extra speed allows the inner core to complete an extra revolution every 140 years; this interaction between the inner and the outer cores may be the cause of the 130 reversals of the north and of the south magnetic poles over the past 65 million years.

This slow relocation means that cartographers must update the angle of declination between magnetic north and geographic north or true north on topographic maps. For some locations in higher latitudes the compass reading for magnetic north can change as much as six minutes per year.

SEASONS

Earth's shape, orbit around the sun, rotation, and tilt combine to cause seasonal variations in the amount of solar energy each latitude receives. Seasonal variations of solar energy govern weather, climate, and vegetation. As Earth travels around the sun, the latitude at which sunlight strikes the surface directly (at 90 degrees) changes slowly and constantly. At one point in Earth's path around the sun, the North Pole has its maximum tilt away from the sun. On this day— the winter *solstice* in the Northern Hemisphere, falling on or around December 21—the solar equator is at the Tropic of Capricorn, at latitude 23.5° south. There is no sunlight north of the Arctic Circle at latitude 66.5° north, and no nighttime south of the Antarctic Circle.

The winter solstice marking the traditional start of winter in the Northern Hemisphere is the summer solstice in the Southern Hemi-

sphere. On the solstice all locations north of the Equator receive the lowest amount of solar energy for the year due, in part, to shortened daylight hours: When the North Pole tilts away from the sun, less of the Northern Hemisphere is sunlit, so days are shorter. In addition, energy is spread over more area. The coldest weather usually follows the winter solstice by about a month due to the resistance of land and water to temperature change, known as *thermal inertia*.

Seasons change as Earth orbits the sun and the North and South Poles slowly continue to reorient relative to the sun. Six months after the winter solstice, on or around June 21, the North Pole tilts its maximum toward the sun. On this day, the summer solstice, the solar equator is the Tropic of Cancer, latitude 23.5° north. The region south of the Antarctic Circle is dark, and all points north of the Arctic Circle have 24 hours of light, which is why

Alaska and other regions north of the Arctic Circle are called lands of the midnight sun. The summer solstice marks the start of summer in the Northern Hemisphere and winter in the Southern Hemisphere.

Between the two solstices, Earth continues to revolve around the sun. At two points in Earth's orbit around the sun, on or around March 21 and September 23, Earth's axis is at a 90-degree angle to the sun: The solar equator is the Equator, and all locations on the globe receive 12 hours of sunlight. Spring and fall begin on these *equinoxes*, when night (Latin *nox*) is equal (Latin *aequus*) to day at all latitudes.

Seasonal changes in the number of hours and angle of sunlight occur around the world, but the impact is greater on the environments at higher latitudes than those near the Equator. Areas between the Tropics of Cancer and Capricorn have minor annual variation in daylight hours and angle of sunlight, while between the tropics and the polar circles seasonal variations increase. Poleward of the Arctic and Antarctic Circles daylight varies in length between zero and 24 hours, and angle of sunlight varies from zero to 23.5 degrees above the horizon. Such seasonal variations are key to global patterns.(See Global Winds and Ocean Currents, page 58.)

LATITUDES AND SUNLIGHT HOURS DURING THE SUMMER SOLSTICE		
LOCATION	LATITUDE	DAYLIGHT HOURS
Guantanamo, Cuba	20 ° N	13 hours
Cairo, Egypt	30 ° N	14 hours
Beijing, China	40 ° N	15 hours
Prague, Czech Republic	49 ° N	17 hours

WEATHER

ATMOSPHERE

Gravity holds Earth's atmosphere. The atmosphere, which is a thin layer of gases, insulates the surface from temperature extremes and protects Earth from most space debris and from dangerous radiation such as gamma rays. Nearly all atmospheric gases are within about 20 miles of Earth; weather occurs within 10 miles.

Composition

Two nonvariable gases, oxygen and nitrogen, make up 99 percent of the volume of dry air, excluding water vapor. Nitrogen comprises 78 percent of the atmosphere, but it rarely reacts with other gases and is insignificant in meteorological processes. Oxygen, at almost 21 percent of the atmosphere, is vital in respiration and is an active agent of the geologic processes of *weathering* (see Weathering and Mass Wasting, page 127), but like nitrogen it is insignificant in meteorological processes.

Far more important to weather and climate are the variable gases water vapor and carbon dioxide. Of the two, water vapor is more variable. Exposed surface water, atmospheric temperature, and altitude are key factors in the amount of vapor

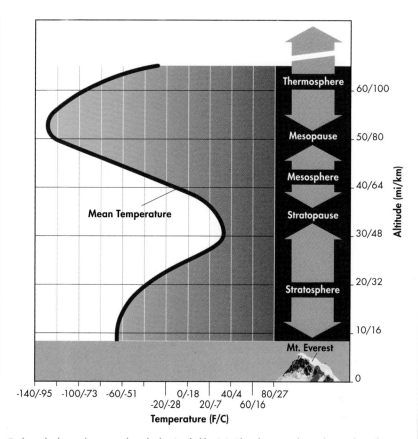

To the naked eye, the atmosphere looks simply like "air," but there are distinct layers, from the stratosphere to the thermosphere. They differ in composition, temperature, and density.

in the air. Above tropical deserts and polar regions water vapor may make up less than 0.1 percent of the lower atmosphere's volume, while water vapor may reach 4 percent over equatorial rain forests. As the atmosphere thins and cools with increasing altitude, there is progressively less water vapor present. Atmospheric water vapor is essential for the processes of *condensation* and precipitation, and it is also the medium by which energy, stored

within water vapor molecules as latent heat, is transferred around the planet. (See Earth's Radiation Balance, page 42.)

Carbon dioxide is the other variable gas important to weather and climate and on average makes up only .035 percent of the atmosphere. The amount of carbon dioxide decreases in summer when increased *photosynthesis* by plants converts it to oxygen and carbon, and it increases in winter when lower temperatures stop plant growth in the midlatitudes and prevent photosynthesis. Atmospheric levels of carbon dioxide and other gases that warm our atmosphere are extremely important in the greenhouse effect that heats Earth. (See Heat and Temperature, page 44.)

In addition to gases, Earth's atmosphere also contains aerosols and particulates that result from natural and industrial processes. Aerosols are light enough to remain suspended in the atmosphere indefinitely and can significantly affect weather and climate. Large concentrations of aerosols, especially sulfurous aerosols from volcanic eruptions, reduce the amount of solar energy that reaches Earth's surface and can cause short-term global cooling. Many smaller particulates—including volcanic dust, sea salt, meteoric dust, ashes from forest fires, and other by-products from combustion—act as nuclei for condensation of cloud droplets.

Aerosols, particulates, and atmospheric gases become pollutants when their concentrations increase to levels that threaten the health of living things or else substantially change existing atmospheric conditions. Aerosol and particulate pollutants include lead and asbestos as well as ash and dust and are generated by internal combustion engines, industrial

Sunsets

At midday, a clear sky appears blue because the sun's rays, especially the shorter wavelength blue rays, are scattered when they contact molecules of atmospheric gases. At dawn and dusk the sun's rays pass through more atmosphere as they approach Earth than they do at midday, and the sky appears orange or red because blue rays are too scattered to be seen. Aerosols and dust increase the scattering and make sunsets red and more colorful.

processes, and waste disposal. Waste incineration and combustion of fossil fuels are primarily responsible for gaseous pollutants such as carbon monoxide, sulfur oxides, nitrogen oxides, and hydrocarbons.

Structure

Earth's atmosphere changes as altitude increases, with significant differences occurring in composition, electrical charge, pressure, and temperature. The atmosphere is divided into layers, from the troposphere, the lowest level, to the exosphere, more than 300 miles above Earth's surface. In the troposphere, from the surface to about six miles above Earth, temperature declines with altitude because the primary heat

source for the air is the ground. (See Earth's Radiation Balance, page 42.) Most weather is confined to the troposphere because it holds more than 95 percent of water vapor and 80 percent of all atmospheric gases. The decline in temperature with altitude results in instability and vertical mixing of the atmosphere. On average, the temperature in the troposphere drops 3.5°F per 1,000 feet: This is called the environmental temperature lapse rate. Occasionally a layer of warmer air will flow over cooler air; the resultant inversion can trap pollutants and particulates as well as cause fogs to form in the lower air.

The top of the troposphere's boundary with the stratosphere is called the tropopause. At the tropopause, temperatures begin to increase because the ozone layer warms the air. The altitude of the tropopause increases slightly in summer because surface temperatures warm the air, which then rises. It also varies year-round by latitude because the centrifugal force that increases the Earth's diameter at the Equator also increases the troposphere to an altitude of 11 miles; at that altitude, the lapse rate drops temperatures to −110°F. Above the Poles, the tropopause averages only five miles high and the lapse rate drops temperatures to −60°F.

The stratosphere, above the

Ozone Hole

By the mid-1980s researchers had confirmed a thinning in the ozone layer above Antarctica—what is now called an ozone hole. For more than a decade, scientists had identified the mechanisms involved in ozone depletion, tracing it to the escape of chlorofluorocarbons (CFCs) from air conditioners and aerosol sprays into the atmosphere. Since 1975, the ozone hole over Antarctica has increased annually, to nine million square miles in 1993; an ozone hole over the Arctic was detected in 1989. Ozone holes shrink in winter, then increase in spring when sunlight reappears and activates chlorine compounds on ice cloud crystals, changing the compounds from inert to active forms capable of destroying ozone. More at: http://www.epa.gov/ozone/science/hole/

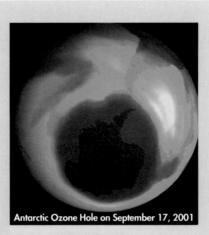

Antarctic Ozone Hole on September 17, 2001

tropopause, extends to about 28 miles above Earth. Temperatures rise with altitude to the stratopause, the upper limit that marks the boundary between the stratosphere and mesosphere. Within the stratosphere, the *ozone layer* warms atmospheric gases by absorbing ultraviolet radiation. Few atmospheric changes can occur in the stratosphere or higher thermal layers because the tropopause, essentially an inversion layer, severely limits mixing between the lower and upper atmospheres. Within the stratosphere, ultraviolet (UV) radiation from the sun creates and perpetuates an ozone layer that protects life on Earth from deadly levels of ultraviolet radiation. Ultraviolet radiation is the energy source for splitting a free oxygen molecule into two ions; each ion then combines with a free oxygen mol-

ecule to form ozone. Ozone is then split by UV radiation into free oxygen and an ion that combines with another ion to form a free oxygen molecule. This cycle of creation and destruction of ozone absorbs most UV radiation. While ozone molecules are found throughout the stratosphere, the concentrated ozone layer occurs near the stratopause. Ultraviolet rays are a high-energy form of radiation that can kill or mutate exposed cells. Life on Earth started 3.5 billion years ago in oceans partially because water was a protection against most UV radiation, and there was no free oxygen to create a protective ozone layer. Ultraviolet rays are divided into ultraviolet A (UVA), ultraviolet B (UVB), and ultraviolet C (UVC), of which UVC has the shortest wavelength and is most danger-

ous to humans and other life. The ozone layer absorbs almost all UVC rays and much UVB radiation.

Temperatures decline in the mesosphere (the layer above the stratosphere, about 28 to 50 miles above Earth) with increased distance from the ozone layer. Few gases are found in the mesosphere and in the thermosphere, about 50 miles to 310 miles above Earth, because the distance from Earth is so great that there is little gravity; thus the atmosphere is thin. Temperatures drop to –130°F at the mesopause, 50 miles high, and begin to increase with altitude in the thermosphere, to perhaps 2,200°F.

The ionization of atmospheric gases creates the ionosphere, a region within the thermosphere where high-frequency radio signals are reflected back to Earth and where solar winds form luminescent atmospheric displays called auroras. Due to Earth's curvature, radio receivers cannot always receive direct signals from a radio station. Different ionized layers form from the sun's radiation; during the nighttime the lowest level (D), which absorbs radio waves, dissipates while beyond D the F layer remains and reflects the waves, especially in the AM radio band. During the day the presence of the D layer interferes with long-distance AM radio transmission and reception.

EARTH'S RADIATION BALANCE

Earth's weather and climate systems involve continual vertical and horizontal movement within the atmosphere's troposphere. Solar radiation provides the energy for these movements. Earth maintains a radiation balance: Energy coming from the sun must eventually be equaled by energy radiated and reflected from Earth. Understanding how the sun's radiant energy is intercepted by Earth and transformed into different forms of energy is necessary to understand the spatial patterns of weather and climate.

Solar Radiation

Radiant energy travels in a wave-like path and extends over a limited portion of the electromagnetic spectrum, which includes all wavelengths and types of electromagnetic radiation. The spectrum ranges from gamma rays and x-rays to radio waves; each type of energy is distinguished by its wavelength—the distance between wave crests, which can range from billionths of an inch to miles long. Visible light, as seen in a rainbow's colors, ranges in wavelength from 0.4 microns (blue light) to 0.7 microns (red light); green light, reflected by chlorophyll in plants, is a combination of yellow and blue light, with wavelengths of 0.4 microns to 0.6 microns.

The temperature of any surface determines what wavelengths and therefore what types of energy are radiated. At 11,000°F, the surface of the sun radiates mostly in the visible light range, although ultraviolet (.01 to 0.1 microns) and near-infrared energy (0.7 to 3 microns) are also radiated. In contrast, radiation from Earth's surface—which averages 60°F, peaks at the 10-micron wavelength, which is in the thermal-infrared range.

The amount of energy radiated from a surface is also determined by its temperature. At 11,000°F, an area on the sun's surface emits about 160,000 times more energy than an equal area on Earth, and Earth, 93 million miles from the sun, inter-

Solar Energy

Radiant energy from the sun can be converted into other useful forms of energy, including electricity. Location on Earth, however, strongly influences the economic viability of using solar energy because the angle and duration of sunlight plus atmospheric interference determine the amount of energy available for use. Cloud-free areas in lower latitudes, such as southern Arizona, have greater amounts of available solar energy than the fog- and cloud-shrouded Pacific coast of Washington. Active and passive solar-energy systems convert radiant energy to sensible heat that warms air or water in a building. A passive solar energy system circulates heat through a structure by convection or conduction, whereas an active solar-energy system requires a pump for circulation. Photovoltaic cells convert radiant energy to electricity. Increased use of solar energy depends on its cost relative to that of other energy sources, especially fossil fuels.

More at:
http://www.eere.energy.gov/RE/solar.html

cepts only a small percentage of this solar energy. The surface temperature of the sun is relatively constant, so the amount of energy emitted by the sun and intercepted by Earth is relatively constant.

Insolation

Of the sun's radiation reaching toward Earth—a stream of radiant energy called *insolation*—only about 52 percent passes through the atmosphere to reach Earth's surface. The atmosphere selectively absorbs, reflects, and scatters the remaining 48 percent. The quantity of solar radiation that reaches Earth also depends on its path through the atmosphere and the angle at which it strikes the surface. The amounts vary with the seasons and with latitude; latitudes farther from the solar equator receive less intense radiation. (See Seasons, page 38.)

Gas molecules in Earth's atmosphere absorb approximately 23 percent of insolation, mostly in the ultraviolet and near-infrared wavelengths. Cloud droplets, ice crystals, and dust in the atmosphere reflect about 17 percent of insolation away from Earth. The amount of reflected insolation varies more than absorption or scatter because clouds are the primary cause of atmospheric reflection, and the amount of cloud cover depends on geographical variations of weather and climate. How

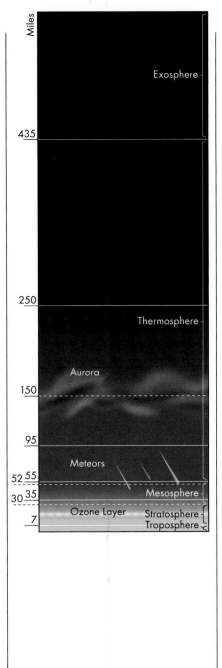

thick a cloud is determines its *albedo,* or reflectivity. A cloud 16,440 feet thick reflects more than 80 percent of insolation, while less than 40 percent is reflected when a cloud is 165 feet thick.

About 8 percent of insolation is scattered when the radiation strikes the atmosphere's gas molecules, dust, and cloud droplets. Some of the scattered energy may reach the Earth's surface, but most is lost to space. On average, 52 percent of insolation reaches Earth's surface, where it is either reflected or absorbed and converted to the kinetic energy that creates and moves weather. Different surfaces on Earth have differing albedos because of their roughness and color. Surface albedos may be as little as 3 percent or as much as 95 percent. The albedo of water is especially important to a clear understanding of Earth's radiation balance because more than 70 percent of Earth's surface is ocean, and the angle of sunlight—which affects albedo—varies a great deal between latitudes. At the solar equator, water has an albedo approaching 3 percent, and the albedo of water also remains low—below 10 percent—within 70° north latitude and south latitude of the solar equator. Near the North Pole and the South Pole, where the angle of insolation can approach

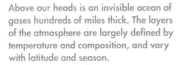

Above our heads is an invisible ocean of gases hundreds of miles thick. The layers of the atmosphere are largely defined by temperature and composition, and vary with latitude and season.

zero, the albedo of water approaches 100 percent. Other factors, such as cloud cover and wave activity, affect the albedo of water, but as a global average the albedo for all of Earth's oceans is about 8 percent.

The albedo of different types of terrestrial surfaces also varies considerably. Roads that are blacktop, with an average albedo of 5 percent to 10 percent, and lighter-colored concrete, with an albedo of 20 to 30 percent, are major reasons why urban areas tend to absorb more insolation than rural areas and heat up, forming *urban heat* islands. (See Urban Heat Islands, page 46.) Snow—especially freshly fallen snow—has an albedo up to 95 percent, which contributes to the lack of absorption and therefore the lack of heat in polar regions. The albedo of chlorophyll in grasses, trees, and crops is usually between 15 and 20 percent; the albedo of deciduous trees increases in winter when the trees lose their leaves.

Of Earth's total insolation, about 46 percent is absorbed by the planet's surface, providing energy for weather systems; 6 percent is reflected by the surface and lost to space, along with 25 percent reflected or scattered by clouds and atmosphere. The remaining energy (23 percent of the total) is absorbed in the atmosphere by gases and dust, but this energy is not concentrated near the surface and does not affect surface weather.

Global Radiation Balance

All solar energy absorbed by Earth's surface and atmosphere is eventually radiated out to space; without this radiation balance energy would accumulate and Earth's temperature would constantly increase. The amounts of incoming and outgoing radiation are not in balance either annually or regionally. Energy from insolation stored in plants by photosynthesis may be released and radiated out to space millions of years later when coal, a fossil fuel formed from swamp vegetation, is burned. Polar regions may annually radiate twice as much energy to space as they receive directly from insolation; this "extra" energy is absorbed between 30° north and south latitudes and is then carried by winds and ocean currents to the polar regions.

While all solar energy absorbed by Earth is eventually lost to space as radiant energy, the atmosphere prevents immediate loss of much of this absorbed energy. Earth radiates energy in thermal infrared wavelengths, of which about 90 percent is absorbed by carbon dioxide and other gases in the atmosphere. This absorbed energy warms Earth's lower atmosphere through the greenhouse effect. (See Greenhouse Effect, page 45.) Ultimately, the energy absorbed by Earth is radiated into space, balancing insolation.

Air Temperature

Air temperature is measured on two major scales: Fahrenheit, used mostly in the United States, and Celsius, once called the centigrade scale and now used by scientists and much of the rest of the world. The scales use the freezing point of water as their base—0 degrees—but the Fahrenheit scale uses saltwater while the Celsius scale uses freshwater. On the Fahrenheit scale, freshwater boils at 212°F and the average temperature of the human body is 98.6°F; on the Celsius scale, freshwater boils at 100°C and average body temperature is 37°C.

Heat and Temperature

Differential heating of Earth's surface and lower atmosphere creates weather—local winds, global winds, ocean currents, and precipitation. Heat is the amount of thermal energy—a form of kinetic energy—that flows among air, water, and rock. Temperature is how much thermal energy a substance contains. Differences in temperatures cause heat to flow between substances, and temperature measures the average, not the total, amount of thermal energy a substance contains. As a result denser substances can contain more energy but have lower temperatures than other substances. Water has a much higher heat capacity than dry beach sand and is transparent, allowing energy to be distributed deeper. As a result, with equal inputs of radiant energy, sand and water may contain equal amounts of thermal energy, but the surface temperature of sand will be higher; more heat will flow from the sand to your feet than from the water.

Thermometers are used to measure temperature. Mercury thermometers have thin, sealed glass tubes containing liquid mercury that expands and rises as it warms and contracts and sinks as it cools. Alcohol is used instead of mercury when especially cold temperatures are to be measured because alcohol freezes at −170°F while mercury freezes at −38°F. Thermostats use bonded bimetallic strips of metals—brass and iron, for example—that expand and contract at different temperatures causing the bonded strip to bend. The temperature is calculated by how much the strip bends.

The U.S. National Weather Service has standardized the collection of temperatures by using shelters to prevent thermometers from absorbing radiant energy that would raise

temperatures and to protect them from precipitation that would cool them. This ensures accurate measurements that are also comparable from one site to another. The shelters are ventilated so that the air inside is not warmed by conduction from the shelter itself; they thus have the same air temperature as outside the shelter. All thermometers are set to measure air temperature at four to six feet above the ground to ensure comparability.

Greenhouse Effect

The warming of Earth's lower atmosphere, the troposphere, is often compared to how air inside a greenhouse warms. Although the analogy is not perfect, there are

important similarities, and the term *greenhouse effect* refers to how Earth's atmosphere warms. Global warming is caused by atmospheric pollutants that have increased the greenhouse effect. *Greenhouse gases*, including carbon dioxide, methane, and nitrous oxide, increased during the twentieth century due to more industrial activity. For example, the concentration of carbon dioxide in the atmosphere is estimated to be increasing by 1.4 percent annually.

The air in greenhouses warms in part because the glass walls and roof of the building, like Earth's atmosphere, allow only certain wavelengths to pass through. Not much solar radiation is absorbed or

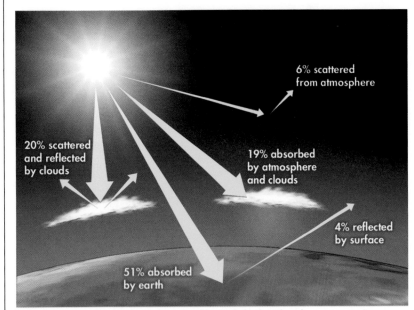

In the greenhouse effect, the sun's short waves hit Earth, but heat from long waves can't escape.

Urban Heat Islands

Large cities tend to be warmer than rural areas; they become urban heat islands. Buildings lose heat, and concrete and asphalt have high heat capacities compared to the lower heat capacities of rural fields and soil. Also, there is little surface water present to absorb thermal energy by evaporation. Temperature contrasts between larger urban areas and their rural surroundings are even greater during nighttime and during winter because urban heating increases as rural temperatures drop.

More at:
http://www.epa.gov/heatisland/

reflected by the glass because it's clear; instead, the glass allows the visible and shorter wavelengths of solar radiation to pass through to be absorbed by the surfaces of the plants and other objects inside. Longer wavelengths, especially thermal infrared energy radiated by the plants and other surfaces, do not pass back through the glass and therefore are absorbed by the air trapped within the greenhouse, warming it. By trapping the air inside, a greenhouse captures *sensible heat*, and the interior stays warm overnight.

The warming of Earth's lower atmosphere is a more complex process than that of a greenhouse. Thermal energy from Earth's surface is transferred to the troposphere, warming it by radiation, conduction, and convection. Water vapor in the air contains its own latent energy that, when released by condensation, also warms the atmosphere. Most of Earth's radiant energy is absorbed by the atmosphere and then reradiated back to the surface, extending the warming process of the lower atmosphere.

More than 90 percent of Earth's surface radiation is intercepted and absorbed by greenhouse gases—water vapor, carbon dioxide, ozone, nitrous oxide, and methane—in the atmosphere, and the rest is lost to space. The greenhouse gases are called that because absorbed radiant energy warms them. Each greenhouse gas takes up specific wavelengths; together these gases take in most terrestrial radiation of wavelengths between 2.5 and 8 microns, as well as most wavelengths greater than 15 microns. Water vapor absorbs the widest range of wavelengths and greatest quantity of radiation. Carbon dioxide absorbs the second largest amount of radiation and contributes to global warming because the burning of *fossil fuels* has increased the atmospheric concentration of the gas. Methane, nitrogen oxide, and other atmospheric gases also absorb surface radiation, although in rela-

tively minor amounts.

Surface radiation, convection, conduction, and the transfer of latent heat in the form of water vapor contribute to the warming of Earth's atmosphere. Of the four mechanisms, latent heat transfers three times more thermal energy to the atmosphere than conduction and convection combined. The energy is ultimately converted to long-wave radiation that is lost to space.

Vertical Temperature Change

As Earth's energy is transferred to the atmosphere, most of the warming occurs nearer the surface. The resulting decline in temperature with increased altitude has important implications for atmospheric mixing and surface conditions. On average, air temperature decreases 3.5°F per 1,000 feet of elevation or altitude—the environmental temperature lapse rate. Two major factors cause this lapse rate: Radiation is more intense and more likely to be intercepted and absorbed by greenhouse gases close to Earth's surface, and because air thins with increasing altitude, the amount of water vapor and other greenhouse gases available to absorb radiation decreases proportionately. Tall mountains, even Kilimanjaro and others near the Equator, are cold at their summits and often snow covered year-round.

Diurnal and Seasonal Temperature Patterns

Atmospheric temperatures and related changes in humidity, condensation, pressure, and winds vary daily and seasonally, as well as geographically. Diurnal heating patterns reflect the time lag between Earth's absorption of solar radiation and the subsequent reradiation of long-wave energy to the atmosphere. Air temperatures usually bottom out and begin to increase at Earth's surface about 30 minutes after sunrise, when the surface has absorbed solar energy and then starts to radiate long-wave energy. Temperatures increase as the sun rises and Earth absorbs and radiates increasing amounts of energy. Peak insolation occurs at noon, but the temperature continues to rise for two to three hours, even though the amount of insolation decreases as the sun drops in the sky. Temperatures peak in mid-afternoon because of the time lag between surface absorption and radiation of energy. After the peak, temperatures fall as the rate of surface radiation decreases. The radiation and temperature cycles begin again at sunrise.

Seasonal heating patterns also reflect a time lag between Earth's reception of insolation and its release of that energy as radiation. In the mid- and higher latitudes, where seasonal variations in insolation are greater than those near the Equator, there is usually a one-month time lag between the amount of insolation and the resulting air temperature. January and July are usually the coolest and hottest months respectively even though insolation is lowest in December and greatest in June. (See Climate Controls and Classification, page 85.)

Compared to soil and rock, water is particularly slow to release heat and thus changes temperature more slowly; water therefore increases the lag time for diurnal and seasonal heating. Water has a high heat capacity. Because of circulation by currents and waves and because sunlight penetrates deeper into water than soil, energy is stored at greater depths in the water and is released more slowly into the atmosphere. As a result, the seasonal time lag for coastal high and low temperatures may be closer to two months, compared with one month for inland locations. Most coastal locations have cooler days and warmer nights than inland locations because air over water heats and cools more slowly than air over land. Areas that have little water or water vapor warm rapidly during the day and cool quickly at night; even the Sahara can have freezing temperatures.

Temperature extremes on Earth, especially record high temperatures, usually occur in drier areas. In North America, the highest temperature yet recorded was 134°F in Death Valley, July 10, 1913; the highest temperature yet recorded on Earth—136.4°F—was at El Azizia, Libya, September 13, 1922. Record low temperatures in winter are found at high latitudes where insolation is minimal and at higher elevations where the air is thin; the lack of water vapor and the high albedo of snow and ice in both locations allows rapid cooling.

Spatial Temperature Patterns

Local surface conditions affect the amounts of insolation absorbed, the timing of long-wave radiation, and therefore the temperatures of limited areas. Terrain is an especially important local factor: Slope aspect, the direction a slope faces, affects the angle and duration of solar radiation and consequently the amount of moisture in the soil. During early summer in the Rockies and other alpine areas in the Northern Hemisphere, north-facing slopes stay

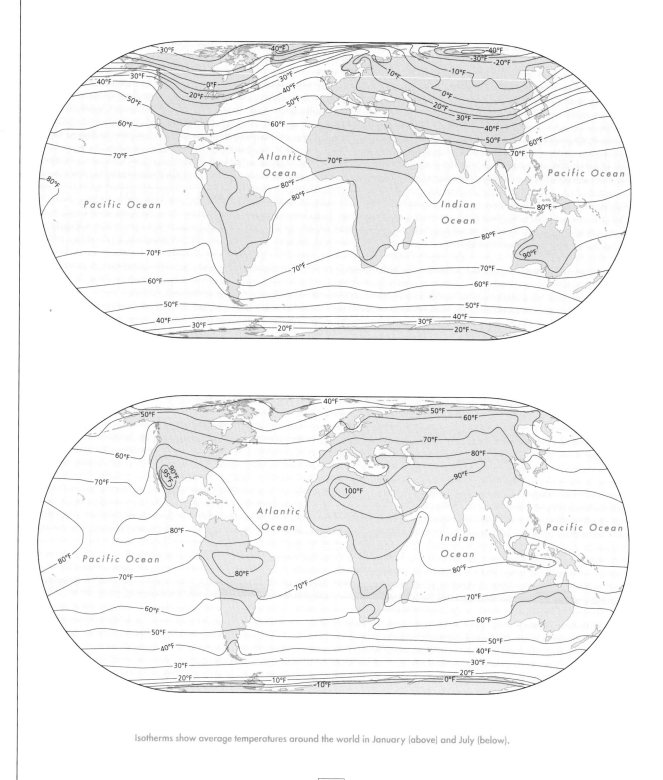

Isotherms show average temperatures around the world in January (above) and July (below).

cooler and retain snow longer than south-facing slopes. Surface moisture and temperature can vary dramatically within short distances, as between a streambed and the prairie through which it runs. Lower areas that accumulate water, such as marshes or swamps, may have smaller daily temperature ranges than drier slopes nearby.

Three global temperature maps can show temperature patterns and reveal main factors affecting average temperatures in January and July, and annual average temperature range. January and July are the coldest and warmest months in the Northern Hemisphere and the warmest and coldest in the Southern Hemisphere. The annual average temperature range is the difference between the January and July averages.

Annual temperature ranges are used instead of just average annual temperatures because the extreme high and low temperatures throughout the year are most limiting to vegetation and ecosystems. For example, the average annual temperatures for Seattle and Kansas City are nearly identical—about 53°F. In Seattle, average January and July temperatures are 41°F and 65°F, an average annual temperature range of 24°F; in Kansas City, the January average is 26°F and the July average 79°F, a range of 53°F. Seattle has far milder temperatures than Kansas City; its annual temperature range puts less environmental stress on plants and people.

Maps of average January and July temperatures use *isotherms,* lines of equal temperature, that show distinct geographic patterns resulting from five primary temperature controls: latitude, elevation, land-water contrast, global wind patterns, and ocean currents.

Latitude, and its correlation to insolation, is clearly shown in the two maps on page 48 by the progressively cooler monthly temperatures farther from the Equator. Ocean temperatures show latitudinal influences more clearly than those of continents because water-surface conditions are much more homogeneous than those on land.

Large water bodies have a strong moderating influence on temperatures because they absorb and release energy slowly. Smaller water bodies, even the large Great Lakes, may freeze in winter and have no moderating influence. Isotherms for oceans show that cooler temperatures exist farther poleward over water than over continents in summer while warmer temperatures prevail farther poleward than on continents in winter. Thermal inertia affects the atmospheric temperatures over coastal areas. Coastal communities tend to have cooler summers and warmer winters than inland areas at the same latitude.

Maritime influences on coasts and inland areas depend partly on ocean current temperatures and on the global winds prevailing at those locations. Ocean currents tend to be warmer off the east coasts of continents and cooler off the west

Elevation and Temperature

Higher elevations are associated with thinner air, and mountains are typically colder than surrounding low-lying areas. Tall mountain ranges such as the Andes and Himalaya are distinctly colder than their foothills, as are highland areas such as the interior plateaus of Mexico and Ethiopia. In Mexico City, at an elevation of 7,570 feet, average July temperatures may be 10°F cooler than in Veracruz, at sea level on the Gulf of Mexico. Elevational changes are often compared to latitudinal changes because the effects on temperature ranges are so similar. Vegetation reflects this vertical distribution of temperatures on mountains; climbing up the east slope of the Sierra Nevada takes a hiker through a sequence of vegetation types from desert sagebrush to alpine tundra.

coasts. (See ocean currents map, page 59.) As a result, summer temperatures are usually cooler on North America's west coast than on its east coast. In San Francisco the warmest month (September, due to the time lag) has an average temperature of 65°F because of the cool California Current; in Norfolk, Virginia, at the same latitude on the East Coast, the warmest month (July) has an average of 78°F because of the warm Gulf Stream.

Global winds, which circle the globe, blow the moderating effects of water onshore until either mountains interfere or distance causes the maritime influence to dissipate. January isotherms for North America and Europe show the different influences that the Rockies, which run north-south, and the Alps, which run east-west, have on the maritime effects blown inland by westerly winds. The Rockies severely limit the inland extent of maritime temperatures from the Pacific Ocean, while the east-west trending Alps do not keep the maritime influences of the Atlantic Ocean from moderating temperature patterns more than a thousand miles inland at Budapest and Bratislava.

Temperatures and Humans

Comfort, and in some cases health, is influenced by ambient air temperature, with air circulation and humidity contributing to its impact on people. High temperatures may be debilitating if heat causes the body's core temperature to rise to 102°F or higher, and colder temperatures that lower body temperature below 95°F may cause hypothermia. In summer, the National Weather Service reports the heat index, which combines temperature and relative humidity to show the apparent temperature felt by humans, in order to warn individuals of possible health threats.

Cold air temperatures cool people far more rapidly when it's windy or breezy than when it's calm. Ordinarily, a warm layer of air near the skin insulates the human body, but wind displaces this layer with cooler air. As a result, more heat is conducted and convected away from the body. A windchill index, sometimes referred to as a windchill equivalent temperature (WET), combines air temperatures and wind speeds to create a measure of how rapidly people lose heat and are chilled at different wind speeds.

ATMOSPHERIC PRESSURE AND WINDS

As solar energy warms Earth, the resulting diurnal and seasonal pat-

Verkhoyansk, Russia

Verkhoyansk, Russia, just north of the Arctic Circle in Siberia, has an average annual temperature range (the difference between highest and lowest average temperatures) of 118°F, one of the largest of any settlement in the world: July averages 60°F and January's average is –58°F. The region around Verkhoyansk is a landlocked part of the world's largest landmass and is almost 4,000 miles from the maritime influences of the Atlantic Ocean to the west and 600 miles from the Sea of Okhtosk to the east. The prevailing winds at this latitude—the westerlies—have little maritime influence on temperatures after traveling so far.

As inhospitable as the temperatures in Siberia are, the former Soviet government and the tsarist government before that used prison labor to develop the region's abundant coal, gold, and molybdenum resources. Russian author Aleksandr Solzhenitsyn's *The Gulag Archipelago* described the Soviet system of prison camps.

WINDCHILL INDEX												
WIND SPEED (MPH)	THERMOMETER READING (°F)											
	50	40	30	20	10	0	–10	–20	-25	–31	–36	–42	–47
Calm	50	40	30	20	10	0	–10	–20	–31	–36	–42	–47	–52
5	48	37	27	16	6	–5	–15	–26	–52	–58	–64	–71	–77
10	40	28	16	4	–9	–24	–33	–46	–65	–72	–78	–85	–92
15	36	22	9	–5	–18	–32	–45	–58	–74	–81	–88	–95	–103
20	32	18	4	–10	–25	–39	–53	–67	–81	–88	–96	–103	–110
25	30	16	0	–15	–29	–44	–59	–74	–86	–93	–101	–109	–116
30	28	13	–2	–18	–33	–48	–63	–79	–89	–97	–105	–113	–120
35	27	11	–4	–20	–35	–51	–67	–82	–92	–100	–107	–115	–123
40	26	10	–6	–21	–37	–53	–69	–86	–93	–102	–109	–117	–125

terns of surface radiation create related patterns of atmospheric warming and atmospheric pressure. Wind, the result of horizontal atmospheric pressure differences, moves clouds, air masses, and particulates locally and around the globe. Understanding the factors affecting atmospheric pressure, wind speed, and direction provides the basis for understanding not only local weather but also global precipitation patterns and climates.

Atmospheric Pressure

Earth's atmosphere, held near the surface by gravity, exerts pressure on all surfaces, much as water does in a pool. The weight of the overlying column of air creates atmospheric pressure, and the amount of pressure is comparable to the height of the column of air above a location. While increases in altitude are matched by decreases in atmospheric pressure, horizontal pressure variations can occur at any given altitude because of changes in temperature and other factors

Atmospheric pressure, the weight of overlying air, decreases with increasing altitude both because there is less overlying air and because gravitational pull decreases with altitude. At Earth's surface, average atmospheric pressure is 14.7 pounds per square inch. In Denver, the Mile High City, atmospheric pressure is 12.2 pounds, or 83 percent, of that at sea level; at 18,500 feet, on the slopes of Mount McKinley in Alaska, pressure is 7.35 pounds, or 50 percent. Lower atmospheric pressure affects respiration because each lungful of air has proportionally fewer oxygen molecules. Mountain climbers who do not take the time to acclimatize to thinner air at higher elevations may experience altitude sickness because less oxygen gets into their blood.

Horizontal pressure variations cause winds and are mapped using isobars; these maps can then be used to predict the movement of storms and weather systems. An *isobar* is a line connecting points of equal atmospheric pressure, so that high-pressure areas (*anticyclones*) and low-pressure areas (cyclones) are shown. To adjust for different surface elevations, the barometric reading for each point is calibrated to a sea-level equivalent. Denver, for example, is at an elevation of nearly 5,300 feet, and under average daily

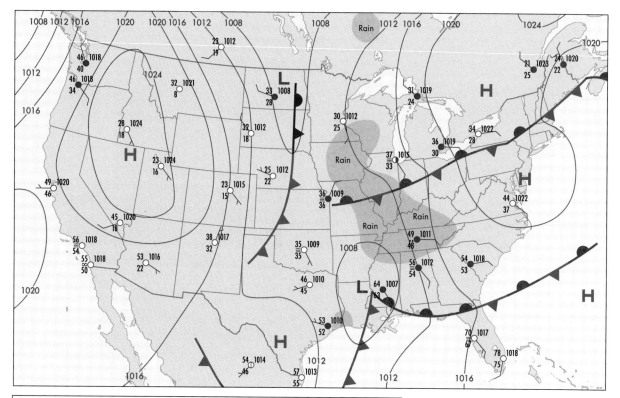

Weather Symbols

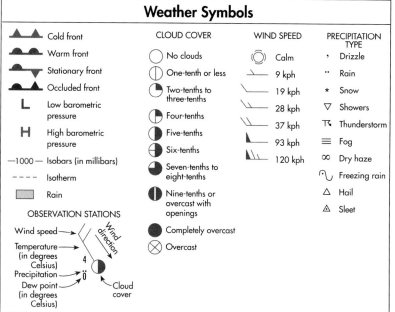

		WIND SPEED		PRECIPITATION TYPE	
Cold front	CLOUD COVER				
Warm front	No clouds	Calm		Drizzle	
Stationary front	One-tenth or less	9 kph		Rain	
Occluded front	Two-tenths to three-tenths	19 kph		Snow	
L Low barometric pressure	Four-tenths	28 kph		Showers	
H High barometric pressure	Five-tenths	37 kph		Thunderstorm	
—1000— Isobars (in millibars)	Six-tenths	93 kph		Fog	
---- Isotherm	Seven-tenths to eight-tenths	120 kph		Dry haze	
Rain	Nine-tenths or overcast with openings			Freezing rain	
	Completely overcast			Hail	
OBSERVATION STATIONS	Overcast			Sleet	

OBSERVATION STATIONS

Wind speed →
Temperature → (in degrees Celsius)
Precipitation →
Dew point → (in degrees Celsius)

Wind direction

4
0̈
Cloud cover

A representative weather map uses commonly accepted conventions to depict warm, cold, and occluded fronts, isobars to join areas of similar air pressure, and a variety of means to show the amount of cloud cover, wind speed, and type of precipitation. Altogether, an amazing amount of weather information is available to knowledgeable weather map readers.

conditions it should have an atmospheric pressure of 24.95 inches. New Orleans, at sea level, has a barometric level of 29.92 inches, and a comparison of the two cities would show the effect of elevation on the change in pressure, not the effect of the horizontal pressure differences that move weather systems.

Altimeters

Altimeters are used to determine the altitude of aircraft above sea level by measuring outside atmospheric pressure and comparing that reading to sea-level pressure. The altitude of the aircraft can be calculated because atmospheric pressure decreases regularly with increasing height above the ground. Radar altimeters determine altitude by measuring the time a radio signal takes to beam down to Earth and return. ("Radar" stands for *ra*dio *d*etection *a*nd *r*anging.) Changes in atmospheric pressure associated with changes in weather, however, can result in incorrect measurements of altitude. When a low-pressure system moves into an area, for example, pressure readings may drop by 2 or 3 percent, and a given reading in an aircraft will incorrectly be translated to a higher altitude. For safety purposes, pilots must adjust their altimeters based on local weather conditions reported by the Federal Aviation Administration.

Measuring Atmospheric Pressure

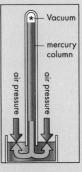

A barometer is used to measure atmospheric pressure. A standard mercury barometer has a glass column about 30 inches long, closed at one end, with a mercury-filled reservoir. Mercury in the tube adjusts until the weight of the mercury column balances the atmospheric force exerted on the reservoir. High atmospheric pressure forces the mercury higher in the column. Low pressure allows the mercury to drop to a lower level in the column. An aneroid barometer uses a small, flexible metal box called an aneroid cell. The box is tightly sealed after some of the air is removed, so that small changes in external air pressure cause the cell to expand or contract.

Pressure Variations

Horizontal atmospheric pressure variations occur primarily because of differences in temperature. When Earth's atmosphere is warmed by the greenhouse effect the kinetic energy of greenhouse gases increases, causing gas molecules to move more rapidly and spread far-ther apart. The result is a decrease in the density of the atmosphere that reduces the atmospheric pressure. Record low pressures are found in cyclones; the lowest yet recorded is 25.69 inches in the eye of a typhoon north of the Pacific Ocean island of Guam in 1979. Atmospheric pressure in the eye of a *tornado* is probably lower, but that has not yet been measured due to the destructive force of the winds.

When air cools, atmospheric pressure increases because the lower kinetic energy causes the gases to compress, increasing air's density. The record high pressure yet measured is 32.01 inches in Siberia during the winter of 1968. Antarctica may have higher atmospheric pressures, but these have not been officially recorded.

While most horizontal variations in atmospheric pressure can be explained by differences in the rates of heating and cooling on Earth's surface, water vapor concentration in the air and the presence of atmospheric convergence or divergence also contribute to pressure differences. Increased amounts of water vapor, a variable gas, tend to lower atmospheric pressure because water vapor is less dense than either the nitrogen or oxygen that it displaces. Atmospheric convergence occurs when winds meet and compress air at their point of convergence.

Atmospheric divergence occurs when winds blow away from each other and leave lower atmospheric pressure at the location of the divergence. (See Wind, this page.)

Patterns of Atmospheric Pressure

An idealized global pattern of surface atmospheric pressure shows five pressure bands circling Earth, with two high-pressure areas centered over the North and South Poles. The *intertropical convergence zone* (ITCZ) along the Equator is a low-pressure zone created by intense solar radiation and heating. Heated air rises at the Equator and moves away from the Equator. By the time the diverging air moves approximately 30 degrees north and south of the Equator, this upper air has cooled and is denser than surrounding air. It then descends, creating high-pressure surface zones near the Tropics of Cancer and Capricorn. At about 60° north and south latitudes—between the high-pressure zones at the tropics and the high-pressure zones over the Poles—lie low-pressure surface zones; these pressure zones are created by the convergence of air moving from the tropical and polar high-pressure zones.

Differences in surface characteristics, especially between water and land, modify this idealized

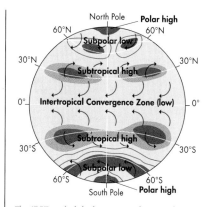

The ITCZ and global patterns of atmospheric pressure are marked above by darker red in the areas of high pressure and lighter red and gray in low-pressure areas; arrows show wind direction.

global pressure pattern. For example, in the case of Eurasia, the continental landmasses of Europe and Asia, continent and ocean contrasts in heating and cooling are also reflected in differences in atmospheric pressure. The Eurasian landmass cools more than its surrounding oceans—the Atlantic, Arctic, Pacific, and Indian Oceans—in winter and consequently tends to have higher surface pressure; in summer the landmass warms more rapidly than the oceans and develops lower surface pressure. The winds over Eurasia reflect this seasonal change in pressure. The Siberian high-pressure area is a well-developed winter phenomenon, with an average pressure of 30.56 inches, and an average pressure below 29.53 inches in summer.

Wind

Wind is the horizontal movement of air in response to differences in atmospheric pressure. If the movement is up or down, it is called an updraft or downdraft. When atmospheric pressure at one location is higher than the pressure at surrounding locations at the same altitude, air flows to equalize this imbalance. Usually the pressure gradient between two locations is the result of differential surface heating, but other factors, such as convergence or divergence, can also affect atmospheric pressure.

All winds, whether local, regional, or global, have two defining characteristics—speed and direction, each of which is the result of several interacting factors. Wind speed is the result of the competing influences of pressure gradient and friction. The greater the pressure differential over a given distance, the faster the wind, but pressure gradients cannot be directly correlated to surface wind speeds because different types of terrain slow wind based on their roughness.

Friction, particularly friction created by large objects on the ground, affects wind velocity and wind flow or turbulence. Buildings, trees, and snow fences can decrease the speed of wind, while the flat surfaces of ice, lakes, oceans, or smooth desert often have faster

BEAUFORT SCALE OF WIND FORCE			
BEAUFORT NUMBER	GENERAL DESCRIPTION	LAND AND SEA OBSERVATIONS FOR ESTIMATING WIND SPEEDS	WIND SPEED 30 FEET ABOVE GROUND (km/hr)
0	Calm	Smoke rises vertically. Sea like mirror.	Less than 1
1	Light air	Smoke, but not wind vane, shows direction of wind. Slight ripples at sea.	1–5
2	Light breeze	Wind felt on face, leaves rustle, wind vanes move. Small, short wavelets.	6–11
3	Gentle breeze	Leaves and small twigs moving constantly, small flags extended. Large wavelets, scattered whitecaps.	12–19
4	Moderate breeze	Dust and loose paper raised, small branches moved. Small waves, frequent whitecaps.	20–28
5	Fresh breeze	Small leafy trees swayed. Moderate waves.	29–38
6	Strong breeze	Large branches in motion, whistling heard in utility wires. Large waves, some spray.	39–49
7	Near gale	Whole trees in motion. White foam from breaking waves.	50–61
8	Gale	Twigs break off trees. Moderately high waves of great length.	62–74
9	Strong gale	Slight structural damage occurs. Crests of waves begin to roll over. Spray may impede visibility.	75–88
10	Storm	Trees uprooted, considerable structural damage. Sea white with foam, heavy tumbling of sea.	89–102
11	Violent storm	Very rare; widespread damage. Unusually high waves.	103–118
12	Hurricane	Very rare; much foam and spray greatly reduce visibility.	119 and up

winds than cities or forests. Wind that is 65 miles per hour over a calm ocean may be only 40 mph over rough terrain, while the same pressure gradient may cause winds to reach speeds of 100 mph at 3,000 feet above the influence of surface friction. A second effect of friction on wind is to increase air turbulence. Instead of flowing smoothly from high to low pressure, wind blows in gusts or eddies when buildings, hills, and other objects slow, channel, or deflect it. Turbulence can also form at higher altitudes when a *jet stream*—an extremely rapid, high-altitude wind—an updraft, or a mountain causes wind shear that interferes with smooth wind flow. Clouds with a wave pattern can be evidence of high-altitude turbulence, although clear-air turbulence (CAT) also occurs and is an occasional hazard for aircraft.

Wind direction is controlled by pressure gradient, friction, and the Coriolis effect. Friction affects wind direction by slowing wind speed and reducing the Coriolis effect. Winds have their direction diverted by the Coriolis effect: to the right of the pressure gradient in the Northern Hemisphere and to the

The Coriolis Effect

The Coriolis effect, first described by French mathematician Gaspard-Gustave de Coriolis (1792–1843) in 1835, influences winds, ocean currents, and even missiles. As Earth rotates, the entire surface travels eastward as part of the rotational motion. The rate of eastward motion at the surface depends on the latitude because Earth is a sphere and different latitudes have different circumferences around Earth's axis. At the Equator the surface rotates at more than a thousand miles per hour, but at 40° north latitude— about the latitude of Denver and Philadelphia—the surface rotates at less than 800 miles per hour. Winds, like all motion parallel to the surface, incorporate the east-

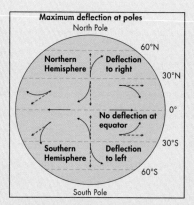

Maximum deflection at poles
North Pole
60°N
Northern Hemisphere : Deflection to right
30°N
No deflection at equator
0°
Southern Hemisphere : Deflection to left
30°S
60°S
South Pole

ward movement from their latitude of origin into their trajectories. As a result the path of wind blowing from one latitude to another will be offset because of the difference between the wind's rate of eastward movement and the surface's rate of eastward movement.

For example, a pressure gradi-

ent from Houston, at 30° north latitude, to Minneapolis, at 45° north latitude, will initiate a southerly wind incorporating an eastward movement. As the wind moves farther north—where the circumference around a latitude is smaller and the surface rotates more slowly—the eastward movement of the wind is faster than that of Earth. As a result, the wind ends up east of Minneapolis—to the right of the original direction. By the same process, a wind blowing from Minneapolis to Houston starts with a smaller eastward component than is present in Houston. When the wind arrives in Houston, the city has rotated farther east than the wind and the wind appears to have curved to its right.

left in the Southern Hemisphere. Winds are named for the direction from which they come.

Wind Speed

Wind speeds are measured in knots (one knot is one nautical mile, or 6,076 feet) or miles per hour by anemometers, which indicate wind speed by how fast an array of cups that catch the wind rotate. The Beaufort scale, a system of descriptions based on wind effects on the surface, is used to describe estimated wind speeds. For example, a moderate breeze—a force four on the

Beaufort scale—moves small tree branches, lifts leaves and paper from the ground, and measures from 13 to 18 miles per hour. The current world record for wind speed is 231 miles per hour, measured on Mount Washington, New Hampshire, in 1934. The reading was for one gust, and winds greater than 185 miles per hour have been recorded there for a five-minute interval. Westerly winds are usually strong in this region, and winds flowing over Mount Washington are constricted between the mountain and overlying air layers, causing an increase in wind speed.

Local and Regional Winds

Local winds are caused by short-term, often diurnal, temperature changes that create short-term pressure gradients. Land breezes and sea or lake breezes are local winds caused by the diurnal differences in temperature and pressure over land and water. During the day, when land warms more rapidly than water, a sea breeze blows from the cooler, higher pressure air over water toward the warmer, lower pressure air over the land. These breezes may reach several hundred yards to several miles inland before their momentum is stopped, partly because of friction over the surface. In addition, the pressure gradient becomes less sharp farther inland from the beach. At night, a land breeze forms when the pressure gradient reverses—after land has cooled more quickly than a lake or ocean.

Differential heating in hilly terrain causes another set of local winds: mountain and valley breezes. In daytime, slopes—especially those facing the sun—warm more and therefore have lower atmospheric pressure than surrounding air at the same altitude. As the less dense air warmed by a slope rises, the cooler and denser surrounding air flows toward the slope, creating a valley breeze. At night, slopes—especially upper slopes in thinner air—radiate energy and cool more rapidly than lower slopes; the cooler, denser air descends from the upper slopes as a mountain breeze.

Regional winds can also be caused by cold, dense air descending from higher elevations. These winds, often more pronounced in the colder temperatures of winter, are the katabatic winds associated with glaciers, high plateaus, and mountains. Katabatic winds in Europe have such names as the mistral of southern France and the bora along the Dalmatian Coast. Other katabatic winds are the tabu of Alaska and the glacial winds in Antarctica.

A second type of regional wind is the *chinook* of the Rockies, called the fohn, in Europe. Chinooks can gust to one hundred miles per hour, and temperature changes can be extremely rapid: In Alberta, Canada, chinook winds raised air temperature 38°F in four minutes on January 6, 1966. Santa Anas are easterly winds blown over the Sierra Nevada from high pressure areas in Utah and Nevada into southern California, where they contribute to the fall fire season by dehydrating vegetation. A comparable wind, the zonda, occurs in Argentina after air has blown over the Andes.

Monsoons

Monsoons are regional winds that reverse direction between summer and winter and dominate coastal Asia from Pakistan to Japan. Weaker versions of monsoons also occur in western Africa and northern Australia. Several billion Asians and Africans rely on the complex monsoon winds because they bring moisture and rainfall essential for growing crops. Even though monsoons are generally seasonally predictable, the duration and amount of monsoonal winds and rain can vary from summer to summer. Occasionally monsoons fail to occur and the resulting summer drought con-

Valley breezes occur in daytime as cool dense air warms and rises up warm mountain slopes (A), while mountain breezes occur at night when the air is cooled and descends downslope (B). Sea breezes occur in daytime when cooler sea air rushes inland to replace sun-warmed land air (C), and land breezes occur at night when the cooled land air rushes back out to sea (D).

tributes to conditions that cause starvation and thousands of deaths.

The Eurasian landmass cools in winter, forming a high-pressure area over Siberia; in summer the landmass warms, resulting in the formation of a low-pressure area. These changing patterns of pressure cause winds to shift. However, recent studies show that the position of a monsoon is also affected by the shifting of the solar equator and ITCZ, as well as by changes in the polar jet stream. Shifting of the ITCZ reinforces the effect of the seasonal pressure changes over Eurasia and its surrounding oceans. More at:

http://www.usda.gov/oce/waob/ja wf/profiles/specials/monsoon/mon-soon.htm

GLOBAL WINDS AND OCEAN CURRENTS

Global atmospheric pressure zones and the Coriolis effect create a global circulation system of surface and high-altitude winds as well as ocean currents moved by surface winds. Global winds move weather systems, and therefore heat energy and moisture, around the world; heat and moisture are also redistributed by ocean currents. Understanding the spatial and temporal patterns of global circulation is key to understanding seasonal patterns of climates.

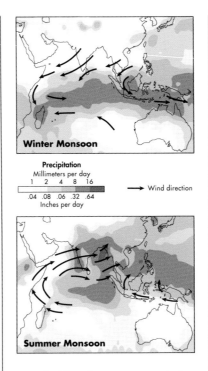

In Asia, land-based winter monsoon winds are generally dry. Summer monsoon winds originate over warm oceans and bring drenching rains.

Causes of Atmospheric Circulation

Unlike local or regional winds, which are seasonal and limited to one continent or region, global winds blow throughout the year and circle Earth. While global winds are consistent features of Earth's environment, their paths shift in response to changes in regional pressure systems and the movement of the solar equator. These shifts, however, are limited, and global winds provide a consistent basis for generating precipitation patterns and global climates.

Surface high- and low-pressure systems create pressure gradients that drive global winds. Air rises at the Equator because of intense solar heating, leaving lower pressure at the surface and increasing air pressure high above the Equator as more air rises to the higher altitude. This rising air will eventually return to the surface when it loses heat by radiation to space and therefore becomes denser than surrounding air. The air descends at 30° north and south latitudes, creating surface high-pressure zones.

High-pressure areas centered on the Poles and low-pressure zones around 60° north and south latitudes are also part of the global pressure system. Polar highs are caused by extremely low temperatures.

In addition to pressure zones, the patterns of global winds reflect the influence of the Coriolis effect. The westerlies circle the globe at more than 40 miles per hour; as a rule, they are diverted about 45 degrees to the right or left from the pressure gradient because of the Coriolis effect. As a result global winds do not blow from north to south (or south to north) between pressure systems. Instead, winds tend to blow from a quadrant, or between two cardinal directions: The westerlies blow from the southwest in the Northern Hemisphere and

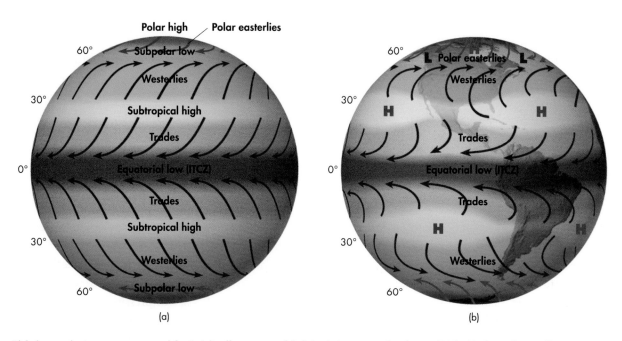

Global atmospheric pressure zones and the Coriolis effect create a global circulation system of surface and high-altitude winds, as well as ocean currents (above) moved by surface winds. Global winds move weather systems, and therfore heat energy and moisture, around the world; heat and moisture are also redistributed by ocean currents. Understanding these patterns is key to understanding seasonal patterns of climates.

from the northwest in the Southern Hemisphere.

Surface Patterns of Atmospheric Circulation

Earth's global atmospheric circulation system is a pattern of wind and pressure zones centered on the doldrums at the ITCZ. Two consistent wind zones are created in each hemisphere: the trade winds, or easterlies, and the westerlies. Near the North and South Poles, winds are often easterly, but they are more inconsistent than winds in other zones of global circulation.

North and south of the Equator are the trade winds, which were known to sailors long before Columbus used them on his first voyage to the Americas in 1492. The trades are more consistent than other global winds; they blow from the eastern quadrant almost 80 percent of the time because the pressure gradient between the tropics and the Equator is so consistent. While the pressure gradient for the trades is from high to low pressure, or north to south in the Northern Hemisphere, the Coriolis effect deflects the wind to a northeast wind. In the Southern Hemisphere, with a south-to-north pressure gradient and a leftward Coriolis effect, the trade winds are southeasterly.

A high-pressure zone with inconsistent winds is centered on 30° north and south latitudes. The surface high pressure along these latitudes causes air to flow outward to the Equator and toward higher latitudes.

Between 35° and 60° north and south of the solar equator are the westerlies. The westerlies are interrupted in the Northern Hemisphere by continents whose pressure systems and mountains interfere with smooth airflow and cause increased variability in wind direction. Nevertheless, most weather systems in these midlatitudes still travel from west to east. In the Southern Hemi-

Global Circulation and Sailing

Sailors named different global winds and pressure systems based on their influence on sailing. The lack of consistent winds at the Equator and along the Tropics of Cancer and Capricorn resulted in sailors being becalmed. If they were immobile for days, they ate or tossed overboard livestock; hence, the name horse latitudes. English poet Samuel Taylor Coleridge (1772–1834) described the experience of wind-lessness in *The Rime of the Ancient Mariner:* "Day after day, day after day/We stuck, nor breath nor motion;/As idle as a painted ship/Upon a painted ocean." Westerly and easterly winds were named for the directions from which the winds consistently blew; easterlies are also called trade winds, a term derived from *trado*, Latin for "direction."

sphere—especially south of South America, Africa, and Australia, and north of Antarctica—there is less land area to impede winds, which allows the westerlies to be more consistent than in the north.

At higher latitudes than the westerlies are polar easterlies. These winds blow from the high pressure caused by extremely cold temperatures at the Poles to the relatively lower pressure at 60° north and south latitudes where the polar fronts are located. Polar easterlies are curved to the west by the Coriolis effect, which is strongest at the Poles. The Antarctic has higher pressure and more consistent winds than the Arctic because of increased cooling from its larger ice-covered surface area. Winds at both Poles, however, are less consistent than those in other global wind zones because of changes in the location and the intensity of the polar pressure systems.

Earth's global wind patterns are far more variable than the system described above. Seasonal shifts in the intensity of insolation and temperature cause the surface pressure conditions to change and therefore affect global winds. The distribution of continents and oceans creates pressure systems that displace global wind patterns; major mountain ranges may also interrupt their flow. Passing storms and frontal activity also disturb wind patterns.

One of the most important influences on global wind patterns is the seasonal shift of the intertropical convergence zone (ITCZ), caused by the movement of the solar equator. During summer in the Northern Hemisphere, when the solar equator coincides with the Tropic of Cancer, the entire global wind system shifts slightly north. In winter, when sunlight shines vertically down on the Tropic of Capricorn, the system shifts slightly south. Precipitation patterns also shift north and south in response to the movement of global pressure zones and winds. Southern California's weather is affected by the doldrums shifting northward in summer, causing hot, dry weather, and by the westerlies shifting south in winter, causing a rainy season with moisture from the Pacific Ocean.

In general, the ITCZ shifts farther north and south between seasons over continents than it does over oceans; water has greater thermal capacity than land, resulting in a slower rate of change in water temperature. Over the Atlantic and central Pacific the ITCZ shifts less than five degrees between seasons, half as much as it shifts over Central America and Africa. Other pressure systems can affect this seasonal shift in the position of the ITCZ: In the monsoon region of South Asia and the Indian Ocean, the ITCZ can shift almost 40 degrees between seasons. The high temperature and low pressure over the land in summer bring the

ITCZ north to the Himalaya; and the low temperature and high pressure over the land in winter force it south of the Seychelles.

Upper-Air Wind Patterns

The upper-air wind component of the global circulation system differs from surface winds in direction and speed because pressure systems and the Coriolis effect vary with altitude. High-altitude winds have greater velocity than surface winds because there is less friction at higher altitudes, and the Coriolis effect increases with speed.

In the midlatitudes, an overall upper-level pressure gradient exists from higher pressure nearer the tropics to lower pressure nearer the Poles. The Coriolis effect curves the resulting upper-air winds toward the east in both hemispheres. Midlatitude westerlies are especially important in directing weather systems over North America and Eurasia.

Midlatitude, upper-air westerlies vary in velocity between seasons because of seasonal changes in temperature and pressure. Polar regions are cool in summer but are extremely cold in winter, while tropical temperatures are consistently warm year-round. As a result there is a greater temperature gradient in winter than in summer; winter, therefore, has greater pressure gra-

The Roaring Forties

During the 1700s and 1800s, before steam power replaced sails, sailing from the Atlantic to the Pacific Ocean around Cape Horn at the southern tip of South America was very hazardous and time-consuming. Rounding the Horn involved tacking into the roaring forties, strong westerlies between 40° and 50° south latitude. Delays of a month or more were possible; some captains chose to sail 8,500 miles eastward across the Atlantic and Indian Oceans to the Pacific to avoid these strong winds.

dients that cause faster winds. Wind speeds in the upper-air westerlies are difficult to specify, but they may average more than 50 miles per hour; jet streams in the westerlies are much faster, up to 150 miles an hour. (See Jet Streams, below.)

Jet Streams

Jet streams are the fastest of the upper-air winds and are influential in directing the movement of surface weather systems. The speed of a jet stream ordinarily varies between 50 and 150 miles per hour, and jet streaks, short sections of jet streams, may top 200 miles per hour as they circle the globe. These speeds are the result of intense pressure gradients in the upper atmosphere and are found only in zones of great temperature contrast. Jet stream speeds are, like those of the upper-air westerlies, faster in winter when temperature contrasts are greater.

While lower-latitude jet streams exist, such as the subtropical jet stream, the polar-front jet stream actually affects more of the world's populated regions. As it travels eastward, the jet stream literally pulls along the top of surface weather systems, while surface friction slows the systems to one-third the speed of the jet stream. During the summer months, the polar-front jet stream is centered at about 45° north and south latitudes, but in winter the cooling of the polar regions causes it to move about 10 or 15 degrees closer to the Equator.

The polar-front jet streams reinforce the effect of the upper-air westerlies on surface weather; together, they increase the eastward motion of weather systems. As a consequence the midlatitudes experience far greater day-to-day weather variations than other latitudes.

Ocean Currents

Earth's global wind system is the driving mechanism for surface ocean currents, which transport heat from the tropics to the polar regions, influence the location of major fisheries, and affect coastal climates around the world. Surface currents are continuous, nearly circular flows of surface water that extend to a depth of about 300 feet and are centered in the horse latitudes. (See Earth's Surface Ocean Currents map, page 63.) Deep-water currents ranging in depth from about 2,000 feet to the ocean floor also flow from Antarctica to Greenland and the Bering *Strait*. These deep-water currents, called density currents, are caused by decreases in water temperature to below 30°F and increases in salinity. They support major fishing grounds and are important in distributing oxygen and nutrients throughout the oceans, but they do not affect climates significantly.

Surface currents are relatively slow, averaging one or two miles per hour, and their motion is the result of frictional drag on the water surface by the global wind system. Waves, the up-and-down motion of the ocean's surface, are also the result of the frictional drag of wind, but their motion is temporary and caused by storms and local or regional winds. *Tides*, the regular rise and fall of ocean levels, are caused by the gravitational pull of the moon and sun. (See Waves and Coastal Landforms, page 156.) Ocean currents move continuously because the trade winds and westerlies are constant.

Weather Prediction

Weather proverbs such as "Red sky in the morning, sailor take warning," have given way to computer modeling for predicting the weather. Accurate forecasting requires predicting how rapidly and from what locations surface and high-altitude winds will bring new weather systems. Modeling and mapping of atmospheric conditions has improved with greater understanding of atmospheric processes, improved data, and faster computers.
More at:
http://cirrus.sprl.umich.edu/wxnet/

Current speeds are far slower than wind speeds because water has great inertia, and the kinetic energy of wind transfers directly only to the ocean surface.

Surface ocean currents flow in a circular pattern because landmasses divert currents sideways and the Coriolis effect causes the currents to flow at a 45-degree angle to the prevailing wind direction. When the trade winds blow in the North Atlantic, the northeasterly winds result in an easterly ocean current.

While ocean currents average a few miles per hour, currents on the west side of oceans—along the east coasts of landmasses—tend to flow

up to several miles per hour faster than currents on the east side of oceans—or along west coasts. For example, in the North Atlantic Ocean, the Gulf Stream on the west side of the ocean flows more than five miles per hour, while the Canary Current on the east side often moves at less than one mile per hour.

Ocean currents affect climates primarily because water temperatures affect atmospheric humidity. When the trade winds cause ocean currents to flow parallel to the Equator, the water warms by absorbing insolation. When the Gulf Stream and Brazil Current curve away from the Equator alongside east coasts because of the Coriolis effect, the water warms the coasts and increases the specific humidity in the area. (See Global Climate Patterns, page 87.) When an ocean current, such as the California Current, heads back to the Equator after traveling to polar regions, it has cooled between 5° and 10°F and, though it still moderates temperature extremes, it cannot increase humidity or precipitation significantly.

Ocean currents have a range of other impacts on humans: Fishing and shipping are affected by upwelling—cold, deep water rises to replace surface water—and by water temperatures. Most of the commercial fisheries in the world's

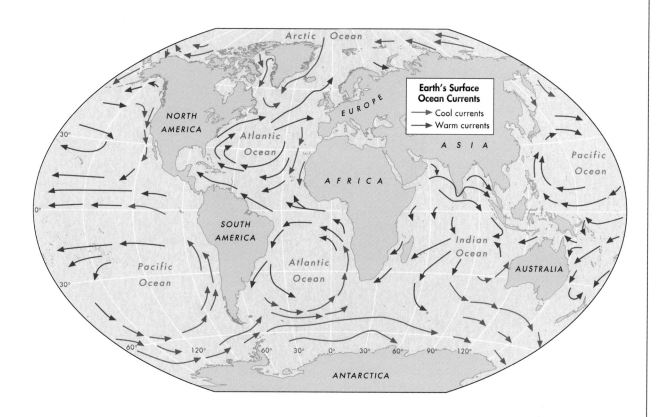

Cool currents convey cold water from the Poles to the tropics; warm currents convey warmer water from the tropics toward the Poles. Ocean currents are profoundly affected by prevailing winds.

oceans are found in waters whose temperatures are below 35°F and where currents cause upwelling. Off the west coast of South America, for example, the cool Peru Current, once called the Humboldt Current, moves north along the coast until it is pulled away from Ecuador as part of the South Pacific circulation pattern, causing nutrient-rich lower waters to upwell and support the food chain for anchovies and other fish. Ocean currents also impact shipping routes: The cooling but still-warm North Atlantic Drift keeps the ports of Norway ice-free during winter, whereas pack ice blocks the coasts of Greenland and Newfoundland.

El Niño and the Southern Oscillation (ENSO)

El Niño, a periodic reversal of the pattern of current flow and water temperatures in the mid-Pacific Ocean, is related to changes in weather around much of the world; *La Niña*, another periodic event in the mid-Pacific, also affects global weather. The exact causes of El Niño have not yet been confirmed, but there seem to be links between changes in the direction of the trade winds and in Pacific Ocean currents. In normal years, the trade winds and the North and South Equatorial Currents flow to the west, and warm water accumulates near Borneo and Indonesia. In this western region of the Pacific Ocean water levels can reach 18 inches higher than levels at the eastern edge, and the accumulation of warm water can cause atmospheric pressure over the western Pacific to drop.

In October or November preceding an El Niño, the low pressure over the western Pacific changes to higher pressure while the higher pressure in the east falls. This southern oscillation of atmospheric pressure causes the trade winds to fail, if not to reverse direction, and the ITCZ to be displaced five degrees or more farther south than is usual. As the trade winds stop, the equatorial currents reverse direction as the accumulation of warm water near Indonesia drifts to the east, bringing warm water to Ecuador, where it spreads north and south along the South American coast.

Ten El Niño events occurred during the last half of the 20th century; most started in December or January—hence the name, El Niño, the Child, referring to Christmas time and Jesus's birth—and lasted from a few months to more than a year. During the more severe El Niños, as in 1992–93 and 1997–98, droughts and floods affected Australia, North America, South America, Asia, and the Pacific Rim. When atmospheric pressure increases in the western Pacific, onshore breezes fail and precipitation drops in Indonesia, Malaysia, and northern

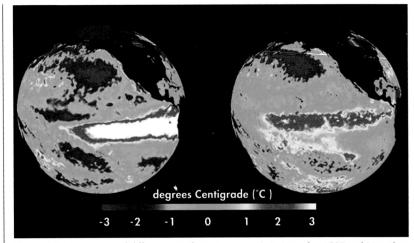

degrees Centigrade (°C)

-3   -2   -1   0   1   2   3

These satellite images reveal differences in El Niño's warm strip in December 1997 and December 2002. Sea-surface temperature anomalies in these maps were computed from measurements of sea-surface temperatures collected by the AVHRR sensor on the NOAA polar orbiting satellites.

Australia; forest fires have been more common during El Niño because vegetation is drier. When the water warms and pressure drops off South America, convection and precipitation increase, occasionally causing flooding from Callao, Peru, to Quito, Ecuador.

El Niño has long-distance effects on air pressure that link weather patterns in the United States, Canada, and elsewhere. Warmer water on the west coast of North America tends to cause milder winters for coastal Alaska and British Columbia, and changing precipitation patterns have been linked to flooding in the southern United States.

La Niña occurs when equatorial waters in the Pacific become colder than normal. Stronger-than-normal trade winds off the west coast of

South America cause more upwelling along the coast, which brings more cold water from the deep to the surface. La Niña sometimes, but not always, alternates with El Niño and causes opposite changes in weather around the world. In India, for example, monsoon rains decrease during an El Niño but increase during a La Niña. More at:

http://www.pmel.noaa.gov/tao/eln ino/el-nino-story.html

ATMOSPHERIC MOISTURE

Water vapor in the atmosphere cools as it rises and condenses into liquid water or ice that forms dew, frost, fog, and clouds. When cloud droplets combine, they form rain and snow that fall to Earth or form dew and frost on the surface.

Hydrologic Cycle

Water vapor is the key element in the global energy and hydrologic systems. Vapor, the gaseous state of water, forms when liquid water or ice absorbs sufficient heat from the sun or Earth to change to a gas. The energy necessary for the change is stored in the water vapor as latent heat, which is released as sensible heat when the vapor changes form again, to either a solid or a liquid. As water vapor moves from one place to another, especially from tropical latitudes to the Polar regions, the stored energy is released as the vapor condenses, thus redistributing heat around Earth.

The movement of water in its different forms around the globe is called the hydrologic cycle, and this circulation connects oceans, continents, groundwater, and the atmosphere. (See hydrologic cycle diagram, below.)

Although the water vapor in the atmosphere amounts to only .001 percent of all water on Earth, the rate of exchange of water between the atmosphere and the surface through condensation and precipitation is rapid, taking an average of ten days. Oceans contribute 85 percent of the water vapor in the atmosphere—almost a hundred quadrillion gallons per year evaporate from ocean surfaces, or almost 3 gallons per square foot of ocean. The rest of the water vapor in the atmosphere comes from surface water evaporation on continents and transpiration from vegetation.

More at:
http://watercycle.gsfc.nasa.gov/

Atmospheric Humidity

The amount of water vapor contained in the atmosphere, or *humidity,* varies over space and time and is measured in terms of absolute and specific humidity. Absolute humidity is the ratio of the weight of water vapor to that of a given volume of air, and specific humidity is the ratio of the weight of water vapor of a given weight of air. A meteorologist can estimate how much potential precipitation exists when the specific humidity is known. The specific humidity can range from less than .02 ounces to nearly one ounce of water vapor per pound of atmosphere. Absolute humidity is not often mentioned in weather reports because the volume of a given mass of air will change as a function of temperature and therefore change the absolute humidity.

Condensation, precipitation, and evaporation result from changes in the *relative humidity* of the atmosphere—the ratio of the water vapor actually present to the maximum capacity at the same temperature. Saturation pressure, the maximum possible amount of water vapor in the air, depends on the temperature of the atmosphere: Warmer air contains more energy and therefore can contain more

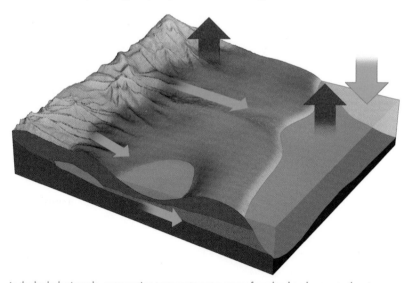

In the hydrologic cycle, evaporation transports water vapor from land and ocean to the atmosphere. Water returns to Earth in the form of precipitation. On land, water constantly seeks lower ground, often flowing into the ocean.

water vapor than cooler air. As air warms, the saturation pressure in the atmosphere increases and more water vapor can be held in the air. If the actual amount of water vapor remains the same, then the ratio between actual vapor pressure and saturation pressure changes and relative humidity drops. Conversely, as air cools, relative humidity increases because the saturation pressure decreases and approaches the actual amount of water vapor in the atmosphere.

Condensation

When the relative humidity of air reaches 100 percent, saturation occurs. At saturation, no more water vapor and therefore no more water-vapor pressure can be added to the air because the gases of the atmosphere do not have enough kinetic energy to evaporate more water. When relative humidity is high, people have difficulty cooling down because evaporation of perspiration, the process that cools the human body by absorbing heat energy, is prevented by the nearly saturated air.

As air cools below the saturation level of the atmosphere, the maximum capacity of the atmosphere drops below the actual amount of water vapor in the air and condensation, the phase change of water from a vapor to a

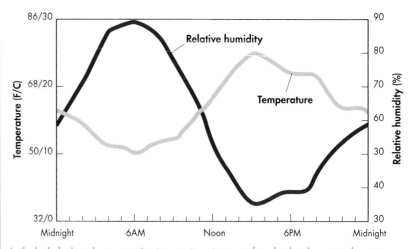

In the hydrologic cycle, evaporation transports water vapor from land and ocean to the atmosphere. Water returns to Earth in the form of precipitation. On land, water constantly seeks lower ground, often flowing into the ocean.

liquid or solid, occurs. Condensation removes water vapor from the air and forms cloud droplets, dew, and frost. Sublimation refers to water vapor that changes directly to ice crystals at subfreezing temperatures. The process of condensation starts to occur at saturation and persists as long as the air continues to cool, during which time the relative humidity remains at 100 percent. Condensation creates clouds and the water droplets that form on a glass holding an icy drink after the cold glass has chilled the air next to it.

The dew point is the temperature at which condensation starts, and it depends on ambient air temperature and the specific humidity of the air. Warm air over a tropical ocean will have higher specific humidity than air the same temperature over a desert because of increased evaporation. The dew point for marine air may be 80°F while the dew point for desert air may be 35°F because its lower specific humidity provides so much less water vapor to start with. During summer in the United States, the lowest dew point temperatures are customarily registered in deserts and the Rocky Mountains, while the highest dew point temperatures are along the Gulf of Mexico.

Condensation of water vapor normally requires a surface on which water or ice—depending upon air temperature—can form. Without a surface, water vapor eventually forms water droplets, but the air must cool below the dew point level, thus becoming supersaturated. While dew and frost form on grass stems and windows, cloud droplets need microscopic particles on which to form. Almost all air contains particles that attract water. Examples include sea salt in the atmosphere over oceans and dust, pollen, smoke, and bacteria in air over land. As a result, even rainwater is a solution containing other substances; no water in nature is pure water. (See Weathering, page 127.)

As air temperature rises and falls throughout the day, relative humidity changes, too. As daily temperature drops after its mid-afternoon maximum, relative humidity begins to rise. When daily temperature reaches its minimum shortly after dawn, relative humidity peaks. Saturation of the air is most likely to be reached in surface air near dawn, and dew or frost tends to form then. Surfaces are also cooler than the atmosphere at dawn because they have cooled by radiation, while the air that absorbed the radiation has been warmed. As a result, dew or frost may form even if air temperatures reported by the Weather Service are not below the dew point.

In addition to changes caused by the daily insolation cycle, air temperature and relative humidity also change when air expands or compresses. When air descends and comes under greater atmospheric pressure, the air always warms. No condensation can occur because as the air warms its capacity for evaporation and water-vapor pressure increases. When absolute humidity is stable and air's capacity to hold water vapor increases, lower relative humidity results.

Areas of downdrafts and the leeward sides of mountains tend to have far fewer clouds and much less precipitation than areas of updrafts and the windward sides of mountains. Even in Hawaii, where rain forests dominate, western leeward sides of the islands have less precipitation. On Hawaii, the shrub and grasslands of the Kona coast on the west receive less than 20 inches of rain per year while Hilo, 60 miles to the east—and windward—receives 100 inches.

Formation of Clouds

Clouds are visible masses of water droplets or ice crystals that form when air rises and cools below saturation level, causing condensation. On average, about 52 percent of Earth is cloud-covered at any moment. Clouds provide rain and snow, the dominant types of precipitation. Dew and frost are also considered precipitation and are virtually the only water available for vegetation in deserts such as the Atacama in northern Chile. Different kinds of clouds result from different movements in Earth's atmosphere and are therefore associated with different types of weather and climates.

While clouds are almost always the result of the cooling of air, the lifting of air is the result of convection, orographic uplift, frontal activity, or convergence. Each process occurs in different spatial and temporal patterns that may coexist and interact in different locations. Different locations have predictable and seasonal types of atmospheric uplift, clouds, and precipitation.

Convection is the spontaneous rise of air after it has been warmed by surface radiation and has become less dense than the surrounding air. Equatorial regions receive more solar energy than elsewhere in the world and are more likely to experience convection consistently through the year. While regions in the midlatitudes also experience convection, it is usually greatest in summer, when land surfaces absorb the largest amounts of solar radiation.

Orographic uplift results when winds force air to rise over elevated land. Hilly islands, for example, cause orographic uplift, but the most important global patterns are associated with the windward coastlines of continents in the path of trade winds or the westerlies.

British Columbia, Norway, and New Zealand have cloudy and wet coasts much of the year because of the dominant westerlies and mountain ranges parallel to their coastlines. The east slopes of the Hawaiian Islands, Madagascar, and the Bahamas also tend to be cloudy and wet because of the dominant trade winds, especially in the afternoons when heat causes convection to increase the rate of uplift.

Frontal activity occurs when two air masses of unequal temperatures meet; the colder air pushes under the warmer air, thereby lifting it. Where fronts, or boundaries, between the air masses exist, precipitation and winds occur. Frontal activity is most likely to take place in the midlatitudes, where warmer tropical and subtropical air masses meet colder subpolar air masses; the weather in North America and Eurasia is dominated by frontal activity. The polar-front jet stream, a high-altitude wind above the boundary of cold polar and warm tropical air, contributes to frontal activity by keeping the air masses moving.

The least common cause of cooling occurs at the Equator, where the northeasterly and southeasterly trade winds come together, forcing the air to rise. Any convergence of air can contribute to rising, cooling, and condensation. Hurricanes and

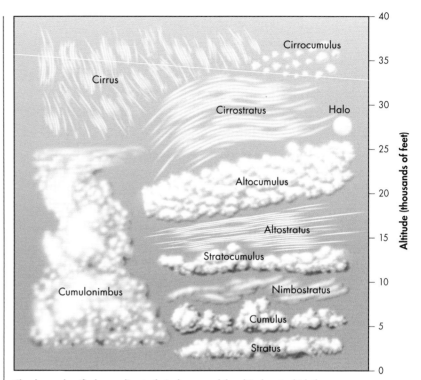

Clouds are classified according to their shapes and the altitudes at which they are commonly found.

other cyclonic storms have a convergence of inflowing air that interacts with convection to cause condensation. (See Weather Systems, page 71.)

Classification of Clouds

Clouds are classified according to their appearance and altitude. Cloud shapes can be categorized as puffy (cumulus), layered (stratus), or wispy (cirrus), and these terms may also refer to their altitudes. Low clouds, or stratus, are below 6,500 feet (but not on the surface); middle clouds (marked by the prefix *alto*) are between 6,500 and 20,000 feet; and high clouds, cirrus, are above 20,000 feet. Cumulus clouds, which have extensive vertical development, are classified separately and may extend from near the surface to more than 50,000 feet. When precipitation falls from clouds, the cloud types have "nimbo" or "nimbus" added as a prefix or suffix to their names, such as nimbostratus and cumulonimbus.

The appearances of clouds provide information on what atmospheric processes are most likely to be at work. Puffy cumulus clouds, including gigantic cumulonimbus storm clouds, are usually initiated by convection, although orographic and frontal activity can contribute to lifting the air. Stratus clouds are most commonly associated with frontal activity, especially warm fronts, although convergence can also be involved. Fog is also classified as a cloud, but it is a surface cloud. Fog is distinguished from mist, a less dense surface cloud, in that fog restricts visibility to 3,250 feet. Radiation fogs result when clear skies allow surface radiation to escape to space and the lowest part of the atmosphere cools rapidly. Valleys, such as coves in the Appalachians and much of California's San Joaquin Valley, often have radiation fogs, especially in winter when the ground is damp and surface humidity is high. Advection fogs occur when humid air advances from a warmer surface to a colder surface. San Francisco and the Grand Banks of Newfoundland experience advection fogs. Avalon, Newfoundland, is the foggiest location in Canada, and Cape Disappointment, Washington, is the foggiest location in the United States; each place has fog nearly 30 percent of the time. Fog occasionally forms along mountain slopes because of orographic uplift and is called upslope fog. If an upslope fog rises above the slope, it becomes a stratus cloud.

More at:
http://www.srh.weather.gov/srh/jetstream/synoptic/clouds.htm

Precipitation

Precipitation returns water from the atmosphere to Earth's surface and continues the hydrologic cycle of water moving from oceans, to land, and to air. Rain and snow support life, carve landforms, and transport energy and material before returning to the atmosphere as water vapor through evaporation and transpiration. Other than dew and frost, which are precipitated directly on Earth, precipitation falls from clouds to the surface.

Rain and drizzle consist of droplets—an average raindrop has approximately one million droplets—that reach the surface. Individual cloud droplets are too small and too light to fall as rain, and each raindrop must be heavy enough to fall through slight updrafts and large enough not to evaporate during the fall through drier air. Virga is rain that evaporates before reaching the surface. In the midlatitudes, most raindrops start out as snowflakes that melt during their fall to Earth.

Snow occurs when snowflakes do not melt or evaporate during their fall to the surface. While snow falls at high elevations along the Equator—Chimborazo in Ecuador and Mount Kenya in Kenya are snowcapped year-round—snow is more common in higher latitudes and in colder seasons.

Sleet and hail are different forms of frozen precipitation. Sleet is rain that freezes during its fall to Earth and requires a cold inversion layer thick enough to allow raindrops to freeze during their fall. Hail is another form of rain that freezes prior to reaching Earth's surface. Unlike sleet's raindrops that freeze as they fall, hail's raindrops freeze as strong updrafts in cumulonimbus clouds carry them up repeatedly into colder parts of the cloud.

Sleet's appearance differs from that of hail: Sleet is a single frozen raindrop, but a hailstone is layers of ice that have frozen onto a central ice pellet during the several cycles of being lifted by updrafts to the freezing tops of clouds.

Global Precipitation Patterns

Global precipitation patterns are often seasonal and are related to global patterns of insolation, temperature, atmospheric pressure, wind, and ocean currents. Precipitation depends upon the processes caused by surface warming (convection), low pressure (convergence), and uplift (orographic uplift

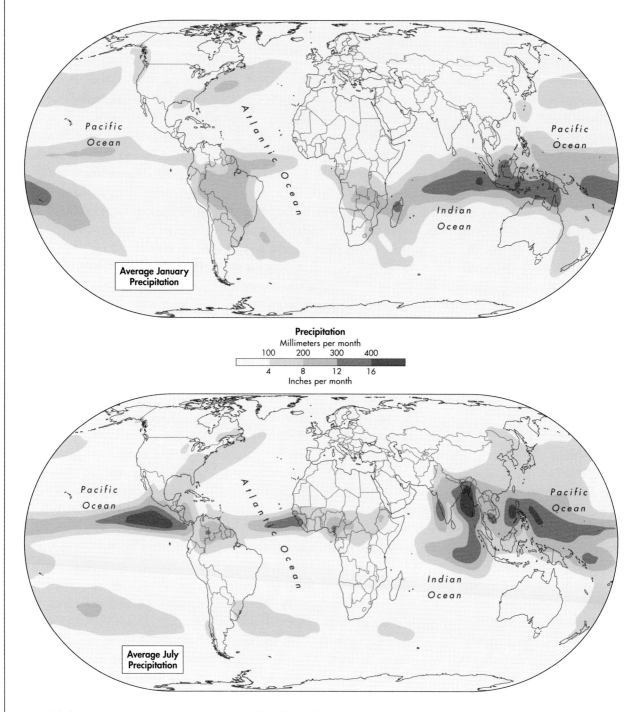

Precipitation

Millimeters per month

100   200   300   400

4   8   12   16

Inches per month

Average January Precipitation

Average July Precipitation

Global precipitation levels vary by season and are affected by insolation, temperature, atmospheric pressure, wind, and ocean currents.

and frontal activity). Ocean currents affect precipitation because the temperature of the water affects the rate of evaporation and the amount of water vapor available for condensation and precipitation. Regional winds and other factors affect precipitation patterns: In South and Southeast Asia, trade winds bring summer precipitation that is dramatically increased by the summer monsoon winds. (See Monsoons, page 57.)

Seasonal precipitation is higher near the solar equator and on the coasts of continents downwind from the trade winds and westerlies. During the Northern Hemisphere's winter, when the solar equator coincides with the Tropic of Capricorn, one band of high precipitation is centered at 5° to 10° south latitude, where convection and convergence from the trade winds occur. Regions at these latitudes, such as central Brazil and the Democratic Republic of the Congo, receive more than 40 inches of rain from January to March. East coasts just north and south of the solar equator, such as northeastern Australia and Mozambique, receive orographic precipitation as the trade winds blow humid marine air over land. Finally, the west coasts struck by the westerlies also receive precipitation from orographic uplift; coastal California and north-

ern Morocco usually receive just over 20 inches of rain in winter, almost all the precipitation for an entire year.

During the Northern Hemisphere's summer, when the solar equator shifts north to the Tropic of Cancer, precipitation patterns also shift north. Convection and convergence at the ITCZ result in a band of precipitation just north of the Equator. Nigeria and the coast of Guyana have their wettest months in June, July, and August. Trade winds also shift farther north in summer, bringing precipitation to Central America and Taiwan. The westerlies and the orographic precipitation caused by the trade winds are also displaced northward. Landforms affect precipitation from the trade winds and westerlies: Eastern Washington State and western Nicaragua are in the *rain shadows* on the leeward side of mountains and are far drier than windward western Washington and eastern Nicaragua.

Global patterns of precipitation are matched by global patterns of nonprecipitation, especially in areas of high atmospheric pressure. Air descending at the tropics warms and has decreasing relative humidity because its specific humidity remains constant. These latitudinal bands north and south of the Equator are therefore kept dry, resulting in the

Sahara, Mojave, and deserts of Australia. West coasts in the latitudes of the trade winds and regions east of the coastlines where the westerlies blow—for example, Montana and interior Siberia—are dry for the same reason: They are far from the source of humidity (oceans). Finally, the Arctic and Antarctic are extremely dry because of high pressure and low absolute humidity caused by low temperatures.

WEATHER SYSTEMS

Weather systems are high- and low-pressure areas, such as air masses and fronts that form and dissolve over time as they move from one location to another. Frontal activity and variable weather are most common in the midlatitudes, where air masses from tropical and polar regions meet: The tropics and polar regions have more consistent weather conditions. Thunderstorms, tornadoes, and hurricanes—cyclonic weather systems that can be natural hazards—occur in distinct patterns on Earth. Understanding weather systems is key to understanding the spatial patterns of climate and atmospheric hazards.

Weather

Weather, in contrast to climate, is the short-term state of the atmosphere for an area, and weather systems are short-lived, regional combinations of temperature, wind, humidity, and precipitation that move as units. Weather systems such as hurricanes may cover thousands of square miles or, as in the case of a tornado, just a few acres; they may travel halfway around the world or dissipate after one mile. The dominant types of weather systems are high-pressure, anticyclonic systems characterized by clear skies and calm weather and low-pressure cyclonic systems characterized by cloudy skies and stormy weather.

Weather systems change as they travel over Earth because they contact different surfaces and encounter different weather systems. Their atmospheric components are related, and components change as systems form and dissipate over time. For example, a weather system that forms over a tropical ocean, such as a *hurricane*, changes rapidly by losing both temperature and wind speed as it moves over relatively cold, upper-latitude landmasses.

Air Masses

Large units of air that may cover a million square miles or more are called air masses; cold air masses may extend only a few thousand feet up, and warm air masses may extend through the troposphere. Atmospheric conditions determine

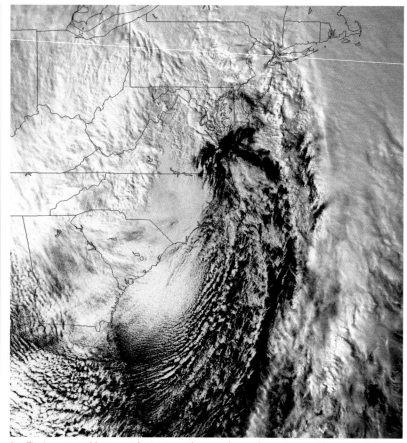

Satellite images enable meteorologists to see large-scale weather patterns, aiding not only immediate weather prediction but also long-term scientific research.

the stability of an air mass, how much precipitation it may have, how dense it is, and how it will interact with another air mass. Each air mass has relatively uniform atmospheric conditions and does not mix easily with different air masses. Most weather systems in the midlatitudes are either air masses or the frontal systems created by the interaction between different air masses. There are few frontal systems in polar and tropical latitudes because these regions have little variation in their atmospheric conditions.

Air masses form when air is stationary for several days over a source region, a relatively homogeneous surface such as an ocean or a landmass. During this time the air takes on the temperature and moisture characteristics of the surface with which it is in contact. Ordinarily there must be high pressure in the region to allow the air mass to remain in contact with the surface and to prevent other systems from moving in and displacing it. Consequently, conditions that are conducive to the formation of stationary or longer term air masses are rare in the midlatitudes because the westerlies, upper-air westerlies, and the polar jet stream are continually moving air, preventing it from having sufficient time to reach equilibrium with the surface. Polar and tropical regions, on the other hand, lack persistent wind systems that bring in contrasting air; thus, most air masses form over these regions.

Air masses are categorized by their source regions and reflect the surface conditions in those regions. Most air masses form either in the tropics and are warm or in polar regions and are cold. For example, they may form at the Equator, in the Arctic, or in the Antarctic; these source regions generate the warmest and coldest air masses. The moisture content or humidity of air masses is related to whether the source region is maritime or continental.

Four types of air masses most often affect locations in North America: maritime-tropical air masses that form over the southern North Atlantic, the Gulf of Mexico, and the southern North Pacific; continental-tropical air masses from Mexico; maritime-polar air masses from the northern Pacific; and continental-polar air masses from Canada. Occasionally, maritime-polar air masses from the northern Atlantic, continental-polar masses from Siberia (the Siberian Express), and continental-arctic air masses from northern Canadian islands also affect the weather in North America. Different types of air masses are more common in different seasons: Continental-tropical air masses rarely develop in winter, and continental-arctic air masses rarely develop in summer.

The characteristics of air masses change as they move from their source region over surfaces that may have different temperatures and amounts of water, say from lakes or oceans to deserts. The changes are greater if air masses move slowly and have extended contact with new surface conditions. After several days or weeks an air mass either dissipates or becomes a different type of air mass based on the characteristics of its new source area. A continental-polar air mass from Canada, for example, will warm up, become more humid, and change into a maritime-tropical air mass if it stays over the Gulf of Mexico. Global winds such as the polar-front jet stream and upper-air winds are most important in moving air masses. Surface winds often reflect the pressure gradients of an individual weather system—the winds that flow inward to the low-pressure centers of a hurricane, for example—and therefore should not be interpreted as moving the weather system.

Fronts

A *front* is a boundary between two air masses with different densities. The front forms when air masses of different densities move and collide with each other. Although different air masses may mix over a number of days, they do not do so easily: Differences in temperature and occasionally humidity cause differences in density that keep the air masses distinct. A front can range in size from hundreds of yards to several miles across, relatively short distances when compared to the size of air masses. A front is tilted up from the surface and moves with the air masses; how much a front tilts and how fast it moves depend upon the characteristics and motions of the air masses. Tilt and speed are responsible for the type of weather experienced on the surface.

Midlatitude Systems

Midlatitude cyclones are low-pressure systems, perhaps a thousand

TYPES OF FRONTS	
COLD	◆ Most common type of front ◆ Result when colder, denser air mass, such as a continental-polar, moves into and under warmer, less dense air mass, such as a maritime-tropical ◆ May move at up to 30 miles per hour because of atmospheric pressure gradients and the pull of upper-air winds ◆ Squall lines, cumulonimbus clouds, thunderstorms, and drops in temperatures of 10°F or more are characteristics of cold fronts.
WARM	◆ Occur when relatively colder and warmer air masses meet ◆ Less dense, warm air mass moves into and over denser, colder air mass ◆ Warmer air overrides the denser cold air mass ◆ Move at approximately half the speed of cold fronts because the less dense, warm air cannot push the cold air mass back easily ◆ Cirrus and stratus clouds precede a warm front, often by hundreds of miles, and rain or sleet can occur
OCCLUDED	◆ Result when faster-moving cold front overtakes slower warm front moving in the same direction ◆ As cold front reaches warm front, warm air mass lifted entirely off Earth's surface ◆ Warm fronts commonly become occluded in midlatitudes because weather systems often include cold fronts following warm fronts ◆ Lifting the warm air front can cause heavy rainfalls lasting a few days or less, depending on how much water vapor in the warm air mass condenses and precipitates
STATIONARY	◆ Occur when boundary between two different air masses does not move at the surface ◆ Neither air mass moves as a unit, but wind still blows from the pressure centers of each air mass into the frontal zone where the air is lifted by convergence ◆ Rain occurs along stationary fronts, often in large amounts

miles across, that form at the surface underneath meanders in the polar-front jet stream. On average, at any given time, ten of these low-pressure weather systems—called midlatitude, frontal, or extratropical cyclones—circle the Northern Hemisphere, and often circle the Southern Hemisphere. Some midlatitude cyclones promote convection that results in thunderstorms and can even generate tornadoes, potentially among the most pow-

erful of the cyclonic storms. High-pressure anticyclones are ordinarily paired with and tend to follow each cyclonic system in the area of the midlatitudes.

Anticyclones usually follow midlatitude cyclones, bringing dry weather to the midlatitudes. Winds blow outward from the high-pressure centers of these anticyclones, preventing convection and keeping skies cloud-free. In winter, cold-core anticyclones form over polar sur-

faces and bring cold, clear weather to the midlatitudes; in summer, warm-core anticyclones from lower latitudes bring warm, clear weather.

Throughout their development, midlatitude cyclones follow a path determined primarily by upper-air winds, although surface features such as mountain ranges also have an effect. Upper-air westerlies dominate the midlatitudes, so that most midlatitude cyclones move eastward. The meandering of the jet stream and surrounding upper-air westerlies causes midlatitude cyclones to move north and south as well as eastward. Seasonal shifts in the global-pressure and wind systems also shift the paths of cyclones poleward during the winter and equatorward in the summer.

The travel speed of midlatitude cyclones varies according to the season. In summer, they travel eastward at speeds averaging 20 to 30 miles per hour; surface friction caused by features such as trees and hills slows the eastward pull of the upper-air wind currents on the cyclones. In winter, midlatitude cyclones tend to travel faster, up to 40 miles per hour, because the cold air masses become even colder, creating greater temperature differences; the increased pressure gradients cause faster winds. Individual fronts associated with a midlatitude cyclone also have different rates of speed; cold fronts, in general, travel faster than warm fronts.

Thunderstorms

The most common weather systems on Earth are thunderstorms, low-pressure systems five to ten miles in diameter and sometimes extending to the tropopause. They are formed by convection, and accompanied by lightning, the electrical discharge that heats the atmosphere and causes thunder. Thunderstorms may also occur in conjunction with frontal or orographic lifting and are also associated with cold fronts, which are often components of midlatitude cyclones. At any given time approximately 2,000 thunderstorms are occurring on Earth. The storms are also accompanied by heavy precipitation, hail, strong winds, and occasionally even tornadoes. Some exceptionally large thunderstorms can last for several hours, but most are short-lived and are over within an hour because precipitation causes downdrafts that interfere with the convectional uplift that creates and sustains the systems.

In order to form, thunderstorms require three conditions: High absolute humidity, instability, and rapid uplift; upper-air divergence and surface convergence of air can also be involved. Thunderstorm formation almost always involves equatorial or maritime-tropical air masses because they contain large amounts of water vapor. This large amount of water vapor releases more latent heat upon condensation and counteracts expansion cooling; this contributes to instability and allows the warm air to rise easily. Rapid uplift can be caused by frontal activity, such as a cold front, an occluded front, orographic uplift, convection, convergence, or a combination of these processes. When upper-air divergence maintains the updraft of rising air, upper-air currents contribute to thunderstorm development by causing lower surface pressure.

Cumulus clouds towering six to nine miles high are the basis of thunderstorms. Surface warming and atmospheric instability initiate the rapid rate of uplift that creates the vertical cloud. The high absolute humidity found in the warm air maintains this uplift by releasing latent heat energy from condensing water and cloud droplets. Heating the air lowers the pressure so that the cumulus cloud extends higher into the atmosphere. At higher altitudes the cloud can meet upper-atmospheric winds that blow the top of the cloud downwind, giving it an anvil-shaped appearance.

The global distribution of thunderstorms shows the importance of warm, tropical air masses and the capacity of land to heat more quickly and cause more convection than water. (See Annual Number of Days with Thunderstorms map, page 76.) Central Africa, Brazil, and Southeast Asia have the highest number of thunderstorm days; in parts of Zaire, Cameroon, and Côte d'Ivoire thunderstorms occur more than 180 days per year. Colder climates have far fewer thunderstorms in general, and very few occur poleward of 60° north and south latitudes or where inland temperatures

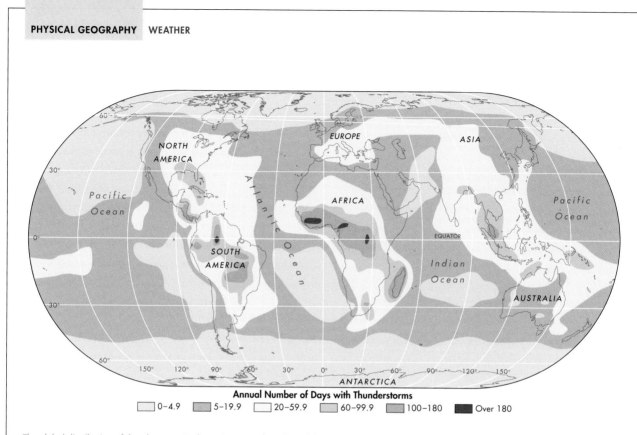

Annual Number of Days with Thunderstorms

☐ 0–4.9 ■ 5–19.9 ☐ 20–59.9 ■ 60–99.9 ■ 100–180 ■ Over 180

The global distribution of thunderstorms is shown in terms of number of days a year a region experiences them. Thunderstorms are the most common weather systems on Earth, with some 2,000 occurring at any given time somewhere on the planet.

are influenced by cold ocean currents. Oceans generally are cooler than land and in summer have fewer thunderstorms.

In the United States, thunderstorms are most common in Florida, but the Front Range on the eastern side of the Rocky Mountains in Colorado is a close second. Warm tropical air converges over Florida from the Atlantic to the east and the Gulf of Mexico to the west; as a result, atmospheric uplift above the central part of the state south of Orlando causes more than a hundred thunderstorm-days per year on average, most in the summer. On summer days in Colorado, south-facing slopes of the Front Range warm rapidly, and the resulting valley winds cause convection. When regional weather systems such as cold fronts intensify this uplift, thunderstorms are most likely to occur.

Midlatitude thunderstorms across North America and Eurasia are often associated with cold fronts or with squall lines, lines of thunderstorms that precede cold fronts. When a cold front moves forward, the warm air mass rises rapidly through frontal uplift and is somewhat compressed by the incoming cold air mass. This convergence of air masses initiates uplift along a line ahead of the cold front and occasionally causes intense thunderstorms.

Thunderstorm Hazards

When heavy rains fall during a thunderstorm, rapid updrafts occur alongside downdrafts. Downdrafts are caused by the fall of rain and snow from the upper reaches of the cumulonimbus cloud. Not only does the falling precipitation cause air to descend, but some of the precipitation evaporates in the lower reaches of the cloud when it reaches air of lower relative humidity. Evaporation absorbs heat energy from the air, causing it to cool, increase in density, and fall toward the surface. As these downdrafts reach Earth's surface, they spread outward and can gust ahead of the storm at speeds up to 60 miles per hour.

Lightning, a giant electrical spark or flow of electrons in the atmosphere, may be a side effect of lighter ice crystals and heavier ice pellets being separated within a cumulonimbus cloud by up- and downdrafts. For lightning to form, positive and negative charges must be separated until the difference in electrical charge is large enough to

overcome air resistance. In thunderstorm clouds, ice pellets and ice crystals develop different charges: Heavy pellets fall faster through a cloud and strike lighter crystals; negatively charged electrons are transferred from the crystals to the pellets, which become negatively charged when temperatures are below 5°F. At higher temperatures, such as those in the lower parts of clouds, the negatively charged electrons transfer from the crystals to the pellets, which then become negatively charged.

As the lighter, positively charged ice crystals are carried to the upper

cloud by updrafts, there is a spatial separation between this positively charged upper cloud and the negatively charged lower cloud as well as between the lower cloud and the positively charged Earth's surface. The surface develops a positive charge to counter the negative field of the lower clouds. When the differences in charge between any two areas is large enough to overcome air's resistance to the flow of electricity, electrons flow as lightning between the charged areas to equalize the number of electrons. The path of the electrons is only an inch or so wide, which is why lightning occurs as a bolt; the presence of clouds and winds can change the appearance of lightning to look like sheets or ribbons. About 80 percent of all lightning discharges occur within clouds, but the 20 percent between clouds and Earth's surface can be a major natural hazard.
More at:
http://www.nws.noaa.gov/om/broc hures/trw.htm

Tornadoes

The most violent cyclonic storm generated by a thunderstorm or by a hurricane is a tornado. Estimated rotational wind speeds may exceed 300 miles per hour, and the central pressure of a tornado's vortex may drop to 27.00 inches or lower, far below the standard 29.92-inch

A single flash of lightning lasting one-quarter of a second may consist of dozens of main strokes. Thunder is produced as lightning instantly heats surrounding air to high temperatures, causing the air molecules to explode or expand rapidly. Sound travels more slowly than light, so thunder is always heard after lightning is seen.

Lightning and People

In an average year lightning in the United States kills nearly a hundred people, starts thousands of forest fires, and damages millions of dollars of electrical equipment. Death rates from lightning are greater in some tropical regions. In Kisii in western Kenya, some 30 people die each year from lightning strikes. Kisii's high rate of lightning fatalities occurs because of the frequency of thunderstorms and because many of the area's structures have metal roofs.

To avoid being struck by lightning, stay away from high areas, tall structures, and metallic objects such as golf clubs; these locations and objects are most likely to build up an electrical charge during a thunderstorm. Lightning can and does strike twice in the same place—the Empire State Building in New York City averages 20 strikes each year—and lightning can occur whether or not it is raining. More at: http://thunder.msfc.nasa.gov/olar.html

---

barometric pressure at sea level. Tornadoes extend from the ground up to about 20,000 feet and are ordinarily small with vortex diameters of about 300 feet, last less than five minutes, and travel less than one mile. A few tornadoes, however, can be nearly one mile in diameter, can last for hours, and can travel more than a hundred miles. Tornadoes may hopscotch from one location to another, but all travel along or near the ground.

Tornadoes are categorized by the F-scale, their rotational wind speed, and the P-scale, the length of their path and width. The intensity of tornadoes is measured on the Fujita scale, which ranges from F-0 (windspeeds from 40 to 72 miles per hour) to F-5 (wind speeds from 261 to 318 miles per hour). Almost 80 percent of all tornadoes are F-0 or F-1 and are considered weak; only one percent are F-4 or F-5 (violent). The P-scale ranks tornadoes into six categories by the length and width of their path; categories range from a tornado 18 feet wide, with a path of 1,600 feet, to about 4,800 feet wide and a path of more than a hundred miles. Rare violent tornadoes, especially those with high P-rankings, cause the most deaths: A 1974 F-5 tornado in Xenia, Ohio, killed 34 people; another F-5 tornado killed 17 people in 1991 in Andover, Kansas.

Less than one percent of thunderstorms generate a strong tornado; extremely rapid updrafts and changes in wind direction with altitude are required. Wind shear, an abrupt change in wind direction or speed, creates a rotating tube of air at an altitude of about 20,000 feet. Thunderstorms, especially the largest—called supercells—can spawn tornadoes from wind shear because updrafts that can reach 150 miles per hour cause the rotating tube of wind to tilt. When the tilt straightens and becomes nearly vertical and reaches the surface, a tornado forms. Dust and water droplets caught in the winds of the vortex cause the tornado to become visible to an observer; tornadoes appear darker when the sun is behind them and lighter as they move away from the sun and reflect sunlight.

Tornadoes travel with the thunderstorm or hurricane that generated them, but accurate path prediction is not possible because they can take any direction, jump around on the surface, and move at speeds up to 75 miles per hour. Midlatitude tornadoes usually travel from southwest to northeast along with the cold front or squall line, at about 30 miles per hour.

Tornadoes occur in dozens of countries around the world, but the United States has by far the largest total number every year; Australia is a distant second. The location of the United States relative to the polar front, where thunderstorms are common and along which midlatitude cyclones travel, is the reason for the large number of tornadoes. Equatorial regions have fewer tornadoes because the intertropical convergence zone (ITCZ) has diverging high-altitude air and therefore lacks the necessary wind shear; Eurasia has few because mountain ranges separate maritime-tropical and continental-polar air masses, resulting in less frontal activity and in fewer midlatitude cyclones than in the United States.

Within the United States, tornadoes are most common in a region known as Tornado Alley, which extends from the Texas Panhandle to southeast Nebraska. All 50 states experience tornadoes, but Tornado Alley averages 20 per year. Despite its large number of thunderstorms, Florida has relatively few tornadoes because its thunderstorms are more often caused by converging air than by frontal activity. Tornado season in the United States varies by latitude; the season usually starts in early March when temperatures rise along the Gulf of Mexico, moves northward along with the polar front, and by midsummer has reached the United States-Canada boundary. Tornado season ends at different latitudes as temperatures

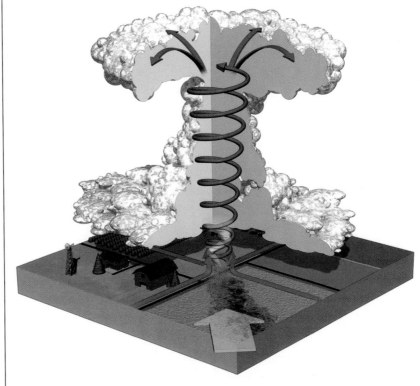

Tornadoes are formed when a strong draft of wind in a thunderstorm catches a rotating tube of air formed by wind shear and lifts it to a vertical position, causing it to twist at the same time. When the tube touches the ground, a tornado forms.

THE FUJITA TORNADO INTENSITY SCALE			
F-SCALE	**CATEGORY**	**(KM/HR)***	**(MI/HR)***
0	Weak	65–116	40–72
1		117–181	73–112
2	Strong	182–253	113–157
3		254–332	158–206
4	Violent	333–419	207–260
5		420–513	261–318
*ESTIMATED WIND SPEED			

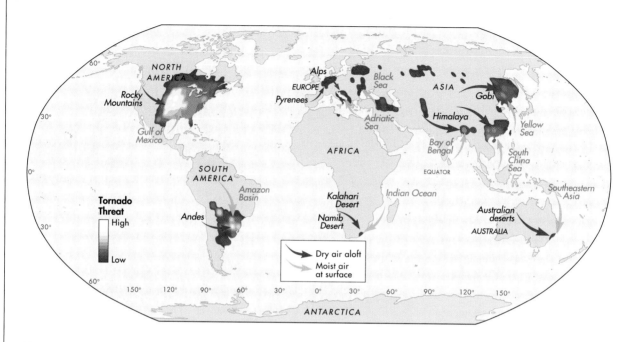

Many countries experience tornadoes, but the United States holds the record with between 700 and 1,000 every year.

cool or as changes in the season cause the polar front and cyclonic storms to shift to different latitudes. More at:

http://www.out:look.noaa.gov/tornadoes/index.html

Tropical Systems and Hurricanes

Weather conditions in the tropics, between 30° north and south, are far more consistent and uniform than those in the midlatitudes, and traveling weather systems are far less common. High pressure protects the tropics from the intrusion of colder polar air masses and the resultant frontal activity that dominates the weather of the midlatitudes. There is also less landmass over which continental air masses can form at these latitudes. Seasonal differences in insolation, temperature, and precipitation do occur, but in general warm tropical and warmer equatorial air masses dominate weather.

The weather systems that do occur in the tropics usually develop from easterly waves, westward moving, low-pressure disturbances in the trade-wind zone between 5° and 30° north and south latitudes from the Equator. During the late summer, as ocean temperatures increase slightly, a low-pressure system (which can appear as a linear trough hundreds of miles long) occasionally forms over tropical oceans, possibly over an island. As this system is blown westward by the trade winds, the normal easterly wind pattern curves into the low-pressure zone (convergence); as the system moves to the west, a wave-like disruption in the trade winds occurs.

A few of these easterly waves develop into larger cyclonic storms. If the low-pressure center intensifies and wind speed increases, then tropical depressions form when winds reach 23 miles per hour. They become tropical storms when wind speeds reach 39 miles per hour, and hurricanes when wind speeds reach 74 miles per hour. For the low pressure to intensify, the middle troposphere must contain sufficient water vapor to allow evaporation to occur, vertical wind shear must be limited, and subsiding air associated with the tropics on the eastern side of the oceans must be restricted to allow vertical development of the system. When these factors are present, convectional uplift increases, lowering surface atmospheric pressure and increasing surface-pressure gradients and cyclonic wind speeds.

Hurricanes, called typhoons in the western Pacific Ocean and cyclones in the Indian Ocean, are tropical cyclonic wind systems that form over warm oceans. On average, the diameter of the area experiencing hurricane-force winds is 150 miles or less, although the overall wind system may be much larger. Winds flow into the low-pressure core of the hurricane (pressure down to 26.22 inches has been measured, far below the 29.92 average) and are curved counter-clockwise in the Northern Hemisphere and clockwise in the Southern Hemisphere. A calm eye is found at the center of the storm; the winds blow into this low-pressure system so that rising air occurs around the eye, creating a wall of clouds but leaving the interior clear.

While hurricanes have traveled as far north as Canada and Hokkaido, Japan, they form only over tropical—but not equatorial—waters. The energy supply for hurricanes is water with a temperature of 80°F or greater; evaporation from this warm water provides the

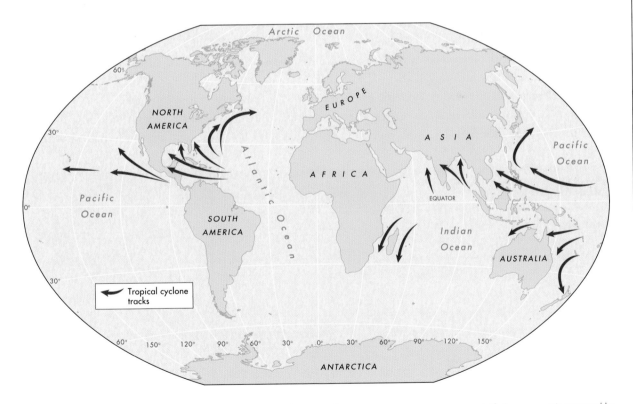

Tropical cyclonic storms develop over tropical waters in summer and autumn when ocean temperatures are warmest. Their movement is governed by the Coriolis effect and global and upper-atmosphere winds. Of the 100 or so storms that form near the west coast of Africa each year, only 10 percent reach Caribbean and American coasts as hurricanes.

latent heat released during condensation, and this heat perpetuates uplift and instability in the center of the hurricane. Most hurricanes are generated between 5° and 20° north and south latitudes in the western North Atlantic, western Pacific, eastern North Pacific, and Indian Oceans; other tropical waters such as those off the west coast of South America and Africa are usually too cold due to the ocean currents that have traveled from polar regions. (See tropical cyclones map, page 81.)

The requirements for hurricane formation also affect their timing and duration. Hurricane season is usually in late summer and fall because ocean temperatures are at their maximum two months or so after the summer solstice. The U.S.

National Hurricane Center, in Miami, Florida, describes the official season for Atlantic hurricanes as June 1 to November 30, while the Pacific season is May 15 to November 30. Most hurricanes lose wind speed and dissipate after one week, although a few may last up to three weeks. When hurricanes travel over colder water or inland, they weaken and die out because surface friction and lack of an energy supply slows wind speeds to below hurricane strength. Hurricanes usually dissipate before traveling to within 120 miles of the coast.

The storms follow a general path shaped by global and upper-atmospheric winds; their travel speed also reflects these influences. Early in their development, hurricanes tend

to travel westward with the trade winds and upper atmospheric winds at speeds between 10 and 20 miles per hour. About two-thirds of all hurricanes eventually curve toward the Poles because of the Coriolis effect and slow down to 5 to 10 miles per hour when they no longer have the trade winds pushing them. The few that reach the zone of the westerlies curve back to the east and may travel more than 30 miles per hour.

Deaths and damages from hurricanes are most severe along coastlines. Hurricanes are destructive because of their wind speeds, accompanying tornadoes, large amounts of rain, and—perhaps most important—because of the storm surges they cause. As a hurricane approaches a coastline, the

SAFFIR-SIMPSON HURRICANE INTENSITY SCALE							
Scale number (category)	Central pressure		Wind speed		Storm Surge		Damage
	mb	in.	mi/hr	km/hr	ft	m	
1	≥980	≥28.94	74–95	119–154	4–5	1–2	Minimal
2	965–979	28.50–28.91	96–110	155–178	6–8	2–3	Moderate
3	945–964	27.91–28.47	111–130	179–210	9–12	3–4	Extensive
4	920–944	27.17–27.88	131–155	211–250	13–18	4–6	Extreme
5	<920	<27.17	>155	>250	>18	>6	Catastrophic

NAMING TROPICAL STORMS

Since 1953, Atlantic tropical storms have been named from lists originated by the National Hurricane Center and now maintained and updated by an international committee of the World Meteorological Organization. The lists featured only women's names until 1979, when men's and women's names were alternated. Six lists are used in rotation. Thus, the 2004 list will be used again in 2010.

2004	2005	2006	2007	2008	2009
Alex	Arlene	Alberto	Andrea	Arthur	Ana
Bonnie	Bret	Beryl	Barry	Bertha	Bill
Charley	Cindy	Chris	Chantal	Cristobal	Claudette
Danielle	Dennis	Debby	Dean	Dolly	Danny
Earl	Emily	Ernesto	Erin	Edouard	Erika
Frances	Franklin	Florence	Felix	Fay	Fred
Gaston	Gert	Gordon	Gabrielle	Gustav	Grace
Hermine	Harvey	Helene	Humberto	Hanna	Henri
Ivan	Irene	Isaac	Ingrid	Ike	Ida
Jeanne	Jose	Joyce	Jerry	Josephine	Joaquin
Karl	Katrina	Kirk	Karen	Kyle	Kate
Lisa	Lee	Leslie	Lorenzo	Laura	Larry
Matthew	Maria	Michael	Melissa	Marco	Mindy
Nicole	Nate	Nadine	Noel	Nana	Nicholas
Otto	Ophelia	Oscar	Olga	Omer	Odette
Paula	Phillippe	Patty	Pablo	Paloma	Peter
Richard	Rita	Rafael	Rebekah	Rene	Rose
Shery	Stan	Sandy	Sebastien	Sally	Sam
Tomas	Tammy	Tony	Tanya	Teddy	Teresa
Virginia	Vince	Valerie	Van	Vicky	Victor
Walter	Wilma	William	Wendy	Wilfred	Wanda

low-pressure center of the storm can cause ocean levels to rise three feet or more. The wind pattern of the hurricane also raises ocean levels by causing storm waves to be blown into the coast along the poleward side of the hurricane. Low pressure and wind can combine to cause storm surges that raise ocean levels an additional 25 feet. In 1900 Galveston, Texas, was flooded by the storm surge of a hurricane; 8,000 people died. The low-lying coast of Bangladesh is subject to cyclone storm surges, and the 1970 and 1991 cyclones each caused hundreds of thousands of deaths.

Hurricanes are classified by the combined factors of wind speeds, atmospheric pressures, storm surges, and potential damage. The United States National Weather Service uses a rating system called the Saffir-Simpson scale to rate hurricanes from one to five, based on likely preperty damage and flooding in coastal areas if a hurricane makes landfall.

While level-five hurricanes are uncommon, they are catastrophic. (See Hurricane Intensity chart, page 82.) Hurricanes originating off the west coast of Africa, called Cape Verde hurricanes, are rare because water temperatures there are usually cool, but especially destructive (frequently level five); they increase their energy supply while traveling 3,000 miles across the tropical Atlantic to the Caribbean.
More at:
http://hurricanes.noaa.gov/

CLIMATE

CLIMATE CLASSIFICATIONS

Climate is the long-term average annual pattern of atmospheric conditions for a location, whereas weather is the short-term state of the atmosphere for the same location. Climate has a tremendous impact on vegetation, soil development, and agriculture, which in turn influence where and how humans live, including energy use and recreation. Classification systems based on meteorological criteria have been created to describe global patterns of climate and enable comparisons to be made between places.

Misinformation and myths about climate have influenced how people have settled and developed the planet. French and British farmers settled Nova Scotia and Newfoundland on Canada's east coast in the 1600s, mistakenly believing that latitudes similar to those of their homelands meant similar climates.

Climates are usually classified according to basic elements of weather. Seasonal patterns of precipitation and temperature, for example, are key to most classification systems because these data are important for the growth of vegetation, and knowing about these conditions enables people to grow crops and raise livestock. Other elements of weather, including wind,

A polar bear and her cubs cross the frozen tundra at Churchill, on Canada's Hudson Bay. Frigid Arctic air masses account for the area's polar climate.

pressure, insolation, and humidity, have also been used to set up classification systems.

Variations from year to year in precipitation and temperature make the averaging of data essential. For example, on July 4 in Kansas City, Missouri, maximum temperatures over a 30-year period have ranged from below 80°F to above 100°F, while precipitation on that date has ranged from zero to three inches. The mean precipitation and mean temperature are determined by adding all values on a date and then dividing by the number of years of data; this figure provides a basis for long-term planning. However, knowledge of the extremes in temperature and precipitation is useful for emergency planning. Means can be misleading: One major rainstorm in a desert can increase the mean annual precipitation even though no rain may fall in nine out of ten years on a given date.

Thirty years of atmospheric data are ordinarily used to prevent shorter-term variations from skewing the averages. In the United States, beginning in 2001, the National Climate Data Center in Asheville, North Carolina, has used weather data from 1971 to 2000 for information about climates. Relying on shorter periods of time can give a misleading picture of climatic patterns.

More at:

http://lwf.ncdc.noaa.gov/oa/climate/climatedata.html

Atmospheric data must be averaged for climatic classification, and monthly averages rather than annual averages are used. Seasonal and monthly variations are extremely important in defining types of climate because temperature and precipitation patterns affect the growing seasons of vegetation. (See Bioregions, page 175.) For example, Vancouver, British Columbia, and Springfield, Massachusetts, have annual average temperatures of 50°F and precipitation amounts of 45 inches. In Vancouver, however, the average January temperature (37°F) is 26° colder than the July average (63°F), far less than the 47° range in Springfield between January (26.5°F) and July (73.5°F). Seasonal precipitation patterns are also different: In Vancouver most rainfall occurs in winter, but precipitation is year-round in Springfield. Vegetation differs too: Douglas fir and hemlock dominate the natural vegetation near Vancouver, while maple and birch are the dominant trees in Springfield.

More at:

http://www.worldclimate.com

Although averaging data for climatic classification is very useful for understanding patterns of vegetation and for planning purposes, such data cannot be used to predict weather accurately. Weather prediction for ten days in advance, much less a year, is unreliable because tiny variations in the levels or locations of atmospheric conditions—conditions too small to be accurately or systematically measured—lead to extremely different results. According to the American Meteorological Society, predictions beyond five days are increasingly limited in usefulness.

CLIMATE CONTROLS AND CLASSIFICATION

Climatic classification systems enable meaningful comparisons to be made between locations; various systems use different criteria to emphasize different sets of weather elements. No single system meets the needs for all possible comparisons and analyses because different systems focus on different aspects and impacts of weather.

Geographic patterns of climates reflect the influences of the same controls that shape weather conditions, especially temperature and precipitation. Latitude, elevation, proximity to water bodies, ocean currents, topography, and prevailing winds are the most important climatic controls. Elevation, proximity to water bodies, ocean currents (despite conditions such as El Niño and La

Ancient Greek Climatic Classification

About 2,500 years ago the Greek philosopher Parmenides (born circa 515 B.C.) developed a climatic classification scheme based on weather conditions. He defined three zones of climate: torrid, for northern Africa; temperate, for southern Europe; and frigid, for northern Europe. Climatic mythology was incorporated into this system because the Greeks believed climates determined personal characteristics: People in the torrid zone were intelligent but lethargic; people in the frigid zone were energetic but unintelligent; and people in the temperate zone (which, coincidentally, included Greece) were intelligent and energetic. This philosophy of *environmental determinism* persisted among many scholars until the mid-1900s when it was fully discredited.

Niña), and topography are relatively unchanging over the centuries, and their impacts on climate are consistent year after year. Seasonal changes in insolation at each latitude and the shifting of prevailing winds at some locations provide the dynamic aspects of climate.

To be useful, climatic classification systems must be based on commonly available weather data, and they must capture differences and similarities between locations that are meaningful for a system's user. In order to create a global classification system, the amount and type of atmospheric data needed must be simplified in order to generalize the information. Simplification reduces the number of possible climatic types so that the geographic patterns of climates are not lost in excessive detail, and generalization is necessary for grouping climates to make comparisons between locations possible. Even though coastal cities in Uruguay, South Carolina, and southern China will have slightly different monthly temperatures and amounts of precipitation, the impacts of their climates on vegetation may be so similar that these areas can be grouped into one climatic category.

Different climates blend into each other because the influence of climatic controls, such as distance from an ocean, changes gradually. Establishing climatic regions includes defining boundaries between locations based on atmospheric data, and this may result in dividing locations with similar atmospheric conditions into different climatic regions. Hence, boundaries are understood to represent transition zones. For example, in many classification systems Lake Charles, Louisiana, and Washington, D.C., are included in the same category. Although Washington, D.C., is much closer to New York, New York City is in a different category because of definitions used to establish regions.

German botanist Wladimir Köppen's climatic classification system, developed in the early 1900s and revised many times since, is one system commonly used in the United States. Köppen's classification system focused on the climatic limits of vegetation and was based on temperature and precipitation patterns. Köppen and other climatologists revised the system numerous times to correspond more closely to vegetation patterns. The climates described below follow a 1957 revision of the Köppen system by University of Wisconsin climatologist Glenn Trewartha.

The Köppen classification system employs temperature as the basis for four moist climatic regions: tropical rainy, humid mesothermal, humid microthermal, and polar. Lack of precipitation designates a fifth region (dry), and topography determines the sixth (highland). There are subclasses within each region, except for highland, that are usually distinguished on the basis of seasonal patterns of precipitation. In all, there are 15 major climates and the highland climate. Global patterns of biomes (regional ecosystems based on life forms) mirror these climatic regions because vegetation reflects the amounts of precipitation, temperature, and sunlight that are available. (See Köppen classification system map, page 88).

GLOBAL CLIMATE PATTERNS
Tropical Rainy Climates (A Climates)
TROPICAL RAIN FOREST (AF)

At the intertropical convergence zone (ITCZ) along the Equator, there are virtually no seasons; each month is hot (averaging 80°F) and wet (averaging eight inches of rain). The consistently high amount of insolation this equatorial region receives throughout the year causes high temperatures and daily precipitation. Precipitation usually peaks in the afternoons, when maximum daily temperatures increase convection.

This climate promotes the growth of trees (rain forests, sometimes called *selvas*, from *silva*, Latin for "wood") because there is no cold season. Tropical *rain forests* are found in the Amazon and Congo River Basins as well as in Southeast Asia, Central America, and Madagascar. Trees dominate this biome and are tall (often more than 200 feet), broad-leaved, and evergreen; the continuous warmth and rain in this biome means no seasonal stress or loss of leaves. Many animal species are arboreal, and insects thrive. Insect-transmitted diseases, such as malaria and yellow fever, are endemic in many of these areas. These forests have the greatest biodiversity of any biome: At least several million species are found in them worldwide, and a single square mile may contain hundreds or even thousands of tree species.

TROPICAL SAVANNA (AW)

Immediately to the north and south of the Equator and the tropical rain forest climate, rainfall becomes seasonal. The ITCZ only dominates during summer, the wet season; high-pressure systems shift slightly into these regions during the winter, the dry season. Temperature remains high year-round; monthly averages are between 70°F and 86°F, with slight cooling during the winter.

Seasonal changes in precipitation result in fewer trees and more grasslands than in tropical rain forest.

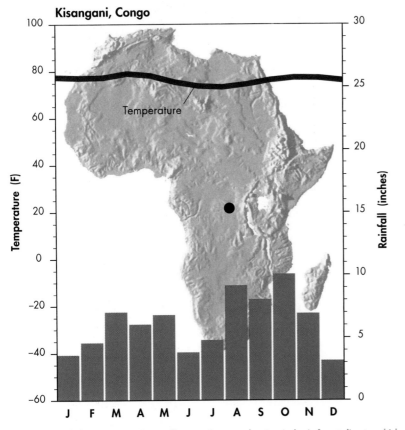

Kisangani, Congo

This climagraph for Kisangani, Congo, illustrates its seasonless tropical rain forest climate, which occurs along the Equator.

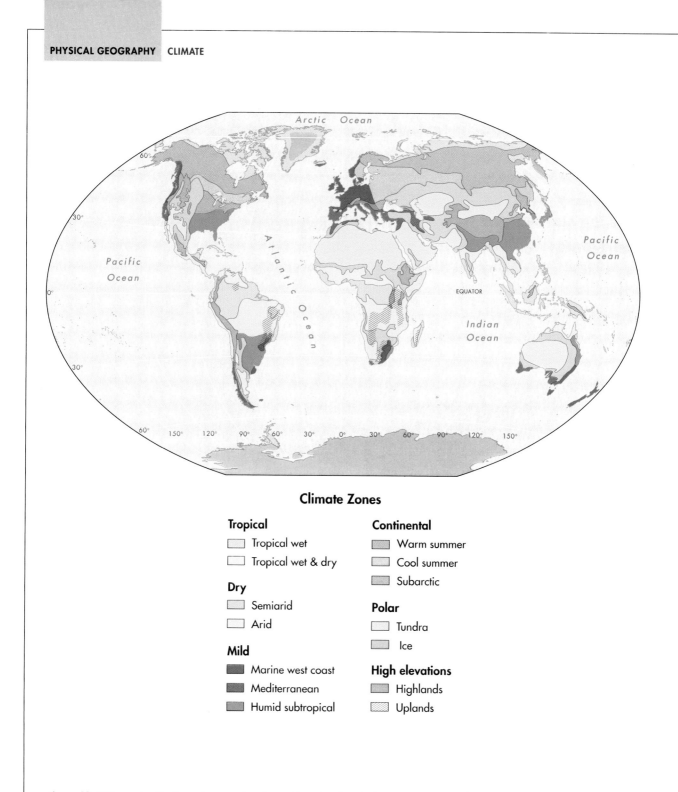

Climate Zones

Tropical
- Tropical wet
- Tropical wet & dry

Dry
- Semiarid
- Arid

Mild
- Marine west coast
- Mediterranean
- Humid subtropical

Continental
- Warm summer
- Cool summer
- Subarctic

Polar
- Tundra
- Ice

High elevations
- Highlands
- Uplands

The modified Köppen classification system organizes the world's regions into six climatic zones: Five are based on precipitation (four moist, one dry), and one is based on topography (highland).

Savanna biomes are usually located near and blend into tropical scrub woodland and tropical *steppe* biomes; much of central Africa, south-central Brazil, and inland Australia are covered by savanna. The lack of precipitation causes most vegetation to become dormant during the dry season, but the availability of grasses provides food for large herbivores, such as giraffes and zebras, plus associated predators.

TROPICAL MONSOON (AM)

In tropical monsoon climates (see Monsoons, page 57) precipitation patterns are intensified forms of those in tropical savanna climates, and wet summers average as much as 15 inches of rain each month. This seasonal precipitation is magnified by wind reversals caused by seasonal heating and cooling of a continental landmass. Monsoons are frequently associated with Asia, but this climate is also found in western Africa and eastern South America. Forest vegetation is most common in these regions; however, in areas where the dry season lasts longer, grasslands will form.

Dry Climates (B Climates)

TROPICAL STEPPE (BSH)

Tropical steppe climates are located poleward of tropical savannas (AW) and often surround the tropical

The Seiplok Sanctuary on Borneo, East Malaysia, contains 10,000 acres of virgin rain forest. Here, animals recovered from poachers are returned to the wild.

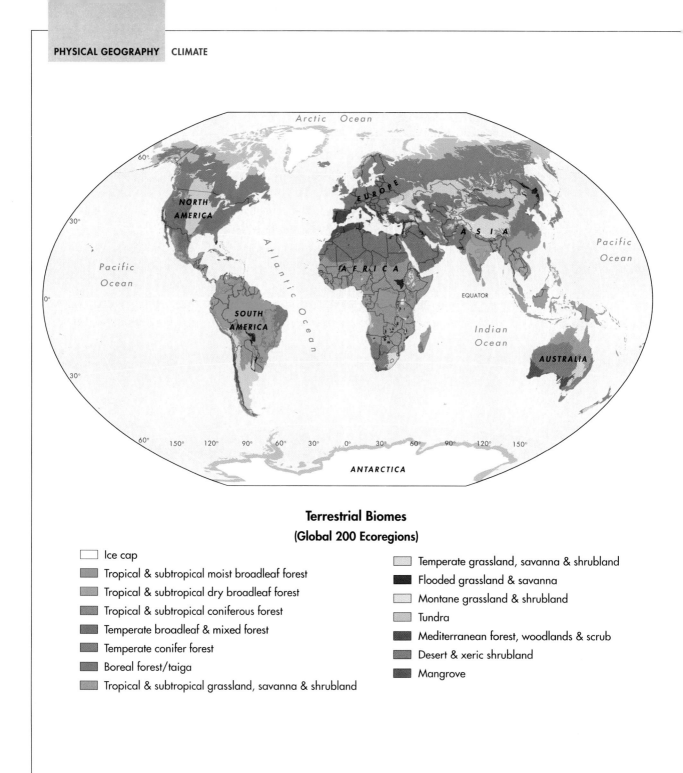

Terrestrial Biomes
(Global 200 Ecoregions)

- ☐ Ice cap
- ▦ Tropical & subtropical moist broadleaf forest
- ▦ Tropical & subtropical dry broadleaf forest
- ▦ Tropical & subtropical coniferous forest
- ▦ Temperate broadleaf & mixed forest
- ▦ Temperate conifer forest
- ▦ Boreal forest/taiga
- ▦ Tropical & subtropical grassland, savanna & shrubland

- ▦ Temperate grassland, savanna & shrubland
- ▦ Flooded grassland & savanna
- ▦ Montane grassland & shrubland
- ▦ Tundra
- ▦ Mediterranean forest, woodlands & scrub
- ▦ Desert & xeric shrubland
- ▦ Mangrove

The U.S. Forest Service system of classifying Earth's biomes consists of four general categories—forest, grassland, desert, and tundra—which are further divided into 13 specific biomes. The list above numbers more than 13 biomes because montane forests include a range of these biomes at different elevations and ice areas lack biomes.

desert (BWH) climate. A brief wet season may occur either in summer, as in southern Sudan, when the ITCZ is present, or in winter, as in coastal Libya, when westerlies bring precipitation.

MIDLATITUDE STEPPE (BSK)

Steppes also exist in the midlatitudes, where subfreezing temperatures are more common than in the tropics and where precipitation levels remain low, between 10 and 25 inches annually. Midlatitude steppes are often found in the rain shadow of a mountain range, as is the case east of the Cascade Range of Washington, Oregon, and northern California, or else so far downwind from an ocean, as in Kazakhstan, that frontal precipitation from maritime air masses is extremely limited.

Steppe biomes, found in either low latitudes or midlatitudes where rain shadows or the high-pressure systems limit precipitation, are dominated by short grass. Low annual precipitation creates steppes in South Africa (the veldt), the Sahel region south of the Sahara, the western Great Plains of North America, the Caucasus of Eurasia, and other locations. Trees are found alongside streams; feather grass and buffalo grass are typical types of vegetation.

TROPICAL DESERT (BWH)

Tropical deserts occur when annual precipitation is minimal (from near

Canyon Prairie, Iowa, is typical of a grassland biome called a steppe. Steppe biomes also occur in South Africa's veldt and in Eurasia's Caucasus.

zero to ten inches per year) in regions centered on the Tropics of Cancer and Capricorn, such as the Sahara in northern Africa, the Atacama in Chile, and the Gibson Desert in Australia. High pressure prevents almost all forms of cooling and precipitation.

MIDLATITUDE DESERT (BWK)

Midlatitude deserts are not as extensive as tropical deserts and are located farther from the Equator,

The vegetation of the Sonoran Desert near Tucson, Arizona, reflects the region's classification as a desert climate, one of the Type B climates.

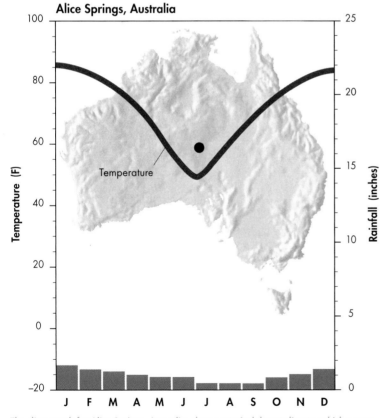

Alice Springs, Australia

The climagraph for Alice Springs, Australia, shows a typical desert climate, which occurs because Australia's mountainous east coast prevents Pacific Ocean winds from penetrating the continent's interior.

where temperatures are lower in wintertime. Precipitation is rare because of rain shadows and distance from maritime air masses, as in the midlatitude steppes. With the exception of Patagonia, Argentina, all midlatitude deserts, such as the Gobi in northern China and southern Mongolia, are in the Northern Hemisphere, where there are continental landmasses at these latitudes.

In desert biomes, only *xerophytes*, plants capable of tolerating annual precipitation levels below ten inches and occasionally prolonged droughts, can survive. In about 75 percent of regions with desert climates, some scattered vegetation exists, such as cactus, thornwood, and tamarisk. Vegetation is usually small and grows slowly, and many plants are ephemerals, grasses or herbs whose life cycles follow the rare rainfalls and include months or even years of dormancy between weeks of blooming.

Humid Mesothermal (C Climates)

MEDITERRANEAN (CS)

Mediterranean climates are found along the Mediterranean Sea and on the west coasts of landmasses bordered by cold ocean currents. They are poleward of tropical dry climates, and temperatures are mild but cooler than in the tropics. Precipitation falls mainly in winter, when the westerlies move toward the Equator, while dry summers are caused by high pressure. Native vegetation, such as chaparral in the U.S. Southwest, olive trees, and cork oaks, has adapted to summer droughts, and irrigation is needed to grow most crops.

The Mediterranean climate creates a Mediterranean woodland and scrub biome. This biome is found in southern California, southern and southwest Australia, and central Chile and on Mediterranean coastlines. Vegetation, called chaparral in California and maquis in southern Europe, is limited to species that can tolerate the stress of high temperatures and summer drought. Other vegetation has small leaves and thick bark for protection from climatic stress; olive trees, cork oaks, and manzanitas are typical species.

HUMID SUBTROPICAL (CA)

Humid subtropical climates are found poleward of the tropics on east coasts bordered by warm ocean currents. In summer, trade winds bring precipitation, while high-pressure causes dry periods. In winter, frontal precipitation occurs when cold polar air masses move into these latitudes.

Vegetation thrives in this temperate, humid climate. Warm ocean currents provide sufficient moisture and temperatures are warm enough to support broadleaf, nondeciduous trees; these subtropical forests grow on the Sea Islands off Georgia, in eastern China, and in southern Japan, for example. Biodiversity is more limited in these forests than in tropical rain forests, and trees are shorter and less dense than they are closer to the Equator. Magnolias,

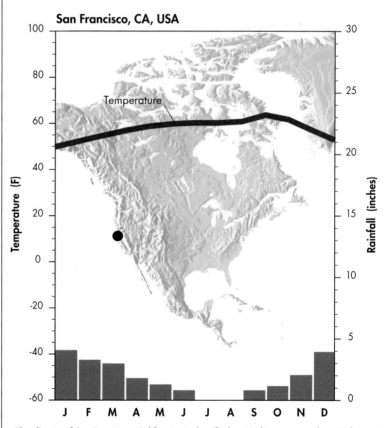

The climate of San Francisco, California, is classified as Mediterranean subtropical. San Francisco's exposed coastal location accounts for its cool summers and mild winters—a combination unusual for this latitude.

A salt marsh typical of the subtropical broadleaf evergreen biome can be seen through oak trees draped with Spanish moss at Fort King George State Historical Park in Darien, Georgia.

live oaks, and Spanish moss are found in this biome in the southeastern United States. Crops such as rice support large numbers of people, as in East Asia.

In the cooler winters of the upper midlatitudes, an increasing number of cold-tolerant trees are found, including needle-leaf evergreens, such as pines, and deciduous broadleaf trees, such as oak, hickory, beech, and ash. Tree species in this biome have more deciduous varieties Equatorward and an increasing percentage of needle-leaf trees toward the Poles; the largest regions with this biome are found in the eastern United States, northern Europe to central Russia, and northeastern China.

MARINE WEST COAST (CB, CC)

Marine west coast climates are found poleward of Mediterranean climates. Westerly winds dominate most of the year, and orographic precipitation, especially on the windward slopes, occurs year-round. Maritime air masses keep temperatures mild, and the climate extends inland to mountains, such as the Cascade Range in the Pacific Northwest of the United States and the Coast Mountains of British Columbia, or to where air masses take on continental characteristics, as in Europe

New Hampshire's White Mountains, part of the Appalachians, are located in a biome region classified as midlatitude deciduous and mixed forest.

where the east-west trending Alps do not block westerly winds.

Midlatitude evergreen forests are found on west coasts that have a marine west-coast climate. The west coast has heavy precipitation and so tends to have large trees; coastal redwoods include the tallest trees on Earth, some more than 300 feet. Mild temperatures, ample rainfall, and a long growing season make farming possible.

Humid Microthermal (D Climates)

HUMID CONTINENTAL–WARM SUMMER (DA)

Humid continental–warm summer climates are the first in a series of climates extending poleward of the humid subtropical climates on the east coasts of continents. Mean July temperatures are 71.6°F or warmer at latitudes about 40° north. Precipitation peaks in summer because the higher temperatures cause con-

vection, although frontal precipitation occurs year-round. Deciduous forests such as those of Ohio and Pennsylvania are common in this climate, which is limited to the Northern Hemisphere because there are few landmasses at these latitudes in the Southern Hemisphere.

Interior locations receive less precipiation and are dominated by grasslands. These are some of the world's major grain producing areas.

HUMID CONTINENTAL–COOL SUMMER (DB)

Poleward of 40° north latitude, summer temperatures less than 71.6°F and humid continental–cool summer climates occur. This climate is absent in the Southern Hemisphere because there is no major landmass at those latitudes. Precipitation is lower than that of humid subtropical California climates because lower summer temperatures reduce convection, and colder air masses with lower specific humidity reduce frontal precipitation. Evergreens are found mixed with deciduous trees at these latitudes because of the cooler temperatures. Although this climate is not found on west coasts, it occurs in much of Sweden because the mountains of Norway exclude the moderating influence of maritime air masses, so continental air masses dominate.

In the interior midlatitudes, tall grass averaging three to four feet is found where precipitation is higher (up to 20 inches) than in the steppes but is still insufficient to support trees. Prairie biomes are limited to the midlatitudes between steppe and forest biomes. In addition to the interior plains of North America, small prairie biomes are found in Uruguay, Manchuria, Hungary, and Ukraine. Bluestem and other grass species that domi-nate this biome once supported huge herds of grazing herbivores, such as bison, but over hunting and agriculture have almost totally eliminated such herds.

SUBARCTIC (DC, DD)

The subarctic climate, also called the boreal forest climate for its type of vegetation, is an extremely cold climate, but it has sufficient precipitation and a long enough growing season to support trees. Arctic and subarctic air masses dominate this area, with frontal activity in summer bringing limited precipitation of two to three inches per month.

Evergreens are the most common trees, and forests can be an important natural resource. Northern coniferous forests are also known by the Russian name "taiga." Vegetation in these areas faces the stress of extremely cold winters. The trees—fir, spruce, and

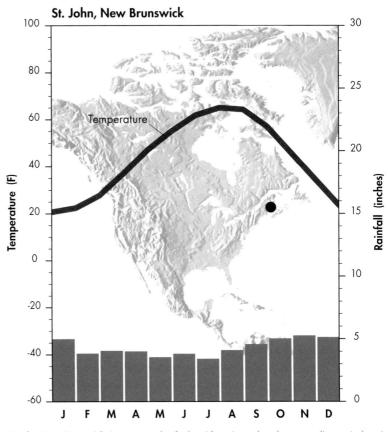

St. John, New Brunswick, is an example of a humid continental-cool summer climate. Its location on the North Atlantic storm track results in frequent heavy snowfalls in the winter months.

pine—mature to be shorter than normal because seasons are short and precipitation is limited. Much of northern North America and northern Eurasia are covered by this biome, which has very limited biodiversity because of climatic limitations on vegetation and therefore on the food web.

Polar (E Climates)

TUNDRA (EET)

Tundra climates are found along the coastlines of the Arctic Ocean, on some coastlines of and islands off Antarctica, and in the highest reaches of mountain ranges such as the Andes and Himalaya. Frigid Arctic and Antarctic air masses dominate these regions but rarely produce precipitation because frontal activity is largely absent. Some climatic classification systems designate this region as a cold desert because it experiences a meteorological drought, receiving less than ten inches of precipitation a year.

Vegetation in the tundra biome is limited to moss, lichen, and other hardy plants. Cold temperatures and two-month-long growing seasons severely limit biodiversity; however, the total number of organisms may be huge during summer, when reindeer, for example, and migratory birds converge and insects such as mosquitoes and

blackflies are active. Vegetation grows slowly, and grasses, such as cotton grass, and lichens, including reindeer moss, are common types of vegetation. Permafrost, where soil temperatures remain below freezing most or all of the year, is common in tundra, affecting soil drainage and plant growth.

ICE CAP (EF)

Freezing temperatures keep the interiors of Greenland, Antarctica,

and some northern Canadian islands ice covered year-round. Ice cap climates are extremely cold, dry, and windy; no vegetation survives in this climate, but bacteria have been found in layers within the ice.

HIGHLAND (H CLIMATE)

Highland climates vary with changes in elevation and slope. Temperatures may be tropical at the base of mountains, such as the

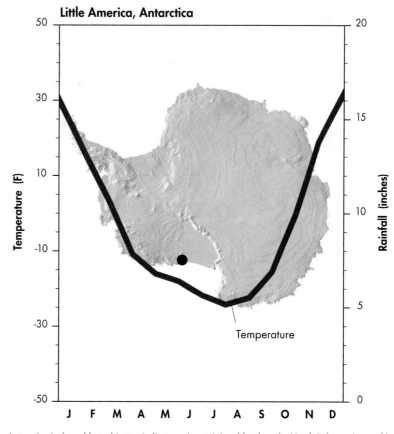

Little America, Antarctica

Temperature

Antarctica is the coldest, driest, windiest continent. It is colder than the North Polar region and is almost completely covered by an ice sheet averaging one mile thick.

Andes in Ecuador, and cold enough for glaciers at the top. Precipitation may exceed 60 inches a year on a windward slope and be less than 10 inches on the leeward side. Highland areas may encompass several of the 15 other climates, but the area covered by each climate is too small to be included on a world map. Therefore, these regions' climates are designated as undifferentiated highlands.

Marine, Maritime, and Continental Climates

The moderating influence of oceans on climates is substantial, and proximity to oceans causes climate to be classified as marine or maritime. Continental climates are either far from or upwind from coastlines. Temperature extremes are moderated in maritime climates, in dramatic contrast to the much larger annual temperature range in continental climates. Even at higher latitudes, where seasonal variations in insolation are great, maritime climates are moderate. Tacoma, Washington, on Puget Sound and about 85 miles from the Pacific Ocean, has a moderate, 20°F annual temperature range, far less than the 56°F range of inland Montreal, Quebec—even though both cities are near the same latitude.

CLIMATIC CHANGE

Climatic change, often associated specifically these days with global warming, includes increases and decreases in temperature, precipitation, and the frequency of extreme weather events such as drought. The exchanges of energy and water between the atmosphere, oceans, and land surfaces, especially ice sheets, make climatic change a complex phenomenon that is difficult to measure and predict accurately.

Most, if not all, climatic changes can be traced to changes in Earth's *radiation balance*. (See Earth's Radiation Balance, page 42.) If the amount of insolation received at a given latitude or in a region increases or if there is an increase in the amount of long-wave energy radiated out to space, surface temperatures will change. Changing temperatures will, in turn, cause changes in atmospheric pressure, wind patterns, humidity, and precipitation. Over millions of years, changes caused by geologic processes—such as plate movement to new latitudes and uplift to new elevations—will also bring about changes in climate.

Changes in Earth's radiation balance can be explained by changes in Earth's orbital patterns or changes in the composition of Earth's atmosphere. There were also measurable changes in the amount of energy emitted from the sun during the 1980s, but this solar constant varies too little to cause climatic change on Earth. In 1893 British scientist E. Walter Maunder (1851–1928) established a relationship between the decreasing frequency of sunspots—large, cool blotches on the solar surface—and colder global temperatures. Two minimums, or periods of low sunspot activity coincided with lower global temperatures; however, no causal relationship between sunspots and Earth's radiation balance has been confirmed.

American scientist, publisher, and diplomat Benjamin Franklin (1706–1790) linked changes in the composition of Earth's atmosphere and climatic changes following the 1783 eruption of Iceland's Laki volcano, arguing that volcanic ash in the atmosphere interfered with Earth's warming. Later volcanic eruptions strengthened the link between eruptions and climatic change. Volcanic dust and ashes spewed into the atmosphere were once thought to reflect sunlight and therefore decrease insolation, but this is now considered an unlikely cause of global cooling because the particles settle out of the atmosphere too quickly to have lasting effects. The eruption of sulfur oxides, which are transformed into sulfurous aerosols, are of much greater concern. These aerosol droplets may stay in the stratosphere for months or years because there is no precipitation at this level, and they scatter and absorb solar radiation. The location of a volcanic eruption and the season when it occurs are also factors in climatic change: Aerosols from low-latitude volcanoes spread farther and with greater effect than those from high-latitude volcanoes, whereas summer wind patterns at lower latitudes disperse aerosols farther than during other seasons.

Human-caused changes to the atmosphere have contributed to global climatic changes. The United Nations Intergovernmental Panel on Climate Change reported in 1995 that global surface temperatures had risen one degree Fahrenheit since 1900 and that increased carbon

Urbanization and Climatic Change

At the local scale, climates in large cities are measurably different from climates in the surrounding countryside. Warmer temperatures, especially at night and during winter, increasing clouds and rainfall (but not snowfall), slower wind speeds, and other changes in urban weather have been documented. The urban heat island phenomenon has been noted for more than a hundred years. (See Urban Heat Islands, page 46.) In cities, humans have changed atmospheric composition through pollution; they have changed ground surfaces with buildings, pavement, and the removal of vegetation; and they have increased the release of heat energy from buildings.

dioxide from burning fossil fuels was partially responsible. Scientists continue to debate how the burning of fossil fuels has damaged Earth's carbon cycle. (See carbon cycle diagram, page 178.)

Earth's climates have changed repeatedly and changes will continue, but predicting change is problematic. The cause-and-effect relationships between the different factors affecting climate—ocean temperatures, cloud cover, and atmospheric carbon dioxide—have not been quantified for modeling climatic change. Scientists continue to work on climatic change, but the scale and complexity of the system make efforts to predict and influence climate unreliable.

More at:

http://climatechange.unep.net/

Milankovitch Theory

In 1938 changes in Earth's orbital patterns were linked to global cooling and to the Pleistocene Ice Age. Milutin Milankovitch, a Serbian mathematician, discovered a statistical relationship among three factors: Earth's tilt, Earth's orbit around the sun, and glacial advances. He noted that three astronomic cycles are involved: Earth's tilt changes from 22.1 degrees to 24.5 degrees and back every 41,000 years; Earth's North and South Poles make one revolution,

a precession, every 25,780 years; and Earth's orbit and distance from the sun change in a 100,000-year pattern, or eccentricity.

John E. Kutzbach of the University of Wisconsin–Madison's Center for Climatic Resources theorized that these cycles could change the amount of insolation received at different latitudes by as much as 5 percent and apparently could allow the northern latitudes to cool sufficiently for glaciers to grow. Evidence in sup-

port of the Milankovitch theory has accumulated in recent years: Fossils found in seafloor cores suggest changing ocean depths and changing glacial coverage of continents during the past 500,000 years. When colder temperatures caused glaciers to expand on the continents, the decrease in runoff from streams and rivers caused ocean levels to drop. No temperature declines sufficient to trigger an ice age are predicted for the next 50,000 years or more.

EARTH MATERIALS AND TECTONIC PROCESSES

MINERALS AND ROCKS

Minerals and the rocks they form combine to provide a history of the processes and patterns of Earth's environments. Minerals and rocks affect landform development and form valuable natural resources such as gold, tin, iron, marble, and granite.

Minerals

So far more than 4,000 naturally occurring minerals have been identified on Earth. A mineral is an inorganic solid that has a characteristic chemical composition and specific crystal structure; these in turn affect its physical characteristics. Quartz, for example, is formed from silicon and oxygen, and halite, or salt, is a combination of sodium and chloride. A mineral's physical characteristics determine how resistant it is to weathering, how it influences landscape development, and how people decide to use it. The composition of quartz makes it very resistant to weathering, transparent, and a source for silicon used in computer chips. Although graphite

Rocks are formed and transformed in a sequence called the rock cycle. (See The Rock Cycle box, page 102.) In a complete cycle, molten rock hardens into igneous rock, which then breaks down and compacts into sedimentary rock. Heat and pressure transform this rock into metamorphic rock, which then melts and becomes molten rock. Not all rocks move through the complete cycle.

and diamonds are composed of carbon atoms, the arrangement of atoms is different, which is why diamonds are the hardest minerals and graphite is one of the softest. Most mineral particles are small because they form within confined areas, such as lava flows and between grains of sediments; large individual crystals found in geodes, pegmatites, and other rocks are relatively uncommon.

Only a few dozen of the 4,000 minerals recognized by the International Commission on New Minerals and Mineral Names are common

on Earth. The most common minerals are combinations of the 112 elements—such as silicon, oxygen, and iron—that are also common on Earth. (See Planet Earth, page 34.) While many combinations of these elements are possible, only a limited number of combinations of minerals form under the environmental conditions found on Earth.

Silicates, the most common class of minerals, are the major components of most rocks and include quartz, mica, olivine, and precious minerals such as emeralds. Other major mineral classes include oxides, sulfides, sulfates, carbonates, halides, and native, or pure, elements such as gold.

Almost all minerals can form in more than one environment, but the existence of a particular mineral is ordinarily evidence of a limited range of conditions such as temperature, pressure, and the presence or absence of oxygen or water. As a result, minerals can be used to map a region's paleogeography, or prehistoric geologic processes. The spatial distribution of minerals, however, may not display easily identifiable patterns because the environmental conditions under which they formed were present over periods of hundreds of millions of years. Halite, for example, forms when salty bodies of water evaporate; salt deposits in Siberia and the midwestern United States are evidence of past oceans and tropical locations. Diamonds form only under extremely high pressure hundreds of miles underground and are exposed on Earth's surface in very few locations, among them central South Africa, Yakutia in Russia, and northern Canada.

Rocks

Mixtures or clusters of minerals form rocks under specific sets of environmental conditions, and rocks are classified igneous, sedimentary, or metamorphic according to how they are formed.

Building Capitols

Some United States capitol buildings have been constructed using stone from in-state quarries, as a matter of state pride and for economic reasons. Granite, an especially strong igneous rock, was used to build numerous capitols, and Barre granite from Barre, Vermont, started an industry. Limestone was also a popular choice in construction: Salem limestone, also called Indiana limestone, from Salem, Indiana, was used to construct not only the capitol building of Indiana but also parts of the capitols of nine other states, including Alaska.

IGNEOUS ROCKS AND RESOURCES

Igneous rocks form when magma, molten material below Earth's surface, or lava, molten material on Earth's surface, cools. Not all minerals in magma or lava solidify at the same temperature when cooling occurs, so different igneous rocks can form from the same molten material. Olivine and augite solidify at warmer temperatures and form gabbro, whereas quartz and muscovite solidify at cooler temperatures and form granite.

Igneous rocks are classified by texture—evidence of their speed of cooling—and color—evidence of their chemical composition. In general, igneous rocks such as granite, with visible mineral crystals, form below the surface; igneous rocks with tiny crystals or a glassy appearance, such as obsidian, form at the surface. Cooling is usually very slow for magma because the rock that surrounds the intrusive magma is good insulation; in some cases, such as in the Sierra Nevada in California, it takes thousands of years for the magma to cool. Slow cooling allows mineral crystals in these rocks that formed below the surface—intrusive or plutonic, rocks such as granite, diorite, and gabbro—to grow large enough to be seen with the naked eye. Lava cools more quickly because it is exposed to air, and minerals in the resulting extrusive, or volcanic, rocks—such as rhyolite, andesite, and basalt—form much smaller crystals than intrusive rocks. Lighter-colored felsic rocks, such as granite, rhyolite, and andesite, have a greater percentage of quartz, micas, and silicate than do darker *mafic rocks* such as gabbro and basalt; mafic rocks have a greater percentage of magnesium and iron. Lavas can be felsic or mafic and their composition affects their rate of flow and how they may be erupted from a volcano. (See Igneous Structures and Volcanoes, page 120.) Despite different textures resulting from different rates of cooling, felsic igneous rocks have comparable mineral compositions. Gabbro is an intrusive rock

The Rock Cycle

Minerals have very specific physical and chemical properties related to their composition; rocks have more diverse mineral components and therefore more variable properties. Sandstone, for example, is most often composed of quartz grains, although any other hard, rounded fragment in a limited size range from .002 to .08 inch in diameter, such as feldspar grains, can form sandstone. More specific names may be applied depending on the size and composition of grains comprising a rock. The environmental conditions in which rocks form are often vastly different from the conditions in which their mineral components formed; the quartz grains of sandstone are usually formed miles below ground, whereas most sandstones are former beaches.

Rocks will not change unless their surroundings change. Their mineral components, stable in the original environment, do react to stresses from changes in temperature, pressure, and the presence of water and air. (See rock cycle diagram, page 100.) All rocks will change when environmental conditions cause sufficient stress, but the mineral composition of different rocks limits what their next forms may be. For example, silicates that form granite cannot change directly into the carbonates that form limestone. Once a rock changes, there is little evidence of the environmental conditions that existed when it was created.

whereas basalt is volcanic, but both are mafic and contain the minerals olivine, pyroxenes, and plagioclase feldspar.

The distribution of igneous rocks reflects Earth's tectonic history. Intrusive rocks dominate continental shields, the ancient cores of continents; the Canadian Shield, core of North America, is granite that is exposed over much of eastern Canada. Mountains, especially older chains such as the Appalachians and Urals, tend to have exposed intrusive rocks, laid bare by erosion. Extrusive rocks such as basalt are usually found where there has been volcanic activity; much of eastern Washington, Iceland, and western India have exposed basalt.

Igneous resources include many of Earth's deposits of diamonds and valuable metallic minerals such as titanium and chromium. As magma cools, heavier mineral crystals sink and concentrate at the bottom of the formation. Some magmas generate superheated water that carry rare metals such as gold, silver, and copper and deposit them as veins in rock fractures.

SEDIMENTARY ROCKS AND RESOURCES

Sedimentary rocks form from the accumulation of different types of sediments, including particles of minerals or rocks, ions in solution, and organic material. The majority of sedimentary rocks are created from particles that settle or precipitate out of lakes, oceans, and seas; however, sand dunes, talus cones at the base of cliffs, and glacial deposits can also become sedimentary rocks. As sediments accumulate, they settle and eventually solidify to form layers, or strata. *Lithification*, the process of becoming rock, includes compaction of strata by burial and cementation by precipitating minerals such as calcite or hematite.

Sedimentary rocks are classified by their composition. Clastic sedimentary rocks are accumulations of particles of other rocks: Clay becomes shale, sand becomes sandstone, and gravel forms conglomerate. Chemical rocks form when ions in solution precipitate or become solid as evaporation or other changes reduce water's ability to hold ions in solution: Calcite forms from solution and combines with clay minerals to form limestone. Organic rocks form when dead plant and animal matter accumulates in bodies of water before it completely decomposes; organic processes are important in the formation of many limestones—especially chalk, which forms from the bodies of microscopic organisms such as foraminifera—and bituminous coal, which is made up of plant materials.

Changes in ocean levels relative to the continents have resulted in sedimentary rocks covering most of the continental surface, including *continental shelves,* the offshore extensions of continents. Ocean floors were originally basalt, an igneous rock, but over time sediments eroded and washed from the continents and eventually became sedimentary rocks that buried the basalt. Many of these sedimentary rocks formed hundreds of millions of years ago; most of the rocks in the midwestern United States were formed in the Paleozoic era, which ended 245 million years ago. In the past four billion years or so, tectonic activity has compressed and folded many former ocean floors into mountains, as evidenced by aquatic plant and animal fossils in sedimentary rocks found on mountain peaks in the Himalaya, Alps, and Rockies.

Many natural resources have their origins in sedimentary rocks;

such as fossil fuels of coal, petroleum, and natural gas; construction materials such as limestone for cement; and metal deposits such as iron and gold. Coal is tropical swamp vegetation that accumulated in acidic water; the acidity killed bacteria that would otherwise have decomposed the vegetation. Most of the world's major coal deposits, including those in the central United States, Europe, and China, date to the late Paleozoic era; their current locations, far from the tropics, are now known to be the result of tectonic plate movements and are evidence to support the theory that continents have shifted over time. (See Plate Tectonics, page 104.)

Petroleum, or crude oil, and natural gas formed in shallow oceans from microscopic organisms that settled and were covered by sediments that prevented decomposition. As the organisms were buried under deeper and deeper layers of sediments, pressure and heat converted the organisms to liquid and gaseous hydrocarbons. Petroleum starts to form at temperatures above 120°F; at temperatures above 212°F natural gas forms. There are major petroleum deposits in the Persian Gulf region and in Venezuela and western Russia where ocean floors have been uplifted and exposed.

METAMORPHIC ROCKS AND RESOURCES

When rocks are subjected to heat, pressure, and chemically active fluids, they can become metamorphic rocks. Any rock can have its texture or its mineral composition changed: Limestone metamorphoses into marble when its texture changes because of increased heat, and basalt metamorphoses into greenstone when low heat and fluids alter its minerals. Shale, a sedimentary rock composed of clay and *silt,* can progressively metamorphose into slate, phyllite, schist, and finally gneiss as an increasing percentage of the original clay changes and more layering occurs.

Two metamorphic processes, regional and contact metamorphism, are responsible for most metamorphism. Regional metamorphism may affect areas of thousands of square miles. When two continents come together, the rocks along the boundary where they converge are heated and compressed, causing sediments and minerals to align at right angles to pressure; rocks such as slate and schist result.

Contact metamorphism occurs when heat from a magma body forms rocks such as marble, greenstone, and hornfels. Pressure is not a factor, and less metamorphism occurs when the heat source is farther away. Garnet and staurolite minerals form where the temperature is highest, and chlorite and biotite form farther away, where the temperature is cooler.

Metamorphic rocks are not common surface rocks because they form far below the surface and are exposed only after substantial erosion. Most often, metamorphic rocks are found alongside igneous rocks where tectonic convergence once occurred. Eroded mountain chains, such as the Appalachians and the Adirondacks in North America and the Urals in Europe, expose belts of metamorphic rocks. Shield areas of continents, such as the Canadian Shield, are the roots of ancient mountains, and they expose gneiss, greenstone, and other metamorphic rocks and granite, an igneous rock.

Most metal deposits created by metamorphism, including copper, lead, and zinc, are created by contact metamorphism involving superheated fluids commonly found where igneous rocks form. Other metamorphic minerals used as resources include graphite, used in pencil lead; talc, used in talcum power; and asbestos, used in insulation. Slate, marble, and gneiss are metamorphic rocks used as construction materials, and precious stones formed by metamorphism include rubies, emeralds, and sapphires.

More at:

http://www.rocks-rock.com/

PLATE TECTONICS

The theory of *plate tectonics* links currents in the mantle to the location and formation of major geologic features on Earth's surface and to earthquakes and volcanic eruptions. Earth's continents, oceans, mountain ranges, and other major surface features change locations as tectonic processes continue.

Plate Tectonics Theory

The theory of plate tectonics that emerged after the 1950s and 1960s addresses many of the weaknesses of *continental drift* theory and explains additional spatial patterns of geologic features. Plates are sections of Earth's lithosphere—its rigid exterior—that form continents and ocean floors; tectonics, from the Greek *tecktonikos* (builder), is the study of Earth's structure. Tectonic processes of convergence and divergence, meeting and separating, create the largest surface features and link the formation of continents and oceans, their major surface features, the global distribution of rock types, and even past climatic changes. Understanding these processes also helps explain how and why earthquakes and volcanoes occur and helps predict geologic events. The theory of plate tectonics continues to be refined as new information gets collected and analyzed.

Geologists have identified several dozen large and small tectonic plates that have stable interiors, move independently, and interact with each other at their boundaries. (See Earth's Tectonic Plates map, pages 108-109.) Plates often are partly continental and partly oceanic; the North American plate extends from the Mid-Atlantic Ridge to the west coast of Canada and the United States (except for part of California's coast). The South American, North American, Eurasian, African, Pacific, Antarctic, Australian, Indian, and Nazca plates are the largest plates. Other plates, such as the Juan de Fuca (off the coast of Oregon and Washington), the Caribbean, the Philippine, and the Adriatic plates, are much smaller but are important tectonically because of their boundary interactions.

During Earth's history the number, size, and location of tectonic plates have changed dramatically. Smaller plates have merged into larger plates, and larger plates have divided into smaller plates. *Pangaea* is the most recent example of a superplate that formed from smaller plates and is now divided into our current tectonic plates. The division of Pangaea into smaller plates started around 225 million years ago and continues today. (See Pangaea and cur-

Continental Drift

In 1620 English philosopher Sir Francis Bacon (1561–1626) noted in *Novanum Organum* that the shapes of eastern South America and western Africa matched one another. Over the years other authors noted the fit of these coastlines and hypothesized that the continents had once been joined and then drifted apart. German meteorologist and geophysicist Alfred Wegener (1880–1930) presented the first well-researched theory of continental drift in *The Origin of the Continents and Oceans* (1915). While his theory offered an explanation for many geographic and geologic patterns around the world, it generated a great debate among scientists because it challenged the accepted idea that continents had fixed locations.

Wegener theorized that all continents on Earth were united as a supercontinent, Pangaea (Greek *pan*, "all," and *ge*, "Earth"), in a single super-ocean, Panthalassa, until the end of the Triassic geologic period, about 208 million years ago. When Pangaea began to drift apart, what is now Africa was at its center, with South America to the west; India, Antarctica, and Australia to the east; and North America and Eurasia to the north. Pangaea split first into two sections: Gondwanaland to the south and Laurasia to the north.

Wegener's evidence that the continents had once been attached included matching spatial patterns of fossils and rock formations between continents. Wegener inferred that rocks with grooves scratched by glaciers in southern India, central Australia, Africa's Kalahari Desert, and Uruguay may once have been attached to Antarctica. The Appalachian Mountains near the east coast of North America and mountains in Greenland, Ireland, and Norway have similar ages, rock layers, and fossils.

Despite the evidence Wegener gathered, most geologists rejected his theory of continental drift because he could not convincingly explain why or how the continents had separated. Parts of his theory were inaccurate: The moon's gravity does cause tides, but it cannot cause continents to move; continents do not, as Wegener proposed, drift over or plow through ocean floors.

In the 1950s, 20 years after Wegener disappeared during an expedition to Greenland, new technology such as magnetometers, instruments that measure the magnetic field of the Earth, was used to examine seafloors. Researchers aboard ships measured and mapped the orientation of Earth's past magnetic fields as preserved in the basalt rocks from the ocean floors; their findings confirmed Wegener's basic claim that continents move. When basalt cools, the iron minerals in the molten rock align to the north and south magnetic poles in such a way that the basalt contains a record of the magnetic poles at the time of the rock's formation. Earth's magnetic poles reverse themselves every million years on average. Mapping the magnetic orientation of the minerals revealed that the sequence of reversing magnetic fields was the same in rocks on both sides of the Mid-Atlantic Ridge.

These matching magnetic patterns are evidence that ocean floors are being created at the ridge and then are spreading slowly outward while new basalt forms at the ridge. In the 1960s, Princeton University geologist Harry Hess proposed the idea of seafloor spreading—that seafloors form at mid-ocean ridges and then move away, thus moving continents farther apart. New drilling technology enabled scientists to sample and date rocks on the ocean floors and to corroborate Hess's hypothesis with paleomagnetic evidence. The youngest rocks were found along the mid-ocean ridges, and samples became progressively older with distance from the ridge. Sediments covering the basalt ocean floor are older, contain fossils from earlier geologic periods, and become thicker with increasing distance from the ridge because there has been more time for these sediments to accumulate.

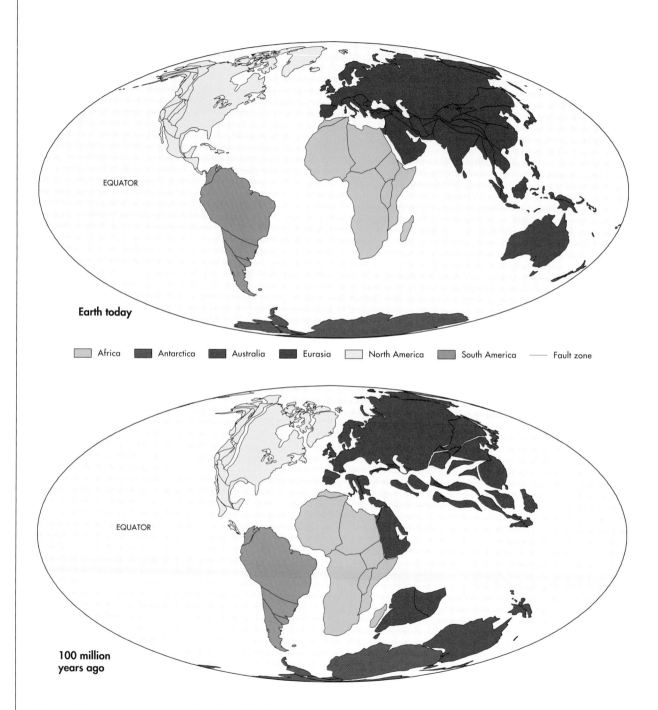

Earth today

Africa   Antarctica   Australia   Eurasia   North America   South America   —— Fault zone

EQUATOR

100 million years ago

Evidence of plate tectonics can be seen in the shifting of land masses over the past 600 million years of Earth's history. The continents found their current positions only about 20 million years ago. Clues to the past positions of these land masses were derived from paleomagnetic information in rocks.

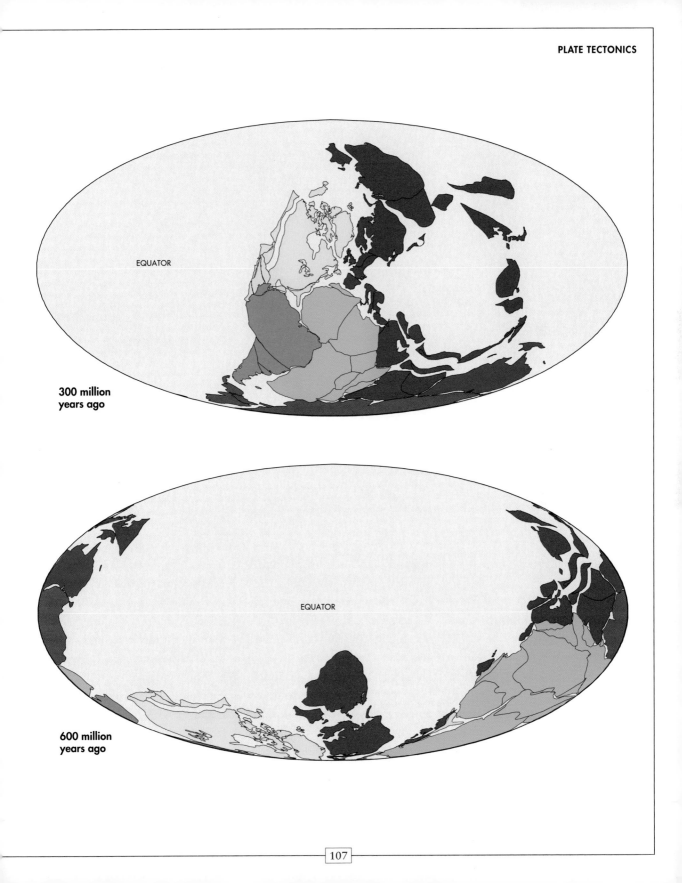

300 million
years ago

600 million
years ago

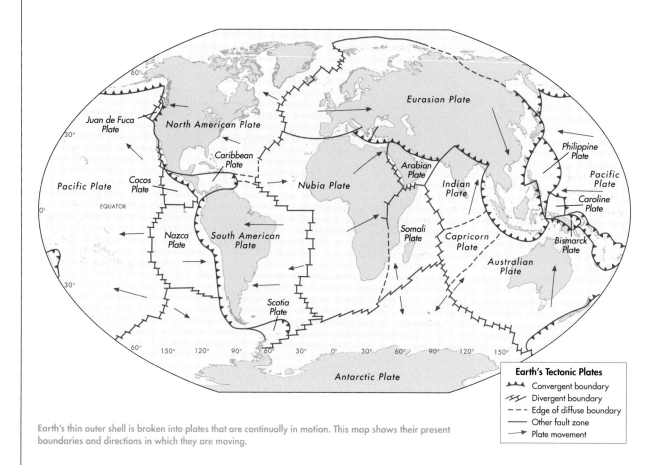

Earth's thin outer shell is broken into plates that are continually in motion. This map shows their present boundaries and directions in which they are moving.

rent continents map sequence, pages 106-107.)

According to the theory of plate tectonics the Earth's *lithosphere,* its rigid outer layer, moves slowly—usually one to two inches a year—over the underlying, softer *asthenosphere.* The lithosphere includes the crust and the uppermost part of the mantle and can be classified as either oceanic or continental. (See Earth Structure, page 35.) The mostly granite continental lithosphere averages 15 miles in

thickness and is more than 100 miles thick under mountain ranges. Basaltic oceanic sections of the lithosphere average 30 miles in thickness and may be only 12 miles thick in places. Oceanic basalt is about 10 percent denser than continental granite; when an oceanic section of a plate meets a continental section of a plate, the denser oceanic plate slides under the continental plate.

Radioactive decay of uranium and radium and the residual heat

from Earth's formation cause temperatures in the asthenosphere to increase to more than 1,000°F. As a result the asthenosphere is softer than the lithosphere, and rock in the asthenosphere can flow, much as hot asphalt does. The lithosphere can glide over this underlying layer with reduced resistance from friction.

The ultimate source of energy for the movement of tectonic plates is heat from Earth's interior, but the exact mechanism for plate movement has not yet been discovered. One mechanism proposed in the 1960s is that heat in the mantle causes convection currents. Some geologists theorize that there may be two other forces related to interactions between tectonic plates that provide mechanisms for plate movement. Slab pull may occur where plates meet: When one plate is pushed down at the boundary, it pulls the rest of the plate behind it. Ridge push may occur where plates are forming with a previously existing plate, and the plates are pushed away from each other.

The rate of movement of tectonic plates varies for different parts of the world and is apparently affected by the relative motions of plates. North America is moving away from Eurasia and South America is moving away from Africa at about 1 to 1.5 inches each year, whereas the Nazca plate, west of South America, pulls away from the Pacific plate at the rate of more than 7 inches per year. The faster speed of the Nazca plate seems to be because it lacks continental lithosphere, and the plate is being subducted under South America. In general, faster tectonic movements occur when one edge of the moving plate is being subducted under another plate and when slab pull is increased.

The Balkans and Middle East

The region extending from the Adriatic Sea to northern Iran and the Sinai Peninsula is a collection of more than a half dozen small tectonic plates, some smaller than South Carolina's area of 31,113 square miles. Boundaries of individual plates—including the Caspian, Ionian, and Levantine—are being mapped based on earthquake locations. As these plates shift, they are being compressed by the African and Saudi plates, which are drifting north into Eurasia, and they have created numerous mountain chains, including the Caucasus, Zagros, and Taurus ranges. The plate movements cause earthquakes to be a continuing hazard.

The differential movement of tectonic plates creates three types of tectonic boundaries: divergent, where plates move apart, or diverge; transform where plates move sideways in relation to each other; and convergent, where plates move into one another, or converge. Earth's surface changes at tectonic boundaries. Over time, however, the convection currents within the mantle change, and boundaries can cease to function or change into different types of boundary. The boundary between Europe and Asia closed when the plates merged to form Eurasia, and the convergent boundary along the coast of California became a transform boundary when the Farallon plate was being subducted under the North American plate. If a boundary ceases to function, the relict boundary, locked within a new plate, can continue to have effects: The New Madrid Fault in southeastern Missouri, a former divergent boundary within the North American plate, remains an earthquake hazard.

More at:

http://www.rocks-rock.com/

DIVERGENT BOUNDARIES AND LANDFORMS

When tectonic plates separate, they create a divergent boundary and new lithosphere. Initially, when a tectonic plate divides, its lithosphere thins from the tension; as a result there is less overlying pressure on the mantle beneath it. This reduced pressure allows part of the mantle to melt, rise through rifts caused by the tension, and fill the gaps between the two plates with basalt, creating basaltic lithosphere. Part of the newly created lithosphere attaches to the boundaries of both tectonic plates, so that each plate grows as it moves away from the line of divergence.

Divergent boundaries also occur in the oceans, and, because magma that rises from the mantle is warmer and less dense than the existing oceanic crust, it creates a bulge, called a mid-oceanic ridge, while adding new tectonic plate material. Today's oceans were created by divergence, and all have mid-oceanic ridges. When Pangaea broke apart, the Atlantic Ocean formed; the Mid-Atlantic Ridge marks the divergent boundary between North America and South America on the west and Eurasia and Africa on the east. The Carlsberg, Mid-Indian, and Southeast Indian Ridges are found in the Indian Ocean and were formed between India, Africa, Australia, and Antarctica.

Not all divergent boundaries continue to spread apart; the driving forces can cease for reasons that are not yet known. The island of Madagascar, now part of the Africa plate, formed a smaller plate that separated from the east coast of Africa but stopped diverging about 90 million years ago; it is now 250 miles off the coast of Mozambique. Several rifts within continents ceased to develop, or failed, because divergence stopped, but thinned, weakened segments of lithosphere remain. In the United States, the lower Mississippi River Valley—including the New Madrid Fault—follows the edge of a failed rift; two other major rivers, the Amazon and the Niger, are also located in failed rifts.

TRANSFORM BOUNDARIES

Transform boundaries occur where two plates slide sideways past each other, along what are called strike-slip faults. (See Faulted Structures, page 114.) The San Andreas Fault in California is a boundary between the North American plate as it slides northwest relative to the Pacific plate as it slides southeast. Transform boundaries create no major landforms, but continual movement causes earthquakes along the entire boundary. (See Earthquakes, page 116.) Transform boundaries, like divergent boundaries, can cease to move. Loch Ness, a linear lake in northern Scotland, is on the Great Glen Fault, a relict transform boundary.

CONVERGENT BOUNDARIES AND LANDFORMS

Convergent boundaries occur when two tectonic plates move toward each other. Differing landforms result from convergent boundaries according to the speed of the plates and whether convergence is between continental plates, continental and oceanic plates, or oceanic plates. When one plate is *subducted*, or forced below the other, it eventually melts into the mantle. Globally, the loss of lithosphere through subduction balances the creation of lithosphere caused by divergence. The surface area of Earth neither grows nor shrinks: As new tectonic plate material is added, an equal area of tectonic plate material is destroyed.

Great Rift Valley

When divergence occurs within a continent, as is now happening in the Great Rift Valley of eastern Africa, a *rift valley* emerges. Rifts slowly widen as the plates pull apart and deepen because basaltic lithosphere at the boundary is thinner than the granitic, continental lithosphere. Gradually, the rift fills with water. At first lakes form, but over time, as the valley widens, the lakes become narrow seas; eventually oceans result if divergence continues. Lake Tanganyika and Lake Nyasa in East Africa have formed because the Great Rift Valley is lower than the surrounding highlands. The Red Sea is on the divergent boundary between the Arabian and African plates, which have been separating for about ten million years; if divergence continues the Red Sea will become an ocean.

Island Terranes

As an oceanic plate is forced under a continental plate, the oceanic plate heats as it is subducted into the mantle. Much of the oceanic plate is eventually reabsorbed into the mantle, but the high pressure and temperatures in the upper mantle release water from the subducted rocks and melt felsic minerals, such as quartz, in the subducted plate and in the mantle above it. The resulting magma is less dense than the surrounding rock and rises toward the surface where it forms intrusive igneous rock bodies, such as the granite of the Sierra Nevada. (See Igneous Structures and Volcanoes, page 120.)

Mountains created by convergence between oceanic and continental plates are partly a result of volcanoes and intrusive igneous rock bodies and partly the result of other tectonic processes, such as compression. A volcanic arc results when a line of volcanoes forms parallel to a convergent boundary; the Cascade Range from Mount Lassen in California north to Mount Garibaldi in British Columbia, is one example. As an oceanic plate is subducted, part of its overlying sediments are scraped off and attached to the continent. In addition, islands that rise from the ocean floor are usually too light or too large to be subducted and therefore also attach or accrete to the continent. These island terranes, distinct geologic regions, make up much of the western edge of North America: There are about 40 or so terranes on the West Coast, including Wrangellia, Stikine, and San Juan. The compression that results from two tectonic plates converging also causes the edge of the continent to fold into mountains. (See Geologic Structures and Earthquakes, page 112.)

When a continental tectonic plate converges with a denser oceanic tectonic plate, ocean floor is destroyed as the oceanic plate slides below the continental plate, and mountains rise along the boundary. As the oceanic plate bends down, the ocean floor angles down and forms a trench. Trenches are the deepest parts of oceans, and their location within a hundred miles of the coastlines of continents rather than in the middle of oceans supports the theory of plate tectonics: The Peru-Chile, Ryukyu, and Java Trenches all parallel coastlines.

Subduction of an oceanic plate beneath a continent can cause the continental plate to stretch like Silly Putty even as its edge compresses. This extension reduces the compressional effect that creates folded mountains at the boundary, making the mountains less prominent than they might have been.

Convergence between two oceanic plates results in subduction of one of the plates. The colder plate subducts beneath the warmer plate and a trench forms. Nearly seven miles deep, the world's deepest trench—the Mariana Trench in the western Pacific—formed when the Pacific plate subducted the Philippine plate.

Volcanic island arcs form parallel to these trenches on the non-subducted oceanic plate. Each island is a composite volcano formed from magma rising above the subducted plate, and is comparable to continental volcanoes forming in volcanic arcs.

Some island arcs, like the Aleutian Islands, are continuations of volcanic arcs found on land, in their case the Alaska Range. Other linear island arcs are the Marianas and Tonga in the Pacific and Lesser Antilles in the Caribbean. Not all volcanic islands are formed by convergence and subduction; like Hawaii, some formed over a *hot spot* of rising magma. (See Igneous Structures and Volcanoes, page 120.)

Convergent boundaries also occur between continental tectonic plates, but subduction cannot take place because neither plate is sufficiently dense to be forced down into the mantle. Continental plates converge only after intervening oceanic plates have been subducted; as a result, when the continents converge, intrusive igneous rock bodies and volcanic landforms will be found in the interior of the landmass.

When the continents meet, the highest mountains in the world form because lateral compression causes folding and therefore vertical uplift. The Himalaya contain the world's highest peaks and are the result of the Indian subcontinent converging with Eurasia about 50 million years ago.

The mountain range created by convergence of continental plates becomes a *continental suture*, marking the former boundary between two plates. The Urals and Appalachians are examples.

The thickest parts of the lithosphere are found at these boundaries while they are active; after convergence stops, erosion eventually exposes the rock that was once buried miles underneath the mountains.

More at:

http://pubs.usgs.gov/publications/text/dynamic.html

GEOLOGIC STRUCTURES AND EARTHQUAKES

Shifting tectonic plates stress rocks along their boundaries, causing folds, domes, and other geologic structures to form in rock layers that influence surface landforms such as mountains, cliffs, and valleys. Earthquakes, hazards to people, are also a result of tectonic shifting and stress. The relationship between tectonic boundaries and tectonic stress explains the spatial patterns of geologic structures, associated landforms, and earthquakes.

Stress, Strain, and Structures

Geologic structures are deformations, or changes, in the arrangement of rock layers. Geologic structures are most easily seen in sedimentary rocks, which are usually deposited in horizontal layers or strata.

Three types of stress occur along tectonic boundaries: compression, tension, and shear or sideways stress. Compression causes rock layers to fold because it reduces surface area but not the amount of rock. Tension deforms rock by stretching it over a greater area. Shear stress causes deformation by applying force in opposing directions to rock layers.

Different types of tectonic boundaries cause different types of stress. Compression occurs at con-

vergent boundaries and is most evident when two continental plates collide and fold rock layers into mountains. Tension results at divergent boundaries and causes faulting along much of the ocean floor near mid-oceanic ridges. Shear stress, also called tangential stress, is caused by transform boundaries and is marked by offset surface features such as stream channels that bend dramatically. (See strike-slip fault diagram, page 115.)

Europe After the Pleistocene Ice Age

Europe, like North America, had huge glaciers covering much of the continent during the Pleistocene Ice Age, an event that lasted 2 million years, from 2 million years ago to about 10,000 years ago. Glaciers about 1.5 miles thick centered on Scandinavia and spread outward to the Bristol Channel on the south coast of Wales, the Harz Mountains in Germany, and the Carpathians in Poland. As the glaciers melted, Scandinavia began to rebound to its preglacial elevation; parts of the northern Baltic Sea continue to rise about four inches every decade, whereas the rebound rate is less than one inch per decade in the southern Baltic Sea.

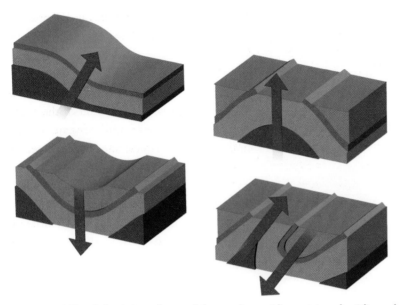

Types of rock folds include, clockwise from top left, monocline, anticline, overturned anticline and syncline, and syncline. Folds result when tectonic plates meet, compressing Earth's surface.

Deformation of rocks may be elastic, plastic, or fractured. Elastic deformation is temporary, and the rocks eventually return to their previous arrangement. Usually, elastic deformation is associated less with tectonic stress than with temporary kinds of stress, such as the weight of glaciers during an ice age. For example, Hudson Bay in Canada is in a temporary downfold that is slowly undergoing isostatic rebound to its above-water level; the glaciers that compressed the region of the bay melted 10,000 years ago. Isostatic rebound occurs because the lithosphere, which essentially floats on the asthenosphere, sinks when a region is weighted by glaciers, for instance, and rebounds upward when the weight is removed, as when glaciers melt.

Plastic deformation is permanent, and folded rock layers are records of the stress involved. As a rule, sedimentary rocks undergo plastic deformation when they are buried because heat and pressure from the burial and the millions of years involved allow the layers to reach equilibrium with the stress. Mountain roots, rock layers buried deep in a folded mountain but exposed by erosion, are excellent examples of plastic deformation.

A third type of deformation—faulting—occurs when rock layers are too brittle to fold. *Faults* are off-set fractures or breaks in rocks where the sides of the break are displaced in any direction relative to each other. Near-surface rocks are generally more susceptible to fracture and faulting because they are colder and under less pressure than buried, hotter rock; however, earthquakes can occur at depths greater than one hundred miles, evidence of faulting in the mantle.

Although geologic structures affect landforms, other factors are involved and the landforms will not always mirror the underlying rock structure. Landforms are also the result of the surface erosion of exposed layers of rock and reflect the type and duration of erosion as well as the types of rock that are present. Similar types of folding may cause different landforms in different regions. For example, when the upper rock layer of a fold is resistant, a ridge forms, but when the upper layer is softer than underlying layers, then the top of the fold erodes away.

FOLDED STRUCTURES

Rock layers that are lifted or pushed down relative to the surrounding area are called *folds*. Linear folds, domes and basins, and plunging folds are variations of folding and reflect different angles of pressure. Folds most often occur when Earth's surface is compressed by tectonic convergence, which causes the leading edges of continental plates to wrinkle like fabric as they contact another tectonic plate. In addition, convergence also causes subducted oceanic plates to melt, rise as magma bodies, and deform overlying rock layers.

Linear folds are the most common of folded structures and form at right angles to the compression of converging tectonic plates. With moderate compression, rock layers form a parallel series of *anticlines,* or upfolds, and synclines, or downfolds. (See types of folds diagram, page 113.)

Almost all major mountain ranges contain complex folded structures and often equally complex surface features. The Alps and Urals of Europe, the Zagros of Iran and Iraq, and Australia's Great Dividing Range are all intensely folded. In the United States, the Ridge and Valley region of the Appalachians developed from a series of folds extending from eastern Pennsylvania to northeastern Alabama.

Two variations of folds—domes and basins—occur when the stress is not linear but centers on a point. Domes are circular or elliptical anticlines and are the result of sedimentary rock layers being warped upward by rising fluids, such as magma or salt, rather than by tectonic compression. When less dense material rises underneath

rock layers toward the surface, it can force the overlying rock layers to rise. The Black Hills of South Dakota and Wyoming, the Ozark Dome of southeastern Missouri, and the Aïr region of central Niger have domal structures.

Basins are circular or elliptical

A fault is a product of tectonic stress and is classified by the relative displacement of its sides. Four categories are (top to bottom): normal, reverse, transform, and overthrust.

downwarps or synclines caused by a loss of supporting pressure, in some cases because the gradual accumulation of sediments on the surface becomes heavy enough to cause underlying rock layers to sink.

There are large structural basins in the midwestern United States, such as the Michigan Basin in Michigan, Williston Basin in North Dakota, and Permian Basin in Texas. In Europe, the Paris Basin extends southwest from the Ardennes in southern Belgium to the Collines de Perch of Normandy in France. Many of these basins are hundreds of miles across and contain thick deposits of sedimentary rock.

FAULTED STRUCTURES

Almost all surface rock contains fractures or joints from expansion as erosion removes overlying layers, but faulted structures represent past or present tectonic stress. The movement that creates fault displacement can be extremely abrupt or very slow depending on the kind of tectonic stress involved, the rock's strength, the presence of groundwater along the fault plane, and the area of contact. Abrupt movement causes earthquakes, but in some cases, the rate of movement along a fault is so imperceptible that it is referred to as fault creep. (See Earthquakes, page 116.) In either case, cumulative movements can extend over hundreds of miles; along California's San Andreas Fault, matching rock layers have been found 350 miles apart.

A fault is classified according to the relative displacement of its two sides: Traditionally the side of a fault that tilts downward is called the footwall and the ceiling is called the hanging wall.

Normal faults, the most common type of fault, occur when the hanging wall slides down relative to the footwall. Tension associated with divergent boundaries and with the stretching of rock layers from warping up, or doming, causes normal faulting. If movements along the fault are slow, there may be no easily noticed resultant landforms; erosion and deposition will hide the displacement. More abrupt movements will create a scarp, or wall, where the upper part of the footwall is exposed. An 1819 earthquake near the mouth of the Indus River in India created a scarp almost 50 miles long and as much as 20 feet high; called the Allah Bund, Dam of Allah, it caused the Indus to flood the region.

Multiple, parallel faults associated with divergent boundaries create horsts, grabens, and fault-block mountains. Horsts are linear ridges or land uplifted between parallel faults, and grabens are linear valleys that slipped down between parallel faults. The Black

Grabens (top) are linear valleys that slipped down between two parallel faults. Horsts (bottom) are linear ridges between parallel faults.

Forest in Germany and the Sinai Peninsula are horsts, and the Rhine Valley in Germany and Death Valley, California, are both grabens. Grabens may become lakes because they drain adjacent areas; Lake Baikal in Siberia, the deepest lake in the world, is a graben, as is the Dead Sea.

Fault-block mountains form when blocks of land are tilted between parallel faults. This occurs when thousands of square miles come under tension by divergence or regional uplifting; the Aberdare

The San Andreas Fault slashes through California for more than 700 miles. The zone marks the boundary between the North American and Pacific tectonic plates.

Strike-slip faults (left) are formed by sideways displacement along vertical faults.

Range in Kenya and the Tetons in the United States are classic examples. Ranges of fault-block mountains have formed in many of the world's large *plateau* areas, including those in Bolivia, Tibet, and Iran. The Basin and Range province in Nevada and Utah is composed of a series of fault-block mountain ranges between basins, or valleys. Uplift from rising magma may have caused the tension; another explanation for this tension lies in the change the west coast of North America underwent 30 million years ago, moving from a convergent boundary to a transform boundary.

Strike-slip faults involve sideways displacement along nearly vertical faults. The stress that causes strike-slip faults is associated primarily with transform boundaries, although divergent boundaries on ocean floors also have numerous strike-slip faults. Different segments of a diverging plate can move at slightly different rates, allowing sideways displacement between faster- and slower-moving segments. The Great Glen Fault in Scotland, Dead Sea Fault in the Middle East, and San Andreas Fault in southern California are strike-slip faults. Displacement offsets drainage patterns and rock formations, but does not create major landforms such as mountains.

Earthquakes

Earthquakes are vibrations caused by movement of rock along a fault. They are unavoidable effects of Earth's tectonic movements, and the pattern of earthquakes around the world nearly matches plate boundaries. More earthquakes occur along convergent and transform boundaries, where rocks are cold and more brittle, than along divergent boundaries, where rising magma warms rocks and makes them more pliant. Most earthquakes not occurring on current plate boundaries—such as the New Madrid earthquakes in Missouri in 1811–12, the Nova Scotia quake in 1929, and the 1886 quake in South Carolina—are responding to stress from a bound-

ary hundreds of miles away or are associated with former tectonic boundaries. Other events cause surface vibrations—including bomb explosions, impacts of large meteorites, and even magma rising toward a volcano—but the release of tectonic stress along faults causes the majority of earthquakes.

More at:

http://hsv.com/genlintr/newmadrd/

Only a limited area of a fault—the focus—is under sufficient stress to move at any one time, causing an earthquake. Most foci are miles underground, so for mapping purposes geologists locate an *epicenter* on the surface directly above the focus. Occasionally the focus of an earthquake has included surface

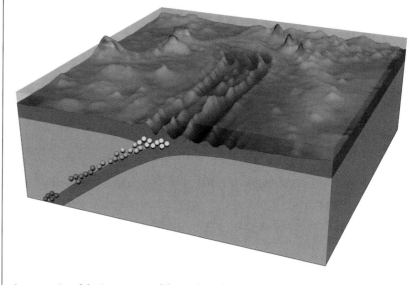

A cross section of the Tonga region of the South Pacific reveals the boundary between the subducting plate and the mantle. The foci of earthquakes follow this boundary. Earthquakes with foci near the surface tend to be the strongest.

A Three-Minute Earthquake: Alaska's Good Friday Earthquake, 1964

On March 28, 1964, Good Friday, southern Alaska suffered one of the largest earthquakes of the 20th century—8.5 on the Richter scale, or 9.2 magnitude on the moment-magnitude scale. Surface faulting caused some areas to rise more than 12 feet and other areas to subside more than 6 feet; landslides, tsunamis, and liquefaction destroyed homes and killed 131 people. Ordinarily, earthquakes last less than 60 seconds—the 1906 San Francisco earthquake, 8.3 on the Richter scale or 7.7 magnitude, lasted 40 seconds—but the Good Friday earthquake lasted more than three minutes, evidence of the large amount of energy that had accumulated.

More at:

http://www.eere.energy.gov/RE/solar.html

faulting: During the 1906 San Francisco earthquake, fences that had been built across the fault line were torn in half. When the focus shifts, the stress that had accumulated there is redistributed to adjacent areas of the fault. Rocks at the focus experience vibrations caused by shifting as they adjust to a loss of stress, and nearby areas undergo small shifts known as aftershocks when they adjust to the redistribution of stress.

Most earthquakes occur in crustal rock within 40 miles of Earth's surface, although a few foci have been recorded as deep as 435 miles. In order for an earthquake to occur, rock must be brittle enough to break and shift along a fault; at greater depths increased heat and pressure cause rock to fold or flow instead of break and fault. Along convergent boundaries involving an oceanic plate, the foci of earthquakes correspond to the depth of the subducted plate. Of the more than one million earthquakes each year, the largest ones tend to occur closer to the surface.

Vibrations in and on Earth during an earthquake are caused by energy waves that radiate from the focus. Earthquake waves are either surface waves, which travel along Earth's surface, or body waves, which travel through the interior of Earth. Two types of surface waves—waves moving sideways and waves moving vertically—are especially important because they cause most of the damage directly associated with earthquakes.

The two types of body waves are also important because seismologists can use them to triangulate earthquake epicenters: Primary or P waves are compressional body waves; they are faster than secondary or S waves, which are shear body waves. Seismographs register P and S waves and record the time that each arrives; the difference in arrival time can be translated into the distance from the seismograph station to the epicenter because the speeds of the two body waves are known. In the granite crust, P waves travel at an average rate of about 13,400 miles per hour, and S waves average about 8,000 miles per hour. Locating the epicenter of an earthquake requires triangulation—using the distances from a minimum of three stations to chart circles and find their intersection.

EARTHQUAKE SCALES

Scientists around the world use various scales to measure the size of earthquakes. The Japanese use the Wadati scale and Russians use the Medvedev scale. In the United States the moment-magnitude scale and the *modified Mercalli scale* are now used more frequently than the *Richter scale*. In the Richter scale each increase in scale value, for example from one to two, denotes a tenfold increase in the amplitude of the S wave. The moment-magnitude scale is based on the seismic

MODIFIED MERCALLI INTENSITY SCALE	
I	Not felt except by a very few under especially favorable circumstances.
II	Felt only by a few persons at rest, especially on upper floors of buildings.
III	Felt quite noticeably indoors, especially on upper floors of buildings, but many people do not recognize it as an earthquake.
IV	During the day felt indoors by many, outdoors by few. Sensation like heavy truck striking building.
V	Felt by nearly everyone, many awakened. Disturbances of trees, poles, and other tall objects sometimes noticed.
VI	Felt by all; many frightened and run outdoors. Some heavy furniture moved; few instances of fallen plaster or damaged chimneys. Damage slight.
VII	Everybody runs outdoors. Damage negligible in buildings of good design and construction; slight to moderate in well-built ordinary structures; considerable in poorly built or badly designed structures.
VIII	Damage slight in specially designed structures; considerable in ordinary substantial buildings with partial collapse; great in poorly built structures. (Fall of chimneys, factory stacks, columns, monuments, walls.)
IX	Damage considerable in specially designed structures. Buildings shifted off foundations. Ground cracked conspicuously.
X	Some well-built wooden structures destroyed. Most masonry and frame structures destroyed. Ground badly cracked.
XI	Few, if any (masonry) structures remain standing. Bridges destroyed. Broad fissures in ground.
XII	Damage total. Waves seen on ground surfaces. Objects thrown upward into air.

moment—the area of rock displaced, the rigidity of that rock, and the average distance of displacement. The Mercalli scale uses a range of I to XII to measure the intensity or damage caused by the earthquake: I, no damage; IV, pictures fall off walls and books fall from shelves, for example. A modification of the Mercalli scale measures a range of values over a region because the degree of damage from an earthquake depends upon distance from the epicenter, type of bedrock, level of groundwater, and type of construction.

EARTHQUAKE HAZARDS

Destruction and deaths from earthquakes are caused by direct and indirect effects of surface vibrations. Love waves collapse rigid structures that are not designed to resist shear stress. Earthquakes in regions of Iran, Afghanistan, Armenia, and Turkey—where mud-brick construction is standard—are responsible for thousands of deaths due to the collapse of structures.

Indirect effects of earthquakes include liquefaction, landslides, *avalanches*, fires, and *tsunamis*. Liquefaction refers to the behavior of water-saturated soils when vibrations cause the soils to lose their weight-bearing capacity, allowing structures built on them to collapse. During the 1964 Good Friday earthquake in Alaska, sediment below Turnagain Heights, an Anchorage residential area, liquefied and 70 homes collapsed. Liquefaction contributed to damage at Vanadzor, Armenia, in 1988 and in San Francisco's Marina district in 1989; both places are built on landfills in marshes.

Landslides and avalanches result when earthquake vibrations upset

Alaska's 1964 Good Friday earthquake caused the ground to liquefy and homes to collapse in an Anchorage residential area.

COMPARISON OF RICHTER AND MERCALLI SCALES, SHOWING ADJUSMENTS FOR FAMOUS EARTHQUAKES		
EARTHQUAKE	RICHTER	MERCALLI
Chile, 1960	8.3	9.5
Alaska, 1964	8.4	9.2
New Madrid, 1812	8.7	8.1
Michoacán, 1985	8.1	8.1
San Francisco, 1906	8.3	7.7
Loma Prieta, 1989	7.1	7.0
Kobe, 1995	6.8	6.9
San Fernando, 1971	6.4	6.7
Northridge, 1994	6.4	6.7

the equilibrium of steep slopes in mountainous regions. In 1970 an earthquake in the Peruvian Andes caused a massive avalanche of ice, snow, and rock to break away high on Huascarán, the country's highest peak. The debris buried the town of Yungay in mud and rock, killing almost all its 20,000 inhabitants.

Earthquakes cause fires when water and natural-gas lines and electrical wires break. An earthquake in Japan in 1923 occurred during the noon lunch hour, when hibachis, small charcoal grills, were lit. They were overturned and more than 140,000 people died in Tokyo and Yokohama, primarily in fires in residential areas where most homes were constructed of wood and paper. Fire also caused much of San Francisco's destruction after the 1906 earthquake, and for decades the disaster was referred to as the Great Fire, not the San Francisco earthquake.

More at:

http://www.sfmuseum.org/1906/06.html

Tsunamis, Japanese for harbor waves, are a series of sea waves caused by the vertical displacement of seafloor during an earthquake or volcanic eruption. The waves are sometimes misleadingly called tidal waves. In the Pacific Ocean, water averaging 3 miles deep allows tsunamis to travel more than 400

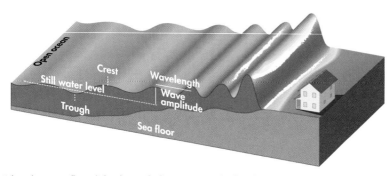

When the ocean floor shifts along a fault, seawater is displaced, generating low, broad waves. As the waves near shore, friction from the shallow seabed slows them down. At the same time the crest surges higher. The faster the depth decreases, the greater the tsunami's force.

miles per hour but have wave heights of only a few feet. When tsunamis reach shallower water, friction from the seabed slows the waves, causing their wavelengths to decrease and heights to increase; when they strike a coastline, tsunamis average 30 feet high, and heights greater than 100 feet have been recorded. Tsunamis are most common in the Pacific Ocean because of convergent boundaries surrounding the Pacific plate. The most vulnerable coastlines are Honshu in Japan, Hawaii, Alaska, Indonesia, Peru, and Ecuador.

Earthquakes are a certainty along tectonic boundaries around the world, and disasters like the July 17, 1998, earthquake and tsunami that killed more than 3,000 in Papua New Guinea are inevitable. In California, the San Andreas Fault in the San Francisco area has not had a substantial earthquake for decades and residents anticipate a big one; unfortunately, the long interval without

significant earthquakes means that the earthquake, when it arrives, is likely to be of greater magnitude because stress has been accumulating. The nearby Hayward Fault did generate a substantial earthquake in 1989, with an epicenter south of San Francisco. In southeastern Missouri, the New Madrid Fault, site of three powerful earthquakes in 1811 and 1812, is also accumulating stress that will eventually cause another large earthquake.

Earthquake prediction, by which the timing as well as the location and magnitude of an earthquake can be identified, is not yet a proven science. While some earthquakes have been successfully predicted, prediction is not consistently accurate. Research focuses on identifying changes that occur prior to an earthquake in precursor phenomena, including animal behavior, radon content in groundwater, and ground tilt.

Recent studies on changes in the electrical resistance of rocks, caused perhaps by stress on quartz grains, are promising.
More at:
http://earthquake.usgs.gov/bytopic/

IGNEOUS STRUCTURES AND VOLCANOES

Magma heated deep within Earth rises and forms igneous structures below the surface and volcanoes on the surface as a result of tectonic processes and hot spots. Igneous structures such as batholiths result from mantle plumes below the surface and also form near convergent boundaries; volcanoes form on the surface near convergent and divergent boundaries and at hot spots and are potentially serious hazards.

Plumes and Plate Tectonics

Magma rises from the upper mantle toward Earth's surface, where it creates and influences landforms. Mantle plumes, convergent boundaries, and divergent boundaries cause magma to form and rise. In order for mantle to melt, either temperatures must increase by about 400°F above the temperature of surrounding rock or pressure must decrease so that the magma will become less dense and rise slowly through surrounding rock. Rock under high pressure remains solid, but once pressure decreases, it melts; on average, partial melting begins at a depth of 15 miles.

Geologists in the 1960s discovered that columns of heated material hundreds of miles in diameter rise from the boundary between the outer core and the mantle. When the molten outer core heats a section of the lower mantle it causes a column of heated rock, a mantle plume, to rise toward the surface at a rate of several inches to several feet each year. As the plume rises, pressure decreases, allowing some of the plume to melt as it continues to rise.

The shape of a mantle plume changes as it rises toward—and eventually contacts—the lithosphere. In the mantle, a plume rises as a column with a slightly larger top. When the plume approaches the rigid lithosphere, the plume's top flattens out to a disklike form that can be 1,000 to 1,500 miles across and 100 miles thick. Below the top, the rising column or tail retains its columnar shape.

When magma from a plume reaches the lithosphere the spreading top of the plume causes the overlying lithosphere to bulge upward as much as 2,000 feet to form a huge plateau in the overlying plate. Faulting or rifting may occur and magma extruded as lava pours from the rifts to cover the plateau with layers of basalt, adding height to the plateau.

Eventually, after perhaps a hundred million years, the heat supply for the plume is exhausted, and magma will cease to rise to the surface.

Hot spots form at Earth's surface above mantle plumes. Mantle plumes and their surface hot spots are fixed in position above their source areas within continental and oceanic tectonic plates. As plates shift, however, landforms such as volcanoes, island arcs, and *seamounts* (underwater volcanoes) created above the hot spots move with the plates much like boxes on a conveyer belt.

The Galápagos Islands in the Pacific and the Azores in the Atlantic are the result of hot-spot activity, and *flood basalts,* huge lava flows formed by past hot spots, exist in Siberia, Brazil, and India. Mantle plumes have created plateaus that cover much of the ocean floor. Active hot spots on continents are rare; the Yellowstone National Park region of Montana, Wyoming, and Idaho is one of the few in North America.

Island chains formed by hot spots can be used to calculate both the speed and the direction of tectonic plate movement; scientists compare the ages and the locations of past hot-spot landforms with the location of an active hot spot today. In the Pacific Ocean, the chain of islands and seamounts, or submerged volcanoes, gets older

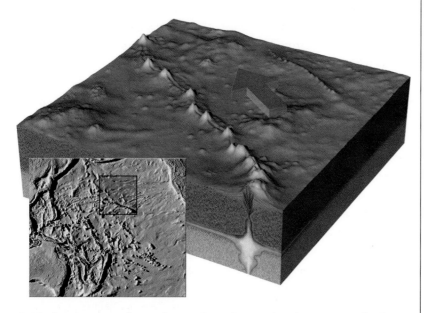

The islands and seamounts that stretch across the Pacific Ocean from the Hawaiian and Midway Islands to the Aleutian Islands increase in age toward the northwest, revealing the movement of the Pacific plate over a hot spot in the Earth's crust.

and older from Hawaii to the northwest, past Midway to the Aleutian Islands. This indicates that the Pacific plate has been moving to the northwest at about four inches a year. Oahu is about 2.5 million years older than the big island of Hawaii. The movement of the Pacific plate was to the north until 43 million years ago, as indicated by the pattern and the ages of islands along the Emperor Seamount.

Plutons

Magma cools in chambers and openings below Earth's surface called *plutons*. Erosion and isostatic uplift may eventually expose the buried plutons. The igneous rocks that form plutons are often more resistant to erosion than surrounding sedimentary rocks, so plutons tend to form mountains, hills, or ridges in the landscape.

Batholiths, the most massive plutons, have a minimum of 40 square miles of exposed surface. They are usually composed of felsic igneous rock, mostly granites and diorites. The Idaho Batholith covers more than 15,000 square miles and may have formed when a number of batholiths coalesced. Batholiths are created by converging boundaries and mantle plumes; the largest batholiths form when continental plates and oceanic plates converge.

Most eroded mountains and

The distinctive granite cliffs of the Yosemite Valley in California are part of the Sierra Nevada batholith.

continental shields include batholiths. Batholiths that formed 100 million years ago on the west coast of North America, from Baja California to the Alaska Panhandle, are now eroded mountain ranges. The Piedmont region of the Appalachians includes batholiths emplaced more than 300 million years ago.

The second type of pluton is tabular, or relatively flat. Most tabular plutons form from the magma of a large chamber, which, when it cools completely, may form a batholith. Massive and tabular plutonic bodies are therefore often located near each other. Laccoliths, lopoliths, and sills are tabular plutons, but differences in their magma composition give

each a different shape. A laccolith forms when viscous felsic magma is forced between rock layers and piles up in a mushroom shape, sometimes even several miles across. Lopoliths are large, concave plutons.

Sills are small and far more common than lopoliths. Unlike lopoliths, they form horizontally, varying in width from inches to hundreds of yards, and form only within 1 to 1.5 miles of the surface. Sills are common throughout mountain and shield regions. The Palisades, which are located along the Hudson River, and the Giant's Causeway in Ireland are classic examples of sills.

Dikes are another type of tabular plutons and may be the most common plutonic bodies. Dikes are likely to form when a magma chamber fractures rocks, especially rocks at the surface. Different magmas can form dikes, but mafic magmas and basalt dikes are most common because felsic magmas are too viscous to flow easily.

Volcanic pipes or necks are cylindrical remnants of conduits from magma chambers to the surface. In some cases, pipes are filled with erosion-resistant igneous rock that becomes isolated peaks in the landscape as weaker rock is eroded away. Devils Tower, Wyoming, and Shiprock, New Mexico, are excellent examples. Other pipes that do

Gettysburg, Pennsylvania

Exposed plutonic bodies became strategically important during the Civil War battle at Gettysburg in 1863. Union troops camped on Cemetery Hill, an exposed sill, while Confederate troops were on Seminary Ridge, an exposed dike. These plutons provided each side with high ground, and Confederate troops led by Maj. Gen. George Pickett had to charge up Cemetery Ridge in their unsuccessful attempt to dislodge Union forces. The ironstone each side used to build walls was basalt from these two plutons.

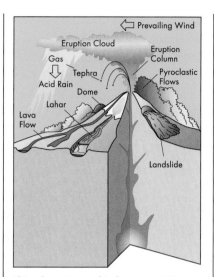

This volcano is typical in the western U.S. Eruptions release lava flows and pyroclastic flows of hot ash, volcanic rock, and gas, which may travel at 150 mph. Mudflows, or lahars, can entomb everything in their path.

not form prominent surface features may instead contain unusual minerals below the surface: Kimberlite pipes, such as those near Kimberly, South Africa, as well as pipes discovered in northern Canada during the 1990s, are sources for diamonds.

Volcanic Landforms

When lava erupts onto Earth's surface from underlying magma chambers, it accumulates into different features or volcanic landforms according to whether the lava is felsic or mafic and whether the shape of the opening through which it emerges is a large, linear fissure or a small, circular vent. Volcanoes are landforms composed of lavas and other volcanic debris that flow or erupt onto Earth's surface; igneous structures such as plutons form beneath the surface and are exposed only by erosion.

Felsic lavas are more viscous and tend to contain more water vapor and other dissolved gases than mafic lavas. As a result, felsic lava does not flow easily; it explodes upon eruption. Mafic lavas are less viscous because they lack silicates and maintain a temperature averaging more than 1,700°F, enabling them to flow more readily and farther; they are less explosive on extrusion than felsic lavas. Linear fissures allow lava to cover larger areas than do vents. Mafic lava in eastern Washington State has been found more than a hundred miles from its source.

When lava, ash, and cinders erupt from a circular vent and accumulate around the opening, they form a volcano. Volcanoes are classified as shield volcanoes, cinder cones, or composite volcanoes, also called strato-volcanoes, depending upon whether the lava is mafic or felsic. The type of lava causes different shapes and sizes of volcanoes, and it also affects how explosive eruptions will be. Volcanoes are also classified as active (expected to erupt), dormant (capable of erupting), and extinct (incapable of erupting). Fewer than 1,500 of the

Cascade Range Volcanoes

Mount St. Helens in Washington, like Mount Rainier, Mount Baker, and Glacier Peak, was one of several Cascade volcanoes scientists considered to be potentially dangerous prior to its dramatic eruption in 1980. St. Helens had last erupted in 1857, and it has been studied intensively since 1980. In November 2004 earthquakes and a growing lava dome inside the crater led scientists to predict another eruption was likely. California and western Oregon also face the threat of eruptions from volcanoes such as Mount Hood, 40 miles southeast of Portland, and Mount Lassen in northern California.
More at:
http://vulcan.wr.usgs.gov/

20,000 or so volcanoes on Earth are considered active; 550 of the active volcanoes are continental and the remainder are oceanic.

Shield volcanoes are the largest volcanoes on Earth and have slopes of only 10 degrees or so, with round bases shaped like shields. When the tail of a mantle plume under an oceanic plate reaches the ocean floor, a smaller column of rising magma may exit through a vent rather than through a fissure. This lava is extremely mafic and flows easily: Thousands of layers may accumulate over time and build up a volcano that rises miles above sea level. Mauna Loa, the largest volcano on the island of Hawaii, rises more than six miles from the ocean floor, and its base is more than a hundred miles wide.

Earth's shield volcanoes have formed primarily in the Pacific and Atlantic Oceans, because the two largest oceans contain the most hot spots, sources for magma. Many of these volcanoes are in linear patterns because, as a tectonic plate moves over a mantle plume, the plate transports its volcanoes away from the plume as the next in the series of volcanoes forms. Easter Island, Samoa, and the Galápagos Islands are Pacific Ocean shield volcanoes; Atlantic Ocean shield volcanoes include St. Helena, the Azores, and Bermuda.

Cinder cones are the smallest volcanoes, usually under a thousand feet high, and form over the tails of mantle plumes beneath continents. As magma melts through continental crust, it becomes more felsic: When silicate-enhanced magma reaches the surface, it has enhanced viscosity and tends to erupt explosively, ejecting cinders, ash, and other fragments. As fragments accumulate near the vent, they form a steep-sided hill with slopes between 30 and 40 degrees. Most cinder cones erode quickly and disappear from the landscape because they are piles of

The Mauna Loa volcano on the island of Hawaii rises more than six miles from the ocean floor. A shield volcano built layer by layer over time, Mauna Loa is still active and growing.

unconsolidated sediment. Paricutín, a cinder cone west of Mexico City, is famous because scientists were able to observe its formation between 1943 to 1952.

Most of the world's best-known volcanoes are composite volcanoes, formed from mafic and felsic lavas; they are large, steep-sided mountains found near coastlines and as islands in oceans. Subduction along convergent boundaries provides the magma that rises to form composite volcanoes. While the magma is mostly felsic, it may also combine melted mafic mantle and melted felsic sediments, and it may erupt as cinders and ash as well as flowing lava. Composite volcanoes tend to have the conical shape of a cinder cone but are larger because of lava flows. Mount Rainier, Washington, is more than 2 miles high and has a base 14 miles in diameter.

Composite volcanoes, like shield volcanoes, often occur in linear patterns because they parallel the convergent boundaries that formed them. There are chains of volcanoes along the west coasts of Central America, South America, and the Pacific Northwest, where the Cocos, Nazca, and Juan de Fuca plates are being subducted. The west side of the Pacific Ocean is also lined with composite volcanoes where the Pacific plate is subducted; the *Ring of Fire,* a nearly complete

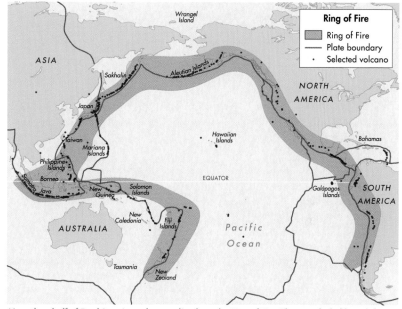

More than half of Earth's active volcanoes lie along the Ring of Fire. They are fueled by subduction: The edges of the plates that cradle the Pacific slide under the plates holding the continents and bend down into the hot mantle. The friction melts rock, which rises and bursts out as lava.

arc of volcanoes, circles much of the Pacific Ocean. Other lines of composite volcanoes are found in Indonesia and the Caribbean, with smaller concentrations in Italy and in the southern Sandwich Islands near the Falklands.

Volcanic Eruptions

Volcanic eruptions pose a serious and increasing hazard to humans, though not all active volcanoes and not all types of eruptions are dangerous. Many volcanoes, such as those in southern Alaska, are far from urban development and pose little threat to populations. Some eruptions, especially those of shield volcanoes, involve slow lava flows

that miss development or can be diverted by barriers.

Eruptions can be classified based on the characteristics of past eruptions of specific volcanoes. Shield volcanoes may have Icelandic- or Hawaiian-type eruptions, which primarily involve lava flows. Hawaiian eruptions release more dissolved gases than Icelandic eruptions and can cause small ash falls. Eruptions of composite volcanoes can range from Strombolian— named for Stromboli, a volcano on an island of the same name off the west coast of Italy—which have moderate explosions and mostly steam eruptions, to ultra-Plinian, such as Vesuvius in A.D. 79, or

The eruption of Mount St. Helens in Washington State in 1980 spread a hazardous cloud of super-heated gases and pyroclastic fragments.

Krakatoan for the Indonesian eruption in 1883, which are the largest types of explosive eruption.

Volcanic hazards include a number of secondary effects, such as tsunamis and landslides, but the primary effects involve the fall or flow of volcanic debris. Lava flows can cause substantial damage to property and land but cause few deaths. Few lava flows travel faster than five or ten miles per hour.

Lahars, volcanic mudflows, occur when water mixes with ash and cinders; the water can be from snow and glaciers melted by an eruption, from rain, or from the eruption of a water-filled crater. Lahars may be responsible for more than 10 percent of all volcano-related deaths; the 1985 lahar associated with the erup-

tion of Nevado del Ruíz in Colombia killed more than 22,000 people. Both shield and composite volcanoes have generated lahars, and deaths have occurred in Iceland, Indonesia (the source of the term), Costa Rica, and the United States.

Pyroclastic falls, including ash, cinders, and larger fragments such as bombs and blocks, are associated with composite volcanoes whose magma is viscous and often sticks in its vent, causing explosive eruptions. The eruption of Vesuvius in A.D. 79 is perhaps the best known example because the ash and cinders preserved Pompeii and the forms of some of the 16,000 victims.

Eruptions have released toxic gases dissolved in magma and lava into the atmosphere and have been

responsible for thousands of deaths this century. Carbon dioxide and sulfur oxides are among the more common gases, although hydrogen sulfide and other gases have also been emitted.

One of the deadliest hazards from erupting volcanoes is a glowing cloud of superheated gases and pyroclastic fragments that can travel more than one hundred miles per hour. The cloud usually explodes from the side of a composite volcano and may travel outward more than ten miles. The 1902 eruption of Mount Pelée on Martinique killed 29,000 people almost instantaneously; other eruptions have occurred in Alaska, the Philippines, and New Guinea, where 3,000 people died in 1951 during the eruption of Mount Lamington.

Tsunamis caused by eruptions have resulted in thousands of deaths, including the coastal population of Crete when Santorini, an island volcano, erupted in 1530 B.C. The eruption that obliterated Krakatoa in 1883 killed more than 32,000 people. Famines may follow eruptions because ashfall may kill livestock and bury crops. Iceland suffered a famine in 1783 after an eruption; 10,000 people and 130,000 head of livestock died.

More at:
http://volcanoes.usgs.gov/
main.html

LANDFORMS AND LANDSCAPES

WEATHERING AND MASS WASTING

Landforms are the visible record of geomorphic processes and part of the landscape, the natural and man-made surface features in a region. *Geomorphology* is the study of Earth's surface features; geomorphic processes include tectonic activity and volcanism, which may uplift land, and gradation, processes of weathering, mass wasting, and erosion that lower land.

Weathering

Weathering is a geomorphic process that disintegrates and decomposes surface rock and soil by physical and chemical processes. When rocks that were formed underground reach the surface, they weather as they are exposed to rain, ice, snow, and wind. The minerals that comprise rocks and the rocks themselves are affected by changes in the environment when they are exposed at the surface. Rocks disintegrate through mechanical processes, such as frost action and root growth; they decompose chemically when mineral components change through *oxidation*, hydration, and solution.

The striking landforms seen in Bryce Canyon National Park, Utah, were formed by weathering, which gradually disintegrates and decomposes surface rocks.

Water that repeatedly freezes and thaws in rock cracks can ultimately break up the rocks, another example of physical weathering.

The products of weathering include fragments of rocks, sediments, and dissolved ions. Talus and scree, boulders and rock fragments that break off from a cliff, accumulate at the cliff's base. Sand, *silt*, and clay-size sediments form from weathered rocks and become the basis of soils. The mineral components of some rocks are soluble and can be dissolved by acidic water and transported in solution elsewhere. They can be precipitated as the basis for chemical sedimentary rocks or withdrawn by plants for nutrients. (See Sedimentary Rocks and Resources, page 102.)

There are two types of weathering: physical (or mechanical) and chemical. Physical weathering involves mechanical forces that break rocks into smaller pieces. Chemical weathering changes the mineral components of rocks into more stable minerals. Chemical weathering also dissolves soluble minerals into ions, which wash away. Both types of weathering affect man-made structures as well as rocks: The outer layer of clay brick eventually flakes off because of physical weathering, and marble gravestones become unreadable over time because of the effects of chemical weathering.

Physical and chemical weathering interact, thus increasing each other's impact. Chemical weathering acts on the surface area of rocks,

and physical weathering increases the surface area subject to weathering. When a block of rock is broken into eighths, the volume of rock remains the same but the surface area doubles. Physical weathering breaks rocks into fragments and sediments more readily when chemical weathering has weakened the rock by creating softer, less-resistant minerals.

PHYSICAL WEATHERING

The expansion of roots, freezing water, and salt crystals within a fracture or pore in a rock is a primary mechanism for breaking the rock into fragments. Roots of trees growing in cracks in cliffs can expand as they grow and exert enough pressure to widen a fracture in a rock and break it, much the same way as tree roots buckle concrete sidewalks and streets.

Frost action is the most common example of expansion as a physical weathering process and occurs wherever there is water and the range of temperatures includes a freeze-thaw cycle. When water freezes, the molecules realign and expand about 9 percent in volume; in rock fractures, as in water pipes where water is confined, this expansion can create a force up to 1,400 pounds per square inch. Rock fractures are initially extremely small, so repeated freezing and thawing must occur before a crack expands enough to break off the outer part of the rock.

Salt-crystal growth is another example of physical weathering by expansion. Salt dissolved in water precipitates as the water evaporates and eventually accumulates as crystals of calcite, gypsum, and halite. Over time, as evaporation continues, the accumulation and growth of salt crystals within rock fractures can exert enough force to break off small fragments.

CHEMICAL WEATHERING

Chemical weathering occurs when water or atmospheric gases come in contact with the surface of a rock and change its mineral composition. Most chemical weathering takes place in the presence of water in which atmospheric gases are dissolved and within a dozen or so feet of Earth's surface. In climates with higher temperatures and higher precipitation, the rate of chemical weathering speeds up because the greater heat energy generally increases the rate of most chemical reactions. In humid tropical climates, chemical weathering can extend several hundred feet below Earth's surface; in deserts, chemical weathering may be limited to a depth of a few feet.

Chemical weathering causes minerals to change into different,

As the massive granite batholith forming Stone Mountain near Atlanta, Georgia, was uncovered by erosion, the granite expanded and eventually formed an exfoliation dome. Stone Mountain rises about 700 feet above the relatively level surrounding landscape.

usually softer, minerals or to divide into soluble ions. Chemical weathering by solution divides soluble minerals into ions, which can be removed by flowing water. Minerals such as gypsum, originally formed by the evaporation of oceans, usually dissolve easily. Solution always involves water and most often acids, which form as water combines with dissolved gases and other ions. (See Groundwater, page 143.)

PATTERNS OF WEATHERING

Weathering varies geographically by type and rate depending upon climate, rock types, vegetation, and topography. Climate is the most significant factor affecting weathering

because of the roles temperature and moisture play in chemical and physical weathering. Surface rocks and minerals vary in resistance to different types of weathering. Limestone, for example, weathers rapidly in moist climates and slowly in dry climates. Vegetation also affects weathering through root growth, which causes physical weathering, and humic acids, which come from the decomposition of vegetation and cause chemical weathering. Topography, especially the angle of slopes, affects runoff and drainage, and therefore the amount of water available for weathering.

Physical weathering occurs in all climates, but different types of weathering are associated with dif-

Gossan Zones

In 1993 two geologists working for a private company were in northern Labrador, Canada, searching for gold and diamonds, and discovered what turned out to be a ten-billion-dollar deposit of copper, nickel, and cobalt. While looking for kimberlite deposits, their serendipitous discovery happened when they recognized a rusty red gossan zone formed from hydrated iron-oxide minerals on a rock outcrop. Gossan zones (from gossen, Cornish for "blood") are associated with the chemical weathering of sulfide minerals such as pentlandite, which is composed of nickel, and iron, and sulfur, and chalcopyrite, which is made of copper, and iron, and sulfur.

ferent climatic patterns. Frost action is most active in subpolar and mountainous regions where freeze-thaw cycles occur daily, at least during some seasons. Salt-crystal buildup is most pronounced in dry climates near water bodies: The Namibian coast on the Atlantic Ocean, the land around the Great Salt Lake in Utah, and the country near the Colorado River—at such places, precipitation is largely absent, but moisture is available for evaporation year-round. Root growth is most common where a warm, humid climate supports forests, as in the Ozarks of Arkansas, Missouri, and Oklahoma. (See Climate Controls and Classification, page 85.)

Chemical weathering is most active in wet, warm climates; humid tropical climates have the highest rates of chemical weathering, mostly due to the climate but partly because of the large amount of vegetation supported by the climate. In contrast, chemical weathering in deserts is extremely limited because of the lack of water. Rock type is also an important factor in chemical weathering rates. Sandstone, for example, is much more resistant to chemical change than is limestone because quartz sand grains are chemically stable, whereas calcite dissolves easily.

Mass Wasting

Mass wasting, also called mass movement or gravity transfer, is the downslope movement of rock, soil, and sediment in response to gravity; it is a component of the gradation process that lowers hills and mountains and works with erosional agents such as streams. All slopes experience mass wasting: Steep slopes may have dramatic landslides and rockfalls, but slopes with angles of only a few degrees also experience downslope movement of rocks and sediments.

For mass wasting to occur, the pull of gravity must be greater than the resistance to gravity of a slope's material. Strong rocks, granite for example, may form cliffs, while unstable sand dunes can have a maximum slope of about 35 degrees. A talus cone at the base of a cliff, consisting of large, angular rock fragments, will have a much steeper slope than will piles of rounded pebbles along a coastline.

Most mass wasting is triggered by a catalyst—some change in the environment that either increases the gravitational force on or reduces the strength of a slope. Adding weight to a slope can cause it to fail. The weight may be heavy rainfall, snowfall, or houses or other structures on a slope. Water can also saturate the underlying rock and soil and trigger mass wasting. The slope's strength is reduced so that gravity causes materials composing the slope to be pulled downward. Earthquakes commonly set off landslides and rockfalls because vibrations unbalance slopes already at their maximum angles; they may also increase the pressure of groundwater and cause liquefaction, which causes soil to act like a liquid.

CLASSIFICATIONS OF MASS WASTING

Mass wasting is classified according to several criteria: motion, material, and speed. The motions for rock and sediment include falls, slides, flow, and creep; materials include rocks, sediments, snow, and mud. Speeds can range from 0.1 inch per year to more than 100 miles per hour. The classification system used here is based on speed of movement to show a continuum of mass wasting and covers creep, solifluction, slumps, mudflows, and landslides.

The slowest and most common type of mass movement is soil creep.

Creep occurs when soil expands outward at right angles to a slope because of the addition or freezing of water. Depending upon the slope's angle, this will cause soil and rock layers within a few feet of the surface to move downslope at a rate of between 0.1 inch and 0.5 inch per year.

Solifluction, the result of meltwater reducing friction and cohesion within the sediments of a slope, occurs in polar regions where the upper soil layer of *permafrost* thaws during the brief summers and then slowly slides downslope. Hillsides in regions near Fairbanks, Alaska, often appear to sag or bulge because of solifluction.

Slumps are faster than creep and

solifluction but occur less often. Most slumps are rotational along a plane of weakness. As the lower section of a slope flows outward and downslope, the upper section slides downward; the entire slope section appears to rotate. Slumps may move a few feet a day and are often wet-season phenomena, occurring after water has infiltrated a slope and reduced friction between the soil layers.

Mudflows, earthflows, and debris flows, like slumps, are faster than creep and solifluction and also involve sediments or rock fragments mixed with varying amounts of water. For flow to occur, there must be little or no vegetation on the slope because root systems increase slope strength and vegetation reduces the rate of water infiltrating the ground. Lahars, the flows of cinders and ash mixed with water from rain or snowmelt, are a volcanic version of mudflows. (See Volcanic Eruptions, page 125.)

Unlike slumps, the material in a mudflow does not move as a unit but has an internal fluid motion because water mixed with the sediments reduces friction throughout the mass of material. The increased weight from the water also increases gravitational pull on the material. Earthflows are viscous, with less water content than a mudflow, and debris flows are mostly rock frag-

Southern California is well-known for its destructive mudflows, which occur during periods of heavy rain following long dry periods or forest fires—both of which reduce protective vegetation cover.

ments, have little water, and are slower than mudflows. The speed of mudflows, the fastest of the three types of flow, varies according to water content, slope, and surface roughness and may reach speeds up to 40 miles per hour.

Dry areas lacking vegetative cover experience mudflows during rare heavy rainfalls. In fall, southern California is subject to forest fires that reduce the amount of vegetation and droughts that precede winter rains; the region is vulnerable to mudflows, especially following forest fires on slopes in the Coast Ranges.

The fastest forms of mass wasting involve abrupt slides or falls of rocks, sediments, and snow. Landslides, rockslides, and avalanches are most common on steep mountainsides and often occur in tectonically active areas where earthquakes unbalance slopes. Slides along planes of weakness between rock layers in slopes and free-falling rockfalls are relatively common. Large landslides that move millions of cubic yards of debris and travel more than a hundred miles an hour are rare. Landslides and avalanches, the general term for extremely rapid slides and falls of snow, rocks, and trees, leave scars and may even move debris many miles downslope.

More at:
http://www.geo.arizona.edu/geo2x/geo218/UNIT6/lecture18.html

Rocky Mountain Slides

Several major avalanches and landslides have occurred in the Rocky Mountains of North America during the 1900s. In 1903 a rock avalanche on Turtle Mountain, Alberta, buried the town of Frank under a hundred feet of debris, killing 70 people. The Gros Ventre Slide in Wyoming, the largest slide in United States history, occurred in 1925 and moved more than 40 million cubic yards of debris.

More at:
http://landslides.usgs.gov/

HUMAN ACTIVITIES AND MASS WASTING

Mudflows, landslides, avalanches, and rockfalls have always been hazardous to people. In 218 B.C. Hannibal, a Carthaginian general, may have lost as many as 18,000 soldiers to avalanches as he traveled across the Alps to attack Rome. People build homes, towns, cities, and roads on slopes that are subject to mass wasting. Such construction may reduce the strength of slopes by changing the angle of slopes, by lengthening slopes and forcing the lowest sections to support greater weight, and by increased soil moisture due to over-watering, leakage from swimming pools, and even, on occasion, septic tank overflow.

Mass wasting is often an unanticipated side effect of resource development that changes the water content of slopes. Clear-cutting in Oregon, Madagascar, and other forested regions has increased the frequency of landslides. The slides may cause few human deaths, but they change the local ecosystems: In Oregon increased mass wasting and erosion have made sediment deposits in streams greater, and the deposits have destroyed downstream salmon spawning grounds. Reservoirs behind dams also cause mass wasting when water infiltrates surrounding rock and reduces friction. In 1963 heavy rains caused a massive landslide into a reservoir on the Vaiont River in northern Italy. The landslide created a huge wave that broke over the dam and surged downstream, washing away towns along the river and killing nearly 2,000 people.

Almost all construction on mountainous terrain involves steepening one slope to create a level area for a building or road. Mountain roads often have retaining walls or wire mesh to prevent rockfalls; avalanche sheds protect travelers from falling and sliding rocks or avalanches. Mining, quarrying, and even ditchdigging can steepen and lengthen slopes, thus increasing stress at the slope's base.

The Cucaracha Formation and the Panama Canal

Construction of the Panama Canal at the turn of the 20th century was complicated by a tremendous number of landslides during excavation. The Cucaracha formation, a layer of clay, contributed to the problem. Clay in general (and wet clay in particular) has little cohesion and is likely to fail and slide. The canal's builders dredged clay to deepen the Chagres Valley and piled the clay along the canal, increasing the angle and the length of the slope. As a result, slopes along the canal failed repeatedly.

STREAMS, FLUVIAL PROCESSES, AND LANDFORMS

Streams and rivers shape more of Earth's land surface than do glaciers or any other geomorphic agent. Even as streams wear away hills and mountains, they create new landforms from the eroded material, and each landform reflects the influences of climate and geology. Flooding, a normal aspect of stream activity, is important in landscape formation and is a hazard to humans. People attempt to control rivers and streams by building dams and levees, which affect wetlands and ecosystems downstream.

Stream Systems

Streams, flows of water in channels, range in size from backyard brooks and creeks to the longest rivers in the world: the 4,238-mile-long Nile and the 3,997-mile-long Amazon. Only the largest streams are called rivers. Landforms created by streams are common throughout the world, except for ice-covered regions and areas where bedrock such as chalk is so porous that water seeps directly into the ground. Streams even shape landforms in deserts where rainfall averages less than ten inches annually.

Streams are a major link between the atmosphere and the oceans as part of the hydrologic cycle. (See Atmospheric Moisture, page 64.) Almost one-third of all the precipitation that falls on land eventually reaches the seas and oceans, in most cases via stream channels; the rest travels through groundwater. Evaporation and transpiration return the remaining two-thirds of the precipitated water directly into the atmosphere.

Most streams flow when precipitation on land collects in channels rather than soaking into the ground or collecting in ponds and lakes. Groundwater flowing onto the surface from springs provides stream flow during drier times. (See Groundwater, page 143.) The readiness with which water sinks into the ground depends upon the texture of soil and nonsoil sediment, the porosity of the surface rock, and the degree to which those surfaces are saturated.

Many separate channels form as water drains off a slope into streams, which then intersect and combine to become a stream system, consisting of a main channel and all the upstream tributaries that contribute water to it. Each system accumulates its runoff from a drainage basin or watershed. Drainage basins may range in size from a fraction of an acre draining into a creek to the 2,722,000-square-mile Amazon Basin, and they will change in size and shape over time.

Elevated land, or divides, separate a watershed and stream system from adjoining watersheds and stream systems. Continental divides separate drainage systems flowing into different oceans and often follow the crests of mountain ranges. In North America the Continental Divide between the Pacific Ocean and Atlantic Ocean drainages follows the crest of the Rocky Mountains, and in South America the Andes mountain range forms the continental divide. Hills and low ridges divide smaller watersheds and subdivide larger watersheds.

Elevation and Temperature

Streams and their tributaries form drainage patterns that reflect the geologic structure of a region. (See Drainage Patterns diagrams, page 136.) Different stream patterns develop depending on the tilting, folding, and fracturing of rock layers and on the presence of glaciers and volcanoes; these factors affect changes in surface slope or how resistant exposed rock layers are to erosion. Dendritic patterns (from the Greek word *dendron*, meaning "tree") are most common, and they form on slightly tilted slopes of equally erosion-resistant rock layers with no significant structural variations, such as folds, that could affect drainage. Trellis patterns form when streams cut through folded rocks of various hardness and follow the path of least resistance. Streams in the Ouachita Mountains of Oklahoma and Arkansas and the Allegheny region of Appalachia follow this pattern. Streams that flow down the slopes of dome-shaped mountains and volcanoes, such as Mount Rainier in Washington or Haleakala in Hawaii, have radial patterns. Rectangular drainage, where tributaries meet at right angles, occurs on surfaces with intersecting rock fractures; deranged drainage, where stream patterns lack a definite pattern or organization, forms on irregular surfaces, such as on glacier-scoured parts of the Canadian Shield. (See Glacial Processes and Landforms, page 149.)

The relationship of streams and stream patterns to the underlying geologic structures in an area can appear to be inconsistent because the structures may change over time. A river may exist for more than 50 million years, long enough for tectonic uplift or subsidence to change the slope of its channel and tributaries. Streams that exist before an area is uplifted or folded may retain their channels despite the new geologic structure because the streams erode their channels faster than the rocks are uplifted by folding. Water gaps form when streams erode through a ridge that is uplifting across their path and so cut through the ridge instead of flowing parallel to it. Wind gaps occur if the stream in a water gap loses its flow to stream piracy (one stream erodes into and inter-sects another, acquiring the upstream reach as a tributary), leaving the valley dry.

Tectonic processes create the geologic structures that influence stream patterns and give rise to the mountains whose elevations determine the location of divides and drainage basins. The headwaters of nearly all major stream systems and the divides between them are in mountains: The major rivers of Asia—the Indus, Ganges, Brahmaputra, and Yangtze (Chang Jiang)—have their headwaters in the Himalaya. Converging tectonic plates are usually associated with one coastline—the west coasts of North America and South America, for example—and cause the world's largest rivers to flow to the opposite coasts of their continents. The 3,708-mile-long Mississippi-Missouri system, with headwaters in the Rocky Mountains, and the 2,634-mile-long Mackenzie-Peace system, with headwaters in the Stikine Mountains, flow to the eastern side of the continental divide. The world's second longest river, the 3,997-mile-long Amazon, flows east from headwaters in the Peruvian Andes.

Stream Flow

A stream's velocity is the key to understanding how land is lowered by erosion, how eroded material is transported to lower elevations, and how and where the eroded material will be deposited. The faster a stream flows, the greater its capacity to erode land and carry sediments and debris; decreased velocity causes the reverse and also increases the amount of deposition.

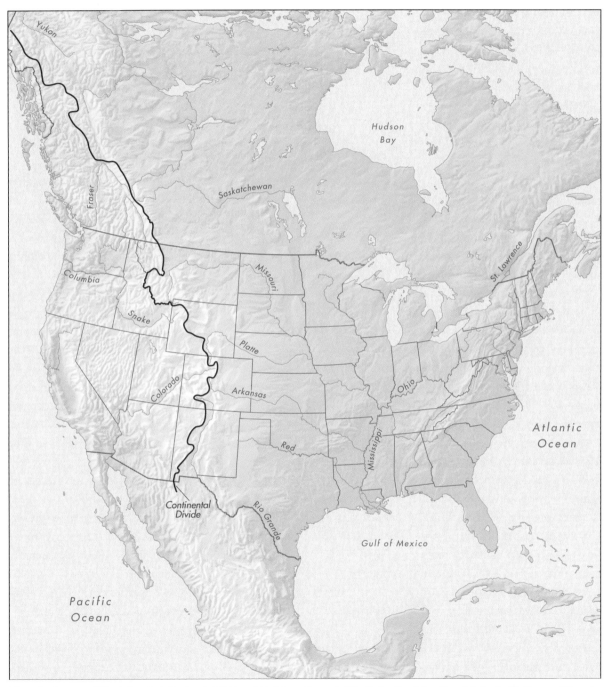

Water flowing west of North America's Continental Divide, which follows the crest of the Rocky Mountains, flows into the Pacific Ocean, while water flowing east drains into the Atlantic Ocean.

Changes in a stream's velocity are a natural part of stream processes but are strongly affected by the building of dams and levees and the straightening of stream channels.

Slope angle, or gradient, is the most important factor affecting stream velocity, and gradient differences are why hilly, upstream areas have fluvial, or stream-formed, landscapes different from more level downstream areas. Streams in level plains areas may slope less than .01 degree, whereas mountain streams can have slopes averaging more than 5 degrees. Greater slope and the total amount of stream flow increase a stream's velocity, but the flow is restrained by friction against the stream's walls and channel bed. Streams on steep slopes may average speeds of five to ten miles per hour; those on plains may flow at 0.5 mile per hour. No equation can calculate velocities for different slopes because of the differing effects of channel shapes and surface conditions.

Most streams are turbulent—with swirls and eddies within the flow—rather than smooth; turbulence increases the erosional and transport capabilities of a stream. Greater velocity increases turbulence, especially near the shoreline, because of friction between the water and the surfaces of the channel. In fast-moving mountain streams and floodwaters, turbu-

Drainage patterns of a stream depend on an area's geologic structure, often due to volcanic or glacial action. Patterns are (from top): dendritic, trellis, rectangular, radial, deranged.

lence may be seen as white water, while in slower water turbulence exists as eddies.

A change in stream velocity also affects the size of particles, based on its diameter, that a stream is capable of transporting. A stream's capacity is the total amount of sediment it can transport. Greater velocity increases capacity, as does increased stream discharge, the total amount of water passing through the channel in a given period of time. Slow-flowing rivers, such as the Platte in Nebraska or the Ob' in Siberia, will transport far more sediment and debris than their faster, smaller tributaries because, despite decreased velocity, the river has much greater discharge.

Changes in stream velocity occur seasonally and within a channel's length and cross section. For all streams, even those receiving water exclusively from springs, water supply is directly or indirectly atmospheric in origin. (See hydrologic cycle diagram, page 65.) The majority of climates have seasonal variations in precipitation, which are reflected in seasonal discharge rates and therefore velocities of stream flow. Streams in regions with humid climates are perennial, or year-round, but still display faster high-flow and slower low-flow seasons. In arid climates, streams may be intermittent, flowing only during the wet season, or ephemeral, flowing only after rainfalls. Flooding, a natural and recurring phenomenon, is a short-term large increase in flow discharge and velocity. (See Flood Processes, page 141.)

Velocity also varies within a stream's channel. As a rule, water flows faster in the headwaters of a stream and slows as the stream nears its base level, the lowest point to which it can erode. Sea level is the ultimate base level because channel flow is dissipated when stream water mingles with seawater. Within any curved section of a stream, velocities vary: Friction slows water on the inside of a curve, and centrifugal force accelerates it on the outside. The velocity of water in a straight portion of a stream is fastest near the surface in the middle of the channel, farthest away from the effects of friction along the streambed and sides.

Stream Processes

Streams create landforms by *erosion*, the removal of sediment, and by *deposition*, the dropping or laying down of sediment. These processes are functions of changes in the velocity and discharge of a stream's flow. When erosion or deposition starts and stops depends on the size of the sediments and at what velocity they can be moved: Clay-size particles and small gravel require a flow velocity of five miles per hour to begin to erode. Clay particles are so flat that they adhere to each other, making them difficult to move, whereas fine sand—sand that is smaller than a pencil dot—is eroded at .75 mile per hour. Clay and fine sand are deposited when flow velocity slows below .25 mile per hour, while small gravel is deposited at velocities below 1.5 miles per hour.

Three processes cause erosion in a stream: hydraulic action, corrosion, and abrasion. Hydraulic action removes sediments when moving water strikes and drags them, usually creating turbulence in the process; small sand grains are most easily eroded by hydraulic action because their cross sections are large enough to receive the force of the flow and their weights are too small to anchor them. Corrosion occurs when flowing water removes the ions of dissolved minerals; in some rivers, especially those flowing through limestone and other soluble rocks, corrosion may be responsible for more than half the total amount of sediments eroded. Abrasion results when particles carried by a stream strike the bedrock or other rocks on the streambed and sides, chipping off particles that are subsequently washed away; potholes in bedrock streambeds are created by abrasion from rocks swirling in eddies. The degree to which each process affects erosion depends upon the type of rock in a region.

Sediments are transported downstream until physical or chemical processes cause them to settle or precipitate out of the water. More than half of all transported sediments (mostly clays and silt) are suspended in the water; this suspended load causes turbidity and is why the Mississippi River is known as the Big Muddy. The Yellow, or Huang, River in China transports yellowish silt from upstream *loess* (wind-deposited silt) deposits to the Yellow Sea. A stream's bed load includes all particles moving along the streambed; larger, heavier particles may move seasonally or intermittently—for example, from an occasional flood—when faster high water is present. Ions transported as dissolved load in water may compose 30 percent of the total sediment load in streams traversing limestone regions.

Transported sediments are deposited when stream velocity falls below their settling velocities; dissolved loads are deposited only when a chemical change occurs in the water. Settling of alluvial deposits (stream-carried deposits) occurs in slower and usually shallower sections of a channel, on floodplains, and in the lakes, ponds, or basins into which a stream flows. If the velocity decreases slowly, as in *estuaries*, transported sediments will be sorted by size. Larger particles, such as coarse sand, will settle when a stream's velocity is still fast enough to transport smaller particles. It may take smaller particles a long time to settle; clay may take a year to settle through a hundred feet of still water.

The dissolved load transported by a stream cannot settle out, even in still water; instead, it precipitates when ions combine and form mineral deposits. Salts such as sodium chloride, or table salt, and other dissolved minerals remain in solution, while alluvial deposits settle in slowed water. Only when water changes chemically, as when it evaporates, does precipitation occur; alkali flats (salt flats) in desert basins form from this process. Calcite, another mineral deposit and the basis of limestone, forms when water releases carbon dioxide to the atmosphere, allowing calcium and bicarbonate ions to combine.

Fluvial Landforms and Landscapes

Fluvial landforms are categorized by their location along a stream channel because slope usually decreases from the headwaters to the base level of a stream.

UPPER-REACH LANDFORMS

In regions where tectonic uplift is relatively recent, streams tend to flow rapidly downslope in straight channels, generating erosional landfoms

The Lower Falls of the Yellowstone River in Wyoming drop 308 feet into the V-shaped Yellowstone Valley below.

such as V-shaped valleys, gorges, waterfalls, and rapids. Streams in strong rocks such as granite, where the sides resist mass wasting and remain nearly vertical, downcut their channels and create gorges. In areas with softer rock, such as shale, V-shaped valleys generally indicate the early stages of valley formation. Where rains are infrequent but cause flash floods, gullies and ravines may dominate the landscape, such as in the Badlands of South Dakota.

Waterfalls and rapids are commonly found in upstream locations. Waterfalls may be *cataracts*, with a single, long drop; the highest is the 3,212-foot cataract of Angel Falls, Venezuela, on a tributary of the Río Caroní. Waterfalls may also occur as a series of stair-step cascades, for

which the Cascade Range in the Pacific Northwest is named.

MID-REACH LANDFORMS

When slope and stream velocity decrease in the middle reaches of a stream, the channel begins to meander, or loop, and form a floodplain. *Meanders* begin when one side of the channel is temporarily blocked, pushing stream flow toward the opposite side of the channel. The force and turbulence of the water cause increased erosion on the opposite side, undercutting the bank and deepening that part of the channel. Stream flow enlarges this curve; water flowing on the outside travels farther and therefore must flow faster than the water flowing on the inside of the curve, much like a runner in the outside lane of a track. The difference in flow rates expands the curve, causing the meander to shift location because the faster, outside flow continues to erode while the slower, inside flow begins to deposit sediment.

Meanders grow outward during the high-flow season of a stream and are relatively stable during the low-flow season. Meanders slowly move downstream because the velocity of a stream increases when water in a curve erodes more of the channel walls that interfere with its downslope movement. As meanders move sideways and downstream, they flat-

The Fall Line

On a line from New Jersey to Alabama, the sedimentary rocks of the coastal plain along the Atlantic Ocean meet the igneous and metamorphic rocks of the Piedmont region of Appalachia. This boundary drops toward the coast because the softer sedimentary rocks have eroded more than the igneous and metamorphic rocks—more resistant to erosion—of the Piedmont. Rapids and waterfalls along the boundary, known as the fall line, block upstream navigation by ships and barges; early European settlements located just downstream of the fall line enjoyed the benefits of ocean transport (limited to calmer and deeper downstream reaches) and used the rapids to drive waterwheels. Philadelphia, Baltimore, and Washington, D.C. are fall line cities; New England mill towns were often located at regional fall lines to obtain waterpower.

ten the terrain on both banks.

The flattened terrain created by meandering is called the floodplain; such areas are susceptible to flooding when heavy precipitation or snowmelt upstream causes a high-flow season. Floodplains may be backed by bluffs on either side of the channel. Floodplains may be less than a mile wide or, for rivers such as the Yangtze and Mississippi, may be more than 50 miles wide.

Along the banks of meandering streams, deposition from flooding creates ridges called natural levees. During floods, when water overflows the banks of a channel, it slows down quickly because of friction from the level terrain of the floodplain. This rapid decrease in velocity causes deposition immediately alongside the channel so that ridges, or levees, are built up; levees can be 20 feet higher than the floodplain and may contain smaller floods. Clays, silt, and other sediments accumulate as alluvium on a floodplain when floodwaters slow.

When streams are shallow, in-channel sediment deposits separate stream flow to form a *braided stream* with channels. Such streams are common downstream from glaciers because the meltwater cannot transport large particles, pebbles, rocks, or boulders very far from their source at the snout of the glacier.

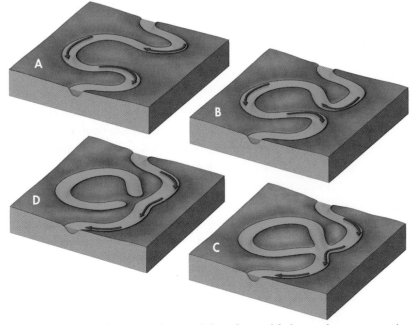

A stream meander develops in several stages. Clockwise from top left, the meander grows outward as centrifugal force deflects water flow toward the outside of the meander (A). Its neck becomes constricted (B). Eventually the meander is cut off from the main stream (C). Finally an oxbow lake forms (D).

DOWN-REACH LANDFORMS

As streams approach their base levels slopes continue to decrease, creating more depositional landforms. Meandering streams form *oxbow lakes*, yazoo streams, backswamps, and wetlands, and when a stream reaches its base level—either a temporary one, such as a lake, or its ultimate base level, the ocean—it may form a delta.

Oxbow lakes, sometimes called bayous in Louisiana and billabongs in Australia, are meander cutoffs. When a stream cuts through a meander neck, the narrow land between two adjoining meander curves, the water in the cutoff meander forms a lake called an oxbow, named for the U-shaped collar that supports an ox yoke. (See the characteristic features of the floodplain of a meandering river diagram, page 139.)

Yazoo streams and backswamps are caused by levees that parallel a stream channel and affect drainage patterns on floodplains. A yazoo stream forms when levees prevent a tributary from entering a river or stream. The tributary flows parallel to the main channel until it reaches a break or opening in the levee that allows it to join the main stream. Yazoo streams are named for the Yazoo River, which parallels the Mississippi River for more than one hundred miles before merging with it at Vicksburg, Mississippi. Backswamps are marshy areas that form in slight depressions on the floodplains behind levees.

When a stream flows into stationary waters, such as a pond, lake, reservoir, or calm ocean or sea, the decrease in velocity lowers the stream's capacity, causing its load to be deposited as a *delta*. Deltas form as a series of beds or layers of sediment; large particles settle first, and lighter sediments travel farther into the calmer water until stream velocity slows to the settling velocity for clays and silt.

Many of the major rivers of the world, including the Nile, Ganges, and Yangtze, form deltas as they reach, respectively, the Mediterranean, Bay of Bengal, and East China Sea; Russia's Volga River forms a wide delta as it flows into the Caspian Sea. In other cases, ocean currents prevent an accumu-

The Mississippi River forms what is called a bird's-foot delta, named because its strings of sediment resemble a bird's foot when viewed from the air.

lation of sediment, or a river lacks sufficient sediment; rivers such as the Amazon, St. Lawrence, and Zambezi do not form deltas. A river's main channel may separate into many branches, or distributaries, and deposit its sediment along smaller channels and breaks in levees to form the larger delta. The shape of a delta, named for the Greek letter D, reflects depositional processes, amounts and types of sediment load, and the influence of waves and tides. The Mississippi River has a bird's-foot delta created by numerous distributaries and shifting channels, while the Nile River has an arcuate delta, a curved delta created by distributaries and modified by wave action.

More at:

http://www.uwsp.edu/geo/ faculty/ritter/geog101/textbook/ fluvial_systems/outline.html

Flood Processes

A *flood* occurs when a stream rises above its banks and covers part of the surrounding land. The water level in a stream usually varies seasonally because of the changes in the atmospheric sources of the water, such as rain or snowmelt. During the high-water season an especially large rainfall or snowmelt can cause water to rise above flood stage, which occurs when the water is level with the stream's banks.

Flooding is a natural event in streams. When not modified by humans, streams naturally flood as often as every other year.

Different sections along the length of a stream may flood for different reasons. In the upper reaches of a stream, flooding is usually caused by a heavy rainfall or sudden snowmelt. In the lower reaches, flooding can result from the downstream movement of upstream floodwaters or from large weather systems that increase runoff over much of the drainage basin. (See Weather Systems, page 71.) Flash floods, rapid rises in stream levels, are limited to small watersheds, and they are brief, usually lasting less than 24 hours.

People build cities, towns, highways, railroads, and farms on fertile deltas and floodplains, thus placing themselves in danger of floods, a major natural hazard worldwide and perhaps the most universally experienced *natural disaster*. In the United States more than 20 million people live on floodplains. In Bangladesh, which has 125 million people, 80 percent of the country's 55,598 square miles are classified as floodplain.

Natural and man-made levees prevent small rises in stream level from flowing onto floodplains, but high waters that overflow levees may become large floods. For this

reason, levees have mixed consequences. The prevention of small floods may cause people to ignore the possibility of larger floods and therefore increase agricultural and residential development on floodplains. In like fashion, artificially increasing the height of natural levees and building artificial levees can prevent small floods but make the area more susceptible to the effects of larger ones.

Floods are described in terms of a number of years, such as a 10-year flood, 25-year flood, or 100-year flood. The terms are based on the frequency that high-water levels were reached in the past; these terms describe the likelihood of a stream reaching a given height during a given year. For example, if there were 100 years of data on stream levels, then the single greatest height reached would be called a 100-year flood, while a height reached on 10 out of 100 years would be called a 10-year flood. There is a 10 percent chance each year that the stream level will rise to the height of the 10-year flood and a one percent chance that stream levels will rise to the 100-year flood level. The levels are not predictions; rather they are expressions of the possibility of certain flood levels. Extremes of weather and events that cause the flood levels cannot be accurately predicted.

The Big Thompson and Nile Floods

Flood danger varies tremendously, depending on the frequency and predictability of the floods. On July 31, 1976, a cloudburst dropped more than ten inches of rain in four hours on the drainage basin of the Big Thompson River in Colorado, causing a flash flood that killed 140 people. The Nile River, prior to completion of the Aswan High Dam in 1971, flooded every fall when summer runoff from the Ethiopian highlands moved downstream. Few lives were lost because the flooding was expected.

More at:
http://www.usgs.gov/themes/flood.html

Human Impact on Streams

Humans intentionally and unintentionally change stream flows when they develop water resources. Controlling stream flow enables humans to manage waterways for flood control, water supply, hydropower generation, recreation, irrigation, and other benefits. Changing stream flow, however, means that velocity and discharge also change; the new erosional and depositional characteristics of a stream can cause problems when they create new landforms. In Washington State, logging companies clear-cut large tracts of forest, exposing the ground to increased erosion and gullying. When runoff reached channels, the increased sediment load was deposited in shallower, slower portions of streams, often on gravel bars where fish such as salmon spawned; the unanticipated channel deposition seriously damaged fishing interests.

Varying surface characteristics within a watershed cause changes in stream flow by altering the amount and type of runoff. Urbanization is especially significant in moderating surfaces, although plowing also causes major surface changes. In cities, pavement may replace vegetation and cover well over 50 percent of the area, thus preventing precipitation from infiltrating the ground and relying on storm sewers and drains to carry runoff directly to nearby streams. In a natural environment, runoff to a stream following precipitation is usually diminished by infiltration of the ground and slowed by vegetation hampering overland flow. After an area has been urbanized, however, the amount of runoff and therefore the stream level—during a flood, the flood crest—increases, and the lag time between precipitation and a rise in stream level decreases.

Water-control structures, especially dams, dramatically alter the flow characteristics of streams. Reservoirs behind dams act as new base levels, causing deposition within reservoirs. When water is released over spillways from dams, it has already deposited its sediment in its reservoir. Thus, because it is underloaded, the stream causes increased erosion below the dam. Damming California's streams and rivers has, in some cases, filled reservoirs to the tops of the dams with sediment and reduced beach size on the Pacific coast because the sand that continually rebuilt the beaches has been trapped behind the dams.

Channelization, the straightening of stream channels for improved navigation and reduced flooding, has a major impact on a stream's channel as well as its surroundings. When streams are channelized by cutting off meanders, the shorter path results in a steeper slope along that stretch. The increased velocity erodes the channel and moves floodwaters downstream more rapidly. When the Blackwater River in Missouri was channelized, it reduced flooding along that stretch of the river but increased the flooding of the river downstream; several downstream bridges collapsed when the increased lateral erosion from the faster flow of the river widened the channel in some places from less than 90 feet to 200 feet.

The Kissimmee River Restoration

Channelization of the Kissimmee River in Florida began in 1961; after the project's completion in 1971 it decreased flooding, drained wetlands and marshes, and provided more water for urban areas. Within 15 years, however, the side effects of channelization became apparent; water quality decreased because fewer wetlands were available to filter water, and wildlife disappeared alongside the river and downstream in the Everglades. The Kissimmee River is now being restored, and the project should be completed by 2009.
More at:
http://www.sfwmd.gov/org/erd/krr/

GROUNDWATER

Groundwater can flow below all types of land surfaces, including deserts and mountains, saturating fractures and pores in rocks and spaces between sediment particles.

Groundwater appears at the surface as springs, seeps, swamps, and thermal features such as geysers. It creates landforms such as caves and sinkholes by slowly dissolving minerals from rock and soil. The flow characteristics of groundwater are key to understanding how springs and caves are formed and how groundwater is an essential natural resource.

Groundwater Systems

Rain and snowmelt are the principal sources of groundwater, and some groundwater comes from surface streams when water seeps directly into rock through a streambed. Like streams on the surface, groundwater is part of the hydrologic cycle. (See hydrologic cycle diagram, page 65.) Not all water that sinks into the ground becomes groundwater: Some water attaches to soil particles and rock fragments, and plants absorb this soil water through their root systems before returning it to the atmosphere through evaporation and transpiration. The remainder of the water continues to sink until it saturates the soil or rock and becomes groundwater.

When raindrops fall through the atmosphere, some of the carbon dioxide in the air dissolves into the raindrops and forms diluted carbonic acid. This water becomes increasingly acidic as it percolates down through the soil layers, where decaying vegetation can increase the carbon dioxide content of the soil to more than a hundred times the normal concentration of the gas in the atmosphere. By the time water from the surface reaches the water table, it can be more than ten times as acidic as surface water; this acidity increases its ability to dissolve soluble rocks such as limestone and marble.

Groundwater accumulates in rocks and sediments because they are porous; there are spaces between sediment particles and within rocks. For example, 60 percent of a pile of gravel, sand, or clay may be solid material, but the remaining 40 percent of the pile's volume consists of gaps between the irregularly shaped individual particles.

Groundwater infiltrates until it reaches an impermeable layer of rock. There are two zones above the impermeable layer where infiltration slows or stops: The lower zone, the zone of saturation, is where groundwater collects, and the upper zone is the zone of aeration, where atmospheric gases occupy some of the space between sediments and in pores in rocks. The *water table* is the upper level of the zone of saturation. Water tables tend to mirror the topography of the surface; they are relatively higher under elevated areas and lower under valleys; they may be at the surface, in the case of swamps, or thousands of feet down in reservoirs under deserts. The level of a water table changes during the year, depending upon how much water is added or subtracted according to the climate or by irrigation. In dry seasons the water table may drop several feet in some areas before rising during the next wet season.

Groundwater Processes

Groundwater usually moves very slowly through rock at rates averaging between five feet per day and five feet per year. A combination of factors affects how groundwater moves: porosity, permeablity, and hydrostatic pressure. Porosity, the amount of space in rock fractures and pores and in gaps between sediments, changes with increasing depth as pressure from overlying rock increases; porosity decreases gradually as the pressure increases and compresses pore spaces, eventually closing them. Permeability is how easily water can travel vertically and horizontally through rock or sediment. To be permeable, pores need to be connected to each other and large enough for water drops to flow through without adhering to sediments or bedrock.

Hydrostatic pressure, the pressure on groundwater from overlying water, is the third factor that determines how groundwater flows. Horizontal differences in hydrostatic pressure control the direction of flow

and—with permeability—the rate of flow. Groundwater flows from areas with higher water tables and greater hydrostatic pressure to areas with lower water tables. The greater the differences between water table levels, the faster the flow; greater permeability also increases the rate of flow. Flow rates range from 800 feet a day in porous limestone to a few inches a year in granites.

Rock layers that are highly permeable can be *aquifers,* or groundwater storage areas; most aquifers form in sandstone and limestone, although some aquifers form in highly fractured basalts. People drill wells to draw water from aquifers. For wells to continue to be productive, the aquifer needs to be sufficiently permeable to allow the wellhead to recharge rapidly.

Surface water flow is turbulent and erodes by physical processes of abrasion and hydraulic force, whereas groundwater flow is smooth and erodes by dissolving limestone, marble, and other soluble rocks. The exception to erosion by solution is when groundwater-fed streams in caves have a turbulent flow; erosion by abrasion and hydraulic force can then take place.

Groundwater transports the ions of dissolved minerals until a change in the water (such as evaporation) causes the carbon dioxide to be removed from the water. The ions

then recombine and precipitate onto solid surfaces as minerals. In caves, where the relative humidity can be 100 percent, precipitation of minerals often occurs as the groundwater drips from the rock ceiling. When the drop falls through the air, carbon dioxide moves from the water to the air because the cave air has less carbon dioxide, leaving minerals to precipitate and accumulate as structures such as stalactites and stalagmites. (See Caves and Karst Topography, page 146.)

Groundwater and Landforms

Landforms created by groundwater may be found on or under 15 percent of the Earth's land surfaces. Springs, thermal springs, swamps, and geysers form when groundwater reaches Earth's surface. Caves and related surface features, such as sinkholes and rock bridges, form *karst* topography in humid regions where there are thick layers of limestone or other soluble rocks.

SPRINGS AND ARTESIAN SPRINGS

Groundwater helps to recharge stream flow in most perennial streams, and it can also flow onto the surface. Springs are surface flows of water caused by a water table intersecting a slope. Spring lines follow faults where permeable and impermeable rocks are next to each other. In many instances,

Hard Water

Hard water develops in a given area because water percolating through limestone and other rocks dissolves calcium, magnesium, and other ions that prevent soap from lathering easily. When hard water drips into a sink or tub, the ions precipitate onto the surface. Water softeners contain insoluble substances, often resins, that attract the ions and remove them from solution and onto the substance by ion exchange.

springs hold only small amounts of water, called *seeps,* which flow slowly out of the ground and eventually into streams. In winter, ice formed on the walls of road cuts shows the locations of seeps.

Groundwater can also flow onto the surface when hydrostatic pressure forces water upward to create *artesian springs.* For an artesian spring to occur, groundwater must exist in a permeable rock layer overlain by an impermeable layer, and both layers of rock must be tilted. Under these conditions, water infiltrates the aquifer at a higher elevation, flows slowly downward, and, because it cannot escape upward due to the overlying impermeable rock layer, is under increasing hydrostatic pressure in the lower end of the aquifer.

When an opening—from faulting, for example—appears near the lower elevations of the tilted rock layers, groundwater rises to the surface because of hydrostatic pressure. Water from artesian springs rises to the height supported by the hydrostatic pressure of the water. Artesian wells are drilled to take advantage of this rise. Artesian springs are found in the Great Plains of the United States, especially near the Black Hills, as well as in Australia and northwestern Africa.

THERMAL SPRINGS AND GEYSERS

Springs can bring groundwater to the surface after it has been heated by contact with magma bodies near the surface or by contact with rock layers warmed at great depths. (See Earth Structure, page 35.) Thermal springs have higher water temperatures, usually more than 10°F warmer, than the average annual air temperature of a region; thermal springs may be called *hot springs* when their temperature is higher than the average human body temperature of 98.6°F. Such springs are found throughout the United States, but the majority are found in the West, where there is more tectonic and volcanic activity. Thermal springs are often used as spas because the heated water contains more dissolved minerals than colder water and is thought to have therapeutic value.

Heated groundwater can produce geysers, springs in which boiling water erupts. Geysers include Old Faithful in Yellowstone National Park and are relatively common in Iceland and Rotorua, New Zealand. For geysers to form, high water tables, groundwater heated to above-surface boiling temperatures, and an aquifer containing a large cavity or connected cavities are required. As ground-

Oases

In deserts adjacent to mountain ranges, precipitation in the mountains can feed artesian springs hundreds of miles away and form oases, local watered areas supporting vegetation. Rainfall in the Atlas Mountains of North Africa supports artesian springs forming oases in the Grand Erg region of the Sahara in Algeria. On Australia's east coast, precipitation on the Great Dividing Range eventually rises inland hundreds of miles away in the Great Artesian Basin, which covers nearly a fifth of the continent.

water fills a cavity, it begins to heat; water at the bottom of the cavity is under greater pressure than that at the top, and therefore its boiling point is higher than that of surface water, which is under less pressure. The same process is at work in a pressure cooker. When groundwater in the cavity approaches the boiling point (212°F), it expands and begins to flow, decreasing pressure on the lower water, thus reducing the boiling point so that it flash-boils and erupts as steam at the surface. Geysers may erupt again and again; Old Faithful in Yellowstone is more predictable than most geysers, erupting every 50 to 100 minutes. Geyser fields may be tapped as an energy supply.

WETLANDS

When the water table intersects basins or lowlands, the depressions can be filled by groundwater, which saturates the ground to form *wetlands*. Floodplains and deltas commonly include wetlands; other low-lying areas are formed by tectonic activity, surface erosion by glaciers, and subsidence. Many lakes in Florida's lime sink region occupy sinkholes created by groundwater erosion. Swamps, such as the Okefenokee and Great Dismal, are often found along coastal lowlands where fresh and saline groundwater intersects the surface.

Old Faithful, in Yellowstone National Park, Wyoming, erupts every 50 to 100 minutes. The eruptions, which last about 4 minutes, send up a spout of hot water and steam 115 to 175 feet high.

More at
http://www.epa.gov/owow/wetlands/

CAVES AND KARST TOPOGRAPHY

Groundwater slowly dissolves limestone, marble, dolomite, and other soluble rocks, creating caves (underground cavities or chambers) usually with access to the surface. Cave formation is a time-consuming process because groundwater moves so slowly. It may take millions of years to form a large cave; Mammoth Cave, Kentucky, has more than 330 miles of passages, and the Good Luck Cave in Sarawak, Indonesia, includes a room that is 2,300 feet long, 1,300 feet wide, and 900 feet high.

Caves begin when groundwater gradually dissolves soluble rock along a main channel that collects groundwater from numerous side channels or tributaries. Most solution takes place within a few feet below the water table because groundwater is still acidic from absorbing carbon dioxide from air in the soil. As long as the water table remains constant, cave passages and side channels continue to expand along joints or faults within the rock. Different spatial patterns of caves can result. The jointed rock near Hannibal, Missouri, caused the cave described in Mark Twain's Tom Sawyer to have a maze pattern, while other Missouri caves have rectilinear patterns.

When the water table moves downward (which occurs when precipitation decreases or surface stream erosion deepens a channel and siphons off groundwater), a cave drains. Water dripping through a cave roof deposits calcite around itself, forming an icicle-like structure, which grows into a stalactite. When water drips from these formations onto the cave floor, mineral precipitation also occurs, and a stalagmite forms; over time the stalactite may meet the stalagmite to form a column.

Caves may last only a few million years because walls and ceilings weaken and are unable to support overlying rock. When a cave collapses, it creates surface features called karst topography. Initially collapse occurs in limited areas. Bowl-shaped depressions called *sinkholes,* or simply sinks, form on the surface above collapsed rooms.

Karst topography is found throughout the world, but it is best developed in humid climates. (See Cave Areas in the U.S. map, above.) In the United States caves and karst topography are found in many states, including Missouri, Kentucky, Indiana, and Florida, and globally karst topography is found in southern Europe, southern Australia, and southeastern Asia. In humid subtropical climates, such as the region near Guilin in south-

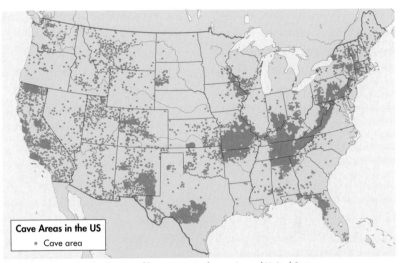

Cave Areas in the US
• Cave area

The dots indicate the distribution of karst caves in the continental United States.

Carlsbad Caverns in New Mexico is an example of a karst cave.

eastern China, towers of more resistant rock, called tower karst, dominate the landscape; the towers are what remain after groundwater removed all collapsed portions of the local cave system.

Groundwater Resources and Problems

Groundwater is a valuable source of potable (safe for drinking) water in the United States. In 1990 about 60 percent of all domestic and commercial users in the United States used groundwater rather than stream water, and more than one-third of the water used for irrigation was pumped from aquifers. The prairie states are most dependent on groundwater: Nebraska, Kansas, Oklahoma, and Texas obtain more of their domestic water supply from groundwater than from streams. California annually withdraws the largest amount of groundwater of all the states—and even more surface water for domestic use.

Groundwater resources throughout the country are vulnerable to depletion, saltwater intrusion, and contamination. The slow flow rate prevents rapid replenishment of aquifers where wells have been drilled and also prevents contaminants from being flushed rapidly out of aquifers. Water percolating into the ground can dissolve harmful

ions from garbage in landfills and carry them into an aquifer.

Groundwater depletion is an increasingly serious problem because populations that are growing in areas such as the U.S. Southwest are dependent on ground-water. Technology has increased our ability to withdraw groundwater; wells can be drilled more than a thousand feet into the ground and can pump out millions of gallons per day, far faster than precipitation can recharge underground supplies.

Surface subsidence is a side effect of groundwater depletion in areas where aquifer materials include unconsolidated sediments. After groundwater has been pumped out of an aquifer, sediments may compact—and the sediments cause the surface to lower. Cities on river deltas, including New Orleans, Venice in Italy, and Bangkok in Thailand, have experienced surface subsidence; Mexico City, built on a former lake bed, has subsided more than 25 feet in some places. Portions of the San Joaquin Valley in California have subsided more than 30 feet due to the withdrawal of groundwater for irrigation.

Groundwater withdrawal by coastal cities can also cause saltwater intrusion into aquifers and therefore into the water supply. Saline ocean water is heavier than

fresh groundwater because salt is dissolved in the seawater, and as a result saline water tends to flow slowly toward the bottom of a coastal aquifer. Eventually, if the water table continues to fall, saltwater reaches the well and contaminates the freshwater supply. Long Island, New York, had severe problems with saltwater intrusion during the 1960s, and many cities on the Gulf of Mexico still face this problem.

Groundwater contamination is a serious problem in the United States as our use of this resource grows. Dumps and landfills used for storing waste, and mining and other industrial activities cause much of the contamination. When water infiltrates the ground and percolates down to the water table, it leaches chemical and biological contaminants from waste deposits, septic tanks, agricultural fields, and other surfaces, polluting the aquifer. Decontaminating a polluted aquifer is very difficult because the water is so inaccessible. Passed in 1980, The Comprehensive Environmental, Response, Compensation, and Liability Act—The Superfund Act—resulted in efforts to decontaminate groundwater at sites such as Cinnaminson, New Jersey, and Endicott, New York.

More at:

http://water.usgs.gov/ogw/

Ogallala Aquifer

Farmers have used wells in West Texas since the 1930s to provide water for irrigation. But some of these wells have since run dry because of dropping water levels in the Ogallala Aquifer, a subterranean ocean of freshwater that lies under eight states in the heart of America's breadbasket. In recent years some Texas farmers have abandoned irrigation on more than two million acres because the water table has dropped as much as 200 feet. Others have switched to plants that require less irrigation. Scientists predict that overuse will cause the water level to fall as much as three feet a year in certain areas.

More at:
http://www.hpwd.com/ogallala/ogallala.asp

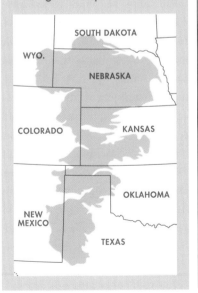

GLACIAL PROCESSES AND LANDFORMS

Glaciers, large accumulations of ice that move on land, cover about 10 percent of Earth's land surface; lakes and even oceans freeze (down to 15 feet or so), but neither is considered a glacier because the ice does not flow as a unit. Glaciers are customarily divided into two types based on their location and size. Valley glaciers, sometimes called alpine glaciers, are found at elevations above 15,000 feet in mountains near the Equator, and at progressively lower elevations at higher latitudes where average annual temperatures are cooler. Alpine glaciers are much smaller than continental glaciers and are tongue-shaped because they follow and expand preexisting stream valleys out of the mountains. Continental glaciers, also called ice sheets or ice caps, are massive. Continental glaciers cover nearly 5 million square miles of Antarctica's 5.1 million square miles, with an average thickness of 7,000 feet. Greenland and Patagonia also have huge expanses of continental glaciers.

In order to form, glaciers require yearly snowfall plus temperatures cold enough throughout the year to allow annual snowfall to exceed annual melting and accumulate for hundreds of thousands of years. Each year's snowfall slowly compresses the previous season's snow, forcing air out to form a granular snow called *firn* (or névé), which is further compacted and over time recrystalizes into ice.

Glaciers move when an accumulation of ice is about 200 feet thick; the overlying weight causes bottom ice to flow. Such an accumulation of ice is only possible on land, where glaciers may be more than two miles thick, as in Greenland and Antarctica. Snow and ice accumulate on the upper two-thirds of a glacier, its zone of accumulation, and then are carried as if on a conveyor belt to the zone of ablation, where the glacier loses its mass by melting, *sublimation* (the process by which solid matter changes directly into a gas), or, in the case of glaciers that flow into the ocean, *calving* (breaking off chunks that float away as icebergs).

Ice in a glacier always moves downslope, away from higher elevations or higher pressure, but the distance the glacier travels and the location of its leading edge, or snout, vary over time. While glacial ice moves continually, the snout may advance, remain in a location, or retreat, depending on whether the climate cools (causing increased accumulation), remains stable, or warms (causing increased ablation).

Glaciers move slowly by sliding along Earth's surface over compressed water. Average speeds of

The movement of glaciers, such as the Ferrar Glacier in Arena Valley, Antarctica, causes erosion and abrasion, sometimes transporting rock and earth for several hundred miles.

glaciers vary between one and ten feet per day—depending upon temperature, thickness of ice, and slope—although faster speeds are possible; glaciers in Greenland have been timed at up to 70 feet per day. Temporary rates up to 300 feet per day have occurred when increased temperature or overlying pressure caused bottom ice to melt and water to accumulate beneath a glacier.

Most large glaciers are found in cold climates near the Arctic and in Antarctica, but Ecuador, Uganda, and Papua New Guinea—all within five degrees of the Equator—have small glaciers on their highest peaks. Not all cold regions have glaciers. Some lack sufficient snowfall, as is the case in parts of Ellesmere Island in the Arctic Ocean and in most of Siberia.

Glacial Processes

As glaciers move, they erode rock and sediment from the walls and beds of their routes. Physical weathering and mass wasting from valley walls above a glacier contribute to erosion by causing rock fragments to fall on the surface of the glacier. Glaciers erode by plucking rocks from the walls or bed (when meltwater freezes onto the rocks) or by abrasion, when the plucked rocks embedded in ice scrape more rock and sediments from the walls and bed. Glacial striations, deep parallel grooves in bedrock, are caused by abrasion and can be found in British Columbia, where they date from the Pleistocene Ice Age; glacial striations can be found in South Africa that date from an ice age 300

The crest of the Sangre de Cristo Mountains in southern Colorado was carved into a continuous arête by Pleistocene glaciation.

million years ago, long before the Pleistocene Ice Age.

Deposition by a glacier occurs at its edge and snout, where warmer air temperatures cause melting. All glacial deposits can be referred to as glacial drift, but deposits that are carried by meltwater and then layered, or sorted, as the streamflow slows are called stratified drift. Deposits that are dropped immediately or pushed aside and not sorted are called *till*. Sand and gravel deposits are often found in stratified drift; till contains a range of sediment sizes from clay to boulders.

Streams, lakes, and wetlands are often relocated when glaciers cover existing drainage patterns and meltwater creates new water features. After the Pleistocene, when some alpine glaciers melted away, water collected in the basins to form *tarns*, or lakes, such as the Enchantment Lakes in Washington State.

Glacial Landscapes—Alpine

Landforms created by alpine glaciers are restricted to higher elevations in mountainous areas where Pleistocene glaciers existed. Snowfall accumulates in upper elevation basins called *cirques*. These basins are in protected areas and form the beginnings of glaciers that flow down preexisting valleys, widening and deepening them through erosion. The erosion creates distinct U-shaped valleys. Thousands of feet of rock can be eroded away in a valley; during the Pleistocene glaciers deepened California's Yosemite Valley more than 3,200 feet, although almost 2,000 feet of glacial deposits now fill the valley. Smaller U-shaped valleys may become tributaries of larger U-shaped valleys. The surfaces of the glaciers may meet at the same elevation, but the larger glacier will erode a deeper valley. When the glaciers melt, the smaller and shallower valley becomes a hanging valley high above the more deeply cut larger valley. Waterfalls such as Bridalveil Fall in Yosemite are commonly found at the mouths of hanging valleys.

Deposition from alpine glaciers is usually limited to the lower valleys but may extend to adjacent flatlands. Conditions suitable for glaciers extend across the higher elevations of a mountain range, so the alpine landscape includes landforms caused by interactions between glaciers. When several glaciers form on different sides of the same peak,

Like most fjords, the Kenai Fjords of Alaska were formed by glacial carving. As glaciers melted and sea levels rose, carved coastal indentations filled with water. The walls of a fjord usually plunge deep below the surface of the water.

they erode cirques, arêtes, and horns into the landscape. Arêtes are steep-sided ridges separating cirques or U-shaped valleys; horns are pyramid-shaped peaks with three or more steep faces that are the back faces of cirques. The Matterhorn in the Swiss Alps is a classic example of a horn, as is Mount Assiniboine in the Canadian Rockies. In other circumstances two cirques may erode backward into each other, forming a col or low point; Tioga Pass in the Sierra Nevada and St. Bernard Pass in the Alps are cols.

Glaciers can erode below sea level on coasts but wave action eventually melts the glaciers. During the Pleistocene Ice Age oceans were lower because more of Earth's water supply was stored in the vast continental glaciers. When the glaciers melted, oceans rose and created *fjords* by flooding U-shaped valleys along many of the world's mountainous coasts—in Alaska, British Columbia, southern Chile, Norway, and New Zealand.

Moraines, the ridgelike till deposits that mark the limits of glacial movement, are common in both alpine and continental glaciers. When alpine glaciers descend to the elevation at which ablation occurs, till is dropped as a terminal moraine at the snout of the glacier and as lateral moraines along the sides of the glacier. As a glacier retreats and the snout recedes up the valley, till covers more of the ground below the snout and, if the retreat is spasmodic, may occur as a series of parallel recessional moraines.

Glacial Landscapes—Continental

Glaciers are now limited to high latitudes and high elevations, but their geographic extent during the Pleistocene and earlier ice ages was far greater and helped create our current landscapes. About two million years ago Earth's temperatures cooled enough for snowfall to accumulate and start forming the continental glaciers that covered North America south to Missouri and blanketed Europe south to London and Krakow; the Andes, the Himalaya, and all of Antarctica were also covered by glaciers. Most of these glaciers have melted, leaving only remnants, but the effects of the Pleistocene Ice Age on the land and water are ongoing and abundantly evident. More than 75 percent of the freshwater on Earth is currently stored as glacial ice, where it remains, on average, for 10,000 years before melting.

Continental glaciers are today limited to Antarctica, Greenland, Patagonia, and larger Arctic islands, but because of their spatial extent during the Pleistocene Ice Age, glacial landforms are found over much of Europe and North America. During the Pleistocene glaciers that

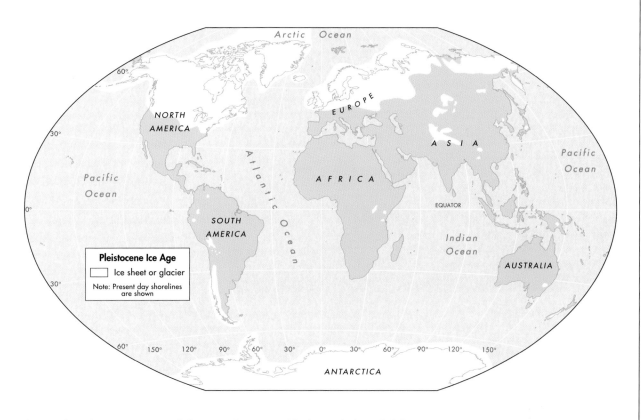

Pleistocene Ice Age

☐ Ice sheet or glacier

Note: Present day shorelines are shown

This map shows the maximum extent of Pleistocene glaciation worldwide, a cycle that ended about 10,000 years ago.

arose from locations such as Hudson Bay and Labrador in North America and Scandinavia in Eurasia advanced and retreated several times. About 10,000 years ago the continental glaciers in the midlatitudes began to melt, ending the Pleistocene Ice Age.

The effects of Pleistocene glaciation are seen in the distribution of erosional and depositional landforms, changes in surface hydrology, and the global distributions of vegetation and animals. Erosional landforms created by continental glaciers are less dramatic in appearance than those of alpine glaciers and are distributed over much of Canada and northern Europe. Like alpine glaciers, continental glaciers alter the surface as they flow outward from their centers, gouging depressions that later filled with water. Minnesota, known as The Land of 10,000 Lakes, was shaped by continental glaciers, as was Finland, with its more than 50,000 lakes.

Glacial deposits cover much of the upper midwestern United States and northern Europe. Moraines were deposited by each glacial advance, and moraines from earlier advances were usually obliterated by later advances. Long Island, New York, and Denmark's Jutland Peninsula are primarily moraines. South of Lake Michigan and southwest of Lake Erie are a series of recessional moraines that can be 20 feet high, 15 miles wide, and 200 miles long in places. Till plains, level deposits of till, are also common in the Great Lakes area behind the terminal and recessional moraines.

Continental glaciers erode the surface as they flow, creating depressions in exposed bedrock that later fill with water to form lakes.

When the continental glaciers advanced over earlier till deposits, the glaciers shaped the till in some areas into low linear hills called *drumlins*, whose long axes paralleled the direction of the glaciers' movement. Bunker Hill in Boston, Massachusetts, is a drumlin, as is Breed's Hill, where Revolutionary War fighting occurred; several islands just off Boston's shore in Massachusetts Bay are drumlins as well.

Glacial erratics, rocks and boulders transported on or in the ice from distant sites, were deposited by Pleistocene glaciers. Large erratics in eastern Washington may be the size of a house and weigh several tons. Other glacial landforms include *eskers* and outwash plains. Eskers are long ridges formed by streams that flowed underneath the glaciers. As the water flows, heavier sediments such as sand and gravel settle in the streambed and may form deposits. Few eskers are

more than a hundred feet high, but some in northern Canada measure more than 500 miles long. Meltwater that flowed from the snouts of continental glaciers formed delta-like deposits called outwash plains.

Depositional landscapes from continental glaciers can be complex, with overlapping outwash plains and moraines and eskers. Small lakes called kettles are common in depositional landscapes; huge blocks of ice would break off the glaciers, become surrounded by accumulating sediments, and eventually melt, leaving a hole in the middle of the sediments. Walden Pond is one of the many kettles in New England.

The Pleistocene Ice Age dramatically changed the hydrology of both North America and Europe. It rerouted streams and changed precipitation patterns, resulting in new lakes. Prior to the Pleistocene, streams north of what is now Nebraska drained northward, and the Missouri River and the Mississippi River did not exist as we know them. During the Pleistocene

The Great Lakes

Before the Pleistocene Ice Age, none of the Great Lakes existed; the entire area was a large drainage basin emptying into the St. Lawrence River. Glaciers moving from Canada into this area followed and deepened the major river channels. As temperatures warmed after each glacial advance, a series of lakes was formed: Lake Maumee was succeeded by Lake Whittlesey, which became Lake Erie. The current pattern of the Great Lakes appeared only in the last 14,000 years.

drainage to the north was halted by the wall of ice, and huge amounts of meltwater had to drain south. The Missouri River's tributary system eventually reached north to Saskatchewan, and the river developed a floodplain several miles across and more than 70 feet deep along much of its course. The Mississippi's sediment load was eventually deposited in the Gulf of Mexico and formed much of what is today's southern Louisiana.

Pluvial lakes, lakes created by rainfall, formed during the Pleistocene through much of what is

Eskers and Arctic Life

Permafrost does not occur in eskers because they are composed of permeable sand and gravel and do not retain water and freeze. They are critical to the ecology of northern Canada. Several plants that grow on eskers, including bear root and cranberries, are important food for bears and migrating waterfowl; animals from grizzly bears to tundra wolves to ground squirrels can burrow into the eskers to survive the long winters.

now the arid American West. Precipitation patterns were different because global climatic patterns had shifted. Lake Bonneville occupied almost one-third of Utah during the Pleistocene, but it has since shrunk and its remnants are present-day Great Salt Lake. The Bonneville Salt Flats are evidence of Lake Bonneville's evaporation, which left behind extensive deposits of salt.

Throughout the northern United States and southern Canada evidence of Pleistocene glaciers is widespread. In addition to the land-

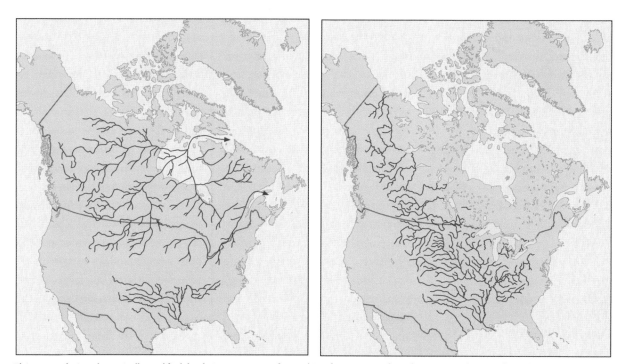

Pleistocene glaciers dramatically modified the drainage systems of central North America. Before glaciation, much water drained northeastward. Afterward, the newly formed Mississippi drainage system channeled water to the south. In the north, meandering streams and numerous lakes characterize a deranged drainage pattern.

forms described above, a number of marshes and swampy areas are found from Saskatchewan to South Dakota. This chain of ponds and marshes provides habitat for migrating waterfowl.

The Pleistocene Ice Age also caused global changes in the distribution of animals and plants. Ocean levels may have dropped more than 400 feet along some coastlines because precipitation was trapped in the continental glaciers and could not flow in streams to the oceans. As a result, several landmasses now separated by water were connected by land bridges across which animals and people could travel; present-day Sri Lanka and India, New Guinea and Australia, and France and the British Isles were each connected. A land bridge between Siberia and Alaska enabled bison and humans to travel to North America; a land bridge connecting present-day Indonesia and the Malay Peninsula enabled elephants and other large mammals to reach the Indonesian archipelago.
More at:
http://nsidc.org/glaciers/

WAVES AND COASTAL LANDFORMS
Shores and Coasts

The boundary or area of contact between continents and islands and seawater can be considered shore, a shoreline, and a coast. A shore is the actual contact area where waves wash over the land surface; the line of contact is the shoreline, which changes with the tides. A coast extends away from the shore as far as landforms created by waves or other marine processes reach inland. Coasts may be less than a half mile wide or dozens of miles wide, depending on past changes in sea level or land elevation. Wave-cut terraces, for example, may be found miles inland if sea levels have dropped since the terraces were formed.

Waves

Landforms along ocean shores and coasts are sculpted primarily by wave action, although in some areas tides can also cause erosion and deposition. Friction and wind pressure create energy that moves through the water, creating waves—undulations in the water surface—that transport kinetic energy to shores. A minimum wind speed of about two miles per hour is needed to create waves; slower breezes cause ripples that die out rapidly.

Wave dimensions reflect wind speed and the amount of energy transferred from the wind. As wind speed increases and waves carry more energy, they steepen and break. When wind speeds reach ten miles per hour, some waves begin to break and form whitecaps.

Undertow and Rip Currents

When a wave's backwash returns to the ocean, it contributes to a *longshore current* and also creates an undertow of offshore flow below incoming breakers. When low spots or undersea canyons are present on the seafloor, undertow can be channeled into fast-moving (more than five miles per hour) rip currents that transport sediments and unwary swimmers offshore.

Waves approach shorelines at various angles because shorelines are irregular. *Wave refraction*, the angle at which most waves and shorelines actually meet, is usually about five degrees from parallel. As each wave reaches shallow water, the friction of the seafloor slows down that part of the wave moving across it, causing it to angle into the shore. Each successive segment of the wave nearing the shoreline is also slowed so that the entire length of a wave meets the shore at almost the same angle. As water washes onto a beach, it does so at an angle of approximately five degrees, but when that water runs back to the ocean, it flows at right angles to the slope of the beach because of the force of gravity. As a result, beach sediment is moved sideways by waves.

Tides

In addition to waves, rising and falling ocean levels, or tides, also affect water along shorelines. Tides are the response of oceans to the gravitational pulls of the moon and sun. When tides rise (flood tides), water extends farther inshore; conversely, falling tides (ebb tides) uncover the shore. The timing of flood and ebb tides on shorelines is influenced by the relative location of the moon and sun and by the Coriolis effect. While few landforms are created directly by tides, erosion and deposition by waves do occur at sea level. The reach of the high and low tides determines what portion of the shoreline is affected by the waves.

Despite the much larger size of the sun, the gravitational pull of the moon is slightly more than twice as strong because the moon is so much closer to Earth. As a result, the locations of high tides follow the path of the moon rather than that of the sun, and tides in general follow a monthly sequence based on the moon's revolution around Earth. It takes a lunar month, or 29.5 Earth days (an Earth day is 24 hours), for the moon to complete one orbit around Earth. During each Earth day the moon travels 12.2 degrees farther in its orbit around Earth. As a result, it takes 24 hours 50 minutes (one lunar day) for the moon to be directly over the same longitude. High and low tides are linked to the location of the moon relative to Earth.

In addition to the moon's gravitational pull, the sun's gravitational pull increases or decreases the levels that high and low tides reach. When the sun, moon, and Earth are in a line, that is in conjunction with one another, the gravitational pull on the oceans is compounded, and spring tides with the highest high tides and lowest low tides result. Spring tides occur during full moons and new moons. At the first and third quarters of the moon, the sun, moon, and Earth form a right angle so that the gravitational pulls on the oceans counteract each other, resulting in neap tides with small ranges.

Tidal ranges and timing are the result of many factors in addition to planetary alignment. For example, the topography of the seafloor and the shape of the coastline affect tidal elevations and timing. In addition, high tides within a body of water such as the Mediterranean Sea move around a central point because the Coriolis effect deflects the movement of the water. Different shorelines along the same body of water can therefore have tremendous differences in the timing and range of tides.

On average, most shorelines have tidal ranges between three and six feet and semidiurnal tides (two high and two low tides) every lunar day. Some locations, such as those on the Mediterranean, may have tidal ranges of one foot, and others, notably the Minas Basin in Nova Scotia's Bay of Fundy, have tidal ranges of more than 50 feet.

Mean Sea Level

The level of the sea changes continuously because tides rise and fall twice daily and waves occur regularly. The average level of the water filling Earth's oceans relative to Earth's center basins and seas is the mean sea level (MSL). It is determined by averaging the heights of high and low tides over a 19-year period. The 19-year time frame is called the Metonic cycle—for Meton, a fifth-century B.C. Greek astronomer—and it averages the effects of periodic changes in the moon's location relative to Earth. Mean sea level is the reference level for elevations on topographic maps, but hydrographic maps (maps of water depths) use a reference level based on average low-tide levels—the mean low water (MLW)—to ensure that ships can clear obstacles on the seafloor.

Coastal Processes

Waves and, to a much lesser extent, tides create coastal landforms by eroding, transporting, and deposit-

Gulf of Saint-Malo

In the Gulf of Saint-Malo off the northwest coast of France, the tidal range reaches 44 feet. The inshore and offshore tidal flows drive the generators of a power plant on the Rance River, which flows into the gulf. Mont-Saint-Michel, site of a medieval Benedictine abbey, lies one mile offshore and is connected to the mainland at low tide by exposed shore, but it is an island at high tide. Outgoing tides in the vicinity expose nearly nine miles of seafloor, and incoming tides move quickly enough to trap the unwary, a phenomenon author Aaron Elkins used as a central part of the plot in his award-winning mystery, *Old Bones*, published in 1987.

ing sediments. A wave's energy is transferred to a shoreline when the wave breaks against it and causes erosion; when the water returns to the ocean, sediment is transported until the water slows and deposition occurs. Large storm waves contain great amounts of energy and therefore increase erosion; human-made changes in the characteristics of a coastline, such as artificial breakwaters offshore, slow wave movement and increase deposition.

EROSION

Waves erode rock along shorelines by abrasion, hydraulic action, and solution. Most wave erosion along steep shorelines is caused by abrasion from gravel and sand carried by waves. Hydraulic action, the physical force of moving water, causes erosion. Large storm waves can crash ashore with more than 2,000 pounds of pressure per square foot. The weight of the water in a wave also compresses air within rock fractures. Erosion by solution along shorelines is most significant where there are limestone and other carbonate rocks.

The rate of erosion is affected by wave size, rock strength, and the shape of the shoreline. Larger waves cause greater abrasion because more and larger particles can be hurled against the shoreline. Harder rocks have greater resistance to erosion than shales and other soft rocks. Erosion is greater on headlands that jut into an ocean than on the headlands of bays or coves because waves are refracted into the headlands.

Sediments that are eroded by waves or introduced into oceans by rivers are transported down the coast. The term "river of sand" is applicable to most beaches because the sand slowly and continuously moves downwave. Sediment is also transported offshore and onshore by seasonal storms. Off the coast of La Jolla, California, for example, winter storms transport sand offshore, where it remains as a sandbar until smaller summer waves transport it back onto the beach.

DEPOSITION

Waves deposit sediment when their speed decreases, which can occur when waves wash inland or when longshore currents reach slower water. Storm waves have greater energy and erosional power than regular waves and will deposit sediments if they wash so far inland that the water sinks into the ground before it flows back to the ocean. The barrier islands off the Atlantic coast of the United States are moving closer to the shore as storms erode their outer beach, overwash the islands, and deposit sediment on the inland side.

BIOLOGIC PROCESSES

Corals, tiny marine animals that live in undersea colonies, create landforms such as reefs. The colonies exist only in clean, shallow tropical seas with water temperatures averaging 64°F or higher. The animals have soft bodies surrounded by hard skeletons; along with deposits of certain algae, coral skeletons accumulate over hundred of years to form reefs.

Coastal Landforms and Landscapes

Coastal landscapes are composed of erosional landforms, depositional landforms, and biologic landforms such as coral reefs. The intensity of wave activity, the type of rock material, the type of tectonic boundary, the water temperature, and the changes in water level all affect the kinds of landforms that occur along a shoreline.

Rising and Falling Sea Levels

Global warming has led to a rise in sea levels, which in turn have increased storm erosion along some of the world's most popular beaches. The Cape Hatteras Lighthouse on the Outer Banks of North Carolina was at risk of toppling into the ocean because more than 1,300 feet of shoreline had eroded since it was built in 1870, leaving just 120 feet between its foundation and the Atlantic. Using a hydraulic lift system, the National Park Service moved the lighthouse inland 2,900 feet in 1999. The lighthouse's beacon will continue to remind us that even the sturdiest man-made structures give way before the sea. More at: http://www.nps.gov/caha/lh.htm

EROSIONAL LANDFORMS

Most shorelines dominated by erosional landforms have experienced geologically recent changes in sea level that caused wave action to strike at a different elevation than previously along the shoreline. Rising or falling ocean levels, tectonic uplift, and rising crust following glacier return can have this effect. Other factors also affect the intensity of wave erosion, however, so that not all shorelines affected by changing ocean levels will be dominated by erosional landforms. For example, the Atlantic shoreline of the U.S. experiences a rise in sea level of about one foot each century, but the shoreline also has many depositional features because of sediments from streams and because of the gradual slope of the continental shelf.

Many of the erosional landforms found on shorelines develop from a wave-cut notch at the base of a cliff, where wave action is concentrated. If the rock forming the cliff is strong enough, the cliff will migrate inland as erosion undercuts it along the shoreline; weak rock, shale, or moraines will erode to form a slope. Fractures or other weak areas at the base of a cliff allow these notches to be expanded by erosion into sea caves, which can extend more than a hundred feet under the cliff. When a sea cave cuts across a narrow headland, the cave can be exposed at both ends until an arch is left. Eventually the roof of the arch will collapse, and the outlying part of the cliff becomes a stack. The west coasts of the United States and Canada and the southern coast of England have cliffs, stacks, and arches.

Other shorelines where wave erosion is (or was) occurring may have wave-cut terraces or an abundance of coves and islands. Waves striking a rocky shoreline leave a flattened offshore area over which waves break and below which waves are unable to erode. When ocean levels fall, these wave-cut platforms, or terraces, are exposed. If past ocean levels were at different heights for long enough periods, then a series of terraces may be eroded; more than ten terraces can be found on San Clemente Island, California.

Glacier- and stream-eroded areas that were flooded by rising ocean levels after the Pleistocene have distinctive erosional landforms created by processes other than wave action. Fjords are found along shorelines that were heavily glaciated during the Pleistocene, including the Panhandle of Alaska and the west coasts of Norway and southern Chile. Some river valleys on the Atlantic coast of North America have drowned since Pleistocene glaciers melted. Drowning of the Potomac and James River Valleys formed the Chesapeake

Bay, while the drowning of the Delaware River Valley formed the Delaware Bay.

DEPOSITIONAL LANDFORMS

When wave velocity slows—as when waves wash inland—sediments are deposited and depositional landforms result. Landforms such as beaches and spits are shaped by depositional processes. Deposition can be seasonal—some beaches are deposited in summer and eroded in winter. Also changes in the amount of available sediment (usually sand) will affect landform development.

Most depositional landforms created by wave action are extensions of beaches. Beaches are accumulations of sediments; the individual particles of a beach are usually sand size but range in size from clay to cobbles. These sediments are deposited by wave action on a shoreline between low- and high-tide marks. Sediment deposits usually extend offshore, occasionally more than a mile. The sandy area above the high-tide mark is the berm or backshore, which often rises to a ridge or dune and is generated by storm waves and storm surges.

Beach drift transports sediment along beaches, parallel to the shoreline; this process creates new landforms when a coastline curves and the drift extends beaches offshore. *Spits* are beach extensions that form along shorelines with bays and other indentations. The growth rate of spits varies widely and depends on the rates of transport and deposition.

Variations of spits occur when they extend across bays, when offshore currents vary, and when nearshore islands exist. Baymouth bars are spits that extend all the way across the mouths of bays: If the bays are not fed by large streams, the bars cut the bays off from the ocean. When offshore currents cause a spit to curve inland, a hook is formed. Nearshore islands change depositional patterns by causing incoming waves to refract and erode the island's seaward sides; at the same time wave speed on the shoreward side is slowed, thus allowing deposition. When a spit connects a beach and island, a tombolo is formed.

Barrier islands are long, linear wave deposits that form parallel to shorelines; Padre Island, Texas, is about a hundred miles long. A variety of geomorphic processes create barrier islands. Some are extended spits, some are berms that were isolated by rising ocean levels, and some are offshore deposits exposed by falling sea levels. The Frisian Islands along the northern coast of the Netherlands are barrier islands. Barrier islands from Cape Cod, Massachusetts, to Brownsville, Texas, protect the East and Gulf coasts from the brunt of storms.

Lagoons are narrow water bodies between barrier islands and the mainland; most are connected to the ocean by tidal inlets. The width of lagoons varies widely, from less than a hundred yards to more than a mile. A very narrow lagoon separates Palm Beach from the Florida

When waves wash inland, they deposit sediments. Depositional landforms result. One kind, a beach, is an accumulation of sediments. A spit is a beach extension along a shoreline with a bay.

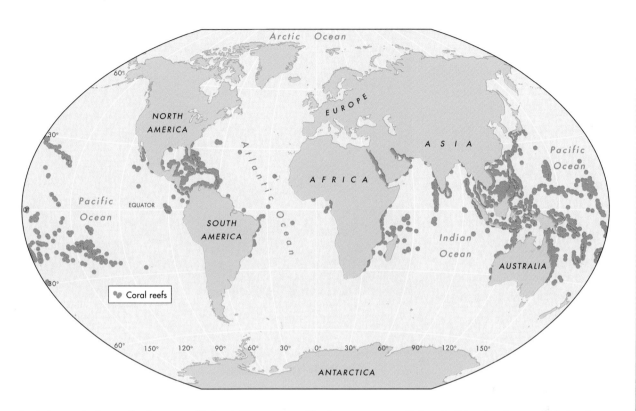

Tropical oceans harbor coral reefs, stony multicolored ridges constructed by tiny sea creatures called corals. Although reefs grow slowly, less than an inch a year, over millions of years they may grow to be hundreds of feet thick.

mainland, while Pamlico Sound separates Cape Hatteras from the North Carolina mainland by more than a mile. Deposition in lagoons along the shoreline and barrier island sides creates marshes, swamps, and tidal flats, and diverse biologic communities thrive in these wetland environments.

BIOLOGIC LANDFORMS

Coral reefs are landforms created by corals, tiny, soft-bodied marine animals that have hard outer skeletons. Corals live in clean, shallow tropical oceans with water temperatures averaging 64°F or higher. Almost all coral reefs are found between 30° north latitude and 30° south latitude, along eastern shorelines that have warm oceanic currents and where sediment deposits from streams are limited.

Corals cannot survive below depths of 500 feet, where the algae on which they depend lack sufficient sunlight to live. As a result, reefs are found along shorelines (fringing reefs), are separated from shorelines by lagoons (*barrier reefs*), or exist as evidence of former shorelines that have since been submerged (*atolls*). Fringing reefs, such as those around the Society Islands in the western Pacific, range from 1,500 to 3,000 feet wide. Barrier reefs form around islands or off continents that are slowly subsiding; as the land sinks, the corals grow upward from one to three feet every hundred years.

Atoll formation, first explained by 19th-century British naturalist Charles Darwin, occurs when a volcanic island subsides below water level. A circular coral reef, paralleling the former shoreline of the island, grows upward at a rate equal to that of subsidence and forms the atoll. The Marshall Islands comprise many atolls in the western Pacific Ocean; one of them is Bikini Atoll, which has coral growing more than three-quarters of a mile above the rock of the former island. This landform implies more than 50 million years of subsidence and coral growth. There are more than 300 atolls worldwide, most of which are in the western Pacific and Indian Oceans.

More at:

http://www.coris.noaa.gov/about/welcome.html

Shorelines and Development

Shorelines are in a constant state of change as cliffs erode, beaches move, and spits are deposited, and they are being increasingly developed for housing, recreation and leisure activities, and shipping. In order to prevent damage to seaside structures and watercraft, people attempt to control or counteract normal shoreline processes, often with limited or temporary success, but sometimes with unanticipated—and often undesirable—results.

Seawalls have been built at the base of cliffs, especially cliffs composed of glacial deposits or soft sedimentary rocks, to prevent wave

Bora-Bora, a volcanic island in French Polynesia, is ringed by a fringing reef that extends more than one thousand feet from shore.

erosion from threatening buildings on top of the cliffs. Unfortunately erosion from wave turbulence increases when seawalls are built; the energy of waves is not distributed on a beach slope but remains undiminished until it strikes the seawall. During strong storms with high waves, erosion around the sides of seawalls may cause them to collapse; it may also undercut the cliffs.

The erosion, transport, and deposition of beach sediment by littoral drift is a perpetual concern for landowners, the recreation industry, and marinas and ports. *Groins*, piles of rock extending outward from a beach, have been used to prevent beach sand from migrating downwave. While a groin will accumulate beach sand on one side, lit-

toral processes still operate down wave so that beaches begin to disappear when erosion is not matched by deposition; as a result, dozens or even hundreds of groins may be built along a shoreline as individual landowners attempt to prevent erosion of their beaches. Pairs of groins, called jetties, built around openings in a spit or barrier island to preserve access, also cause up wave deposition and down wave erosion.

Breakwaters, piles of rock built parallel to a shore to prevent wave damage to watercraft or construction, have mixed results. Waves are reflected off breakwaters, and marinas built on the landward side are protected from all but the highest storm waves. Slowing the waves, however, can cause deposition behind the breakwater of sediments being transported by littoral drift along the shoreline. In some coastal locations, such as southern California, where longshore currents transport more than 250,000,000 cubic yards of sediment annually, deposition can nearly fill a marina: The breakwater at Santa Monica has caused the beach behind the breakwater to expand tremendously.

In other coastal locations, cities have attempted to manage the gain or loss of beach sediments either by pumping out the sand or by trucking in sand to increase beach size. To prevent beach drift from closing its harbor, Santa Barbara, California, must regularly dredge the accumulation of sand, which averages 280,000 cubic yards a year. Miami Beach, Florida, and Ocean City, Maryland, have the opposite problem and bring in sand to replace eroded beaches, but this is also not a permanent solution.

WINDS AND DRYLAND LANDFORMS
Dryland Landscapes

Regions with desert and steppe climates have limited precipitation and sparse vegetation; as a result, rates of gradational processes are different from those operating in regions where surface water supports more extensive plant cover.

Many Chicago beaches have been re-formed with the help of beach groins. These groins prevent sand from migrating south, promoting beach growth on the north side of the groins.

There are also variations within dryland regions because of differences in wind patterns, abundance and type of surface material, and amounts and timing of precipitation. Landforms created by wind erosion—called eolian erosion and deposition are almost exclusive to deserts; fluvial landforms in drylands are the result of occasional rainfall that creates ephemeral streams and temporary lakes that evaporate quickly.

Deserts, regions that average less than 10 inches of precipitation annually, are found on every continent except Antarctica and cover about 20 percent of Earth's land surface. A reg is a stony desert while an erg is a sandy desert. Steppes, semiarid regions that receive between 10 and 20 inches of rain per year, cover about 15 percent of Earth's surface. (See Climate Zones map, page 88.) Deserts grade into steppes, which in turn grade into more humid climates, and the landforms associated with these climates also grade into each other. About 75 percent of the world's desert regions are rocky or covered with thin soils and, along with steppe regions, are dominated by fluvial landforms; the remaining 25 percent of the world's deserts are covered by sand and dominated by eolian landforms.

Today's desert and steppe climatic regions, like other climatic regions, are not permanent but may shift when climates change. During the Pleistocene Ice Age, wetter cli-

Africa's Sahara, the largest desert in the world, is mostly covered by dunes. The volume of sand, geology of the landscape, and wind direction and strength determine dune size and shape.

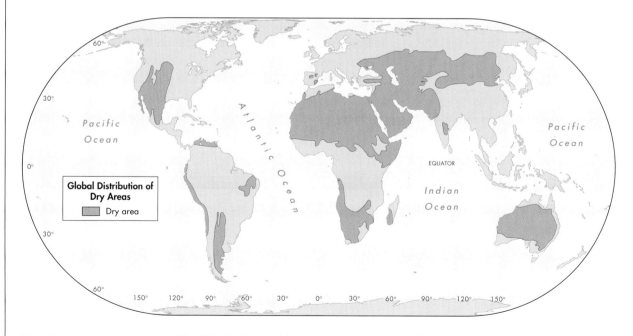

Dry regions cover a large percentage of Earth's land surface, and that percentage may increase in the future due to human impact.

The Aral Sea

In an attempt to increase crop yields, particularly cotton, the Soviet Union diverted waters of the Syr Darya and Amu Darya—rivers that fed the Aral Sea—for irrigation in the 1960s. Since then, the sea, which is really a large lake, has lost nearly 80 percent of its volume and more than 60 percent of its area. Desertification plagues the area around the lake, with salt buildup occurring on the former bed and dust storms feeding off eroding sediments. In 1997 the people of Kazakhstan, recognizing that their portion of the lake was fed by the cleaner Syr Darya River, built a 12-mile earthen dike that sealed off their healthier portion of water from the dangerously polluted southern sections of the lake. In 2001 the World Bank funded a project to improve water management by building a dike intended to rehabilitate the Northern Aral Sea.

More at: http://www.dfd.dlr.de/app/land/aralsee/

mates created lakes in what are now the Sahara, Mojave, and Australian deserts. Today the impact of human activity on the environment is forcing a shift in the other direction—overgrazing, overfarming, and overlumbering are causing existing deserts to expand in a process called *desertification*. The effects of droughts in the Sahel region of northern Africa during the 1970s and 1980s were exacerbated by human activity and have resulted in desertification; Mauritania, Mali, Ethiopia, and Niger were especially hard hit.

Dryland Gradational Processes

WEATHERING AND MASS WASTING

Rates of weathering and mass wasting are far slower in drylands than in areas where water is more abundant. Landforms tend to be more angular than those in humid areas. Because there is less water, less chemical weathering by solution occurs. Limestone, for example, which is highly soluble and forms caves in humid environments, resists gradation in deserts; limestone strata can form caprock on cliffs over shales and softer sedimentary rocks. Physical weathering of surface rocks from salt-crystal growth does occur in dryland areas when saline groundwater rises and evaporates; the lack of surface water and vegetation means that there is reduced opportunity for frost action and for fracturing by growth of roots.

Mass wasting is particularly slow in deserts, where only thin soil layers can develop due to the lack of moisture and vegetation. In steppes, however, it is more common. Occasional thunderstorms can cause mudslides, and small rockfalls loosened by physical weathering and minor downslope movements cause some accumulation of rocks and debris on talus slopes at the base of cliffs.

EOLIAN PROCESSES

Eolian processes include erosion by deflation and abrasion, transportation, and deposition of sediments. Wind erosion by deflation of sediments occurs after wind reaches the critical threshold speed for a sediment size. Small sand grains, for example, begin to be blown away when wind speed reaches about 11 miles per hour. Small, flat particles such as clay are not easily blown away because they lie flat on the surface, protected from wind by rocks and larger objects. Gravel and larger sediments are too heavy for deflation except when winds reach gale force, or at least 32 miles per hour. Deflation can lower land and create basins by blowing away sedi-

Spires in Bryce Canyon National Park, Utah, were created by weathering processes that produced uneven erosion between the interbedding of soft siltstone and hard limestone.

ments until the water table is reached; at this depth moistened sediments adhere to each other and resist deflation.

Abrasion involves windblown sediments (almost always sand grains) that strike rocks or other objects and chip away at them. Abrasion is only operative within a few feet of the surface because most sand grains are too heavy to be lifted very high by the wind. As a result, rocks undercut by abrasion form pedestal rocks. In parts of Antarctica wind blown snowflakes that are technically sediments have also caused abrasion of exposed rocks.

Sediments transported by winds may be suspended or bounced, depending upon wind speed and their weight. Lighter sediments such as clay and silt are suspended by updrafts at altitudes that may reach 10,000 feet, and they may travel thousands of miles before they settle out or are rained out. Dust storms are evidence of sediments transported by wind. Heavier sediments such as sand and occasionally gravel are transported by skipping, rolling, and sliding of the sediments along the surface.

When wind velocity slows, sediments are sorted and deposited on the surface. Sand and larger particles transported near the surface are deposited first and often collect in mounds or dunes near their source. Suspended dust particles require far lower wind speeds to stay aloft and travel farther, settling out weeks or months later over large areas.

Dust Storms

Dust storms can carry more than 6,000 tons of sediment in one cubic mile of air. During the Dust Bowl years of the 1930s in the United States, winds blew clay and silt from midwestern states as far as the East Coast. Southeastern Australia is subject to brickfielders—dust storms named for winds that kicked up dust in brickfields near Sydney—from the interior deserts. Northern Africa has several winds, including khamsins, haboobs, siroccos, and simooms, that carry dust from the Sahara.
More at:
http://www.weru.ksu.edu/

FLUVIAL PROCESSES

Most streamflow in drylands occurs rapidly and very sporadically. Thunderstorms and brief, heavy rains are characteristic of desert climates; though high temperatures cause convection, low absolute humidity reduces the number of convection events that can produce precipitation. Precipitation may cause flash flooding because the thin soil and lack of vegetation reduce infiltration and increase runoff and erosion.

More than 90 percent of streams in deserts are ephemeral, and their channels contain water only a few days or hours each year. Fluvial processes in ephemeral streams are brief but intense. Streams can move large rocks and boulders down valleys during the high flow of flash floods; the average transport distance is shorter than in humid climates because the amount of water flowing in the channel is usually limited. Deposition occurs shortly after rainfalls, and most deposits are left within the stream channel, at the base of hills and mountains, and in depressions within the desert. Sediments are not sorted well because they are deposited abruptly. Internal drainage, where streams flow into landlocked basins within drylands, and deposition within dryland areas are common, occurring in the Mojave Desert, the Sahara, and the Great Sandy and Gibson Deserts of Australia.

Dryland Landforms and Landscapes

EOLIAN LANDFORMS

Wind erosion creates landforms such as desert pavement and blowouts. Desert pavement, found in the Sinai Peninsula and Death Valley, forms where all small surface sediments, including sand, have been blown away, leaving only pebbles and larger rocks on the surface. This surface layer of rock may take centuries to form; it is rarely more than the thickness of one pebble or cobble because underlying sediments are protected from the wind.

When wind turbulence removes exposed sediment, it may form shallow surface depressions called blowouts or deflation hollows. During the Dust Bowl years of the 1930s, blowouts were common in eastern Colorado and western Kansas, where drought had killed much of the vegetation and dried the soil before windstorms swept across the region. Most of these Great Plains blowouts are like those in other deserts and on sandy coastlines—less than three feet deep and ten feet across. Larger deflation hollows as much as several hundred feet deep have been found in the northern Sahara.

After a dust storm's winds slow, dust composed of mostly silt and clay may be deposited as loess over extensive areas. Many loess deposits are found in steppes, including the Palouse of eastern Washington and the Pampas of Argentina. Deserts, especially the Gobi, are common sources for the dust that forms loess, and the sediments are transported downwind hundreds or even thousands of miles from their sources. Areas alongside glaciers are also major sources for loess sediments; much of the loess deposited in the midwestern U.S. and in Ukraine is composed of dust eroded from the outwash plains of Pleistocene glaciers. (See Glacial Processes and Landforms, page 149.)

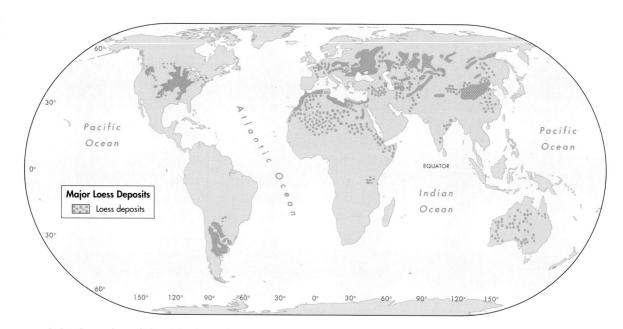

Loess, which is dust made up of silt and clay, has settled over large areas after being blown in by wind or deposited by glaciers in regions such as the midwestern United States.

Loess deposits can cover hundreds of thousands of square miles; particles suspended in the wind take weeks or months to settle out and land at increasing distances from the source areas. The thickness of loess varies as a function of the amount of source material, its weight, and wind persistence. There are deposits a hundred feet thick in the Mississippi Valley and several hundred feet thick in northern China.

Sand dunes are found in almost all deserts, although the sand-covered portion of major deserts is rarely higher than 30 percent of the total area, as in the Arabian Desert, and may be less than 5 percent of

the total area, as in the southwestern United States. Active dunes migrate downwind as sand on the windward face is blown up and over the top of the dune. Migrating dunes usually travel less than 50 feet per year—although some walk 200 feet or more per year—and can be a major inconvenience; U.S. Highway 95 north of Winnemucca, Nevada, runs through the Black Rock and Owyhee Deserts, and numerous dunes have crossed it in past decades.

Different types of sand dunes form depending on the direction of the wind, the amount of sand available, and the presence of vegetation. Barchan dunes form when

wind blows constantly from one general direction, sand is limited, and vegetation is absent; if more sand is available, barchans merge to form transverse dunes. When wind blows from more than one quadrant—for example if it varies between a westerly and a northerly wind—then a longitudinal dune forms; central Australia is noted for its longitudinal dunes. A star dune, with multiple arms radiating from a central mound, forms when winds are highly variable in direction, as is the case in Egypt and on the Arabian Peninsula, especially in the Rub al Khali, the Empty Quarter, in southeastern Saudi Arabia. Parabolic dunes form when some vegetation is present to anchor the upwind horns of the dune; they are associated with blowouts on sandy coastlines.

The dimensions of sand dunes are highly variable because of differences in wind speed and the amounts of available sand. Barchans, the smallest dunes, are rarely higher than a hundred feet; longitudinal dunes, which require far greater amounts of sand, may be more than 600 feet high and more than 50 miles long. Wind speed must be strong enough to blow sand upward to the tops of the dunes but not so strong and persistent that sand is unable to settle; wind speed increases with altitude and eventually is fast enough that sand at the crest does not accumulate.

FLUVIAL LANDFORMS

Stream erosion in deserts creates channels that are dry throughout much of the year. Dry channels are called arroyos or washes in the United States; in Arabic-speaking countries they are called wadi, and in South Africa, donga. Closely spaced, deeply eroded stream channels cut through areas of shale or clay to create badlands. Badlands National Park in southwestern South Dakota is a well-known example in the United States, and badlands topography is also found in southern Israel.

Stream erosion at the base of cliffs topped by resistant caprocks of limestone, for example, causes the cliffs (scarps) to retreat and remain steep. Plateaus, extensive elevated areas, will over time have progressively smaller areas because fluvial erosion will undercut the surrounding cliffs; the cliffs will slowly retreat and eventually converge. *Mesas*, Spanish for "tables," are plateaus surrounded on all sides by scarps; mesas will eventu-ally turn into buttes as they become smaller. Southern Argentina, western China, and the Four Corners area of the United States all contain these landforms. Over hundreds and thousands of years, fluvial erosion can also expose huge, resistant rock masses, such as Ayers Rock located in central Australia; these prominent features are called inselbergs.

Stream deposition in and near channels emerging from the base of mountains, where channel gradients level out, creates *alluvial fans,* the desert equivalent of deltas.

Deposition also occurs in depressions or basins in deserts where suspended sediments settle out in still water. The temporary, or ephemeral, lakes that form after heavy rainfalls disappear rapidly as evaporation and infiltration remove the water, leaving *playas*. Salts dissolved in the lake water are precipitated onto the sediments during evaporation and can form valuable mineral deposits.

Several lakes now present in desert basins are remnants of much larger lakes that formed in wetter climates during the Pleistocene Ice Age. In Utah, for example, the Great Salt Lake is a remnant of the former Lake Bonneville.

More at:
http://pubs.usgs.gov/gip/deserts/contents/

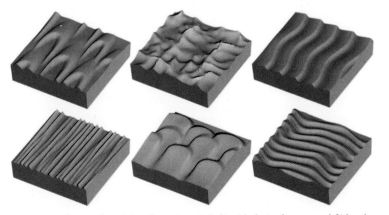

There are several types of sand dune formations, including (clockwise from upper left) barchan, star, longitudinal, transverse, parabolic, and linear.

SOILS AND BIOREGIONS

SOILS

Soils are surface layers of rock material and minerals that contain organic material and are capable of supporting the growth of rooted plants. Soils support vegetation that people and animals depend upon for survival, and historically the presence of fertile soils has strongly influenced settlement locations and the numbers of people. Plants, and therefore soils, also provide the chemical energy that supports Earth's animal populations, either directly in the case of herbivores or indirectly in the case of carnivores. Soil scientists, or *pedologists*, focus increasingly on understand soil-forming processes rather than on soil qualities and characteristics. These processes produce all soil characteristics in a region, and the spatial patterns of these processes are the basis for the distribution of soil types.

Soil Characteristics

The weathering processes and the available rock material in an area play a major role in the develop-

This alluvial fan in California's Death Valley is a classic example of how stream depositions near the base of mountains create the desert equivalent of river deltas.

ment of the physical, chemical, and biologic characteristics of soil. Physical characteristics of soil include composition, texture, and structure; chemical characteristics include acidity/alkalinity and nutrient exchange capacity; and biologic characteristics include the number and types of organisms in the soil and the amount of *humus*. All the processes producing these characteristics are interrelated.

PHYSICAL CHARACTERISTICS

All soils are composed of weathered rock material or sediment, organic matter, water, and air. In a midlatitude forest with a cool, moist climate, 45 percent of the volume of the soil might be sediment, 25 percent water, 25 percent air, and 5 percent organic matter. Soil in a tropical rain forest near the Equator might be 60 percent sediment, 30 percent water, 9 percent air, and one percent organic matter. Water and air fill up the pore spaces in the gaps between the sediments and organic particles and replace each other as water percolates into the soil or as low humidity and drought evaporate the water.

The parent material of soil, the rock debris from which sediment derives, is either the decomposed and disintegrated bedrock in an area or rock debris or sediments moved into the area by water, wind, or glaciers;

or it can be a combination of the two. Different kinds of parent material contribute different minerals and different particle sizes to the soil. Certain types of transported sediment, such as loess deposited by winds and alluvium deposited by floodwaters, can form soils very rapidly; in contrast is the slow soil formation on glacial debris.

The physical and chemical weathering of bedrock and rock debris creates sediment particles of different sizes that determine soil texture. Texture affects the size of the pores between sediments and the ease with which water (and therefore dissolved ions and suspended particles) can move. When sand dominates a soil, the large pores that result allow water to flow easily through to the water table, where it is not readily available for plant use. Clay has such small pores that water is retained, but the water will not sink easily into or flow through the soil. (See Groundwater, page 143.)

Soil structure is defined as the way soil particles clump together; it is a function of the amount of clay present, biotic activity, the mineral composition of the soil, and climate. A soil clump can be flat, linear, blocky, or crumblike. If soils do not clump, such as loess, they are described as massive and lacking in structure. Soil water moves along the planes of weakness between soil clumps; for

example, a columnar structure allows rapid drainage between columns. Thus, soil structure affects a soil's *fertility* by determining how easily dissolved nutrients in soil water flow through it.

CHEMICAL CHARACTERISTICS

The chemical composition and solubility of the minerals within soil affect and are affected by chemical weathering and organic activity, including agriculture. Chemical weathering occurs within soil because air (including oxygen) and water (which is often slightly acidic after absorbing carbon dioxide from soil air) are present in pores within the soil. In addition to forming new minerals in soil chemical weathering creates ions, meaning atoms or molecules with a positive or negative charge. Positively charged ions include calcium, magnesium, and aluminum and are especially important to plant life; they are plant nutrients that can attach to negatively charged *colloids* and move with the flow of soil water.

Soil color, once used to classify soils, derives from the chemical as well as the biologic characteristics of soil. Iron oxides commonly form in areas where warm, humid climates cause oxidation and hydration. In these areas soils are often red or yellow. In cooler areas with poor drainage, iron minerals change chemically to give a bluish color to the soil. White soil may be caused by the chemical precipitation of calcite or gypsum from water within the soil. Humus, decomposed vegetation, is black or dark brown, and therefore soils rich in humus are also dark.

Soil fertility depends on the number of nutrients available to plants in soil water. Calcium, magnesium, potassium, and sodium ions are nutrients that plants can absorb through their root systems. Nutrients can be dissolved from sediments or rock, such as calcium from limestone, or obtained from decomposing vegetation. They are attached to tiny particles in soil water; however, large amounts of water can leach nutrients from soil by washing out these particles.

Hydrogen and aluminum also attach to colloids, and their relative presence makes soil either acidic or alkaline. The number of hydrogen ions present in a unit of soil water is measured on the *pH* scale, the "power of hydrogen" scale. The scale ranges from zero to 14, and a value of 7 is considered neutral. Values below 7 have greater amounts of hydrogen and are acidic; values above 7 are alkaline. Highly acidic soils are usually infertile; highly alkaline soils may also be infertile because there are too few hydrogen ions present.

BIOLOGIC CHARACTERISTICS

Humus, the organic component of soil, can vary from nearly zero percent of soil volume in very dry areas where vegetation is scarce to more than 50 percent of soil volume in swampy areas where vegetation is plentiful. Vegetation that is alive or decaying usually composes the greatest volume of organic material in soil, but animals are also present: More than one million earthworms can inhabit an acre of topsoil. Bacteria, which are also present, decompose vegetation; 15 million bacteria can live in a cubic inch of soil.

Greater amounts of humus are present in midlatitude moist soils where vegetation is lush. In hot, rainy equatorial regions, bacteria rapidly decompose vegetative organic compounds that are then washed from the soil, so that in those tropical regions only minimal humus can accumulate. The soils in peat bogs, called muskegs in Canada, are mucks composed primarily of humus from moss, sedges, and other vegetation from surrounding pine and fir forests; cold temperatures and acidic water slow the decomposition process so that layers of humus accumulate.

Soil Formation

As soils form in a region, soil characteristics vary with depth, and creating soil layers or horizons, which can be seen in a soil cross section or profile. The vertical movement of material dissolved or suspended in soil water is most important in the formation of horizons; downward movement can remove or leach nutrients from the area if the water reaches the groundwater. Surface material—volcanic ash, for example—may be added to soils in an area or removed by wind or stream erosion; at different depths within the soil, different types of chemical weathering can change minerals and organic debris.

Soil formation is controlled by five interacting factors: climate, organic activity, relief, parent material, and time. Soils may be described as being in dynamic equilibrium with their environments because when the factors controlling soil formation change, so do the soils. Parent material, whether bedrock or transported rock debris, has decreasing influence on soil characteristics over time. Initially, parent material supplies the minerals to the soil and therefore influences the rate of weathering and the amount of sand, silt, clay, and other particles that will be produced.

Time is critical because soil formation requires time to change parent material to soil and to create horizons in the soil. The rates of these processes vary considerably around the world, and average rates for soil formation—for example, one inch per century—are misleading. In deserts almost no soil may form in a century, whereas soil may develop at the rate of 0.4 inch per year on volcanic ash in humid tropical locations.

Over time, climate becomes the most important influence on the rate and type of soil formation because rates of weathering, levels of organic activity, and the transport of material within the soil are affected by the levels and timing of precipitation and temperature. Warm and wet climates experience more rapid weathering than do cold or dry climates and have more vegetation as well; bacteria are also increasingly active as temperatures rise. As a result, different kinds of parent material in similar climates tend to form similar soils over time, although factors such as flooding or a volcanic eruption can modify these processes in localized areas. Changes in climate affect soils too, although centuries or even millennia may pass before soils reach equilibrium with a new climate and soil horizons stabilize.

Organic activity and relief also affect processes of soil formation. Organic activity, especially the amount of humus available, is closely tied to climate. Relief, including a slope's angle and its aspect (the direction the slope faces), affects the rates of the soil formation process by affecting drainage, soil erosion, mass wasting, and *microclimates* (the near-surface atmospheric conditions for a local area). As a rule, steeper slopes have thinner soils because of erosion and mass wasting; slopes with angles greater than 45 degrees usually allow no soil to form.

Aspect also affects microclimate and therefore the soil; south-facing slopes in the Northern Hemisphere are warmer and drier than north-facing slopes because of the greater intensity of sunlight.

HORIZONS

As soil water is drawn deeper into the soil by gravity or as it rises because surface evaporation causes water to ascend by capillary action, it transports ions and clay particles within the soil. As a result of these materials being lost from one soil level or gained by another, *soil horizons* develop: These are layers with different physical, chemical, and biologic characteristics. Climatic influences dominate the development of soil horizons, although not all soils have all horizons because of lack of time, moisture, or other factors.

Soil Classification and Patterns

Soils, like climates, vary gradually between locations as the factors controlling rates and types of soil formation change. Soil classification systems are therefore generalized so that the major characteristics of soils can be captured without being obscured by minor variations. One of the simplest systems is based on climates (humid temperate, dry temperate, and tropical), and soils are classified as

SIX HORIZONS HAVE BEEN DEFINED BY THE U.S. DEPARTMENT OF AGRICULTURE	
O Horizon	Nearest the surface; contains the highest percentage of organic debris, whether decomposed or unaltered; little sediment; may be extremely thin, as in deserts, or several yards thick, as in peat bogs
A Horizon	Commonly referred to as topsoil; consists mostly of sediment from weathered parent material; may include considerable amounts of humus from the O horizon
E Horizon	Distinguished by its loss of clay particles and ions to downward-percolating soil water, a process called leaching if the material reaches the groundwater and is lost from the soil
B Horizon	Characterized by accumulation of material from overlying surface layers; clays or silica from the E horizon accumulate here; minerals such as calcite in the B horizon are lost to the A horizon when they rise because of capillary action
C Horizon	Consists of weathered parent material or rock debris, either transported or local; little if any organic content; just beginning to be affected by soil forming processes
R Horizon	Consists of under-lying, unweathered bedrock

pedalfers (dominated by ions of aluminum and iron), pedocals (dominated by nutrients such as calcium), or laterites (highly weathered pedalfers).

The soil classification system used in the United States divides soils into 11 orders, the largest groupings of soils, then subdivides each order into suborders, great groups, subgroups, families, and series. With each level, soils are divided into classes on the basis of more specific information about the soil. At the series level, more than 13,000 different kinds of soils have been identified; more will undoubtedly be identified as soils in different countries are analyzed.

In an idealized diagram of a soil profile, the true soil is composed of the O, A, E, and B horizons. The C horizon is parent material and the R horizon is bedrock.

The 11 orders, based on both the physical and the chemical characteristics of different soil horizons, are entisols, inceptisols, andisols, histosols, spodosols, aridisols, mollisols, alfisols, vertisols, ultisols, and oxisols. (The suffix, "-sol," is from the Latin *solum*, which means "soil.")

Soils are not categorized specifically on the process that formed them, but soil characteristics provide clues to their history of formation. Specifications for each of the 11 orders are quantified where possible, so that a certain soil falls into only one class.

More at:

http://soils.ag.uidaho.edu/soilorders

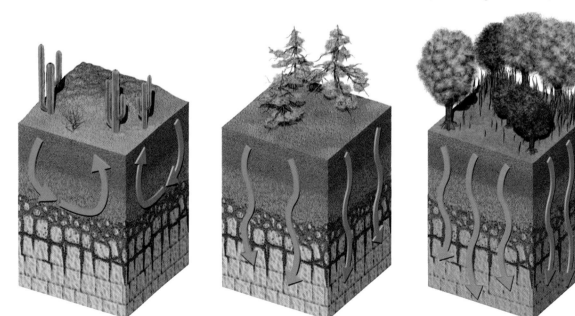

Different soil types occur in different climates. From left to right, podocal soils form in dry climates; pedalfer soils, in moist climates; and laterite soils, in very wet climates, where considerable leaching takes place.

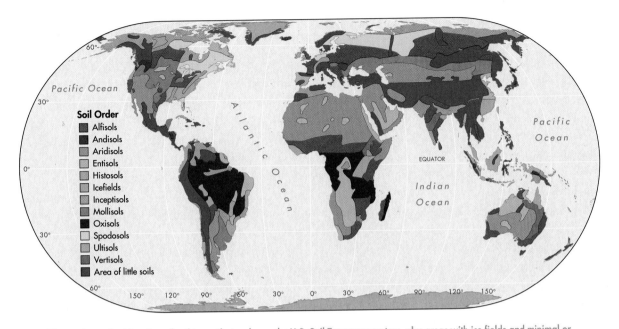

This world map shows the 11 orders of soil types that make up the U.S. Soil Taxonomy system, plus areas with ice fields and minimal or missing soil cover.

BIOREGIONS
Life on Earth

Biogeographers study the spatial distribution of individual organisms in *biotic communities*—composed of plants and animals—and of *ecosystems*—environmental systems that are associations of biotic communities interacting with their *environment*. An ecosystem may be defined and studied at scales ranging from a local pond to a global biome, such as prairies or tropical rain forests.

Organisms can have significantly different environmental requirements. A cactus, for example, can thrive in Arizona's Painted Desert in a climate that averages less than ten inches of rain per year. A man-grove tree, on the other hand, requires the hot, humid climate and huge amounts of water found in swamps such as Florida's Everglades. A cactus would not thrive in a swamp, nor would a mangrove thrive in a desert because their requirements for life are different. Different organisms such as pine trees, gray squirrels, and bacteria are

Salinization

In areas where evaporation rates are high, as in arid and semiarid regions, minerals naturally found in water become concentrated in surface soils. In extreme cases, irrigated fields may even look "snow-covered" as the salts cover the surface with a white crust. The accumulation of salts in surface soils, known as salinization, can be toxic to plants because it limits their ability to take up water.

Salinization can occur naturally, but is most common in areas practicing extensive irrigation, especially in semiarid parts of the Middle East, northern China, central Asia, and the western United States. Up to 20 percent of irrigated land worldwide experiences salinization.

able to share a single environment because they use different sources of energy, water, and nutrients.

Organisms can be classified as producers, consumers, or decomposers, according to how they acquire their energy. Green plants and some algae are producers; they use radiant energy from the sun to produce chemical energy by photosynthesis. Herbivores, animals such as rabbits and cattle that eat plants, and carnivores, animals such as lynx and hyenas that eat herbivores, are consumers that are directly or indirectly dependent upon plants to form energy. Bacteria and worms comprise another classification, decomposers, because they acquire energy by breaking down organic waste such as dead plants and animals, which also provide a supply of nutrients that may be used by other organisms.

Organisms can also be classified based on differences or similarities of anatomy, bodily functions and processes, and shared ancestry with other organisms; physiology and genetics are the basis for understanding how organisms have changed over time to adapt to different and changing environments. The biologic classification of organisms is based on the Linnaean system developed in the 1730s by Carolus Linnaeus (1707–1778), a Swedish naturalist. The Linnaean system has been revised over time and is hierarchical: At the broadest scale, organisms are grouped into one of five kingdoms: plant, animal, fungus, moneran (such as cyanobacteria), and protoctista (such as amoebas). Kingdoms are progressively subclassified into phyla, classes, orders, and families, and more narrowly by genera and species. Organisms of the same species are able to reproduce and have fertile offspring. For example, horses and donkeys are classified in the same genus (Equus) and can mate, but their offspring (mules) are not fertile. Estimates of the total number of species on Earth vary widely, from 4 million to more than 100 million.

More at:

http://en.wikipedia.org/wiki/Biological_classification

Different species of plants and animals have adapted to diverse environmental conditions; organisms now inhabit nearly all Earth's surface environments, except the vents of active volcanoes. Bacteria are found in bedrock hundreds of feet below the surface, in Antarctic ice, and in ocean water miles deep. Animals, plants, and microscopic organisms have adapted to changing environmental conditions within an area: Peppered moths adapted to England's industrialization by gradually becoming darker, a camouflage that helps them avoid predators on soot-covered tree trunks. In other cases, organisms have adapted to new areas and environments. Organisms either adapt to changes, migrate to a more hospitable environment, or eventually become extinct.

Ecosystems

Organisms interact with other organisms and the physical environment in an area, forming an ecological system or ecosystem. Within an ecosystem each organism obtains the energy, water, nutrients, and space necessary for survival, growth, and reproduction. Nutrients required by organisms and the chemical energy stored within organisms flow through an ecosystem, cycling from producers to consumers to decomposers and back to producers. A *food chain* or food web exists within the ecosystem as the energy and nutrients move between organisms as they are eaten or decomposed.

The *carbon cycle* demonstrates how carbon, one of the basic nutrients required by living things, flows through an ecosystem. Carbon, in the form of carbon dioxide, is removed from the atmosphere by plants during photosynthesis and changed into carbohydrates in plant tissues. From plants the carbon trav-

Years ago (millions)	Geologic Time			Biologic Events
			Cenozoic	Age of mammals
65			Cretaceous	Massive extinctions First flowering plants Climax of dinosaurs and ammonites
	Messzoic		Jurassic	First birds Abundant dinosaurs and ammonites
			Triassic	First dinosaurs First mammals Abundant cycads
248	Late Paleozoic		Permian	Massive extinctions (including trilobites) Mammal-like reptiles
			Pennsylvanian	Great coal forests Conifers First reptiles
			Mississippian	Abundant amphibians and sharks Scale trees Seed ferns
			Devonian	Extinctions First insects First amphibians First forests First sharks
408	Early Paleozoic		Silurian	First jawed fishes First air-breathing arthropods
			Ordovician	Extinctions First land plants Expansion of marine shelled invertebrates
			Cambrian	First fishes Abundant shell bearing marine invertebrates Trilobites
570			Neoproterozoic	Rise of the metozoans

This geologic time scale relates the development of life forms on Earth with the corresponding geologic periods.

plants during photosynthesis and changed into carbohydrates in plant tissues. From plants the carbon travels through consumers and decomposers, cycling back to the air by respiration or oxidation; carbon from plants is also stored in fossil fuels such as coal, petroleum, and natural gas before returning to the environment after combustion. Carbon dioxide can also be removed from the air when it dissolves in water bodies, which act as carbon sinks or storage areas; this carbon, too, will eventually return to the active carbon cycle. (See carbon cycle diagram, page 178.)

The environmental conditions within an ecosystem, including the presence of other organisms, form *habitats* for those species residing there. Each habitat has a specific physical location within an ecosystem, such as shallow water at the edge of a pond; the specific environmental conditions of each habitat overlap with bordering habitats.

Field mice and great horned owls inhabit the same forest ecosystem and interact when owls hunt mice, but their habitats differ: Field mice occupy underground nests, whereas owls live in trees.

Climate and topography are the most significant environmental factors that determine an ecosystem. Climate has the greatest influence on an ecosystem because it controls

Carbon, in the form of carbon dioxide, is removed from the air by plants during photosynthesis and converted into carbohydrates. From plants, the carbon travels through consumers and decomposers, returning to the air via respiration or oxidation. Carbon is also stored in fossil fuels before returning to the environment after combustion. Some carbon dioxide is removed from the air when it dissolves in water bodies. This carbon, too, will eventually return to the active cycle.

one species through drought, for example, or predation by animals or humans may not significantly interfere with the functioning of the entire ecosystem. Systems with limited biodiversity, such as deserts and tundra, are more vulnerable than tropical rain forests to disruption from the loss of a single species and its function in the ecosystem.

Ecosystems are geographically defined by the environmental conditions needed by members of the biological community; cave ecosystems, for example, provide the atmospheric conditions required by bats, blind cave fish, and other animals adapted to cave environments to survive. Each species has an optimum level of requirements, such as the amount of water, and also a

tosynthesis, precipitation, and soil characteristics are controlled principally by climatic factors, and plants in particular depend on climatic conditions for survival.

Topography influences environmental conditions because the angle and aspect of slopes, drainage patterns, and relief also affect temperatures, moisture, and sunlight. North- and south-facing slopes of mountains, for example, have slightly different temperatures and precipitation regimes and form microclimates; each, therefore, supports different ecosystems with different habitats.

Ecosystems vary in the quantity of organic matter or *biomass* they produce (productivity) and in the number of species present (*biodiversity*). Warm, wet environments tend to have greater productivity and greater biodiversity than cooler, drier environments; plants that grow and reproduce more easily generate more chemical energy that can be used by other species in the ecosystem. Greater biodiversity, with more species interacting within an ecosystem, makes the food web more complex so that the loss of

Locoweed and Selenium

Species have different nutrient requirements and tolerance levels for chemical elements, such as sodium or calcium, above and below which they cannot survive. Locoweed is one of the few plant species that tolerate soil with high concentrations of selenium, a rare element. Selenium is often associated with uranium so that geologists regard locoweed as an indicator species for uranium deposits.

Adirondack Lakes

In upstate New York effects of changes in the environment were noticed by fishermen in lakes of the Adirondacks in the 1960s. Fewer fish were being caught, and all the fish that were being caught were older fish. A similar pattern had occurred in Scandinavia a decade earlier. The National Academy of Sciences in the United States eventually linked the change in fishing to increased acidity of lake water and, in turn, to acid precipitation. Acid levels in lakes in springtime were especially high because of the influx of snowmelt; many female fish were unable to produce eggs, and only mature fish could survive.

range of tolerance, extremes of requirements within which the species can survive and reproduce and beyond which it cannot survive. As a species, the saguaro cactus can survive extremely low levels of moisture in southern Arizona, but it is particularly sensitive to cold temperatures: All new growth can be killed after 12 hours of temperatures below 32°F. Ranges of tolerance are usually greater for mature members of a species than for the young.

Ecosystems are transformed into other ecosystems when environmental conditions, especially those associated with climate, change. Marshy areas, for instance, often surround ponds and are an example of a zone between an aquatic ecosystem and a terrestrial ecosystem; raccoons, snakes, and other animals will cross from the pond to the surrounding forest and back in search of food and water.

Ecosystems are not static; short-term disturbances, such as fires, droughts, and insect infestations, and long-term changes, such as climatic shifts, affect the flow of energy and nutrients through ecosystems. Temporary disturbances can change the availability of energy, water, or nutrients so that new species will expand their range into the area. If the newer species are more competitive than the resident species, the newcomers may displace them. Over time, however, if environmental conditions return to predisturbance levels, a sequence of plant and animal species more competitive in those conditions will return (ecological succession) until a stable ecosystem or climax community is reached. In midlatitude forests, for example, a field cleared within the forest may take 200 years to become a climax community of oak and hickory trees and associated animals. The sequence is that first the clearing fills with fast-growing weeds and grasses, then shrubs and fir trees move in, shading out the smaller plants; shrubs and firs are then replaced by oaks and hickories, tree species that germinate better in shade.

Climatic changes can have longer-term effects on an ecosystem than a forest fire or volcanic eruption, changing it for hundreds of thousands of years or longer. During the Pleistocene Ice Age woodland and grassland ecosystems existed in the Sahara and the southwestern United States; tundra ecosystems thrived just south of the Great Lakes. If global warming continues at the current rate of about one degree Fahrenheit per ten years for several more centuries, the range of hemlock and sugar maples may reach several hundred miles farther north into Canada from their present limit at about 44° north latitude.

Humans affect ecosystems when they remove or add species to a region, as in the case of agriculture, or when they change the availability of water, nutrients, or sunlight. Commercial farming attempts to replace a diverse ecosystem with one crop (monoculture), and forestry may remove all mature tree species. Both of these activities change the food web in a region and alter the amounts of moisture and sunlight available for surface plants. In Australia indigenous species such as the Javan tiger and Tasmanian wolf became extinct, and exotic, nonindigenous species such as rabbits were introduced; local ecosystems were changed by adding and removing consumers and producers from the food web.

More at:

http://www.invasive.org/

Biomes: Global Distribution

Global patterns of mature ecosystems are studied at the biome scale—macroregional ecosystems based on life-forms (the form, structure, and function of the organisms) rather than on sets of particular species. Rain forests in Africa and South America, for example, contain vines, arboreal animals such as monkeys, and reptiles although the particular species present will be different on the two continents. Biomes enable comparisons between regions to improve the understanding of ecosystem processes and to establish management practices when species (crops, for example) are introduced from elsewhere.

Plant communities are more often used to identify biomes than are animals because vegetation is more visible, constitutes more of the biomass, and produces the chemical energy that flows through the food web. Although agriculture, forestry, and urban development have displaced native plant and animal species throughout much of the world, biomes are mapped based on the mature ecosystems that the environmental conditions would produce without human intervention. Biomes mirror climatic regions because vegetation reflects the amounts of precipitation, temperature, and sunlight that are available. Soil patterns, also strongly affected by climates, are similar to biome patterns. (See Climate Controls and Climatic Classifications, page 85.)

Several classification systems for biomes have been developed, none of which are universally accepted. In 1976 the U.S. Forest Service mapped biomes, using the term "ecoregions," and began to use the concepts of biomes to manage ecosystems within national forests in 1993. The U.S. Forest Service classification system, as revised by Ralph Scott, organizes biomes into four general categories: forest, grassland, desert, and tundra. The forest and grassland categories are divided into specific biomes, such as tropical rain forests in the forest category and prairies in the grassland category, based on the life-forms of the vegetation; these biomes display significant differences in ecosystem productivity and biodiversity.

More at:

http://www.blueplanetbiomes.org/world_biomes.htm

UNDERSTANDING PHYSICAL GEOGRAPHY

Geographers focus on spatial patterns from a global to a local scale, and the key to understanding environments lies in understanding the physical processes that shape them. Physical geography studies past, present, and future environmental patterns on Earth, which are linked because physical processes act over space and time: Hurricanes do not form in a day, canyons in a year, or oceans in a century.

One of the most important applications of physical geography is to be able to predict future patterns and processes on Earth's surface. Many of the processes appear to be slow—the Atlantic Ocean widens one inch per year, and global temperatures rise one degree per century—but other processes are far faster. Understanding wave erosion helps to predict the retreat of coastal bluffs on the coast of the Atlantic Ocean, and an understanding of tectonic stress buildup may help to predict sites of future earthquakes.

The Biologic Invasion of Hawaii

Ecosystems in the Hawaiian Islands are especially vulnerable to introduced species because the islands have a limited area, which also limits the complexity of their food webs. Since 1778, when Capt. James Cook first visited the islands, nearly 4,000 species of plants and animals—including wild blackberry bushes, mongooses, Argentine ants, and cannibal snails—have been introduced, usually unintentionally, into the Hawaiian Islands. Native species have often been unable to compete with aggressive and prolific new species, and now some 20 percent of native plant species and almost 50 percent of the remaining bird species in Hawaii are endangered and at risk of extinction.
More at:
http://endangered.fws.gov/

The natural environments in which we live do not control our behavior, but they influence and constrain many of our choices, such as where to live, what crops to grow, where to raise livestock, and where to build cities or seaports. Physical geography is basic to understanding how humans interact with their environments. Despite their relative disadvantages, such environments as Siberia, the Mosquito Coast in Central America, and the Kalahari in southern Africa have been successfully occupied by different cultures for many centuries; in none of these areas, however, is urban development or commercial agriculture extensive. Perceived environmental advantages, on the other hand, attract people. Many older Americans are relocating from colder northern states to warmer Sunbelt states, while the beauty of coastal regions continues to attract more people than do nearby inland areas.

The geographer's concern with how humans interact with their environments relies on not only predicting changes in those environments but also anticipating the effects on human societies and their responses to those threats—from coastal erosion's effect on beachfront houses to the threat posed by earthquakes on densely populated metropolitan areas. Physical geography provides the knowledge of where, how, and at what rate physical processes are likely to cause environmental changes that will necessitate societal adjustment. This basic knowledge is essential to all societies preparing for the future, even though their policies and responses may be very different.

The knowledge of physical processes and patterns is also essential to understanding and predicting the effects of resource development on Earth's systems. Humans have tremendous ability to affect physical processes and change the face of Earth, but they have not alway accurately predicted the consequences of their actions.

Physical geography is a challenge for the intellectually curious. Each individual can enjoy observing the world, appreciate Earth's diversity and the beauty and complexity of its environment, prepare for changes, and take measures to protects its environments. Understanding Earth is the challenge and reward of physical geography.

The Jungle

Jungles are extremely dense thickets that occur within tropical rain forests after some disruption, such as fire, creates a clearing. Pioneer species, the first to sprout in clearings, include low shrubs and thickets that grow so densely that they create a nearly impassable mass of vegetation. Ordinarily the lack of sunlight on the rain forest floor results in plenty of open ground under the canopy.

SOURCES OF FURTHER INFORMATION

The Audubon Society Field Guide to North American Rocks and Minerals. New York: Alfred A. Knopf, 1995. Comprehensive review with good photos.

The Audubon Society Field Guide to North American Weather. New York: Alfred A. Knopf, 1997. Good photos and explanations of events.

Elsom, Derek. Earth: *The Making, Shaping and Workings of a Planet*. New York: Macmillan, 1992. Good visuals with less detailed text.

Gates, David. *Climate Change and Its Biological Consequences*. Sunderland, MA: Sinauer Associates, 1993. Specialized look at climate.

Junger, Sebastian. *The Perfect Storm*. New York: W. W. Norton, 1997. Best-selling case study.

Lamb, Simon and David Sington. *Earth Story: The Shaping of Our World*. Princeton, NJ: Princeton University Press, 1998. Good geologic history with illustrations.

Lambert, David and the Diagram Group. *The Field Guide to Geology*. New York: Facts on File, 1988. Designed for younger readers.

Levy, Matthys and Mario Salvadori. *Why the Earth Quakes: The Story of Earthquakes and Volcanoes*. New York: W. W. Norton, 1995. Good background and illustrations.

McPhee, John. *Assembling California*. New York: Farrar Straus & Giroux, 1993.

_____. *Basin and Range*. New York: Farrar Straus & Giroux, 1981.

_____. *In Suspect Terrain*. New York: Farrar Straus & Giroux, 1983.

Rees, Robin, ed. *The Way Nature Works*. New York: Macmillan, 1998. Colorful and reader-friendly overview of Earth.

Restless Earth: Disasters of Nature. Washington, D.C.: National Geographic Society, 1997. Case studies; excellent photos.

Simon & Schuster's Guide to Rocks and Minerals. New York: Simon & Schuster, 1978. Includes worldwide locations.

Roadside Geology Series. Missoula, MT: Mountain Press Publishing Company. Descriptions and explanations of geology viewed from highways in a number of states.

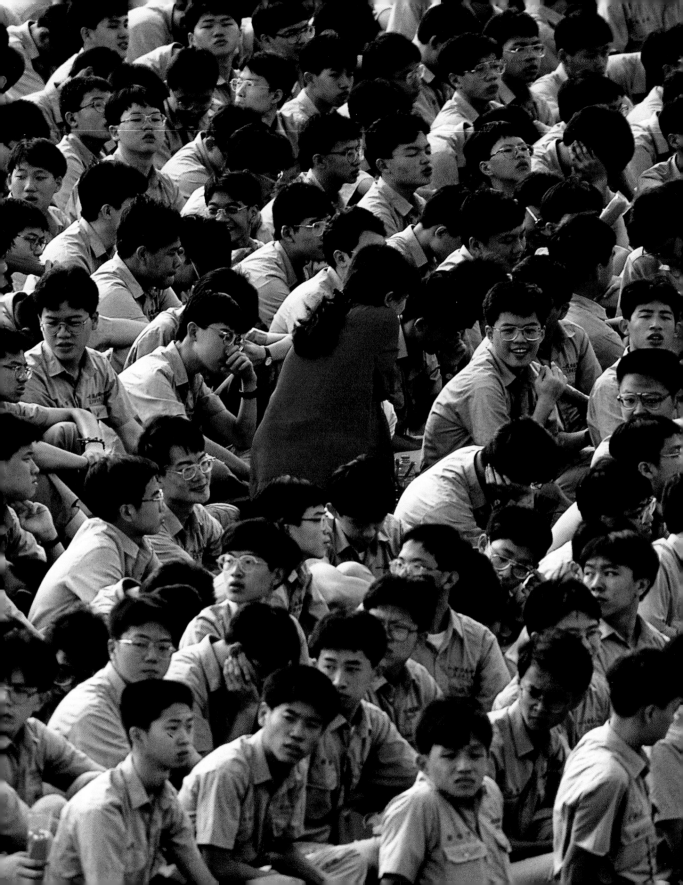

PART III

HUMAN GEOGRAPHY

"Double Ten," or Taiwan's National Day, marks the October 10, 1911, anniversary of the Chinese Revolution. Taipei is the capital and largest city in Taiwan.

POPULATION

PEOPLE AND THE ENVIRONMENT

Geographers approach the study of human populations with a spatial perspective—an eye to the available land on Earth. They try to understand why births, deaths, and migration vary from place to place. They study the changes in the physical environment prompted by changes in population size, composition, and patterns of consumption. Because of their perspective, some geographers express concern for Earth's ability to accommodate a population that is approaching the seven billion mark.

The rate of world population growth has declined since the late 1960s, but the population continues to grow by 77 million people each year. The developing countries, where a larger proportion of the population is in the reproductive years, account for most of the world's population growth. But industrialized countries, with their affluent populations, account for high rates of consumption of Earth's resources.

Geographers also examine interrelationships between people and their environment. In recent years, geographers have studied the effect that large, dense populations have on the quality of the environment and, by extension, the quality of life. The connections between the environment and population vary considerably from region to region. Many countries of sub-Saharan Africa are faced with rapidly growing populations and dwindling resources, while Japan is planning for an aging population—25 percent of the population is expected to be over 65 years old by 2010—and a dwindling work force. Persian Gulf countries, with small populations, have encouraged immigration of a large semi-permanent work force—called guest workers—to exploit bountiful sources of energy.

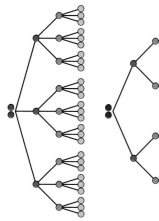

Three-child families-Two-child families: In just three generations, three-child families produce more than triple the number of kinsmen generated by two-child families.

With the advance of industrialization the world has witnessed an ever accelerating population growth.

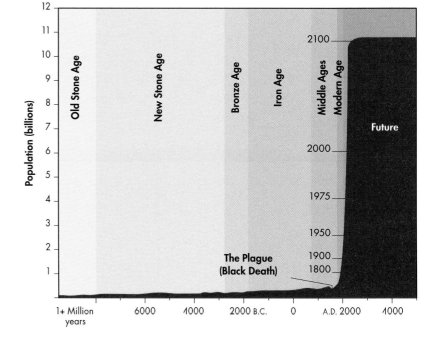

Doubling Time

For most of human existence, the natural increase rate of the world's population has hovered near zero —births roughly equal to deaths. Historically, a rise in population entailed higher standards of living, larger markets, a finer division of labor, and economies of scale in the production process.

The engines of population change are natural increase and net migration. Natural increase is defined as the difference between birth and death rates. (While crude rates, i.e., births or deaths per 1,000 people per year, are commonly used, demographers prefer age-specific rates, which are more accurate predictors.) Net migration is the difference between in-migration (immigration) and out-migration (emigration). The world's natural growth rate in 2004 was roughly 1.3 percent per year, which amounts to approximately 134 million births and 57 million deaths for a net gain of 77 million people per year, or about 211,000 per day. Growth rate is the sum of natural increase and net migration.

Growth rates can be used to estimate doubling times, the number of years a population requires to double in size. With a growth rate of 1 percent, a population would double in about 70 years, or 35 years at 2 percent growth rate. Doubling time of a population, therefore, can be calculated by dividing 70 by the growth rate. If the world's annual natural increase rate remains at 1.3 percent, the population will double in 54 years to almost 13 billion, far more and slightly faster than present projections. (70/1.3 = 54) Doubling times tend to be long in the developed countries—an average of 583 years—where growth rates are low. Doubling times are short in the developing countries—an average of 40 years—where growth rates are generally high.

THE WORLD'S CHANGING POPULATION

The growth of the world's population over the last two centuries has been dramatic. Early human populations were restricted to the number of people who could be supported by the available land, called carrying capacity. Since recorded history, Earth has supported a small population, estimated at up to 10 million around 8000 B.C., increasing, as humans began farming and raising animals, to 200 million about 2,000 years ago. Tribal hostilities, famine, and changes in weather all kept human population in check, leading to a life expectancy at birth of only 10 years! With such drastic "culling" of the population, the species could survive only with a very high *birth rate*—much higher than that found in high fertility countries today.

In tracing the history of population growth, demographers have weighed dozens of variables and estimated that the total number of humans born since 50,000 B.C. is 105.7 billion. The human population at the beginning of the 21st century is about 6.4 billion, or 6 percent of the total number of people who have ever lived on Earth.

Historically, regional population growth often was associated with technological advances, especially in agriculture. Population fluctuations in the Middle Ages, due largely to wars and pandemics, held world population to about 500 million in the mid-17th century. But from that time on, with the advance of industrialization, the world witnessed an ever-accelerating population growth. The landmark number of one billion people was reached in the early 1800s. A population of two billion was attained in 1927. By 1960, the population had increased to three billion, then to four billion only 14 years later, in 1974. The five billion landmark was reached in 1987, the six billion

Two Approaches to Measuring Birth Rates

The two most common methods of measuring births are the crude birth rate and the total fertility rate. Both have distinct uses. The crude birth rate—given as the number of live births per 1,000 population—indicates the number of births relative to the entire population. Crude birth rates for the world range from about 9 in many European countries to about 55 in Niger in West Africa. The rate helps in assessing the impact of births on population growth of an area.

The crude birth rate does not take the age or sex structure of the population into account. Age and sex, on the other hand, are crucial in computing the total fertility rate. This rate is the average number of children a woman would have in her lifetime given prevailing birth rates for each of her child-bearing years. The total fertility rate ranges from 1.2 children in many of the formerly communist countries of Eastern Europe to 8 children in Niger. Fertility rate is the best birth indicator for comparing countries or places, because it considers births relative to the population of women of reproductive age, 15 to 44.

mark in 1999. Note how less and less time passed between each billion, even though the world's rate of *natural increase* (the difference between births and deaths) began declining in the late 1960s.

What underlies the population increase? First, small percentage increases in a population over a long period lead to large increases in actual numbers of people. This is a compounding effect. Greater and greater numbers of women reach childbearing age, then, their children grow up and produce a larger total number of children. Second, growth contains a built-in *population momentum*: Higher birth rates at a previous time and lower death rates at the present time keep populations growing even after families have achieved replacement fertility, defined as the level at which each person on average has a single succes-

sor in the next generation. For most countries this is an average of 2.1 children per woman over her reproductive years. Countries with high *total fertility rates* have births of up to seven children per woman over the woman's reproductive years. Countries with low total fertility rates have births of fewer than two children per woman over the woman's reproductive years.

THE DEMOGRAPHIC EQUATION

Population change results from three components—births, deaths, and migration. The relationship among these components can be expressed in the form of the following equation:

Population Change =
Births – Deaths +
Immigration – Emigration

The equation calculates changes in

a population's natural increase (birth minus deaths), changes in *growth rates* (natural increase +/- migration), changes in the size of populations, and projections of future populations. A benefit of the *demographic equation* is its simplicity—only three components are needed to calculate population change.

WHY AND HOW POPULATIONS CHANGE OVER TIME

Earth's population is currently (2004) estimated to be increasing at a rate of 1.3 percent each year, down from a peak rate of a little over 2 percent in the late 1960s. But population momentum continues to increase the number of people in the world. The predominance of young people in most South American, African, and Asian countries ensures that populations there will continue to increase even though families bear fewer children. Countries such as Taiwan, Thailand, South Korea, and China still will add appreciably to their populations in coming years, despite being at or below the replacement fertility level of 2.1 children per couple.

The United Nations 2002 population projections were revised downward from the 2000 projections by 0.4 billion people, to 8.9 billion for the year 2050. The downward revisions occurred mainly in developing countries, primarily Africa where it is now assumed that declines in the birth rate and rising *mortality* due to AIDS will be more dramatic than previously expected.

Despite a falling rate of increase, Earth's population continues to grow in absolute numbers. Two divergent views have proposed possible outcomes. One, the Malthusian theory of "overshoot and collapse," is pessimistic. The other, known as the cornucopian view,

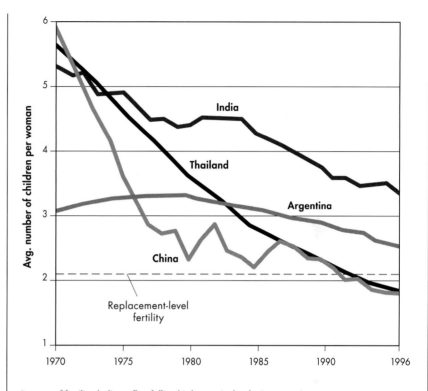

Patterns of fertility decline reflect falling birth rates in developing countries.

POPULATION AGE DISTRIBUTION (%), 2004		
REGION	<15 YEARS OLD	>15 YEARS OLD
World	30%	7%
Europe	17%	15%
North America	21%	12%
Oceania	25%	10%
Asia	30%	6%
Latin America/ Caribbean	32%	6%
Africa	42%	3%

which predicts natural deceleration of population growth, is optimistic.

Thomas Malthus (1766–1834), an English economist, was alarmed by contemporary population data from France and England. He concluded that world population would inevitably increase to a point that would surpass Earth's capacity to sustain it. (See box, page 190.) He said that population increased at an exponential rate, and that it would be only a matter of time until an "unholy trinity" of war, pestilence, and famine came onto the scene to control population.

In the cornucopian view, forces in human society are expected to retard population growth before it exceeds Earth's carrying capacity. Human adaptability and human aspirations for wealth and prestige, and for a better life will prompt people to limit family size. In this scenario, technology will continue to squeeze more and more from Earth's resources, so that the carrying capacity of Earth will actually be increased. This so-called cornucopian view is predicated on the demographic transition model of population change.

Malthus: A Controversial Prophet

The causes of population change are linked to economic, social, and political circumstances. A pioneer in explaining the link was the English economist Thomas Malthus, known for his 1798 *Essay on the Principles of Population*. Malthus argued that the population of countries tends to grow geometrically. The term used today is "exponential" growth, such as 4, 16, 64, 256, etc. In other words, Malthus postulated that given a constant rate of change, larger and larger absolute numbers of people will be added to the population each year. Malthus said that the poor, because of their generally large families, were mainly responsible for spiraling growth in population. He further argued that food supplies increase arithmetically, that is, at a constant absolute rate over time, such as 4, 8, 12, 16, etc.

Malthus said the future holds famine, unless population is held in check. He argued that catastrophe was avoidable only through negative checks, such as war and disease, which increase the death rate, and through positive checks, such as abstinence, which reduces the birth rate.

Malthus failed to foresee the enormous impact of the industrial revolution and great gains in agricultural productivity. Indeed, other than the Irish potato famine of the 1840s, large-scale famine in Europe ended with industrializa-

tion. Furthermore, the fertile lands of North America and Australia provided vast new food sources. More important, he did not foresee the drop in birth rates and family size associated with industrialization and the growth of urban society. Birth rates reflect economic and social circumstances, not biological inevitabilities. Malthus also could have little idea of the effect that advanced methods of contraception would have on the birth rate. Consistent with the main method of lowering fertility in his time, Malthus advocated that people delay marriage.

Malthus's theories helped mold public policies after World War II, particularly birth control programs in developing countries. Low rates of population growth were assumed to be a prerequisite to economic growth. Birth control became integral to foreign aid, but with minimal success because its proponents generally failed to take into account the cultural and economic motivations that contribute to high birth rates.

Malthus's notions have been revived several times since his death. Most recently, high rates of natural increase in the 1950s and 1960s led to worldwide concern about overpopulation. Two influential books appeared at that time: Paul Ehrlich's 1968 *The Population Bomb, and The Limits to Growth* by the Club of Rome in 1972. These neo-Malthusians

believed that time will validate Malthus's theory of overpopulation, arguing that population growth eventually will be constrained by finite resources because ecosystems cannot withstand the accumulated assaults of industrialization, deforestation, and pollution. Others, loosely labeled "technological optimists," hold that new production techniques, such as aquaculture and genetic engineering to boost plant yields, will allow food production to keep pace with population growth. World grain yields have jumped in the last three decades. High-yield rice and wheat, products of the "green revolution," have significantly increased food supplies in India, China, and Mexico, where populations long have pushed the limits of the land's capacity. Many developed countries, which typically have slow-growing populations, produce food in excess of their needs. The excess becomes a valuable export, helping to ease world hunger. In many developing countries, with the notable exception of the region of sub-Saharan Africa, food supply has begun to keep pace with or exceed population growth, tending to validate, over the short haul, the cornucopian argument. But the question remains: Can increases in food production continue to keep pace with population growth?

INDUSTRIALIZATION AND POPULATION DYNAMICS

An alternative to Malthusian interpretations of population growth is the demographic transition model, made famous by demographers after World War II. Based on the historical experience of Western Europe as it industrialized, the demographic transition model ties birth, death, and natural increase rates to the changing socioeconomic circumstances faced by households in the midst of the shift from a predominantly rural, impoverished society to an urbanized, more prosperous society.

Societies typically pass through four stages en route to industrialization. The first stage involves an agrarian economy. Birth rates tend to be high and family size large, often with ten or more births per family. Children provide labor for farm families. High birth rates also reflect the reliance on children as a form of social insurance in old age, for which organized programs are generally lacking in poorer countries. Finally, high infant mortality rates lead families to bear large numbers of children in the hopes that a substantial proportion of them will survive to adulthood. Thus, poverty is a major contributing factor to high fertility rates: "overpopulation" may be the consequence, not simply the cause,

of low standards of living. In poor areas, death rates are high and life expectancies are low, largely due to poor nutrition, tainted water supplies, high infant mortality, and widespread prevalence of infectious bacterial diseases. Although birth and death rates are high, the difference between them—natural increase—is low. Pre-industrial societies grow slowly.

The second stage, often associated with early industrialization, includes a gradual decline in death rates and a gradual rise in life expectancy. Improved medical care is generally thought to be the primary cause. However, historical evidence suggests that improved diets as a result of mechanized agriculture are a major factor. Antibiotics and public health measures, such as improved sanitation and potable water, also play a role, particularly in lowering infant mortality rates. Birth rates remain high; natural increase rises.

The third stage witnesses a decline in birth rates, usually after the death rate has declined. The reasons for declining birth rates in the face of industrialization, contrary to Malthus's theory, lies largely in the changing cost-benefit ratio of children. As mothers join the labor force, the utility of child labor declines and the cost of raising children rises. Birth rates

typically decline as per capita incomes rise and the natural increase rate falls.

The final stage, stage four, commonly occurs with advanced urbanization and industrialization, accompanied by low death rates and long life expectancies. Modern populations, appreciably older than earlier ones, die most commonly from degenerative diseases such as heart disease, cancer, and stroke. While both birth and death rates are low, the difference is negligible. Therefore, natural increase rates in the developed world are uniformly low. Contrary to developing countries, where population policies emphasize birth control, demographic incentives in industrialized countries, faced with aging populations and shrinking labor forces, often seek to increase birth rates through subsidies for childbirth and child rearing.

Newly industrialized countries are going through the early stages of the demographic transition far faster than countries in Europe did, resulting in a much greater gap between birth and death rates and therefore a higher population increase rate. The demographic transition postulates a fall in rates of natural increase, and it also acknowledges the social, rather than purely biological, determinants of fertility.

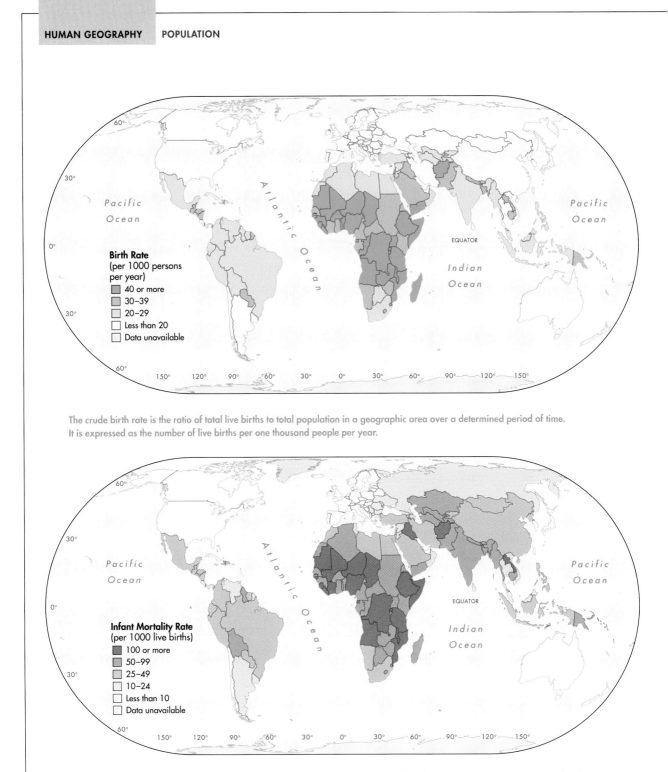

Birth Rate
(per 1000 persons per year)
- 40 or more
- 30–39
- 20–29
- Less than 20
- Data unavailable

The crude birth rate is the ratio of total live births to total population in a geographic area over a determined period of time. It is expressed as the number of live births per one thousand people per year.

Infant Mortality Rate
(per 1000 live births)
- 100 or more
- 50–99
- 25–49
- 10–24
- Less than 10
- Data unavailable

Health education, prenatal care, good diet and hygiene, and the presence of trained medical personnel contribute to lower levels of infant mortality.

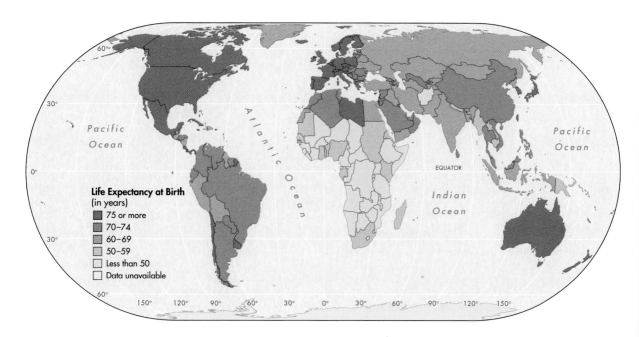

Much of the world benefits from improved medical care, and in many industrialized countries extensive systems of social welfare have improved the quality of life for elderly citizens.

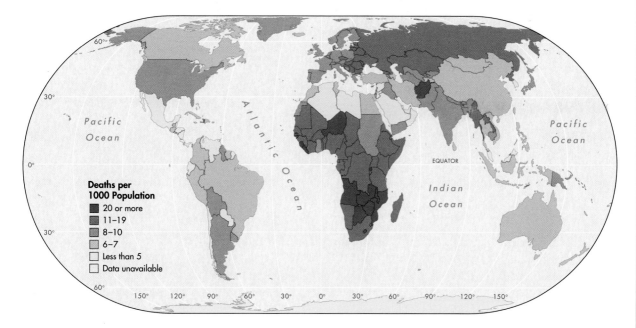

The crude death rate is the ratio of total deaths to total population in a geographic area over a determined period of time. It is expressed as the number of deaths per one thousand people per year.

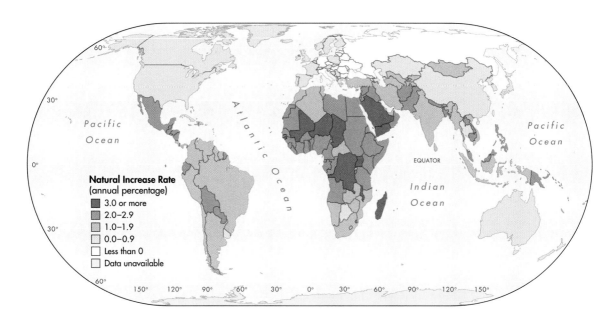

Natural growth rates are determined by subtracting death rates from birth rates. Growth rates are used to predict future population numbers and requirements for social services and food.

Population Projections

A population projection is a numerically derived estimate of a future population. Projections commonly cover up to 30 years. Beyond that the chance of error increases significantly. Accuracy generally increases with large populations projected over short periods, say ten years. Relatively accurate projections are possible if the calculation begins with recent and precise counts of the population, broken down by such demographic components as numbers in each age and sex group, plus rates of birth, death, and migration. More recent estimates from the time of the latest census can also be based on changes of address (often easily traced through drivers' licenses), new utility hookups, school enrollments, and the like.

Of the three demographic components (births, deaths, and migration) used in projections, migration is by far the trickiest to gauge. Fluctuations in the local economy—such as new jobs created by the opening of a major industrial plant, or drought and the consequent monetary loss to farmers—can cause wild fluctuations in migration. Mortality schedules for most populations, on the other hand, are fairly well established; and births can be estimated with adequate information on age structure and other vital statistics of a population. More at:
http://www.census.gov/population/www/projections/popproj.html

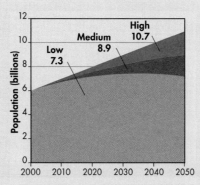

The UN projections for global population depend on anticipated total fertility rates (TFR) or the average number of children born to a woman in her reproductive years.

POPULATION: ASSET OR LIABILITY?

Population growth often brings with it economic growth, specialization, and material gain. However, it can also bring with it major problems. Social services—roads, schools, housing, law enforcement—often fail to keep pace with population growth, especially with the increasingly dense populations in cities of developing countries. Rapid growth and high density can also exacerbate many environmental problems.

Is population growth the primary cause of these problems? Or is the culprit a complex combination of population size, affluence, and technology? Paul Ehrlich, author of the *Population Bomb,* posits that the adverse impact of population on Earth relates to a population's size, affluence, and technological sophistication. Environmentalist Barry Commoner agrees that technology continues to ravage the environment with nitrogen fertilizers, throwaway plastic containers, and the like. However, he also argues that technology has the potential to dampen a population's negative impact on Earth. He notes that technological advances in the 1970s helped improve fuel efficiency, which in turn reduced car emissions and improved air quality.

Some people maintain that population growth is beneficial to economic growth. Most notably, Julian Simon, author of *Population: The Ultimate Resource,* argues that the great advances of civilizations, such as ancient Rome and classical Greece, were propelled by technological innovation, which was in turn fueled by increased population size. Anthropologist Ester Boserup cites evidence from 16th- and 17th-century Tokugawa, Japan, and 15th- and 16th-century Holland that supports the "beneficial" argument. She showed that a doubling of the population of Holland from 1550 to 1650 and a concomitant increase in demands on productivity were met with more efficient agricultural practices, including the expansion of the country's dikes to open more land to farming. Somewhat contradicting these findings, anthropologist Clifford Geertz found that on the densely populated island of Java, Indonesia, food per person decreased even though more and more people worked the land as the population increased in the 20th century.

WORLD POPULATION DISTRIBUTIONS

Exact records are lacking for the spatial distribution of world population in ancient times. Studies indicate that preindustrial urban centers were very small, compared to today's metropolitan areas. The Mesopotamian metropolis of Ur had up to 50,000 people in the 22nd century B.C., while Thebes, capital of Egypt at its zenith in the 14th century B.C., may have contained as many as 225,000 inhabitants. Rome, the largest city of antiquity, may have reached 350,000 around A.D. 200. Medieval Paris contained a mere 30,000 to 60,000 people, and commercially vibrant Venice probably counted no more than 70,000 in the 14th century.

Empires outside Europe could claim notable populations. China, which routinely conducted a census, beginning with the Han dynasty (206 B.C. to A.D. 220), grew to almost 60 million around the time of Christ, more than tripling to over 200 million by 1760. The 17th-century Thai capital of Ayutthaya contained hundreds of thousands of people before its destruction by Burmese armies in 1767. By the 18th century the thriving commercial city of Edo (precursor of modern Tokyo) was one of the largest cities of the world with more than a million people.

In the 19th century Europe's population burgeoned because of industrialization and because medical breakthroughs, better diets, and improved sanitation contributed to a reduction in the death rate. Paris, London, and Berlin

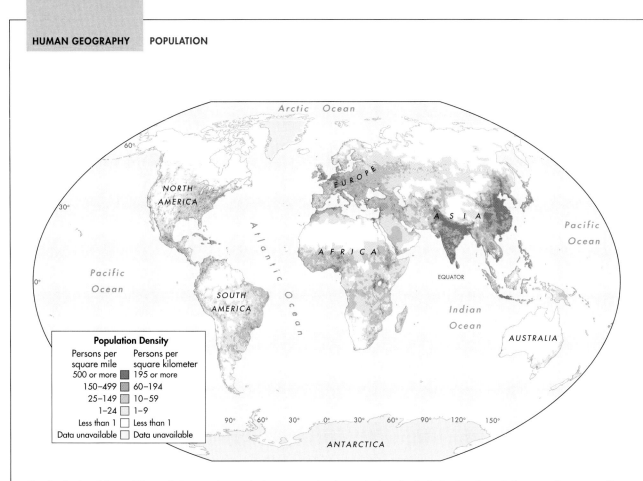

Population Density

Persons per square mile	Persons per square kilometer
500 or more	195 or more
150–499	60–194
25–149	10–59
1–24	1–9
Less than 1	Less than 1
Data unavailable	Data unavailable

The distribution of the world's population reveals not only the most populated countries but also the inclination of populations to settle near coastlines and significant resources such as fertile soil or industrial minerals.

emerged as the world's leading urban centers, and their culture was spread throughout the world by colonization. National populations of developing countries grew rapidly in the 20th century, initiated by a rapid drop in the death rate, first in South America, later in Asia and Africa. The onslaught of AIDS, however, has begun to push East and Southern Africa's death rates back up in the last two decades.

Pushed by generally precipitous drops in the death rate, the population growth of the developing countries in the 20th century far surpassed growth in the developed countries, which experienced an earlier, more drawn out, demographic transition. World population reached more than 6.4 billion by 2004, increasing by 77 million annually, or roughly 1.3 percent per year. Asia accounts for 61 percent of the world's population. China, with 1.3 billion, and India, with 1.09 billion, together account for about 37 percent of the world's people. Distribution by other regions is 11 percent for Europe, 5 percent for the United States and Canada together, 9 percent for Latin America, and 14 percent for Africa. The population division breaks down to 5.2 billion in the developing countries and 1.2 billion in the developed countries in 2004.

A quick glance at the world population distribution map, above, reveals that Earth's population is very unevenly distributed. Major agglomerations include East Asia and South Asia, followed by Western Europe, Southeast Asia, and

Northeastern North America. More localized concentrations are visible along North America's west coast, in central Mexico, and in West Africa and the Nile River valley. Population concentrations generally reflect areas with a high carrying capacity, either fertile land for agriculture or resources that support industry and trade. Vast "empty" regions, with very low population densities, include areas that are too dry, too wet, or too cold to support large populations, such as the Sahara, the Amazon rain forest, or the Arctic tundra.

During the last 100 years, Europe, the U.S., and Canada have sharply diminished in their relative share of the world's population, from 30 percent to 16 percent. Population increases have been most notable in the last several decades in Indonesia, Brazil, Pakistan, and Bangladesh. Several mid-size Asian countries—for example, Vietnam, the Philippines, and Thailand—with populations exceeding 60 million have overtaken all the major European countries in size, with the exception of Germany and Russia.

The future will most likely witness increasing relative importance of African populations. Africa is the only region of developing countries not yet experiencing a major decline in birth rates. The United Nations has indicated that despite increases in deaths and falling life expectancies in some countries due to the spread of AIDS, Africa still will go from 9 percent of the world's population in 1950 to almost 21 percent by the year 2050. With 98 percent of the current population growth occurring in developing countries, this region will account for an increasingly large proportion of the world's total population by the middle of the 21st century.

UNDERSTANDING THE DEMOGRAPHIC TRANSITION

The *demographic transition* model of population change postulates that a decline in mortality, not an increase in fertility, has produced the rapid population growth of the last two centuries. In the demographic transition, the decline in birth rates lags behind the drop in death rates. Several European countries, such as England and Sweden, experienced the transition described by the model and reached low birth rates by the mid-20th century. The demographic transition can both accompany and precede industrialization. In France, for example, the onset of a decline in the birth rates in the 1820s preceded the industrial revolution. On the other hand, Thailand, a predominantly rural country, has witnessed in the last 30 years a rapid drop in birth rates to replacement levels without experiencing widespread industrialization.

DISTRIBUTION AND PROJECTED CHANGE (%) OF WORLD POPULATION			
AREA	2004 (MILLIONS)	2050 (MILLIONS)	% CHANGE
World	6395	9276	45
More Developed Countries	1206	1257	4
Less Developed Countries	5190	8019	55
Africa	885	1941	119
Asia	3875	5385	39
Europe	728	668	-8
Latin America/Caribbean	549	778	42
North America	326	457	40
Oceania	33	47	43

As indicated in the graph (at right), as a country develops, high birth and death rates give way to a drop in death rates followed by a drop in birth rates. In developed countries, both rates are low.

In Myanmar (below), a developing country, women give birth to an average of 3.8 children in their lifetimes. A full 33 percent of the population is under 15 years of age. Children are assets. They provide families with farm labor, security in old age, and a chance to get ahead if there is success in school.

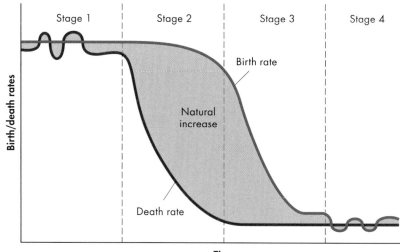

Lowering death rates is largely a matter of access to technology and medical resources. Lowering birth rates, however, requires a shift in cultural values and generally takes much longer to achieve. The speed at which a country can lower its birth rate is critical to closing the gap between mortality and fertility. Otherwise, such countries are saddled with disproportionate numbers of children, rapid rural to urban migration, and continued strain on limited resources—all the socioeconomic problems associated with population growth at the turn of the century.

Places with high infant and child death rates tend to have high birth rates. For example, in rural India, where infant death rates remain high, couples must bear several children to insure that at least two sons reach adulthood. Allowing for infant and child mortality, this could result in the birth of up to seven children. Children provide families with farm labor, security in old age, and a chance to get ahead if the child succeeds in school and in work outside the home. Despite cultural demands for large families, the transition to smaller families is under way. The fertility rate, the average number of live births per woman during her childbearing years, in rural India has dropped from 4.7 in 1973 to 3.1 in 2004.

Fertility rates continue to decline in most regions of the world in the early 21st century. A number of countries have completed the demographic transition, or are close to it, but others lag behind.

Women's changing role in societies is key to the change in fertility rates. In societies that emphasize family and women's role in the home, fertility rates have remained high. For example, Islamic countries, where women generally are restricted to domestic life, have some of the highest birth rates in the world. In the West, where women generally participate extensively in work outside the home, the fertility rates are low and continue to drop. A woman's educational level is also closely associated with her family size. Increased education has changed women's values and heightened women's awareness of opportunities outside the home. In countries with greater educational opportunities, women delay marriage and childbearing. They typically find better jobs, which in turn lessens their incentive to adopt the domestic life of house and children. Indications are, therefore, that as women continue to gain equality with men, birth rates will decline. Where women have been allowed a place in the work force, yet given little support in their roles as homemakers, the drop in fertility can be dramatic. The low birth rates in Rus-

WORLD'S HIGHEST FERTILITY RATES	
Niger	8.0
Guinea-Bissau	7.1
Somalia	7.1
Mali	7.0
Yemen	7.0
Uganda	6.9
Afghanistan	6.8
Angola	6.8
Comoros	6.8
Congo, Dem. Rep. of	6.8
Liberia	6.8
Chad	6.6
Malawi	6.6
Sierra Leone	6.5
Congo	6.3

sia today are due to changing attitudes. Russian women with demanding jobs and domestic responsibilities are finding that one child is enough. More at:

http://www.uwmc.uwc.edu/geography/Demotrans/demtran.htm

REGIONAL TRENDS IN POPULATION CHANGE

The 21st century sees the world in various stages of demographic transition—from agrarian economies with high birth rates at one end to economies of advanced urbanization and industrialization with

Population Pyramids

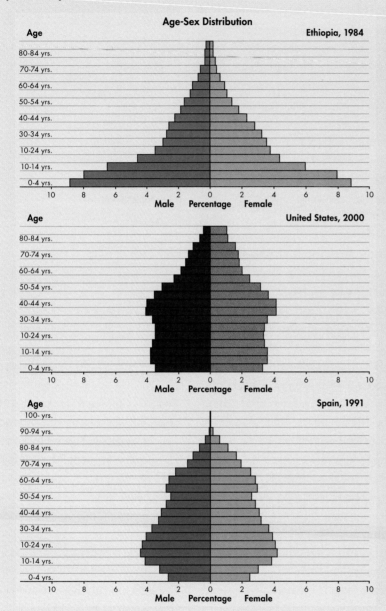

Age-Sex Distribution

Ethiopia, 1984

Age

80-84 yrs.	
70-74 yrs.	
60-64 yrs.	
50-54 yrs.	
40-44 yrs.	
30-34 yrs.	
10-24 yrs.	
10-14 yrs.	
0-4 yrs.	

10 8 6 4 2 0 2 4 6 8 10
Male   Percentage   Female

Age

United States, 2000

80-84 yrs.	
70-74 yrs.	
60-64 yrs.	
50-54 yrs.	
40-44 yrs.	
30-34 yrs.	
10-24 yrs.	
10-14 yrs.	
0-4 yrs.	

10 8 6 4 2 0 2 4 6 8 10
Male   Percentage   Female

Age

Spain, 1991

100- yrs.	
90-94 yrs.	
80-84 yrs.	
70-74 yrs.	
60-64 yrs.	
50-54 yrs.	
40-44 yrs.	
30-34 yrs.	
10-24 yrs.	
10-14 yrs.	
0-4 yrs.	

10 8 6 4 2 0 2 4 6 8 10
Male   Percentage   Female

The age structure of any population can quickly be understood by looking at its pyramid, three of which are shown here. A pyramid is a complex bar graph that shows the proportions of men and women in each age group. The shape of the pyramid can show whether the population is predominately young, as in Ethiopia (top), middle-age as in the United States (center), or older, as in Spain (bottom). The pyramid also suggests the demographic history of a country, typically over 80 years, the span of a few generations. It is possible to make inferences about countries with low birth rates, about proportion of the sexes, most notably the preponderance of older women in most societies and the significantly greater numbers of boys in certain East Asian countries.

The three pyramids give insight into the populations of each nation. Spain's pyramid is representative of Europe's population. Associated with Europe's decline in fertility has been a progressive aging of the population and the need for foreign workers. Although favorable levels of mortality have been achieved in Western Europe, life expectancy recently has dropped in Eastern Europe and Russia.

low birth rates and long life expectancies at the other end.

Europe

Europe, with a population of 728 million, includes some of the more densely populated areas in the world outside of South and East Asia. The Europeans were the first to complete the transition from high birth and death rates to low birth and death rates. Europe went through the demographic transition during the last two centuries while building up substantial numbers of people, despite losing many millions as emigrants, primarily to North America.

Europe's decline in the death rate began about 200 years ago. Total fertility rates continued to decline into the Depression era to *replacement level*, fluctuated in the 1940s and 1950s, and then began a persistent decline after the mid-1960s to below replacement levels. Rates in Northern and Western Europe declined to below replacement levels in the 1970s, while Eastern Europe experienced a precipitous drop in the 1990s with a regional fertility rate of 1.3.

Associated with Europe's decline in fertility has been a progressive aging of the population and the need for foreign workers. Although favorable levels of mortality have been achieved in Western Europe, life expectancy recently has

dropped in Eastern Europe and Russia, particularly for males in their later working years, a trend partly blamed on alcohol consumption and industrial pollution.

South Asia

Second only to China in population, India had grown to over one billion people in 2004. The populations of Pakistan and Bangladesh are more than 159 million and 141 million, respectively, making these two countries among the world's ten most populous. India's population has grown from 250 million at the beginning of the 20th century to 350 million at independence in 1947, and 996 million in 1999, almost a fourfold increase during the last 100 years. Both rapid population growth and high densities prompted the government of India in 1952 to establish a national family planning program, the first in the world. Fertility, however, remained high until the 1960s, then gradually decreased from 5.7 children per woman in the latter part of that decade to 3.1 in 2004. Bangladesh also has witnessed a transition to lower fertility (3.3), but fertility rates remain high in Pakistan (4.4).

East Asia

China's population has been historically large, with almost 60 million during the Han dynasty 2,000

years ago, more than 200 million by the mid-18th century, and one billion in 1981. After the Communist takeover in 1949, China's population increased by more than 145 percent in 55 years, from 529 million in 1950 to 1.3 billion in 2004.

The government, concerned with the country's ability to support such rapid growth, adopted pervasive birth control in 1970. Family planning and abortion services were widely accessible and inexpensive. The new policies included neighborhood goals (much like production goals for factories, only in reverse), incentives and disincentives for couples, massive propaganda efforts, and open doors to education and health care. All this and the flood of affordable new consumer items for the average Chinese set the stage for a drop in the fertility rate from 6.06 children per woman between 1965 and 1970 to 3.32 children per woman between 1975 and 1980, a monumental social change in but a dozen years.

China's successful birth planning program, however, has not been without side effects. China started to experience "missing females." The ratio of male babies to female babies, which typically has been about 105 boys to 100 girls worldwide, had risen from 106 in 1979 to 114 by 1989. The ratio—boys to

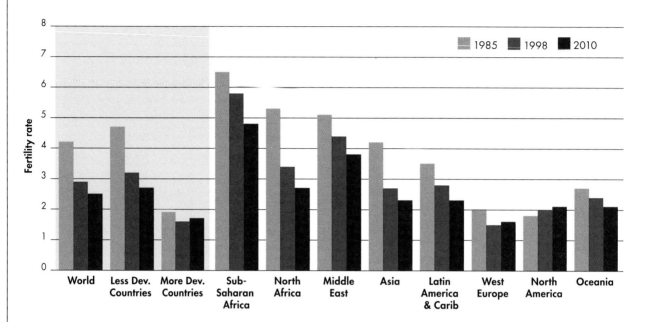

In most of the world the trend is toward lower fertility rates, with some of the most dramatic decreases occurring in less developed countries. In North America and Europe, where fertility rates are near or below replacement levels, a rise is predicted in the next decade.

girls—for the fourth born was 132 to 100. Although stories of female infanticide proliferated, most likely the major reason for the unusual ratio was abortions of female fetuses and unreported female births. Women eligible for marriage soon will become a premium, which should help control China's population growth, but could also lead to another form of social disruption. China likely will have too few workers to support the aging "baby boom" generation in its later years. Now, however, a large proportion of the Chinese population is in the prime labor-force ages, auguring well for continued growth of China's economy.

Japan's 127.6 million people are highly concentrated; less than a fifth of the land is suitable for settlement. Between 1726 and 1852 Japan's population remained close to 25 million, with lower birth rates partly realized through infanticide and by Japanese women delaying marriage until their mid-20s. During Japan's initial period of rapid industrialization, from the 1870s to the end of World War I, the population grew by 60 percent, from 35 million to 56 million, an increase of 21 million. Birth rates in the period of maturation of the Japanese economy, the 1920s, actually increased, contrary to the model of demographic transition and the usual

effects of industrialization. Government propaganda exhorted Japanese women to reproduce. They responded in the 1930s by pushing the birth rate to a high of about 31 children per 1,000 women.

Before the onset of World War II in 1939, Japan's population stood at 72 million. After the war, Japan quickly reduced its fertility from 4.5 children per woman in 1946 to 2.0 children in the late 1950s, primarily through the use of condoms and abortion. In 1990 the Japanese media created a stir with the expression "1.57 shock," reminding the country that Japanese women, on average in their lifetimes, were bearing children at a rate far below replacement fertility. Young Japanese women, now delaying marriage for several years and becoming less accepting of traditional domestic roles, have responded to such talk of "shock" by asserting that they are not "baby machines."

Given that Japan has among the highest life expectancy rates in the world (78 years for men; 85 years for women) and that the age groups born between 1930 and 1950 are analogous to the U.S. "baby boomers" of post World War II, Japan will not only have the world's

oldest population in the next few decades, but it also will have a smaller labor base to support the disproportionately large number of elderly. The large aging population soon will increase the country's death rate and initiate a natural decrease in the population beginning about 2006.

THE MIDDLE EAST AND NORTH AFRICA

North Africa and the Middle East together contain 400 million people, about 6 percent of the world's population. Predominantly Islamic, these regions have experienced high

fertility rates (3.4 and 3.7, respectively in 2004) since World War II. Contributing factors are early marriage and strong family customs. Despite the wealth gained from oil revenues during recent decades and improvements in health and literacy, the fertility rate has held steady. Many countries of the region, especially the oil-rich ones, such as Saudi Arabia and Kuwait, have relatively small populations, requiring migrants, known as guest workers, to supplement the labor force. Guest workers, typically men between the ages of 20 and 50, are

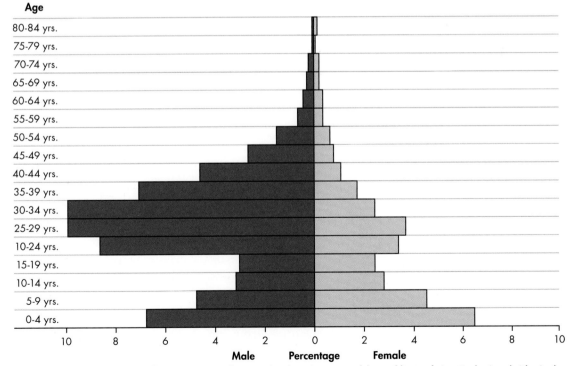

United Arab Emirates: Age-Sex Distribution, 1990

The Middle East and North Africa together contain 400 million people, about six percent of the world's population. Predominantly Islamic, these regions have experienced high fertility rates since World War II. Contributing factors are early marriage and strong family customs.

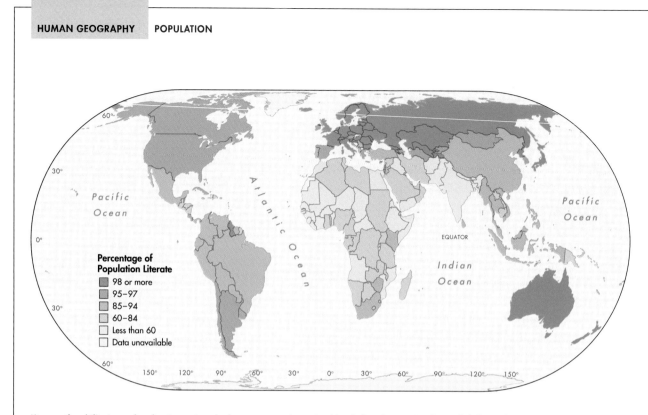

Literacy, the ability to read and write one's native language at a determined level of proficiency, is a learned skill usually obtained in a formal educational setting. Illiteracy is more prevalent in poor and less developed countries, and may affect women more than men in some societies.

visible as an unusual bulge in the age-sex structure of these countries. (See the population pyramid for the United Arab Emirates, p. 203.)

SUB-SAHARAN AFRICA

Sub-Saharan Africa shares with its neighboring region to the north the distinction of the world's most rapidly growing population. Its 48 countries cover less than 20 percent of the Earth's land surface, yet possess 11 percent of its population. Given the rapid rate of population growth in recent decades and the cultural preference for large families, regional growth rates probably will remain high for some time to come. From the 16th to the 19th centuries, Africa lost up to 15 million people, primarily in their reproductive years, to the slave trade. As recently as 1950, the sub-Saharan population stood at only 177 million, more than quadrupling by 2004 to about 733 million. Projections indicate the population may rise to 1.12 billion by the year 2025.

Use of contraceptives is up from 5 percent of married women ages 15–49 years, to 21 percent, dropping the region's fertility rate from 6.7 in the early 1960s to about 5.6 percent in 2004. Although significant declines in the fertility rate do not appear imminent, population growth will be slowed somewhat by a rise in the death rate because of the region's high incidence of AIDS. In the 37 sub-Saharan countries highly affected by autoimmune deficiency syndrome, the average life expectancy at birth has decreased by two years and is expected to continue to decline. The population is projected to be lower by 91 million people, or 10 percent, than it would have been without AIDS.

LATIN AMERICA

Early in the 21st century, Latin America had more than 549 million people, about 9 percent of the world's total. The fertility rate has decreased from 5.9 in the early 1960s to 2.6 percent in 2004. Four major groups account for the cultural origins in the region. The indigenous Indian population is most prevalent in the highlands, particularly in the Andes. Blacks are numerous in Brazil and parts of the Caribbean. European stock predominates in southern Brazil, Uruguay, Argentina, and Chile. People of mixed ethnic background are common throughout the region. Indian and black populations have distinctly higher fertility rates, while South American whites have fertility rates similar to those in European countries.

Mexico and Brazil together contain more than half of the region's population. Both experienced rapid declines in fertility rates, from around 6 in the 1960s to 2.8 for Mexico and 2.2 for Brazil in 2004. In Argentina, Costa Rica, and Colombia total fertility rates have remained near replacement levels (2.1 to 2.6), for the last decade. The rapid decline in Latin America's crude birth rates since the 1950s has been somewhat unexpected, given the pronatalist governments that existed before the 1980s and the influence of the Catholic Church. In terms of population growth, the region's rate increased from 2.6 percent in the early 1950s to 2.75 percent a decade later. In the late 1970s,

COUNTRIES OF AFRICA WITH THE HIGHEST PREVALENCE OF HIV (>20% OF TOTAL POPULATION)
Botswana
Lesotho
Namibia
South Africa
Swaziland
Zambia
Zimbabwe

the rate had dropped to 2.3 percent. By 2004, Latin America's rate of natural increase was 1.6 percent.

ANGLO AMERICA

Canada and the United States have completed the demographic transition and have low birth and death rates. Their combined population in 2004 is 326 million, 5 percent of the world's population. Life expectancy at birth is 79 years for Canada and 77 for the United States, compared with an average life expectancy of 67 years for the world. Fertility is below replacement level, 2.0 in the United States and 1.5 in Canada. Growth will depend on population momentum (caused by the larger proportions of the population presently in the reproductive ages) and immigration. At least two major demographic issues face Anglo America. First, society must accommodate baby boomers as they retire and begin to collect Social Security. Baby

boomers, born between 1946 and 1964, make up almost 30 percent of the population in the United States. Second, both countries must adjust to the shift in ethnicity as the European stock begins to yield population dominance, particularly to Asians, Hispanics, and African Americans. As interaction increases between Latin America and Anglo America, the Hispanic population is becoming increasingly influential, supplanting African Americans as the most numerous.

OCEANIA

Throughout history, Oceania's relative share of the world's population has been only about 0.5 percent, and today, with approximately 33 million people, this large, spread-out region of islands contains only three countries of significant population size.

Australia, with 20.1 million people—about two-thirds of the region's population—and New Zealand, with 4.1 million people, have both completed the demographic transition. Historically, their changes in birth and death rates have generally been similar to those of Western Europe, consistent with the European, especially British, origins of most of their people. Australia augmented its population earlier in this century through immigration from the British Isles, and more recently from a diversity of places. In the early 1990s, Australia was the world's fifth most important country in net international migration. The Polynesian, Melanesian, and Micronesian island countries in the region have higher birth and death rates than Australia and New Zealand, and population growth rates ranging from 3.7% in the Marshall Islands to 0.8 in Palau. Of these, only Papua New Guinea, with approximately 5.7 million people, has a statistically notable national population size.

MINORITY POPULATION IN THE UNITED STATES, 2000	
Total Population	281,421,906
Black/African American	33,947,837
Asian	10,123,169
American Indian/Alaska Native	2,268,883
Native Hawaiian/Pacific Islander	353,509
Other, or two or more races	5,069,916
Hispanic*	35,305,818

*Hispanic is an ethnicity, not a race, and includes persons who self-identify as either white or black, and therefore are already part of the racial count.

Sources of Population Data

The field of population geography is rich with data. The most comprehensive sources are national censuses, taken every ten years in most countries. A census contains information on housing, race, economic status, sex, age, education, and more. The ancient Romans and the Chinese dynasties of thousands of years ago conducted a regular census. Sweden has taken a census since 1749—the longest continual demographic count for a major modern country. In the United States, a census every ten years is mandated by the Constitution. The first census was taken in 1790 and data is maintained by the U.S. Census Bureau.

A wealth of population data is available through government records of vital statistics—births, deaths, marriages, diseases, etc. The extent of vital statistics varies considerably among countries. Contributing to the study of human populations are sample surveys, covering a myriad of topics, such as personal decisions behind migrations, the size of families, the reasons for marriage, choice of job, income related to time of retirement. Governments, universities, and private organizations often finance institutes devoted to conducting surveys, most carried out repeatedly to monitor change. Princeton University researchers, for example, have traced the change in attitudes on birth control and reproduction behavior in the American family. The University of Chicago has a continuous study on the changes in health and medical care practices in the United States. The World Fertility Survey, conducted in the 1970s by the International Statistical Institute, London, measured attitudes and behavior on family planning, interviewing more than 400,000 women in 43 developing countries and 20 developed countries.

Technology is rapidly changing the form in which statistical reports appear. The U.S. census for the year 2000 has only one printed summary per state whereas previously, thousands of volumes were printed. However, the entire census is accessible on the Internet. Some countries, Singapore, for example, are considering conducting and publishing their next census entirely via the Internet.
More at: http://www.census.gov

MIGRATION

CHANGING PLACES

Migration has redistributed people over the Earth for millions of years. Prehistoric humans migrated far in search of food, found favorable conditions, settled, multiplied, and then depleted their sources of meat and fur. One and a half million Irish migrated to the United States in the late 1840s when a potato blight destroyed their main source of food. Migration of industrialized people has led to the leveling of mountains in West Virginia for the purpose of extracting coal, while in Brazil migrants have cleared rain forests for farms and ranches and diverted rivers for crops and cattle in the last several decades.

Migration is the changing of one's place of habitation for a substantial time, normally across a political boundary. People react to the push of natural disasters such as war, overpopulation, religious persecution, politics, and slavery and respond to the pull of economic opportunity, religious freedom, social equality, democracy, safety, food sources, and open land.

During the great age of exploration, European navigators and adventurers of the 15th, 16th, and 17th centuries expanded the frontiers of Europe and diffused European culture and peoples to the Western and Eastern Hemispheres and to Australia. The 20th century has experienced several forced mass migrations, including the transplanting of tens of millions of urban dwellers to the countryside

In the first half of the 19th century, Tasmania received about 68,000 British convicts. Agriculture became their primary means of subsistence.

during China's Great Leap Forward and Cultural Revolution, and the exodus of ethnic Albanians from Kosovo in 1999 in the ethnic cleansing of Yugoslavia. On another scale, one of the most significant migration patterns of the 20th century has been the enormous rural to urban migration that has occurred in countries around the world.

MODELS FOR THE STUDY OF MIGRATION

The most popular explanation of human migration is the *push-pull model*: Migrants are pushed out of one place and pulled to another. People's decisions are based on a calculation of the pluses and minuses of staying put and the pluses and minuses of moving. Travel costs also figures into the calculations. Terrain, cultural barriers, distance, modes of transportation, and time of travel all stand between migrants and their destination, and all usually weigh heavily on the decision to move or to stay.

Many geographers and others have tried to explain migration patterns. For example, E. G. Ravenstein attempted to identify social laws governing migration. (See box below.) Migration tends to be selective; that is, people who move often have characteristics in common. Better-educated, white-collar, and military personnel tend to move more often. Historically, women have moved less often than men and have moved shorter distances, on average, but now have approximately the same migration rates as men in most countries. People in formative stages of life—graduation, marriage, birth, separation, divorce—are associated with a greater likelihood of moving. Migration is selective for certain ages, most notably younger adults, 18 to 30 years old, and their young children, who must move with them.

Demographer William Petersen has classified migration into five categories on the basis of degree of choice.

1) Primitive migration is the movement of preindustrial peoples in response to the physical environment, e.g., the effort to find sufficient land for hunting or farming.

2) Impelled migration involves relatively powerless people, such as indentured servants of the 18th century and workers under the pejorative title "coolie contracts" of the 19th century.

3) Forced migration includes people who are completely pow-

Ravenstein's Laws of Migration

E. G. Ravenstein is credited as the first to attempt to explain patterns of migration. His celebrated paper on the "Laws of Migration," presented to the Royal Statistical Society on March 17, 1885, maintains that pull factors take precedence in the decision to migrate. "Bad or oppressive laws, heavy taxation, an unattractive climate, uncongenial social surroundings, and even compulsion—all have produced and are still producing currents of migration, but none of those currents can compare in volume to that which arises from the desire inherent in most men to 'better' themselves in material respects."

Ravenstein's Migration Laws (1870s–1880s):

1. Most migrants go only a short distance.

2. Longer-distance migration favors big-city destinations.

3. Most migration proceeds step by step.

4. Most migration is rural to urban.

5. Each migration flow produces a counter flow (i.e., return to place of birth).

6. Most migrants are adults–families are less likely to make international moves.

7. Most international migrants are young males.

More at: http://csiss.ncgia.ucsb.edu/classics/content/90

Flight of the Snowbird

Retired Americans who seasonally migrate to the Sunbelt—often called "snowbirds"—represent a recent new migration trend. Migration, normally associated with displacement and hardship, takes on an entirely different meaning for snowbirds. Typically, they are an Anglo couple from the northern United States or Canada.

With retirement, snowbirds flee their family home of several decades, flee the cold winters of the north, leave their grown children, jump in an RV (recreation vehicle), and drive the highways to Arizona, Texas, and other warm regions of the South and Southwest. Along their leisurely way, snowbirds stop at recre-

ational areas and trailer parks to meet old friends. They continue on to their winter home for sunshine and more social activities. The migration is reversed in summer. Their extended trips back and forth are recreation all the way. Indeed, snowbirds celebrate a culture of migration. Movement is their life.

erless, such as African slaves, Native Americans who were forced onto reservations, and the Jews of Nazi Germany.

4) Free migration, in which the unforced will of the migrant is the decisive factor, describes early pioneer movements in the settlement of the American West and contemporary movement of Americans to the Sunbelt states of the South and Southwest.

5) Mass migration often has involved persecuted minorities who are deported from their homes. Such was the case of the Poles, Germans, Crimean Tatars, and many others in Stalinist Soviet Union. Mass migration also finds impetus in politically and economically disenfranchised peoples. Millions of Germans migrated to the United States in the 18th and 19th centuries seeking religious and political freedom and the chance to own land.

MEASURING MIGRATION

A rich source of migration data is a recurrent national census, highlighting information on origins and destinations. Population registries also provide valuable statistics in the study of changes in residence by requiring migrants to transfer their records from one local registry office to another. In Japan the resident registration system contains monthly migration statistics dating to 1954. In China, the household registration system initiated in the early 1950s has provided important annual data for the world's most populous country. (The system, however, was used primarily to deny rural Chinese access to government subsidies for such standards as rice, cooking oil, housing, and urban transportation.) Place-of-birth statistics can serve as the best available movement data in some of the developing countries.

In the United States, migration data is recorded as part of the decennial census. According to *Census 2000,* more than 22 million people in the United States changed their state of residence between 1995 and 2000. Following the predictions of Ravenstein's Laws, aboout half of these migrants relocated to places within the same region of the country. The region with the highest levels of in-migration was the South; the greatest out-migration was from the Northeast.

In addition to domestic migration, the United States is a destination for international migration. The United States Census Bureau records data on the foreign-born population. In 2003, 33.5 million foreign-born persons lived in the U.S., more than half of whom were from Latin America.

More at:

http://www.census.gov/population/ socdemo/migrate.html

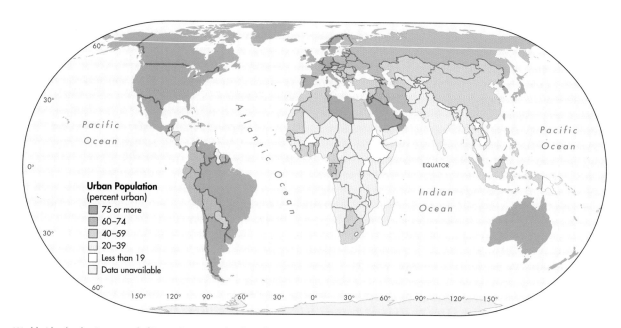

Worldwide, the dominant trend of internal migration has been from rural to urban locations. Experts estimate that almost half of the world's population live in urban areas.

Urban Population
(percent urban)
- 75 or more
- 60–74
- 40–59
- 20–39
- Less than 19
- Data unavailable

Surveys provide insight into cultural and behavioral influences on decisions to migrate. In measuring migrations, geographers work with numbers and rates, including in-migration and out-migration, gross migration (the total number of in-migrants and out-migrants) and net migration (the difference between the number of in-migrants and out-migrants). Geographers also distinguish between internal and international migration. Indirect methods for estimating migrations involve birth and death statistics. More at:

http://www.un.org/esa/population/publications/ittmig2002/ittmigrep2002.htm

DISTANCE, TIME, AND BOUNDARY CONSIDERATIONS

Distance is a major factor in migration flows. Short-distance moves are more common than long-distance moves, and a short-distance move across town has a minor impact on a person's life compared with a long-distance move to a new country.

Decisions based on economic factors predominate in the regional relocation of people in the United States. The push of job dissatisfaction or unemployment and the pull of good or higher paying jobs often entice people to move long distances. Migrants must also decide on a new home, taking into consideration such factors as distance for commuting to the job, housing costs, social groups, and quality of schools.

Migration can be either temporary or permanent. Demographers generally define migration as a permanent change in residence. Although daily commuting is not counted as migration, demographers do count commuting as migration when workers change residence for several weeks in response to work conditions. Seasonal migration, including the annual winter flight of snowbirds to warmer southern climates of the United States, usually involves two residences. Such seasonal migrations are sometimes referred to as

COUNTRIES WITH THE LARGEST INTERNATIONAL MIGRANT POPULATIONS (THOUSANDS) 2000		COUNTRIES WITH THE HIGHEST PERCENT OF INTERNATIONAL MIGRANTS IN THE POPULATION, 2000	
United States	34,988	United Arab Emirates	73.8%
Russian Federation	13,295	Kuwait	57.9%
Germany	7,349	Jordan	39.6%
Ukraine	6,947	Israel	37.4%
France	6,277	Singapore	33.6%
India	6,271	Oman	26.9%
Canada	5,826	Estonia	26.2%
Saudi Arabia	5,255	Saudi Arabia	25.8%
Australia	4,705	Latvia	25.3%
Pakistan	4,243	Switzerland	25.1%
United Kingdom	4,029	Australia	24.6%
Kazakhstan	3,028	New Zealand	22.5%

Migration: Moving Up—or Out?

Demographers study migrations between neighborhoods, counties, regions, and countries. The scale of migrations differs widely. Studies of migration in the United States between city neighborhoods have shown the importance of socioeconomic and educational considerations in decisions to move. In the 19th and 20th centuries, various ethnic groups migrated to areas of inexpensive housing near city centers. Over time, and with increasing affluence, the groups moved progressively outward, triggering an overlay of cultures throughout the cities. "Moving up" in the United States was often equated with "moving out," resulting in the socioeconomic homogeneity of suburban neighborhoods.

Study of county-to-county migration in the United States has been facilitated by extensive data of the Census Bureau. Summarized on maps, the information traces the redistribution of Americans since the 1960s toward the South and the West, pushed by retirement and drawn by coastal counties rich in amenities.

Geographers also have analyzed state-to-state and region-to-region movements. Studies at these larger scales have provided considerable insight into the relationship of regional population flows and economic cycles. In the United States, the exodus of migrants from the Northeast and Midwest during the last few decades can be traced to a concentration of outmoded infrastructure, lack of new, high-tech industries, shift in job openings, and appreciation of the dollar in the early 1980s. Between 1970 and 1985, the Northeast lost more than a million jobs; the Midwest lost 700,000. Economic restructuring and downsizing in the Midwest in the late 1980s, however, led to a balance of migration.

The South, with lower taxes, wages, and land costs, continued to prosper, recording its largest population growth in the late 1980s. Trends in the 1990s show the continuing economic and demographic strength of the South. California started to lose migrants in the late 1980s and early 1990s partly because of inflated housing prices. More at:
http://www.census.gov/population/www/cen2000/migration.html

circular movements. In a broad sense, circular migrants also include those who return to their homeland, even after years of absence. History is replete with stories about the vast numbers of European immigrants who passed through Ellis Island; however, little is written about the fact that nearly a third of these immigrants returned to their homelands later in life.

Migration involves the crossing of a political, or some other, definite boundary. If, in the study of migration, the sampled geographic units are small, relatively few migrants will be missed in the counting. However, chances of missing migration streams increases when the sampled geographic units are large.

REFUGEES AND TEMPORARY LABOR MIGRANTS

A refugee is a special type of migrant. War and persecution have swelled the ranks of refugees. Host countries that have signed the 1951 Geneva Convention may not force refugees back to their homeland. Therefore, the host country must determine whether migrants qualify as refugees, that is, whether they, in the words of the Geneva Convention, have "a well founded fear of being persecuted for reasons of *race*, religion, nationality, membership of a particular social group or political opin-

ion." Determining who qualifies as a refugee is a daunting task. Many migrants claim to be refugees in order to stay in the host country, not for fear of harm back home but for a desire to find a better job, or better housing, or for another one of the various "pull" factors.

More at:
http://www.unhchr.ch/html/menu3/b/o_c_ref.htm

According to the UN High Commission for Refugees, there were more than 17 million refugees, inter-

FOREIGN WORKERS IN SELECTED EUROPEAN COUNTRIES, 1999 LABOR FORCE (1,000)			
Country	Total	Foreign	% Foreign
Total	136,294	8,161	6.0
Austria	3,177	30S	9.6
Belgium	4,096	340	8.3
Denmark	2,842	54	1.9
France	24,903	1,544	6.2
Germany	39,000	3,432	8.8
Ireland	1,333	40	3.0
Luxembourg	168	65	38.6
Netherlands	7,128	278	3.9
Norway	1,067	48	4.5
Spain	16,400	82	0.5
Sweden	4,333	221	5.1
Switzerland	3,346	726	21.7
United Kingdom	28,500	1,026	3.6

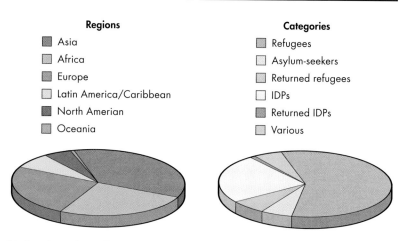

Regions
- Asia
- Africa
- Europe
- Latin America/Caribbean
- North Amerian
- Oceania

Categories
- Refugees
- Asylum-seekers
- Returned refugees
- IDPs
- Returned IDPs
- Various

This chart shows the number of migrants—many refugees and asylum-seekers—in major regions around the world. Often war and persecution account for migration to foreign soil.

ORIGINS OF MAJOR REFUGEE POPULATIONS IN 2003		
Country of Origin	**Main Countries of Asylum**	**Total**
Afghanistan	Pakistan, Iran	2,136,000
Sudan	Uganda, Chad, Ethiopia, Kenya, D.R. Congo, Central Afr. Rep.	606,200
Burundi	Tanzania, D.R. Congo, Zambia, South Africa, Rwanda	531,600
Dem. Rep. of the Congo	Tanzania, Congo, Zambia, Burundi, Rwanda, Angola, Uganda	453,400
Palestinians	Saudi Arabia, Iraq, Egypt, Libya, Algeria	427,900
Somalia	Kenya, Yemen, United Kingdom, Ethiopia, Djibouti, USA	402,200
Iraq	Iran, Germany, Netherlands, Sweden, United Kingdom	368,500
Vietnam	China, Germany, USA, France	363,200
Liberia	Guinea, Côte d'Ivoire, Sierra Leone, Ghana, USA	353,300
Angola	Zambia, D.R. Congo, Namibia, South Africa	329,600
MAJOR REFUGEE ARRIVALS DURING 2003		
Country of Origin	**Main Countries of Asylum**	**Total**
Sudan	Chad, Uganda, Kenya, Ethiopia	112,200
Liberia	Côte d'Ivoire, Guinea, Sierra Leone, Ghana	86,800
Dem. Rep. of the Congo	Burundi, Zambia, Tanzania, Rwanda, Uganda	30,000
Côte d'Ivoire (Ivory Coast)	Liberia, Guinea	22,200
Somalia	Yemen, Kenya, Tanzania	14,800
Central African Rep.	Chad	13,000
Burundi	Tanzania, Zambia, Rwanda	8,100
Angola	D.R. Congo, Namibia	1,500
Russian Federation	Georgia	390
Rwanda	Zambia, Uganda	360

nally displaced persons, and stateless persons worldwide in 2003. Unlike migrants, who are stimulated by economic factors, refugees are often fleeing from war, civil unrest, or religious or ethnic persecution. Most refugees hope to return to their home country, but many end up staying permanently—either legally or illegally—in the host country. Yet another special group of migrants—temporary laborers known as guest workers—have become common in a world that depends on a flexible labor force. In some oil-rich Middle Eastern countries, foreign workers form a majority of the labor force; in South Africa they make up about 15 percent of the labor force; and Germany, France, Israel, and the United States also relied heavily on guest workers to bring in the harvest and to take over many low or unskilled jobs that the local population has abandoned.

At the beginning of World War II, U.S. farmers were asked to step up food production. Pressure on the farmers increased when the United States continued to expand armed forces as the war continued, leading to a shortage of farm labor. In response, the United States recruited Mexican farm workers under the so-called *bracero* program of 1942.

Mexico and the United States agreed on rules governing transportation, pay, housing, working conditions, and medical care for the contract workers, who were called braceros. The program was resurrected in 1951 when the Korean War again took workers from U.S. farms, but was ended in 1964. More at: http://www.farmworkers.org/bracer op.html

MAJOR MIGRATION PATTERNS OF WORLD REGIONS

International migration today involves greater numbers than ever before. About 120 million migrants, slightly more than the combined population of Italy and France, lived outside their country of citizenship or birth at the end of the 20th century. Migrants are not spread evenly in the world: One third live in seven of the world's most affluent countries—Canada, United States, France, Italy, Germany, United Kingdom, and Japan.

Europe

Europe has absorbed millions of immigrants, France and Germany the most, with close to 45 percent of Europe's almost 25 million immigrants in the mid-1990s. Albania accepted hundreds of thousands of Kosovo refugees during the ethnic cleansing by the Serbs and NATO bombing of Yugoslavia in 1999. Europe has seen four stages of international migration in the period since World War II:

1) Stage One (1945 to 1960) involved displaced persons relocating or returning home, e.g., 13 million ethnic Germans resettled to West Germany in 1945 to 1950.

2) Stage Two (1961 to 1974) resulted from the opening of borders and the rapid economic growth associated with the formation of the European Economic Community in 1957. Migrants from southern Europe and outside Europe formed a large cadre of foreign workers, most notably Turks in West Germany and Algerians in France. Many Slavs from Yugoslavia also migrated north to work in several European countries, most often on assembly lines, or in construction.

3) Stage Three (1975 to 1985) involved the slowdown of the European economy and the consequent public backlash against guest workers, which lead to government attempts at repatriation.

4) Stage Four (late 1980s) occurred with the breakdown of communist regimes in Eastern Europe when people sought economic opportunity beyond their borders. The country most affected in this stage was Germany, which initially allowed free return to all ethnic Germans.

South and East Asia

In Asia five major displacements of peoples have resulted from regional conflict since World War II:

1) With the surrender of Japan in 1945, eight million Japanese were repatriated from the far stretches of the empire.

2) A massive refugee movement in Asia followed the partition of the British Indian Empire. In 1947, with civil war brewing, some 15 million people fled to the land of their religion, Hindus and Sikhs to India and Muslims to Pakistan.

3) Millions of North Koreans fled to South Korea after the Korean War in 1953.

4) The Vietnam War, 1950s to 1970s, accelerated the movement of refugees, especially Catholics and ethnic Chinese, first southward within Vietnam and later abroad. Huge resettlement camps were set up on the Thai border to help relocate the refugees. About a million Southeast Asians were admitted to the United States between 1980 and 1994.

5) The Soviet occupation of Afghanistan in 1979 and the sub-

sequent civil war created vast movements across national borders. Five million Afghanis fled to Iran and to Pakistan. When the Soviets withdrew and an Islamic government again took power in 1992, about half of the refugees returned to Afghanistan.

The most notable migration of Asians has been of laborers attracted by governments and business opportunities. South and East Asian workers responded to labor demands in the Middle East in the 1970s to help build the infrastructure to support the emerging oil industry. South Korea, with the maturation of an economy based on exports, has evolved from a labor exporter in the 1970s and early 1980s to a labor importer in the 1990s. Several other Asian countries, Thailand and Malaysia included, import and export labor

Migration Statistics

◆ The number of Mexicans who died trying to cross illegally into the United States from 1993 to 1996 was 1,185, according to Worldwatch Institute.

◆ Foreign workers form the majority of the labor force in several Middle Eastern countries, and about one-seventh of the labor force in South America.

◆ Africa and western Asia contain more than half of the world total of 15 million refugees and displaced persons.

◆ Newly arriving immigrants account for all the population growth in Germany and about a third of the annual growth in the United States.

◆ The United States had a foreign-born population of almost 33.5 million in 2003, about 11.7 percent of the national population.

◆ In the 1990s Europe, North America, Australia, and New Zealand had net population gain from migration. Africa, Asia, and Latin America experienced a net loss.

◆ Countries hosting the most refugees include:

1. Iran, 2.2 million, mostly Afghanis.
2. Democratic Republic of Congo, 1.5 million, mostly Rwandans.
3. Pakistan, 1.2 million, mostly Afghanis.
4. Jordan, 1.2 million Palestinians.

◆ Refugees from three countries or geographic areas accounted for nearly half the world's refugees in 1994: Palestinians 3.1 million, Afghanis 2.8 million, and Rwandans 1.7 million.

◆ Although their population is only about 5 percent of the world population, the United States and Canada contain about 20 percent of the world's migrant stock.

◆ More than 50 million Europeans immigrated, primarily to North America, South America, and Australia, between the mid-1800s and the outbreak of World War I. The trend was reversed in the 1960s, when millions of immigrants flowed into the countries of Northern and Western Europe as guest workers, asylum seekers from former communist countries, and unauthorized refugees from the violence in former Yugoslavia.

◆ In much of Europe, a foreigner is defined not by birthplace, but by ethnicity or ancestry.

◆ The high-income industrial democracies contain about 60 million immigrants, refugees and asylees, and authorized and unauthorized migrant workers.

◆ The percentage of countries whose policies in 1976 were aimed at reducing immigration was 6, increasing to 33 in 1995, according to Worldwatch Institute.

◆ Since 1960 the percentage of people living in urban areas has gone up about 30 percent.

◆ In developing countries, urban populations have more than doubled since 1950 to 39 percent.

◆ Oceania has the highest percentage of migrants (17.8 percent). The 4.7 million are only one-tenth the migrants in Asia (43 million).

◆ The industrialized countries' native labor pool is expected to shrink as the developing world's workforce doubles.

and appear to be approaching a "migration transition" similar to that experienced by South Korea. North America has accepted many migrants from Asian countries. Filipinos, Indians, Chinese, and Koreans recently have grown to significant minorities in large cities and on the west coasts of the United States and Canada.

The Middle East

Political refugees and economic migrants have played a major role in the politics of the Middle East. The Palestinians, disenfranchised after the Middle Eastern wars of 1948 and 1967, arguably are the world's best-known refugee group. According to United Nations estimates, they number over 4 million, some spending up to 50 years as refugees in Arab countries surrounding Israel. With the evolution of autonomy and with the semblance of a homeland in the West Bank and Gaza, the Palestinian refugee problem may be on the way to resolution. The Palestinians, arguably the best educated of the Arab peoples, have prospered as foreign workers in the oil-rich countries of the Persian Gulf. Indeed, the group in Kuwait most devastated by the 1991 Kuwait-Iraq War was not the ethnic Kuwaitis but the Palestinians, numbering about 300,000, who lost their homes and

much of their life savings. They were not allowed back into Kuwait after the war because a small number of the Palestinians were suspected of sympathizing with the Iraqis.

The Middle East is a magnet for workers. Petrodollars from OPEC price increases of 1973 and 1979 spawned prodigious economic growth and a concomitant flow of foreign workers into countries of the Persian Gulf. Different nationals often filled different job classifications. It is not uncommon to find Yemeni day laborers, Egyptian teachers and college professors, and Palestinian bureaucrats. Pakistanis make up a large number of the construction laborers. Koreans and Thais have tended to work in semiskilled jobs. Filipinos are popular as entertainers and musicians. Bengalis have worked as domestics and hotel employees. Large numbers of Egyptian farmers migrated to Iraq, partly because of a familiarity with irrigated river-basin agriculture. In the less populated countries of the Persian Gulf, such as Kuwait and the United Arab Emirates, foreign workers often make up the bulk of the labor force.

Africa

Wars, ethnic hatred, and economic disparities have stimulated much of the African international migration

since World War II. Independence movements of the 1950s up until the 1970s prompted the small British, French, and Portuguese colonial populations to return to their native countries.

Independence also led to civil strife, which frequently turned a large portion of the population into refugees. In the last two decades, millions of refugees from Uganda, Rwanda, Democratic Republic of the Congo, Liberia, Sierra Leone, Sudan, and Angola have fled to neighboring countries to escape civil wars. Guinea, with a population of only 7.5 million, harbored 700,000 refugees in 1999 from civil wars in four neighboring states; about 350,000 of the refugees are from Sierra Leone. Kenya provided haven for 420,000 refugees during the worst of the civil war and famine in Somalia in 1992. In 1994 more than two million Rwandans fled to Tanzania and Zaire (now the Democratic Republic of the Congo) to escape the genocidal war between the Hutus and Tutsis. By 2003 Africa, with but 14 percent of the world's population, harbored more than 25 percent of the world's refugees. Drought and degradation of farm and grazing lands also have driven large numbers of Africans from the Sahel, the southern fringe of the Sahara, to overcrowded cities in search of jobs.

Latin America

The earliest migrants to Latin America were native peoples who journeyed to the Western Hemisphere from Siberia by way of the Bering Strait land bridge as early as 10,000 years ago. The first migrants to Latin America in the modern era were primarily Spanish and Portuguese. The European colonial era brought disease, starvation, and war to the hemisphere, devastating indigenous populations. When the Spanish arrived early in the 16th century, Mexico and Central America contained an estimated 25 million people, but this population was reduced to 2.5 million only one century later. Large numbers of African slaves were imported to the Caribbean and to Brazil for work on sugar plantations. Portuguese slavers brought between three million and four million Africans into Brazil, accounting for about a fifth of all African slave labor in the Western Hemisphere. Later movements added other Europeans, primarily Germans and Italians, to the dominant classes, especially in Argentina and Brazil.

The most important stream of international migrants in the region in recent times has been from Mexico and Central America to the United States. Jobs, higher wages, and better living conditions have been the major draws. Civil wars in El Salvador and Nicaragua have turned thousands of residents into refugees. Almost 600,000 Central Americans sought political asylum in the United States between 1987 and 1996. Several hundred thousand Cubans and Haitians have fled to the United States to escape political and economic oppression. The refugees often have been contained in camps and then subsequently deported. More than 150,000 Hondurans and Nicaraguans were allowed to work

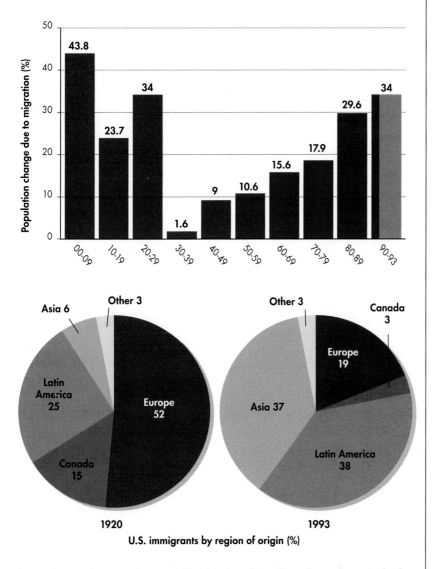

U.S. immigrants by region of origin (%)

Perceived economic opportunity, war, and immigration policies all contribute to fluctuating levels in the flow of migrants.

in the United States for up to 18 months to earn money to help impoverished people in their homeland after the devastation of Hurricane Mitch in 1998.

United States

Land, resources, political and religious tolerance, and economic opportunity all draw immigrants to the United States. Net flows into the country during the 1990s were at historical highs, with somewhat less than one million persons per year added to the U.S. population by migration. As birthrates continue to hover around replacement and as baby boomers move beyond the parenting years, the United States, like Germany, will depend primarily on migration for population growth.

The historical migrations to the United States comprise four major waves, mainly based on the home region of the immigrants:

1) Northwest Europe and Africa—1600s to the 1850s. Northwestern Europeans and Africans built the American economy during its first 200 years of foreign settlement. Africans came as slaves, a trade abolished in 1807. The Native American population, decimated by wars and European diseases, was pushed west onto marginal lands in Oklahoma, the northern Great Plains, and the desert Southwest. The population of Native Americans dwindled from about 7.5 million at the time of the first European settlements at the end of the 16th and the beginning of the 17th century to less than 250,000 by 1900. Settlers from the British Isles and Germany account for the largest number of European migrants to North America during the late 18th and early 19th centuries. Irish, as well as Germans, and Scandinavians, were pushed out of their homeland in the 1840s potato blight. Ireland's population in 1844 was 8.4 million. Today, it is close to 4 million, while about 23 million people of direct Irish descent live in the United States.

Hispanics in the United States

Hispanics, accounting for over 13 percent of the U.S. population in 2002, have become the largest minority in the United States. About two-thirds of Hispanics in the U.S. are from Mexico. In recent years the U.S. government has tried to stem migration from Latin America, primarily Mexico, with tough border controls, but areas of the Southwest have been unable to provide the infrastructure to keep up with the influx. Migrations from Latin America to the United States will continue as long as stark wage disparities exist between the regions.

More at:

http://www.census.gov/prod/2003pubs/p20-545.pdf

2) Southeast Europe—1880s to 1914. The second major wave of migrants to the United States comprised mainly southeastern Europeans. Unlike earlier migrants who came to a land scarcely settled, southeastern Europeans arrived in the country in the midst of intense industrial and agricultural growth. Slavs and southern Italians accounted for most of the immigrants to the United States from the 1880s to the start of World War I. They worked in the fields, in mines, and in factories of a growing America. The first decade of the 20th century witnessed the greatest impact of migrants on U.S. demographics. The number of immigrants was highest relative to the resident population. The 1920 census found that 58 percent of residents in big cities (100,000 plus) were of foreign birth or foreign parents. Immigrants played a role in population increase approaching that of natural increase of the resident population.

3) The 20th-century Trough—From 1914 to 1964, international migrations slowed appreciably, particularly to the United States. During this time, migration was slowed by World War I, but regained momentum nearly 30 years later with the end of World War II. The depression years of the 1930s brought immigration almost to a standstill; in fact, during the 1930s,

Near El Paso, Texas, farm workers from both sides of the Mexico-U.S. border harvest peppers. At least 13 percent of the U.S. population is Hispanic.

earlier immigrants to the United States, especially those from Southeast Europe, chose to return to their homeland. In 1929 the United States placed quotas on immigrants based on each ethnic group's share of the U.S. population in 1920. The quota system and other exclusionary laws effectively prevented the entrance into the country of anyone not of European, particularly northwestern European, origins. During the time when foreign immigration was low, the Great Migration of African Americans from the South to the North took place—a huge movement that contributed much to the redistribution of African Americans.

4) Asia and Latin America—1965 to the present. The liberalization of immigration laws in 1965 opened the door to more migrants from outside Europe. This most recent wave of immigrants from developing countries is changing the cultural fabric of the United States. Asians, such as Chinese, Filipinos, and Koreans, have become influential minorities, most notably in California and major urban centers. Texas and California are home to a recent and large influx of Hispanic immigrants. Asians and Latin Americans, predominantly Mexicans, account for 40 percent each—that is, a full 80 percent together—of all immigrants to the United States. Indications are that this trend will hold for the immediate future.

More at:

http://www.census.gov/prod/2004pubs/p20-549.pdf

IMMIGRANTS ADMITTED TO THE U.S.: TOP 10 COUNTRIES OF BIRTH—2003		
COUNTRY	NUMBER OF IMMIGRANTS	PERCENT OF IMMIGRANTS
Mexico	115,864	16.4
India	50,372	7.1
Philippines	45,397	6.4
China	40,659	5.8
El Salvador	28,296	4.0
Dominican Republic	26,205	3.7
Vietnam	22,133	3.1
Colombia	14,777	2.1
Guatemala	14,415	2.0
Russia	13,951	2.0

IMPACT OF MIGRATION ON SENDING COUNTRIES

In the 1950s and 1960s, the loss of educated citizens—referred to as the "brain drain"—primarily to Europe and the United States was commonly cited as a major problem for developing countries. High wages, modern lifestyles, and other amenities drew professionals and skilled workers to western countries, where they often had received their education. It appears, however, that the negative impact of the brain drain has been attenuated. Many of the highly educated return to their home countries. For example, the vast majority of Thai professionals choose to return home after receiving education in western countries because they prefer their homeland. The Korean government, through generous incentives, has successfully enticed many of its scientists and engineers to return home.

Countries may also benefit when their citizens emigrate. In recent years emigrant countries have earned substantial income from remittances, money sent by guest workers and immigrants to their homeland. In 1990 annual worldwide remittances amounted to $71 billion. During this decade, Egyptians working abroad sent home what amounted to almost a third of their homeland's foreign earnings. Some 4.2 million Filipinos, working overseas, sent home remittances estimated at eight billion dollars annually, almost three times the amount of foreign aid received by the Philippines.

Immigration tends to be selective of age and gender. It is often the young and educated who strike out for opportunities elsewhere.

People with upward social mobility often have little choice but to migrate to cities. For example, Bangkok and Rangoon in Thailand and Myanmar are urban demographic magnets. Likewise, selective emigration from the American Midwest has created devastating demographic effects on the small towns and rural counties there .

Urban growth strategies have been quite popular in national and regional planning schemes, but sometimes unsuccessful in practice. South Korea's investment in industry away from Seoul, primarily for national defense, has led to the growth of a vibrant network of cities of intermediate size throughout the country. Brazil's capital, Brasilia, was founded to divert population growth from coastal Rio de Janeiro and São Paulo. Abuja was dedicated as the Nigerian capital to lessen the seeming intractable urban problems of the overgrown metropolis of Lagos. Several other countries have relocated national capitals as a way of directing urban growth. India has developed two state capitals at Chandigarh and Bhubaneswar. Planned satellite cities are common around the world. Shenzhen, for example, has grown in China over the last 20 years from rice fields to a metropolis that is home to several million people.

Government Programs and Policies Affecting Population Redistribution

Many countries have found it in their best interests either to promote or to discourage migration. For three centuries, Russian czars fostered frontier settlement in Siberia. The U.S. Homestead Act of 1862 promoted settlement of the Midwest by guaranteeing permanent settlers 160 acres. On the other hand, the Great Wall of China is a classic example of a government's use of a physical barrier to discourage migrations—in this case movement of the Mongols from the north to the south. The Great Wall also had the effect of stemming the flow of Chinese northward.

In the 20th century, China, Brazil, the Philippines, Indonesia, and Malaysia implemented major rural settlement programs. Perhaps best known is Indonesia's effort to resettle residents of densely populated Java to the outer islands. Migrants were provided with free transportation, land, and provisions for their first harvest. But the program was plagued with administrative problems as well as political suspicions of a Javanese attempt to dominate the outer islands. While Indonesia relocated more than a million people, a comparable government program in Malaysia, started in 1957, was more successful, although it was accused of ignoring the needs of the rural population.

In Brazil the Transamazonian highway was intended to help settle western Amazonia. However, the more than 200,000 annual migrants to this territory have settled mostly in the southwestern rim of the Amazon basin. Furthermore, the lure of riches, such as the discovery of gold in the northern Amazon in the 1980s, provided a stronger pull for migrants to the interior than the convenience of modern highways and other government incentives. The Mahaweli Dam project in Sri Lanka, started in the 1960s, involved building up a rural region in the center of the island, intended eventually to hold one million people.

Large settlement projects on China's frontier have moved several million people to Xinjiang in the northeast and other sparsely populated regions during the last four decades. China's largest resettlement program, however, has been in the region around the Three Gorges Dam on the Yangtze River in central China. Abut 40 percent of the resettled population is rural. Farmers have been promised new agricultural land, but arable land is in short supply. About 10 percent of the people are likely to be resettled away from the region.

Some countries want to staunch the flow of people from the countryside to urban areas. Often, these attempts are unpopular and ineffective. Before the economic reforms of the late 1970s, China tried to restrict migration to the cities by putting up railroad checkpoints where officials would turn back unauthorized migrants. The government also established a population register and rationed food. During the same period, the Philippine government tried to stop the rush to the cities by withholding residence certificates, which were necessary for free schooling. South Korea recently went after the pocketbook when it applied a graduated residence tax based on city size.

Governments have razed urban slums and kicked squatters and indigents out of cities. In Bangkok and Jakarta, officials have loaded hawkers and tricycle taxi drivers onto trucks and shipped them off to the countryside. But such efforts have been in vain because governments are treating the symptoms of overcrowding instead of treating the real causes—rural poverty and urban opportunity.

One further consideration in population resettlement is the impact on people already living in the resettlement area. Tensions may arise over resources and jobs. Most countries and international agencies prefer to improve living standards in rural areas. Although these intentions are good, programs that attempt to do such a thing often fall short because of insufficient financing and organization.

CULTURAL GEOGRAPHY

TRAITS OF CULTURE

Culture is a group's way of life, including the shared system of social meanings, values, and relations that is transmitted between generations. Culture incorporates such traits or distinguishable attributes as language, religion, clothing, music, courtesy, legal systems, sports, tools, and other material and non-material components. Culture includes all learned behavior. Visually, one can distinguish between Mennonites and Sikhs as well as the landscapes in which they typically live. However, the values and beliefs that underlie the formation of these landscapes are important to understanding cultural identity, as well as conflicts between cultures. Even as a culture unites members of a community, it can also separate communities.

Cultural geography focuses on understanding the formation, transformation, and significance of spatial patterns of cultures. Cultural geographers focus on cultural meanings and values, as well as on the spread of modern cultures at the expense of traditional ones. They study landscapes, for example, not just for the artifacts created by residents, but also for the symbolic meanings that provide a sense of place to residents and that perpetuate cultural values. For example, log cabins may convey a rural, family-oriented lifestyle, whereas a football stadium conveys competitiveness.

Cultural identity shared by members of a culture forms in a

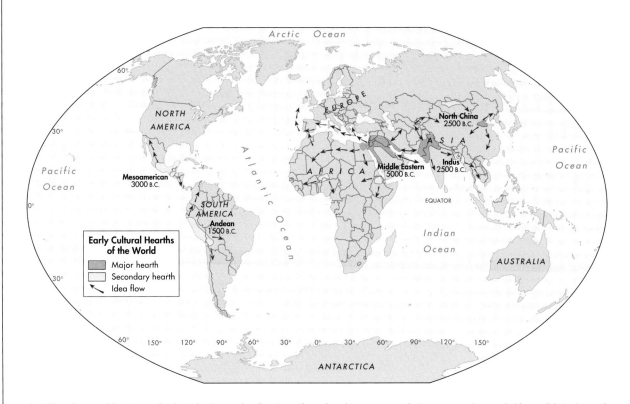

Cultural hearths reveal the origins of today's dominant cultural regions. Throughout history, as populations grew and expanded beyond their places of origin, contact with other cultures meant exchange of ideas and materials. Modern communications and transportation rapidly accelerate this process.

Mount Kilimanjaro looms behind Masai in traditional dress. Such regional cultural identities remain important despite increasing globalization.

region as members of the group interact with each other and with their shared environment; cultures change with exposure to other cultures. Spatial patterns of culture continually evolve as contact between cultures causes mixing as well as conflicts between them. Today, the process of globalization is creating more and different economic and political links between different regions and cultures, yet regional cultural identities remain extremely important: Serbians remain different from Bosnians, Hutus from Tutsis, and Canadians from Americans. The idea of a global village sharing a global culture may be a useful and appropriate metaphor for understanding the spread of Western culture around the world by mass media, but local, regional, and national cultural identities remain strong.

Three concepts—race, ethnicity, and society—are often linked to culture. Race is used to refer to biological differences between humans that are not genetically significant, but are often visible, such as skin color, eye shape, and hair color. While race is biological and therefore not cultural, it is associated with culture and ethnicity because racism—the social practice of discrimination based on appearance—can be a value of one cultural group and can change the way of life of another group, creating a separate identity for it. Americans of African ancestry, Britons of Pakistani ancestry, and Indonesians of Chinese ancestry have all faced discrimination and therefore may not share all the cultural values and traits of their larger cultures.

Ethnicity refers to a minority group with a collective self-identity within a larger host population. Within the United States, ethnic groups—often composed of recent immigrants—exist in most large cities, and some ethnic groups preserve their cultural traditions to form ethnic communities, such as Italians in Little Italy in New York and Cubans in Little Havana in Miami. While these examples of ethnic communities are culturally based, African Americans have been described as forming an ethnic group whose group identity is based on appearance. Ethnic communities often reside apart from the host population, by choice or by coercion, and may be found in ethnic islands. Jews, for example, were forced to live in segregated neighborhoods, or ghettos in 16th-century Italy, and today they may choose to live in Jewish neighborhoods in the United States.

A society is a system of interactions, including rules of conduct and class distinctions, among individuals as well as among groups. Every culture can be viewed as a society because status and conduct are based on cultural values. The status and role of women or of royalty, for example, are components of society. Societies, however, are not limited to members of one cultural group but can include members of numerous cultures and ethnic groups. American society incorporates members of Navajo, Jewish, and Amish ethnic groups as well as members of the majority Euro-American, Christian, capitalistic culture. The distinctions between society and culture are increasingly fuzzy and vary widely from place to place. Cultural geographers study both.

The societal role, status, and opportunities established for men and women create gender differences that are closely associated with culture and cultural values. Differences in level of education, infant mortality rates, income, and political power between men and women are described in terms of a gender gap, with women almost always at a disadvantage. Overall, African and Latin American countries rank lowest in terms of gender equality and Scandinavian countries rank highest. Religion can be extremely important in defining gender roles, but generalizations can be misleading. For example, women in Muslim Afghanistan have been compelled by the ruling Taliban to cease economic activities outside their homes and to wear shape-concealing clothing, while in Pakistan and Bangladesh, also Muslim countries, women have been able to hold public office, even serve as prime ministers.

Geographic aspects of gender roles are also evident within societies. Households remain the primary responsibility of women worldwide and usually serve as the geographic focus of their activities.

Culture has a strong spatial dimension because members of cultural groups often share a physical location and attribute similar meanings and values to it. Cultural regions based on these cultural groups range from a small neighborhood to much larger areas that cultures claim as homelands. Ordinarily, cultural similarities between people and ties to place are stronger at local scales where interactions among members of the culture are more frequent and face-to-face; group identity requires this type of interaction. *Cyberspace*, the computer-linked global network, may create future placeless cultures whose members share cultural values and are linked electronically rather than by geographic proximity.

Neighborhoods are typically differentiated by cultural groups, by social (class) divisions, or by both. In inner cities of developed countries, neighborhoods are often inhabited by recent immigrants and by the poor, often poor minorities; ethnic segregation remains the rule in neighborhoods. Culturally based neighborhoods, such as those of Koreans and Vietnamese in Los Angeles, can provide residents with support groups, economic opportunities, and a defense against threats by other cultural groups.

To be useful in understanding the development of geographic patterns of cultural identity and conflicts, geographers often generalize culture at different spatial scales. At a national scale, individuals often share enough culture traits so that a national culture—such as Turkish, French, or Japanese—is a valid description. In countries where cultural diversity results in far less sharing of traits—as in Sudan—the idea of a national culture would be misleading: Sudan's population is most accurately generalized into two major groups—Muslims in the north and Christians and animists in the south.

Regional cultures exist within many countries and may be in conflict with the larger, dominating culture. Corsicans in France and Kurds in Turkey are seeking autonomy. Subcultures, variations of the dominant culture, also exist, such as the Bible Belt region of the southeastern United States. Ethnic groups and subcultures can also be distinguished at more local scales. Greenwich Village in New York City, for example, is well-known as a residential area for subcultures of artists and bohemians.

Culture Hearths

Dominant cultures may expand outward from their hearths, regions where cultural traits such as religion and agriculture originate, transforming other cultures as well as themselves. The region around Mecca, for example, is the hearth of Islamic culture from which Muslims traveled to convert hundreds of millions of people to Islam around the world. Other culture hearths are found in northern China, the Indus River Valley, the Middle East, and Middle America. Cultural changes are usually the result of contact and sharing ideas, practices, and artifacts with other cultures.

CULTURE DIFFUSION
Innovations and Divergence

Cultures form as groups, become separated spatially or socially, adapt to a new physical and human environment, and develop customs, values, and ways of life that are passed on to succeeding generations.

The concept of *cultural divergence* explains the formation of new cultures as the result of groups dividing, migrating to seek resources, territory, or other advantages, and slowly changing in response to environmental stresses and new ideas. During prehistory, small tribes are believed to have emerged from Africa to wander across Eurasia and later to Australia and the Americas, encountering a variety of environments as they traveled. Over many generations changes accumulated, such as new vocabularies, beliefs, and technologies, that were handed down in each tribe until new cultures had evolved. The ancestors of modern Iranians, Greeks, Spaniards, and Swedes, for example, are believed to have emerged initially from what is now Anatolia, Turkey, around 7,000 years ago; similarities in their modern languages provide evidence that they evolved from a single language and therefore a single cultural group.

Those cultural groups with the populations, the attitudes, and the resources most conducive to generating and fostering new ideas and other innovations, especially in agriculture, established Earth's major civilizations.

In the *culture hearth* areas for civilizations, such as the Nile and Indus Valleys, new weapons, tools, and social structures increased the ability of the civilizations to further develop their own environments and to dominate neighboring cultures. Iron tools and weapons, for example, were first used perhaps 3,500 years ago in what are now Turkey and Iran and proved superior to bronze weapons then in use.

Biologist and anthropologist Jared Diamond argues in *Guns, Germs, and Steel,* published in 1997, that the development of early

Environmental Determinism

The idea that the natural environment determines cultural development was used by Plato and Aristotle to explain why Greeks were so much more developed than cultures in colder or hotter climates. This concept of environmental determinism was revived by German geographer Friedrich Rätzel (1844–1904), following Charles Darwin's *Origin of Species* (1859). It became a major theme of geography when his students, including Americans Ellen Churchill Semple (1863–1932) and Ellsworth Huntington (1876–1947), expanded on Rätzel's work. By the 1950s, however, environmental determinism was replaced by environmental possibilism as geographers, following the work of Paul Vidal de la Blache (1845-1917), recognized that the environment provides opportunities to which groups respond differently depending upon their cultures and access to technology.

dominant civilizations was dependent on the presence of certain locational and environmental factors, including climates favorable to agriculture, the presence of domesticable animals (which were disease reservoirs and, as such, helped build up the population's immunity), and regions easy to traverse.

Ancient culture hearths, based on agricultural productivity, were able to conquer and dominate surrounding areas, thereby increasing their access to labor and resources; early hearths in the Middle East relied on wheat as their food staple, and those in Middle America relied on maize (corn). Increased food supplies allowed greater numbers of people to specialize in activities other than food production, such as arts and crafts. Armies as well as innovators were supported by having individuals freed from concern over sustenance so as to focus energy on specialized activities.

Cultural Diffusion

Cultures and cultural traits spread and interact with other cultures through the process of diffusion, while barriers—both physical and cultural—prevent or channel diffusion. Historically, the spread of culture relied on direct contact between members of different cultures and on the migration of members of a culture. Modern technology has not replaced direct human contact as the means of diffusion, but mass communication and rapid transportation allow cultural diffusion to occur much faster and over far greater areas. One result of technological improvements has been to reduce spatial and temporal constraints on diffusion. Television, radio, and the Internet make new ideas more accessible to more people. A culture can spread into new regions by expansion or relocation diffusion.

Expansion diffusion occurs where contact causes cultural traits to be adopted. For example, when the Spanish converted indigenous Middle Americans to Roman Catholicism in the 16th and 17th centuries, the result was an increase in the population sharing those religious traits. Relocation diffusion is the result of migration, where the members of one culture change locations, taking their culture with them, but the total number of members does not increase. For example, settling English prisoners in Australia beginning in 1788 expanded the English cultural region, but it did not expand the number of Englishmen; over a number of generations the culture of the descendants of these expatriates changed and a distinctly Australian culture evolved.

Conquest and trade have undoubtedly been the causes of most expansion diffusion over history, and trade—as part of the global economy—is now increasingly important in spreading culture. The spread of McDonald's restaurants around the world is an example of how a concept and products developed within one culture can be accepted or imposed elsewhere. Blue jeans, Coca-Cola and Pepsi, jazz, and numerous other components of American culture have spread around the world, evidence of the role of the United States as a culture hearth.

While physical barriers such as mountains and deserts may hinder cultural diffusion, those barriers are more easily overcome with improved technology than are cultural barriers such as traditions and values. In some cases, members of a culture may intentionally bar innovations from outside in a deliberate attempt to remain unchanged by outside influences. Mennonites, for example, reside in ethnic islands, in contact with but mostly unchanged by the larger American culture. Cultural barriers may be relatively permeable where members of a culture accept some new cultural traits, but not others.

Rural cultural values often dictate how neighboring farmers cooperate. For example, several in a community may work together to contour their farmlands, to trap rainwater, and check erosion. The result can be both economically profitable and visually rewarding, as shown here.

Canadians and Americans

Americans and Canadians are often thought by Americans to share a single, American culture. However, Canadians—even Anglo-Canadians—have distinctive societal values, heroes, and traditions despite sharing language, religion, technology, and other cultural traits with the United States.

Canadians are in some ways the original anti-Americans, because they refused to support American independence in 1776. The American principles of "life, liberty, and the pursuit of happiness" are a strong contrast to the Canadian principles of "peace, order, and good government."

McDonald's hamburgers are not welcomed by members of India's mostly vegetarian culture, although blue jeans are.

As members of one culture interact with members of another, cultural traits may be adopted on both sides. When individuals enter into a new society and culture, as when migrants move to a new country, they—or at least their children—may undergo assimilation, adopting the identity—the values, meanings, and way of life—of the new culture. Some cultural traits brought by migrants may be adopted, at least in a revised version, by members of the dominant culture and result in syncretism, a fusion between the two cultures. As Mexicans migrate to the United States, Tex-Mex food and music have become incorporated into American culture.

Culture is transmitted or passed on to individuals from a number of sources, but primarily it is transmitted from family, peers, and social institutions. Governments have a vested interest in their citizens sharing cultural values in order to reduce the potential for cultural conflicts. Social institutions, such as schools, are especially important in transmitting culture. Large numbers of young, receptive individuals are introduced simultaneously to the same information and in the same language. The importance of public schooling in promoting cultural values is evident when there are disagreements within state and local school boards in the United States over the choice of textbooks, and how subjects such as the Vietnam War, evolution, and the legacy of past Presidents, are to be presented.

In some cases where members of one culture have been overwhelmed by a dominant culture, *cultural revivalism* may emerge decades or centuries afterward as those members seek to regain economic or political status. The rediscovery of former cultural identities may include celebrations and festivals or have revolutionary overtones as members of the resurgent culture seek to establish a new relationship with the larger society. In New Zealand, for example, the integration of the indigenous Maoris into the dominant culture has progressed slowly, and a Maori revival has taken place during the 20th century.

Cultural Regions and Landscapes

Cultural region is a general term for areas where some portion of the population shares some degree of cultural identity. Cultural regions based on the Basque language or on blowgun usage—found in Brazil, New Guinea, and the Caroline Islands—are valid means of locating specific cultures, because the Basque language and the blowgun are both limited to small, traditional societies sharing a common history, ancestry, and culture. In other cases a single criterion, such as the use of the English language, is useful only at the broadest scale because of the diversity of peoples—including Scots, Nigerians, Jamaicans, and Indians—for whom English is the primary language.

The members of most cultures tend to be concentrated within a *core area* and have a decreasing presence in peripheral areas. Peripheral areas may be divided into a surrounding cultural domain where the cultural identity is still prominent and a cultural sphere farther away, where it is still influential. In a 1965 study of the spatial extent of the Mormon cultural region, geographer Donald Meinig examined population characteristics (for example, the percentage that are Mormon) and placed the regional core around Salt Lake City, the domain over most of Utah and southeastern Idaho, and the sphere south to Mexico and north to eastern Oregon.

The spatial scale of cultural regions can be expanded to the global perspective of realms, based on generalized perceptions of cultural similarities rather than on the spatial distribution of specific cultural traits. The usefulness of realms in understanding cultural variations is limited, but they can be used to show historical and societal linkages among populations.

Members of cultures interact with and transform their environments through agriculture, land demarcation, construction, and transport systems to create cultural landscapes. Housing styles, street patterns, crop choices, and land ownership patterns are all components of the cultural landscape and reflections of cultural values. In the United States, for example, superhighways, front lawns, skyscrapers, and football stadiums are part of and reinforce the dominant cultural preferences, but regional differences in culture are also visible. Landscapes not only reveal current cultural traits and values, such as choice of sport, religion, and transport, they also provide a partial history of previous cultural preferences in the area. Larger buildings, often with religious significance, may last centuries as evidence of an earlier culture. For example, Angkor Wat and other Hindu temples in Buddhist Cambodia and the Alhambra and other Muslim buildings in Catholic Spain attest to religious changes.

Land use patterns also reflect past cultural values. In the United States, the 1800s division and sale of land in the West by 40-acre and 160-acre lots—township and range—for farming remains evident in today's grid patterns of county roads. On the other hand, contrasting land use patterns can be found in the East where metes-and-bounds surveying, i.e., following streams and other natural features, was used. In Quebec the long-lots system, which established narrow fields, all with access to navigable rivers, was used.

CULTURAL IDENTITIES

Modern-day cultural identities are the result of centuries of cultural divergence and diffusion. Two components of culture—language and religion—are particularly important in understanding cultural identities and appreciating the processes that have led to existing cultural patterns.

Language

Language is the key means by which culture is transmitted and cultural identity gained. Differences in language mean differences in culture and are a potential source of misunderstanding and conflict between peoples.

The geographic study of lan-

Esperanto

One solution to the problems of communicating between cultures is to create an entirely new language.

Esperanto, one such created language, was developed in 1887 by L. L. Zamenhof (1859–1917), a Polish philologist, from a number of European languages.

Despite being endorsed for use by the League of Nations, Esperanto failed to gain wide acceptance and now has only about two million speakers.

More at:
http://www.esperanto.net/info/index_en.html

THE TEN LEADING LANGUAGES IN NUMBERS OF NATIVE SPEAKERS			
Language	**Family**	**Speakers (in millions)**	**Main Areas Where Spoken**
Han Chinese (Mandarin)	Sino-Tibetan	874	China, Taiwan, Singapore
Hindi	Indo-European	366	Northern India
Spanish	Indo-European	358	Spain, Latin America, southwestern United States
English	Indo-European	341	British Isles, Anglo-America, Australia, New Zealand, South Africa, former British colonies in tropical Asia and Africa, Philippines
Bengali	Indo-European	207	Bangladesh, eastern India
Arabic	Afro-Asiatic	206	Middle East, North Africa
Portuguese	Indo-European	176	Portugal, Brazil, southern Africa
Russian	Indo-European	167	Russia, Kazakhstan, parts of Ukraine and other former Soviet republics
Japanese	Nipponese	125	Japan
German	Indo-European	100	Germany, Austria, Switzerland, Luxembourg, eastern France, northern Italy

guage focuses on current distributions of language and on the evolution of languages and language patterns over time and space. Studies focus especially on source areas and paths of diffusion.

SIGNIFICANCE OF LANGUAGE

Language is a system of spoken communication using sounds to transmit meanings. Language is almost always coupled with a written system, and language is a critical component of cultural identity. Translation between languages can be extremely difficult. One reason is that meanings are often contained in phrases rather than single words. For example, literal translation of the words in the phrase "out in left field" to another language would miss the connotative meaning entirely. Values, experiences, and meanings of a cultural group are contained in its language. Finding similar words for similar meanings in different languages is one key to determining if they evolved from a single root language. (See The World's Language Families, page 233.)

Estimates of the number of languages spoken today range from about 3,000 up to 6,500 because of differences in differentiating distinct languages from the dialects of one single language. *Dialects* are versions of one language with different vocabularies, pronunciations, accents, and sometimes syntaxes; most dialects are regionally based, but dialects can also be class- or gender-based. Mandarin Chinese has the most native speakers, those for whom it is their first language, but is not the most commonly spoken language. English is the primary or second language of choice for more people because of the legacy of the British Empire and American dominance in economics, military power, and computers.

More at:

http://www.ethnologue.com/web.asp

During the last several centuries, hundreds of languages and the knowledge they contain have become extinct as the last speakers have died. Younger generations learned other languages so they could participate in the larger society and consequently adopted different cultural identities. The loss of these languages means the loss of meanings and of experiences that were contained in their own individual vocabularies.

Dialects evolve from a single language as populations separate and the frequency and intensity of interactions decline. Over time, if the vocabulary, pronunciations, and accents of the dialects become sufficiently different, then they evolve into related but different languages, and the speakers into members of related but different cultures. British English and American English have different accents and use different vocabularies—British say "lift" instead of "elevator," for example. But within both the United Kingdom and the U. S. regional dialects exist. East and West Midland are two dialects in the United Kingdom, and Southern and Bronx dialects are found in the United States.

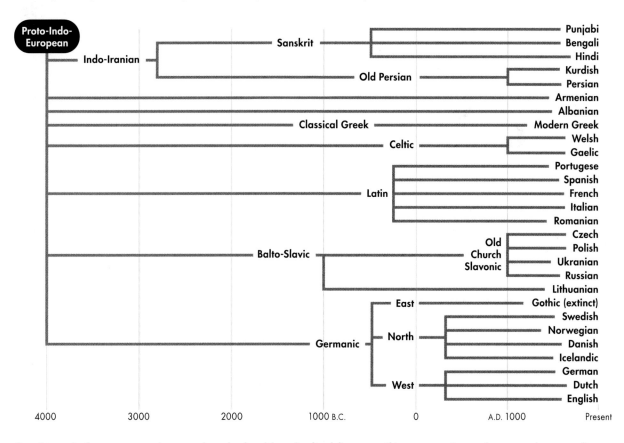

Over thousands of years numerous languages have developed through cultural divergence. This process continues today as some languages disappear—when the cultural group diminishes or is absorbed by another group—and others grow.

Contact between speakers of different languages—or in some cases between different dialects—creates difficulties in communication and generates a need for a mutually intelligible language. A simplified version of one language, called a *pidgin* language, may be used during exchanges between different cultures, but not among members of the same culture. If a pidgin is used enough to become the primary language of a population, then it has become a *creole*. Swahili, which evolved from Arabic and Bantu languages, in eastern Africa, and Bazaar Malay in Malaysia are creoles. *Lingua francas*—literally "Frankish languages"—are existing languages adopted by members of different cultures, and are an alternative to pidgins and creoles. Several lingua francas have been used in different regions and at different times. Latin, Arabic, and Hindi were or are lingua francas. Lingala is now used in the western Congo, and English is the most commonly used lingua franca around the world today.

When a country is formed from a mixture of cultural groups with different languages, a *polyglot state* is created and an official language may be needed. In a number of former British colonies, such as Fiji, Ghana, and India, English has become an official language to enable communication among the diverse groups and to avoid recognizing one group's language above others. French, Dutch, and Portuguese are used as official languages in many of the former colonies of those countries.

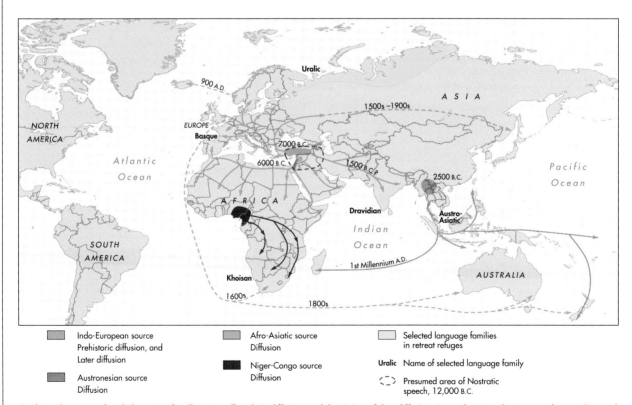

Indo-European source
Prehistoric diffusion, and
Later diffusion

Austronesian source
Diffusion

Afro-Asiatic source
Diffusion

Niger-Congo source
Diffusion

Selected language families
in retreat refuges

Uralic Name of selected language family

⟨‾⟩ Presumed area of Nostratic
speech, 12,000 B.C.

Studying the origin of early language families, as well as their diffusion and the timing of that diffusion, is another way that geographers understand the movement of people and cultures.

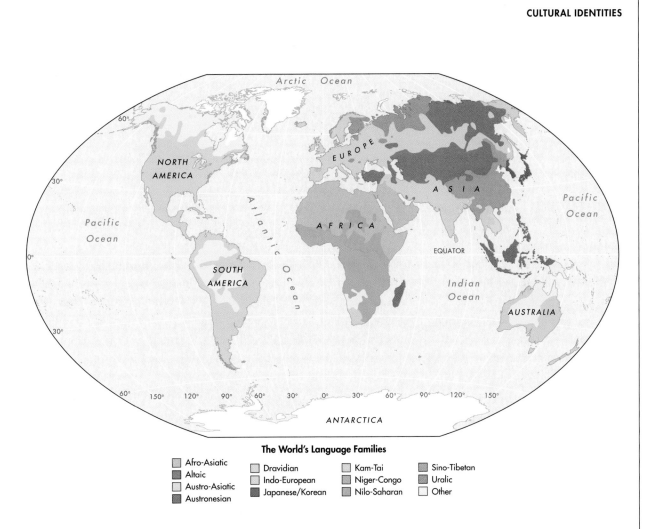

The World's Language Families

- Afro-Asiatic
- Altaic
- Austro-Asiatic
- Austronesian
- Dravidian
- Indo-European
- Japanese/Korean
- Kam-Tai
- Niger-Congo
- Nilo-Saharan
- Sino-Tibetan
- Uralic
- Other

It is estimated that there are up to 6,500 languages spoken in the world today. Mandarin Chinese, Hindi, Spanish, English, Bengali, Arabic, Portuguese, Russian, Japanese, German, French, and Malay-Indonesian are thought to be the 12 most widely spoken.

Language is also evident in the cultural landscape as place-names, and the choice of language on signs is another visible symbol of culture on the landscape. *Toponymy,* the study of place-names, can reveal the history of a region and the values of the people. Throughout the United States places have been named by immigrants in honor of their native lands, such as New Rochelle, New York (France) and New Bern, North Carolina (Switzerland). Pakistan's capital of Islamabad, translated as "the place of Islam," is clear evidence of that country's Muslim strong religious values.

LANGUAGE FAMILIES

Different languages evolved over perhaps the last 200,000 years as members of cultural groups separated and *language divergence* occurred. Languages that diverged from a single ancestral language are related and can be grouped into *language families.* For example, English, Russian, Hindi, and Greek are all members of the Indo-European language family. Different languages diverge at different times, so some languages are more closely

Spread of Indo-European Languages

Two major theories have been used to explain the spread of Indo-European languages from their core area: the conquest theory and the agriculture theory. According to the conquest theory, Indo-European evolved among pastoral nomads in the steppes of what is now Ukraine and spread west into Europe as the nomads conquered neighboring populations. The agriculture theory argues that Indo-European started among farmers in the Caucasus Mountain region who migrated slowly outward to Iran and India as well as to Russia and Europe. The linguistic, genetic, and geographic data available are not yet conclusive as to which, if either, theory is more correct.

related than others. For example, English and German, both of which evolved from an early Germanic language, are more closely related to each other than to Spanish, Italian, and other Romance languages, which evolved from Latin.

The emergence of new languages was probably never a smooth transition but involved a number of competing influences. The English language can be traced to Germanic tribes—especially the Angles, Saxons, and Jutes—who invaded Britain some 1,500 years ago, but it has also been strongly influenced by Latin, which was brought along with Christianity about 1,400 years ago, and by French, brought by William the Conqueror in 1066. The Vikings also influenced the development of the language during their invasions of Britain from the 9th to the 11th centuries. The English of today is the result of 1,500 years of linguistic development and of cultural evolution.

The current spatial distribution of language families reveals the global spread of Indo-European—the most widespread of all language families—and a set of large and small regions dominated by about 20 other language families. Minor language families also exist but do not appear on global scale maps. Within Europe, for example, the Basque language is found in a small region from Biarritz, France to Bilbao, Spain, and is unrelated to any other known language, suggesting long isolation of this cultural group.

The ability to trace languages back to earlier forms enables geographers and linguists also to trace populations back to earlier locations. Linking modern populations to an ancestral group provides evidence of earlier migrations and may be useful in suggesting genetic similarities. Natives of Madagascar, for example, speak an Austronesian language that diffused westward from the Pacific and links them with Filipino, Vietnamese, and some Pacific Island populations. In recent centuries, however, the diffusion of languages has become increasingly associated with global economic and political influences rather than with the relocation of cultural groups.

Language Conflicts

The importance of language as a cultural identifier involves language issues in numerous political conflicts, ranging from regional autonomy to the selection of languages in which school classes should be taught. As a symbol of their culture, members of an ethnic group may seek to protect their language from being overwhelmed by the language of the dominant society. Dominant societies, on the other hand, may try to reduce the use and importance of minority languages in order to acculturate members of minority groups and strengthen the larger, national identity. Thus, before the breakup of the Soviet Union learning the Russian language was compulsory in all Soviet republics.

In the United States, which has as yet no official language, numerous language-based political issues have arisen in recent years. The choice of which language(s) to use in grade schools has become contentious as the number of Spanish-speaking residents has increased by about a half million per year for the last two decades, and as some schools have defined Black English Vernacular, referred to as Ebonics, as a separate language necessitating bilingual education. Several groups, such as English First and English Only oppose the use of other languages and seek to have national legislation passed to establish English as the sole official language of the United States.
More at:

http://www.us-english.org/inc/

In a number of countries, languages that had been suppressed are now being revived for future generations. In Wales ("Cymru" in Welsh) and Ireland ("Eire" in Gaelic) the number of speakers of Welsh and Irish Gaelic is growing because of compulsory language education. In the 1990s a number of cities in western India were renamed in the regional Gujarati language in response to a Hindi cultural revival; Bombay, for example, is now Mumbai. The struggle for greater adoption of French within Quebec resulted in the passage of Bill 101 in 1977, requiring education of most immigrant children in French, and the use of French in workplaces, on signs, and in public places.
More at:

http://www.uni.ca/history.html

Language is critical in perpetuating a culture and its values, symbolism, and meanings. Many key values of cultural identity, however, derive from religious beliefs held by most members of that culture. Cultural identities thus nearly always include both language and religion.

GEOGRAPHY OF RELIGION

Religion, like language, can be a defining component of cultural identity and one that may influence choice of clothing, food, tools, and occupation. Religion may also be more or less replaced by secularism in some modern societies.

As a social system based on a concept of the divine and involving beliefs, values, and behaviors, religion organizes many aspects of a culture. Shared beliefs and values establish strong group identification, and, because some groups can also be linked to specific areas, geographic patterns of different religions result. Conflicts between religious groups may also emerge, some because of differing religious beliefs and others due to economic or political relations.

The global distribution of religious groups reflects past migrations of peoples and the diffusion of their religions. Religions originate in localized areas and expand outward. Islam, for example, started in western Saudi Arabia around A.D. 610, but within three centuries was dominant in Spain,

Disappearing Languages

Globalization in communication and economics is increasing the rate at which world languages are disappearing. In 1995 linguistic experts predicted that as many as half of the 3,000 to 6,500 languages would disappear by 2100 because children are learning other languages—such as English, Spanish, and Arabic—to communicate and interact with larger populations than their ethnic associations. Most of these losses will come in tropical regions where cultural groups have lost their isolation. California, however, will also lose languages as the remaining speakers of perhaps 50 Native-American languages die.
More at:
http://www.globalpolicy.org/globaliz/cultural/2002/0425fast.htm

Mecca is the birthplace of the Prophet Muhammad, founder of Islam, and therefore the most sacred of Islam's holy cities. In the courtyard of the city's great mosque stands the Kaaba, a windowless cube-shaped structure. The faithful turn toward Mecca each day to pray.

Turkmenistan, and northern Sudan as Arabs carried this new religion outward. More recently, following the collapse of communist governments in Eastern Europe and Russia, representatives from numerous Christian denominations have traveled as missionaries to this region to try to reawaken religious systems that had been suppressed under communism.

Spatial patterns of religion are also apparent in the location of sacred spaces, sites associated with holy or divine events. Sacred spaces may be either features in the natural environment or structures designed to honor a religion or deity.

Rivers and mountains are commonly designated as sacred spaces by religions: the Jordan River is sacred to Christians and the Ganges River is sacred to Hindus; Mount Fuji is holy to Shintoists and Uluru (Ayers Rock) is sacred to Australian animists. Structures such as the Western Wall, or Wailing Wall, in Jerusalem, the Dhamek pagoda in Sarnath, India, and the Kaaba in Mecca, Saudi Arabia, are sacred to Jews, Buddhists, and Muslims, respectively.

Pilgrimages are often made by members of a religion to visit its sacred spaces. All Muslims, for example, are expected to make one pilgrimage to Mecca during their life if possible, and so each year more than a million Muslim pilgrims make this journey. Many Hindus travel to Varanasi (Banaras), India, and Shintoists to Ise, Japan, as part of their religious observances.

Minor sacred spaces also inspire pilgrimages; hundreds of shrines are located in Western Europe—primarily in Roman Catholic regions—some of which are visited by hundreds of thousands of pilgrims each year.

Religion changes landscapes through the construction of religious buildings and patterns of land use. Mosques with their towering minarets are visibly different from churches and contribute to the difference between Muslim and Christian landscapes. Bo trees, symbolic of the tree under which the Buddha reached enlightenment, are found throughout Buddhist regions. The absence of particular types of buildings can also be part of a religious landscape: Muslim communities lack both taverns and hog farms because alcohol and pork are taboo in Islam. Clothing, such as the concealing chador worn by Muslim women, and headwear, such as the turbans worn by Sikh men, are personal signs of those religions.

Types and Diffusion of Religion

Religions, and therefore religious patterns, can be differentiated in a number of ways. Monotheistic religions worship one deity; Judaism, Christianity, and Islam are all monotheistic and are related in their worship of the same God. Polytheistic religions, including many animist religions, worship multiple deities, but this may be a misleading description of some religious beliefs. Hinduism is often described as polytheistic. Vishnu, Siva, and Brahma are all major deities, but Hindus also believe in one supreme consciousness of which these three are simply different aspects.

One important distinction among religions that affects their spatial distributions and the potential for conflicts between them is the role that proselytizing plays in the religion. Universalizing or proselytizing religions actively seek converts, and the social structure of such religions often includes missionary activity. Conflicts may arise when other religious groups object to this activity.

Christianity, Islam, and Buddhism are the three major universalizing religions and together have more than three billion followers. Christianity claims the largest number of followers worldwide at nearly two billion, of whom some one billion are Roman Catholic.

The numbers of followers are affected by how each religion defines membership: Roman Catholicism counts all who are baptized in the church, and some Protestant churches count only those who are baptized as adults.

Membership in non-proselytizing religions is limited mostly to individuals raised within the specific cultural group practicing that religion, because conversion of outsiders is not actively sought. Larger religions of this kind, those with millions of followers, are labeled cultural, regional, or ethnic religions

Feng Shui

Some religions view the physical environment as containing divine energy or spirits that must be respected. Traditional Chinese and Korean religious beliefs include the principle of *feng shui*, whereby settlements, buildings, and even graves must be located and oriented in harmony with the surrounding environment. As Chinese have migrated to North America, their belief in feng shui has affected location and placement of factories, restaurants, and residences.

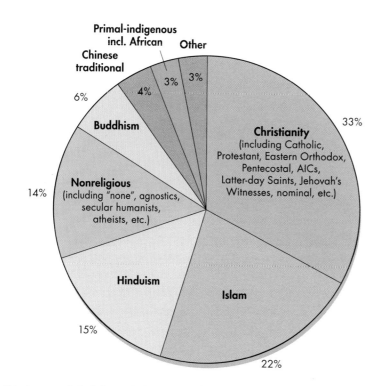

Primal-indigenous incl. African — 3%

Other — 3%

Chinese traditional — 6%

Buddhism — 4%

Nonreligious (including "none", agnostics, secular humanists, atheists, etc.) — 14%

Hinduism — 15%

Islam — 22%

Christianity (including Catholic, Protestant, Eastern Orthodox, Pentecostal, AICs, Latter-day Saints, Jehovah's Witnesses, nominal, etc.) — 33%

This chart reveals the balance of religions around the globe. Christianity, Islam, and Hinduism have the largest presence. Distribution of a religion is largely affected by how active a group is in seeking converts. The social structure of widely spread religions often includes missionary activity.

because they are associated with a single distinct cultural group, such as Hinduism, Sikhism, and Judaism. Religions associated with small and often isolated groups are termed traditional or local religions, such as Totemism and Druidism. Many traditional religions are animist, with followers believing in the presence of divine forces throughout nature; Central Africans, Native Americans, and residents of Oceania are often animists.

Not all religions are easily placed in one category, nor do all religions remain in one category through his-tory. Syncretic religions meld beliefs from different religions; voodoo in Haiti and the Caribbean includes traditional beliefs from African and American local religions with beliefs from Roman Catholicism.

Hinduism, though now a cultural religion that is found almost exclusively among Indians, was a proselytizing religion thousands of years ago as it spread eastward from the upper Indus Valley and converted local populations.

The diffusion or spread of religion occurs outward from the hearth area where the religion orig-inated. All religions spread through relocation diffusion when followers relocate and bring their religion with them. British settlers in North America, South Africa, and Australia brought Christianity to those areas.

Universalizing religions also undergo expansion diffusion as additional populations are converted to the religion, adding to or expanding the total number of followers. The Spanish converted natives of Mexico, the Philippines, and other Spanish colonies to Roman Catholicism.

As religions moved outward from their hearths, they were occasionally replaced in those hearths. Christianity started in what is now Israel, but has relatively few followers in the Middle East. Buddhism started in northeastern India near Nepal, but it is now nearly absent in India because Hindus regained their influence in this region. Islam and Judaism are still found in their hearth areas, but while Islam has remained dominant in the Arabian Peninsula, Judaism and Jews returned to Israel in the 20th century after being absent for nearly 2,000 years.

Trade and war have been important historically in the expansion of religions. Arab traders, who were Muslim, brought Islam to the eastern coast of Africa, to Indonesia, and to the southern Philippines by

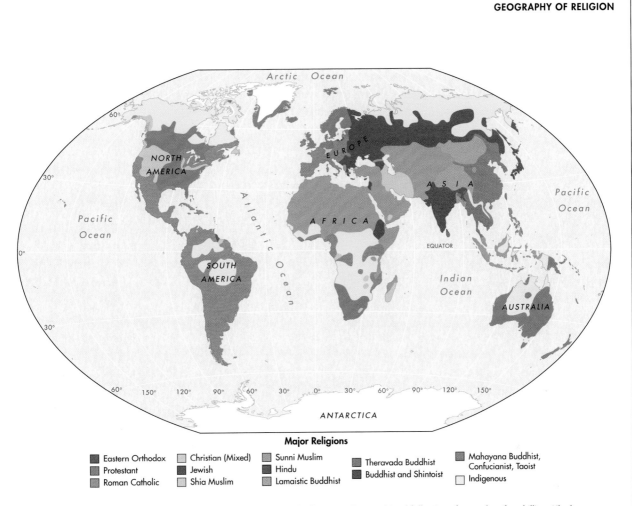

Major Religions

- Eastern Orthodox
- Protestant
- Roman Catholic
- Christian (Mixed)
- Jewish
- Shia Muslim
- Sunni Muslim
- Hindu
- Lamaistic Buddhist
- Theravada Buddhist
- Buddhist and Shintoist
- Mahayana Buddhist, Confucianist, Taoist
- Indigenous

Christianity, Islam, and Hinduism are the world's most widely practiced religions with a combined following of more than four billion. Like languages and regional dialects, major religions and branches can be cultural indicators of the origin and history of an area's population.

1200, and Muslim armies captured and converted populations in northern Africa, the Iberian Peninsula, and parts of the Balkan Peninsula.

European colonialism involved both trade and conquest and has resulted in Christianity being the dominant religion over the largest area of Earth.

The spatial pattern of religions is made more complex by the divisions that exist within every major religion. The largest divisions are branches, such as the division of Christianity into Roman Catholic, Eastern Orthodox, and Protestant faiths, and are often based on major differences among followers over who should head the religion and interpret the faith.

As followers of the different branches migrate, different spatial patterns develop. For example, Roman Catholicism rather than Protestantism diffused to Latin America because Spain and Portugal, both Roman Catholic countries, colonized this area.

Branches can be further subdivided into denominations, such as the Baptist, Congregationalist, and Lutheran denominations of the Protestant branch, which may also have different spatial patterns. Lutherans are concentrated in the upper Great Plains region, where immigrants from Norway and Sweden, two Lutheran countries, settled.

Smaller divisions are often termed sects, and if dominated by a single personality may be referred to as cults.

Distributions of Religions

Although numerous religions can be found within any region, including Jews in Iraq and Buddhists in Peru, generalized global patterns of the major religions and their branches exist. Each pattern reflects centuries of diffusion and will probably continue to change in centuries to come. The current diffusion of Protestantism into traditionally Roman Catholic Latin America is one example of ongoing change.

Two major source areas account for all major religions—the Middle East and South Asia. The three major monotheistic religions—Judaism, Christianity, and Islam—started in the Middle East and, despite their differences, share many key beliefs as well as some prophets. Hinduism

evolved into a major religion in South Asia, after being brought there millennia ago from Central Asia, and Buddhism originated as an offshoot of Hinduism.

MIDDLE EASTERN RELIGIONS

Judaism, the oldest of the three major monotheistic religions originating in the Middle East, is concentrated in Israel, but large Jewish populations exist in many major European and American cities. The religious hearth of Judaism is centered in Jerusalem and includes the area Jews consider their Promised Land from God. When the Jews were driven from this area by the Romans in the Diaspora of the first century, other groups inhabited the area. For centuries, until the 1900s, the region was called Palestine and was populated by Muslims (Palestinians). In the 1880s, the Zionist movement started whereby Jews began returning in steadily increasing numbers to their ancestral home, eventually gaining control

over the area when Israel was recognized in 1948. The Palestinian issue was created by this return of the Jews and the displacement of the Palestinians; there are currently about 1.4 million Palestinians in the Gaza Strip, 2.4 million in the West Bank, 2.8 million in Jordan, and 850,000 in Lebanon and Syria.

Christianity evolved from Judaism with the teachings of Jesus, a Jew born in Bethlehem, raised in Nazareth, and crucified in Jerusalem—three locations in Israel and the West Bank. The spread of Christianity to Europe was due to the proselytizing of the Apostle Paul and other missionaries and later to the influence of the Roman Empire after the Emperor Constantine converted to Christianity in 313. The global diffusion of this religion was due primarily to the establishment of empires by European countries.

Each of the three major branches of Christianity—Roman Catholicism, Eastern Orthodox, and

The Pentecostal Movement

A new Protestant sect based on the religious experience of "spirit baptism" or being "born again" started in the United States in the early 20th century.

In 1901 students at Bethel Bible College in Topeka, Kansas,

began speaking in tongues. This action was attributed to Christ's original 12 disciples during Pentecost, the seventh day after Easter, according to the Christian calendar. Five years after the 1901 event, a mission was established

on Azusa Street in Los Angeles.

As the evangelical movement of Pentecostalism began to receive greater exposure, it expanded rapidly. Now the church counts more than a hundred million members worldwide.

Within the city limits of Rome is Vatican City, the smallest independent state in the world. The most important edifice is St. Peter's Basilica, world center of Roman Catholic worship.

Protestantism—tends to dominate certain regions within the global Christian realm. Roman Catholicism dominates Southern Europe, where Roman influence was strongest, and the countries in Latin America colonized by Spain and Portugal. The Eastern Orthodox Church, separated from Roman Catholicism in the Great Schism of 1054, dominates Eastern Europe and the former Russian Empire of the tsars. Protestantism, which originated in the Protestant movement of the 16th century that separated from Roman Catholicism during the Reformation, dominates Northern European countries and their former colonies.

In recent polls, about one-quarter billion North Americans define themselves as Christians; numerous other religions are also present, including more than four million Muslims and nearly six million Jews. Regional concentrations of different Christian branches and denominations have developed: In the U.S., Baptists dominate the Southeast. Roman Catholics dominate New England and areas bordering Mexico, which is nearly 90 percent Roman Catholic, Luther-

A crowd gathers before the Western Wall in the Old City, Jerusalem, to celebrate the Jewish festival of Shauvot. Jerusalem's greatest concentration of religious and historical sites is in the Old City, which is surrounded by modern Jerusalem.

ans the upper Great Plains, and Latter Day Saints—Mormons—Utah and counties in bordering states. Migrations have changed these distributions over the years; the Latter Day Saints originated in New York, moved to Ohio, Missouri, and Illinois before relocating to Utah to escape religious persecution and find their promised land.

Islam, the third major religion originating in the Middle East, started in Mecca in western Saudi Arabia with the revelations of Mohammed around 610. Mecca, Medina, where Mohammed found sanctuary after being driven from Mecca in 622, and Jerusalem are the three holiest cities of Islam. A proselytizing religion, Islam was spread by Muslim armies outward from Saudi Arabia as part of a jihad, holy war, to secure the peace of Islam. At one point, Islam domi-

nated much of India as well as Spain, Portugal, and Yugoslavia. Islam is currently the fastest growing major religion, due more to contact conversion by individuals rather than from organized missionary efforts.

Islam has two major branches, Sunni and Shiite, which separated during the seventh century in a dispute over leadership of the religion. Sunnis compose about 84 percent of all Muslims and dominate the Arabian Peninsula, northern Africa, and most Muslim countries around the world. Shiites are the majority of the population in Iran, which is about 90 percent Shiite, and in Iraq, which is some 60 to 65 percent Shiite.

SOUTH ASIAN RELIGIONS

Hinduism, the oldest major religion, originated in Central Asia before being brought by Aryan tribes migrating to the Indus Valley perhaps 3,500 years ago. Although now a cultural religion, Hinduism diffused eastward toward the Ganges Valley and to much of Southeast Asia, including Malaysia and Indonesia. Bali in Indonesia is still predominantly Hindu and Angkor Wat is one of several ancient Hindu temples in what is now Cambodia. Regional variations in Hinduism are not as spatially discrete as in Islam and Christianity, but they do exist; for example, the deities Siva and Shakti are venerated more in the north of India and Vishnu more in the west. Hinduism has spread over the last few centuries, primarily as Indians have migrated to other countries, and sizable Hindu populations are found in Fiji, northern Sri Lanka, and South Africa. About 80 percent of India's one billion people are Hindu, which is why Hinduism is the third largest religion despite no longer being universalizing.

Indian society has been very much shaped by Hinduism, especially its caste system, which defines social and religious status. Hindus believe in reincarnation in which an individual is reborn into a caste based on the karma or totality of actions by the individual during his or her prior life. There are four major castes—priests, warriors, tradesmen, and laborers—and interactions between the castes, such as marriage, are religiously defined. In recent years, riots have erupted in protest against laws enforcing the rights of people without caste, called pariahs or untouchables, to obtain higher education, hold government positions, and seek other opportunities.

Buddhism is an offshoot of Hinduism based on the teachings of Siddhartha Gautama, born a member of the Hindu priest caste about 563 B.C. Buddhism started near the border of what are now India and Nepal where Siddhartha was born, reached enlightenment, and died. Lumbini, now in Nepal, and Bodh Gaya and Kusinagara in India are sacred places in this religious hearth based on these three events.

Missionaries carried Buddhism to eastern Asia following the conversion of Asoka, Emperor of India, around 261 B.C. Different branches developed, and Buddhism also merged with other religions. The number of Buddhists worldwide is consequently uncertain because individuals often practice Buddhism along with cultural Chinese and Japanese religions. Theravada Buddhism, the more conservative branch of the religion, diverged from Mahayana Buddhism over different interpretations of Buddha's teachings. Theravada now dominates Southeast Asia. Mahayana is most common in China, in Korea, and in Japan. Lamaism, which also evolved from Mahayana, dominates in Tibet and Mongolia.

Traditional East Asian Religions

Three East Asian religions are closely associated with Buddhist practices. Confucianism and Taoism, two philosophical religions that arose in the sixth century, were dominant in China until the 20th century. Their influences have declined markedly since the country became communist in 1949. Shinto is an ethnic Japanese religion in which ancestors are venerated, but it lost influence following World War II and has become less popular in recent decades. Mahayana Buddhism merged with Taoism and Confucianism in China and with Shinto in Japan, where a form of Buddhism referred to as Zen emerged.

Sikhism, a second offshoot of Hinduism, has relatively few followers—around 22 million worldwide—but it is important because of its conflicts with Hinduism. Founded in the 1500s and based on the teachings of Guru Nanak, the religious hearth of Sikhism is in the Punjab region on the border of India and Pakistan.   Sikhism attempts to reconcile Hinduism and Islam. Currently, many Sikhs are seeking an independent homeland—Khalistan—separate from Hindu-dominated India and have been involved in a sometimes violent campaign to secede.

More at:

http://www.adherents.com

Religious Conflicts

Members of different religions coexist without problems in many areas of the world, but in other areas religious conflicts arise where different religious communities come into contact. Many conflicts between members of different religions, such as that between Buddhist Sinhalese and Hindu Tamils in Sri Lanka, are struggles over economic opportunities and political justice between groups identified by their religions. Other conflicts are primarily over beliefs and values. In 1989 more than 400 people died in Ayodhya, India, when Muslims and Hindus fought over whether a mosque or a Hindu shrine should be located on a site sacred to both religions. In 1992 a Hindu mob destroyed the mosque, causing numerous riots in India. Disputes between religious and secular groups are also common; Islamic fundamentalists are currently battling the government of Egypt on the grounds that it is not adequately following the principles of Islam.

Religious conflicts usually occur when boundaries between religious regions do not match political boundaries, thus allowing one religious group to dominate another economically and politically. Interfaith boundaries are those between different religions. Within Chad, Ghana, and many other countries in the Sahel region of northern Africa, interfaith boundaries separate Muslims in the northern areas from animists and Christians in the southern areas. Muslims often are the national majority and dominate the governments. When the Sudanese government implemented Sharia, the Islamic rule of law, in the 1980s, southern Sudanese, who are Christians and animists, protested and the ensuing conflict has since claimed thousands of lives.

The ongoing dispute between Jews and Muslim Palestinians in Israel has clear economic and political components, but to a large degree is based on different beliefs about who should possess this territory. Muslims claim the land based on centuries of possession, and Jews claim that God promised it to them. Despite this ongoing conflict, Jerusalem remains a city sacred to both Jews and Muslims—as well as to Christians—and mosques, churches, and temples are located close to each other in the Old Town.

Territorial claims by religious groups are the basis for numerous conflicts around the world. Fighting between Sikhs and Hindus in the Indian state of Punjab is over political control of territory as is the fighting between Muslim Kosovars and Christian Serbians in Yugoslavia. Ethnic cleansing, the removal of one ethnic—often religious—group from an area, was used by the Serbians against Kosovars in 1999, much as the Serbians and Croatians removed Muslim Bosnians from territory during the early 1990s.

Political control by one religious group often leads to conflict when minority religious groups seek power. In theocracies, where the leader of the dominant religion is also the political leader, members of minority religions may be persecuted or reject the imposition of laws from another religion. When the Ayatollah Khomeini came to political and religious power in Iran in 1979, Iranians of the Baha'i reli-

A serene icon for pilgrims and tourists, a Buddha's head presides over a cliff in Lu Shan, China. The Min River roils at the statue's base 233 feet below. By the time Buddhism arrived in China, during the first century A.D., two other organized religions and philosophies were already woven into China's rich indigenous culture—Confucianism and Daoism. By 350 A.D., Buddhist texts had been translated, and the religion had taken hold.

gion faced tremendous discrimination. Baha'i originated in Iran (Persia) in 1844 and teaches that all religions are in ultimate agreement.

In some countries, including Norway, Ireland, and Peru, there may be an established church or an official state religion, but religious and government authorities remain separate. In these countries political discrimination against minority religions can still occur. Judaism is the state religion of Israel, and Muslims who live in Israel complain about limitations on voting and other civil rights.

In a number of countries fundamentalists, those who believe in literal interpretation of religious writings, have come into political conflict with the less fundamentalist and more secular members of the larger society. In Algeria, a civil war has been waged for years between fundamentalist Muslims and the government. Not all disputes between fundamentalist and moderate members of a religion involve civil war. In the United States, protests, which sometimes become violent, are the dominant form of dispute between fundamentalist Christians and other groups of the population over issues such as women's rights and abortion.

Intrafaith Conflict

Many current religious disputes have long histories and consequently cannot be resolved easily or quickly. Some conflicts, called intrafaith conflicts, occur between branches of the same religion, such as the conflict between branches of Christianity in Ireland. Centuries of economic and political dominance by Protestants, mostly Scottish and English immigrants, created the conflict between Roman Catholics and Protestants in Northern Ireland. The island on which Ireland (Eire) and Northern Ireland (Ulster) coexist was divided in 1922 after counties voted individually on whether to become self-governing or to remain fully part of the United Kingdom. Irish Catholics were the majority in most counties and voted for dominion status, but the majority populations in the northern counties were Protestant and voted to stay in the United Kingdom. Establishing a lasting peace in Ulster is complicated by the continuing desire of most Roman Catholics to join Ireland, now a sovereign country, and the desire of most Protestants to remain part of the United Kingdom; armed groups still support each of these positions.

Religion and language remain vital components of cultural identity and societal relations, but increasing contacts among members of the global community have led to changes in all aspects of culture. The spread of new cultural traits and values is part of a larger process of modernization, in which the political, economic, social, and cultural values of the world's more developed societies diffuse to its less developed societies. Members of formerly isolated folk cultures are increasingly adopting aspects of more modern popular cultures.

FOLK AND POPULAR CULTURE

Improvements in communications, transportation, and production have increasingly exposed populations to new products, new ideas, and new cultures during the 20th century. As a result, many people have moved from rural conservative, local folk communities into urban, dynamic, popular societies. While no global culture has yet emerged, cultures from around the world share some material components as well as cultural values; blue jeans, computers, and democratic ideals—all associated with American culture—are increasingly common, part of what some cultures complain is the process of Americanization.

Folk culture refers to isolated populations whose cultures are traditionally based on kinship groups or clans, and which utilize artifacts and mentifacts—beliefs, ideas, and their expressions such as religion, music, and folklore handed down over generations. The spread of folk cultures has traditionally been associated with migrations. For example, the diffusion of the Mennonite culture in the United States has been restricted almost exclusively to relocation of members of this ethnic group. Individual components of folk culture, such as music, myths, and medical knowledge are passed down orally between generations.

Popular culture, on the other hand, refers to large urban populations, mass production of artifacts, and change. Ideas, activities, fashions, and other cultural traits diffuse rapidly through the mass media and can be adopted and discarded rapidly as well; fads and trends are considered components of popular culture.

More at:

http://www.utexas.edu/depts/grg/adams/305/305lect05a.ppt

Modern technology allows a far greater number of connections among individuals within their larger culture, but most of these connections are less developed than in a folk culture, thus making urban society more impersonal.

Much of popular culture derives from the United States, the world leader in technology and economics. The latest rap and rock music, colorful hairstyles, newest computer games, and hippest cuisines are all components of popular culture that diffuse through the United States and often to the rest of the world. At the same time as cowboy boots and beepers appear in Russia, however, Japanese appliances and Parisian fashions appear in the United States. Popular culture is often seen as a threat to folk values, and therefore cultures attempting to protect established beliefs may erect barriers to the diffusion of new ideas. Satellite television dishes have been seen as a direct threat by the Chinese, Saudi Arabians, and Iranians because of the type of American programming available. Iran outlawed satellite dishes in 1995, China banned private individuals from owning them, and Saudi Arabia dismantled 150,000 dishes.

The contrast between folk and popular cultures is easily visible in the material aspects of each one. Houses, barns, and other structures, as well as instruments, furniture,

and household items are usually handcrafted in folk cultures rather than mass-produced or purchased. The resulting cultural landscapes of folk cultures are more strongly influenced by environmental considerations, such as resource availability, than are the landscapes of popular cultures, which are more reflective of franchised styles, standardized designs, and architectural fashions.
More at:
http://www.geog.okstate.edu/users/lightfoot/lfoot.htm

In remote areas of the world, such as in Nunavut (Canada) or Kalimantan (Indonesia), folk cultures continue to exist, though they do not thrive. There are often development pressures on these areas, and folk societies such as the Yanomami in the Amazon (Brazil) are losing territory and the security of isolation to prospectors and other settlers. As members of these cultures change in the face of modernity, the folk cultures increasingly incorporate popular items and activities and folk items and activities are slowly retired. In Nunavut and other Inuit (Eskimo) societies, sled dogs are being replaced by snowmobiles and skin tents by prefabricated houses.

Pure folk cultures may not exist anymore within the United States because our economic and social

systems make isolation nearly impossible. Amish teenagers, for example, still do not drive cars, but some do use in-line skates. Isolated regions, which at one time sheltered folk cultures, have been invaded by telecommunications and highways so that former folk cultures, such as that of the hill people in the Ozarks, are in steady decline. Small communities and individual members of these cultures still exist, but younger generations are discovering new options.

As the same fads and trends become available to populations throughout most countries, if not the world, a standardization of culture can occur. As various cultural landscapes converge, the result is termed placelessness, in which places become indistinguishable because malls, fast-food restaurants, and service stations are mass-produced and standardized. Despite the standardization implicit in many aspects of popular culture, regional cultural identities still exist because not all folk values or traits are replaced by popular ones. Even in the United States, there remain distinct regional differences in choice of language, magazines, fast food, and sports, among other things; traditional values and interests derived from folk cultures have not disappeared.

ECONOMIC GEOGRAPHY

DEFINITION

Economic geography concerns the ways societies create economic landscapes and generate spatial patterns of economic activity at scales ranging from the local to the global. This set of topics involves not only traditional issues such as resource use and population growth but also the way in which businesses decide where to locate, how governments and private firms form both regional and national development strategies, international trade, technological change, public policy, and the world economy.

This section provides an overview of economic geography, focusing primarily on global patterns of production, trade, and development, with particular reference to the United States and its capitalistic system. Under *capitalism*, a type of political economy under which the means of production are privately owned, all production is organized around the maximization of profits. Although there have been many types of economic systems throughout history, capitalism has largely eclipsed other forms of economic organization, including such early forms as hunting and gathering, feudalism, or more recently, socialism.

The preindustrial system of terraced rice farming, such as this one in Bhutan, has been practiced for thousands of years in much of Asia. Terraced fields demand intense labor, a relatively inexpensive commodity, but produce high yields, currently feeding nearly two billion people worldwide.

Hunting and Gathering

Hunting and gathering was the mode of production that sustained human beings for more than 95 percent of the human presence on Earth. Characteristically organized around a division of labor in which men hunt game and fish and women gather edible plants, this form of economy involved frequent migration, virtually no private property, kinship bonds as the essential social organization, and low population densities. (See Migration, page 207.) Almost extinct except in remote areas of rain forest and tundra, hunting and gathering began to be displaced by agriculture following the Neolithic Revolution that began around 10,000 B.C.

ECONOMIC SECTORS AND ECONOMIC GEOGRAPHY

Economic activity is frequently classified into four sectors: the primary, secondary, tertiary, and quaternary sectors. In general, primary sector activities pertain to extracting raw materials from Earth's surface, including agriculture, forestry, fishing, and mining. In preindustrial economies this sector accounts for the bulk of employment; in economically developed countries, however, such activities tend to be highly capital-intensive and account for less than 5 percent of employment.

Secondary sector activities include manufacturing and construction industries that transform raw materials into finished goods. *Tertiary sector* activities revolve around the production of intangibles such as services, including finance, producer and consumer services such as wholesale and retail trade, transportation and communications, health care, nonprofit organizations, and the public sector. Tertiary sector activities compose most of the economic activity in developed countries as well as in many developing ones. Quaternary sector activities are a result of the technology revolution and involve management and processing of information and knowledge

While there is considerable overlap among these sectors and enormous diversity within them, economic theory has traditionally viewed development as a sequential movement from the primary (preindustrial) to the secondary (industrialization) to the tertiary and quaternary (postindustrial). However, this approach has questionable application in the developing world today, given that the employment structure of many developing countries moves directly from the primary to the tertiary sector. Developing, or *Third World*, countries include the majority of the world's population and comprise, essentially, former European colonies in Latin America, Africa, and Asia, with the exception of Japan.

Primary Sector Activities

The *primary* economic *sector* involves extracting raw materials from the surface of Earth and includes agriculture, fishing, forestry, and mining. Historically, primary sector activities, particularly agriculture, involved the majority of the labor force of most societies, a reflection of the labor-intensive nature of the production process. In many less developed parts of the world today, these activities still compose a large share, often the majority, of employment, much of it organized around local subsistence systems.

AGRICULTURE

One of the most important facets of economic geography is the way societies procure food supplies for themselves. For the last several millennia, agriculture has been the primary means by and for which people transformed the natural environment. In many places, rising population levels necessitated the creation of new farmland by clearing forests and draining wetlands.

The Origins of Agriculture: The origins of agriculture date back to a period near the end of the Stone Age, usually called the Neolithic Revolution, beginning about 10,000 B.C., in which several societies independently discovered the process of domestication, the selective breeding of plants and animals such that they require human intervention in order to reproduce. Because agriculture allowed more calories to be harvested per unit area, thus sustaining denser population levels, it eventually displaced hunting and gathering, the mode of production that had endured for tens of thousands of years before.

Different crops and animals were domesticated in different parts of the world, some in more than one place simultaneously. (See Primary Areas of Crop and Livestock Domestication chart, page 251.)

These societies were located in the Fertile Crescent of Southwest Asia and in Egypt, China, Southeast Asia, the Indus River Valley, parts of western and southern Africa and, somewhat later, among cultures such as the Olmec and Maya in Central America, the mound building Mississippian culture of the Mississippi River Valley, and the Incas of the Andean region of South America. From these culture hearths, farming diffused to surrounding regions, and different cultures based their agricultural systems around different complexes of crops. In Southwest Asia, for example, wheat and barley dominated; in East Asia, rice and millet; in Africa, diets centered on a millet–yam–sorghum combination; in Central America, maize, beans, and squash were most important; in South America, the potato was the primary staple.

The importance of agriculture in the historic emergence of human civilization cannot be overstated: Farming provided the surplus that allowed the subsequent development of a class society, cities, the state, a supply of products that exceeded demand, and innovations such as writing and metalworking. Population levels, densities, standards of living (including free time and experimentation), and life expectancies rose markedly during this first agricultural revolution.

More at:

http://www.ipgri.cgiar.org/publications/HTMLPublications/47/index.htm

Preindustrial Agriculture: Despite the worldwide dominance of commercial farming today (farming that is industrialized and for profit), preindustrial agricultural systems continue to persist in several forms. Most common among these is slash-and-burn, which is found in the humid tropics of central and western Africa, Central America and Brazilian Amazonia, and parts of Southeast Asia. Roughly 40 million people live today on the basis of this technique. Because tropical soils are very low in nutrients—most of the nutrients are contained in the living biomass—slash-and-burn agriculture is well developed for maximizing short-term yields in such environments. As a general rule, the sequence employed in this type of farming involves cutting down much of the vegetation in small clearings, burning it to release the nutrients into the soil and to clear farmland, then planting a mixed group of crops. Such a system may work for up to four or five years in a given site, until the crops exhaust the available nutrient supply. Then it becomes necessary to move on and clear new land through the same means. When population levels and available land are constant, this form of agriculture is quite effective.

Another preindustrial agricultural system is terraced rice farming, which is widely practiced across Asia from eastern India through southern China, Japan, Korea,

PRIMARY AREAS OF CROP AND LIVESTOCK DOMESTICATION

SOUTHWEST ASIA	MEDITERRANEAN	SOUTHEAST ASIA	CHINA	WEST AFRICA	CENTRAL AMERICA	SOUTH AMERICA
Barley	Celery	Bananas	Apricots	Cotton	Beans	Alpaca
Beans	Dates	Chicken	Barley	Coffee	Chilis	Guinea pig
Carrots	Garlic	Citrus fruits	Cabbage	Melons	Cotton	Llamas
Cattle	Grapes	Cucumbers	Peaches	Millet	Dogs	Papayas
Ducks	Lentils	Coconuts	Plums	Oil palm	Maize	Pineapples
Goats	Lettuce	Dogs	Soybeans	Okra	Manioc	Potatoes
Horses	Olives	Eggplant		Pigs	Squash	Pumpkins
Oats		Hemp		Sorghum	Sweet potatoes	Tobacco
Onions		Pigs		Yams	Turkeys	Tomatoes
Rye		Rice				
Sheep		Sugar cane				
Wheat		Tea				
		Water buffalo				

Indochina, and Indonesia, and currently feeds almost two billion people. Terraced rice paddies are carved into hillsides as a means of creating arable land and involve intricate irrigation systems to control the flow of water, especially when rice plants are young. The system of dikes and levees used for this purpose, as well as the terraces themselves, have been constructed over millennia. This method is labor-intensive and involves working relatively small plots of land. Today rice is grown widely in East Asia, a prime example of intensive agriculture with high inputs per unit area, resulting in high outputs, as opposed to extensive agriculture, which involves few inputs scattered over large areas. (See World Corn, Rice, and Wheat Production map, page 253.)

Finally, preindustrial agriculture includes various forms of pastoral nomadic herding generally practiced in arid or semiarid regions in Central Asia and Africa in which livestock rather than crops are the primary form of sustenance. The Masai of East Africa, for example raise cattle, and the Kirghiz of Central Asia migrate with herds of goats. Frequently such cultures, which are often in areas not suitable for growing crops, may follow natural migratory cycles of their animals and trade with local settled communities.

Historically, agriculture was organized around small-plot subsistence systems relying on animate sources of energy, such as animals to pull plows and people to plant seeds, harvest crops, and fetch water. European agriculture was gradually transformed, starting in the sixth century, because of a number of factors: rising population levels, technological changes such as the plow and stirrup, the replacement of oxen with horses, and the open-field system, in which several large fields surrounding a village were farmed.

Commercialized Agriculture: By the 17th century, commercial agriculture began to make its appearance in Europe: The motive for profit displaced subsistence production. Techniques included growing crops for profit, enclosing fields by fences and hedgerows, improvements in breeding stock and seeds, crop rotation rather than fallow fields. Commercial agriculture was also boosted by the diffusion of food crops such as the potato from South America to Spain, then Britain and Ireland, and European cereals to the grasslands of North America and Australia.

The worldwide expansion of commercial agriculture was intricately linked to the plantation system, the primary institution through which foods and commodities grown or produced for profit were transplanted to European colonies elsewhere. Examples include the fruit and sugar plantations of the

In a preindustrial form of agriculture—pastoral nomadic herding—a young shepherd tends his flock of sheep near Irbil in northern Iraq.

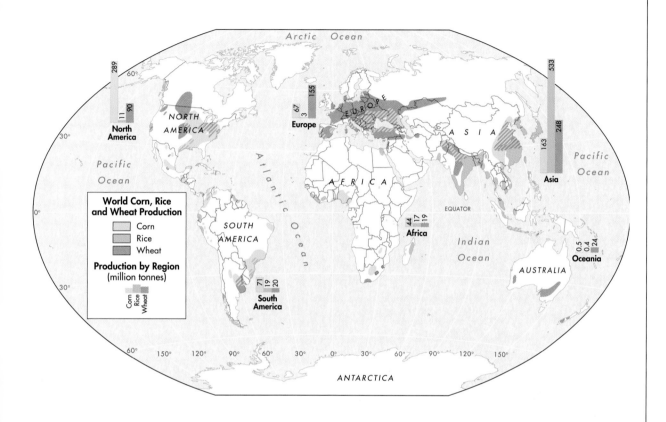

The principal growing areas of wheat, found at cooler latitudes, and rice, found at very warm, humid latitudes, supply most of the world with grain. While wheat cultivation has become largely commercial and highly mechanized, rice still involves labor-intensive, hands-on cultivation in most places. In the U.S. and Japan, rice is grown with little human labor.

Caribbean and Latin America, cotton and tobacco in the southern United States, groundnuts (peanuts) and cocoa in west Africa, cotton and tea in east Africa, cotton in Egypt, tea in India and Ceylon (Sri Lanka), and rubber, tobacco, sugar, and other crops in Southeast Asia.

Low-cost food imports from European colonies accelerated a transformation in agriculture in Europe, driving many laborers out of work and into cities, and encour-aging a wholesale reconstruction in agricultural organization and in productivity.

Capitalist agriculture—farming for a profit—is typified by mech-anized production, making con-temporary agriculture among the most capital-intensive industries in the world. Mechanization throughout the 19th century dras-tically reduced the number of workers needed per unit area of land, driving countless rural fam-ilies off farms and freeing up a labor supply that was quickly absorbed by industrializing cities.

The 19th and 20th centuries experienced the widespread use of technological innovations such as the cotton gin and threshing machine, which resulted in increases in productivity. High yields per unit of land or labor are also largely attributable to vast expenditures of energy through petroleum-dependent farming sys-

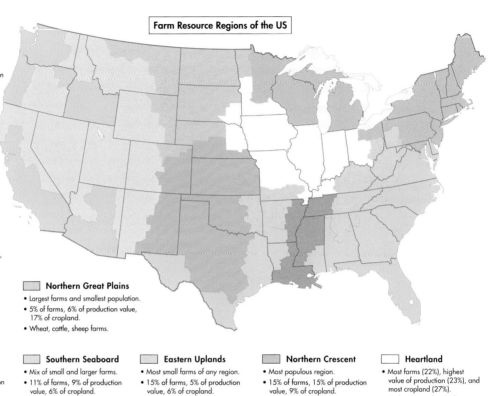

Farm Resource Regions of the US

Basin and Range
- Largest share of nonfamily farms, smallest share of US cropland.
- 4% of farms, 4% of production value, 4% of cropland.
- Cattle, wheat, and sorghum farms.

Fruitful Rim
- Largest share of large and very large family farms and nonfamily farms.
- 10% of farms, 22% of production value, 8% of cropland.
- Fruit, vegetable, nursery, and cotton farms.

Prairie Gateway
- Second in wheat, oat, barley, rice, and cotton production.
- 13% of farms, 12% of production value, 17% of cropland.
- Cattle, wheat, sorghum, cotton, and rice farms.

Northern Great Plains
- Largest farms and smallest population.
- 5% of farms, 6% of production value, 17% of cropland.
- Wheat, cattle, sheep farms.

Mississippi Portal
- Higher proportions of both small and larger farms than elsewhere.
- 5% of farms, 4% of production value, 5% of cropland.
- Cotton, rice, poultry, and hog farms.

Southern Seaboard
- Mix of small and larger farms.
- 11% of farms, 9% of production value, 6% of cropland.
- Part-time cattle, general field crop, and poultry farms.

Eastern Uplands
- Most small farms of any region.
- 15% of farms, 5% of production value, 6% of cropland.
- Part-time cattle, tobacco, and poultry farms.

Northern Crescent
- Most populous region.
- 15% of farms, 15% of production value, 9% of cropland.
- Dairy, general crop, and cash grain farms.

Heartland
- Most farms (22%), highest value of production (23%), and most cropland (27%).
- Cash grain and cattle farms.

Approximately 44 percent of U.S. land area, more than a million square miles, is used to produce crops and livestock. Agricultural patterns depend largely on geography: Soil, slope of the terrain, climate, distance to market, and storage and marketing facilities all play significant roles.

tems, including the production and operation of farming machinery such as tractors, combines, harvesters, trucks, diesel pumps, and other equipment, as well as petroleum-derived fertilizers. Linkage to the petrochemical industry has made industrialized agriculture especially sensitive to world fluctuations in the price of petroleum.

In Japan agriculture today employs only 5 percent of the labor force, whereas in China 50 percent of the population is employed in the production of food.

Several subsystems characterize commercial agriculture in the U.S. including wheat, corn and hogs, fruits and vegetables, dairy farming, and cattle ranching, each of which exhibits a unique cluster of labor and management practices, land uses, geographic location, technologies, and market structure.

Recently, agriculture, particularly in the United States, has been steadily reshaped by new technologies. The computer revolution introduced digital controls for irrigation, the application of pesticides, and harvesting. Innovations, which include fertilizers, hybrid seeds, agrochemicals, and biotechnologies in pharmaceuticals, recombinant DNA techniques, and the genetic alteration of crops—sometimes referred to as the Third Agricultural Revolution— have all been responsible for increasing productivity even further. Conversely, industrial substitutes, such as artificial sweeteners for sugar, have dampened the demand for some farm goods.

Today, the U.S. is by far the world's largest producer of agricultural commodities, exporting half the world's volume of traded foodstuffs. U.S. output in agriculture exceeds most industries, including construction and automobiles. Exports, primarily wheat, rice, corn, and soybeans, are plant rather than animal products, and are concentrated at the bottom of the food chain. In contrast, the U.S. is a net importer of meats, which are located higher in the food chain and require far more calories to produce, even though the U.S. has an enormous domestic meat-processing industry. This is a reflection of the nation's high demand for meat;

further, a large share of cereals output, for example half of corn output, is dedicated to feed for cattle, pigs, and chickens.

More at:

http://www.usda.gov/nass/

Due in part to its high levels of efficiency, U.S. agricultural output frequently exceeds demand, creating a chronic surplus. This situation is also due in part to a series of federal government incentives implemented during the Depression of the 1930s, particularly price supports that encourage farmers to grow more than the market can absorb, thus creating a surplus the government is obligated to purchase. Some of this surplus is used as foreign aid or in domestic U.S. Department of Agriculture subsidized food programs.

The U.S. is not the only country to subsidize its farmers. Japan, for

example, with very limited quantities of arable land, is self-sufficient in rice largely because of subsidies, and most of the budget for the European Union (EU) goes to agricultural subsidies. Persistent surpluses have led to long-term declines in prices.

The combination of low prices, the advantages enjoyed by large firms, and technological change has steadily eroded the number of family farms. In the 20th century, the number of farms in the U.S. has declined from 5.5 million in 1950 to 2.2 million in 2002, while the average farm size has risen from 200 acres in 1950 to 436 in 2002. Long-term challenges to American farming include shortages of groundwater, soil erosion, and rural-to-urban land conversion.

Globally, concerns about agriculture are tied to impending limits to the world supply of arable land and fears that global food stocks will fail to keep pace with rising population levels. Large parts of Earth's surface are too dry, too steep, too wet, or too cold to sustain agriculture. The supply of arable land varies widely around the world and changes over time as prices and technologies allow crops to be grown in marginal environments. As farming encroaches upon such environments, however, *diminishing returns* make harvests there costly and inefficient to sustain in

WORLD'S TOP 15 FOOD AND BEVERAGE COMPANIES			
RANK	COMPANY	HEADQUARTERS	FOOD SALES (U.S.$ MILLIONS)
1	Nestlé S.A.	Switzerland	$46,628
2	Kraft Foods Inc.	USA (Illinois)	$38,119
3	ConAgra Inc.	USA (Nebraska)	$27,630
4	PepsiCo Inc.	USA (New York)	$26,935
5	Unilever plc	UK & Netherlands	$26,672
6	Archer Daniels Midland Co.	USA (Illinois)	$23,454
7	Cargill Inc.	USA (Minnesota)	$21,500
8	The Coca-Cola Co.	USA (Georgia)	$20,092
9	Diageo plc	UK	$16,644
10	Mars Inc.	USA (Virginia)	$15,300
11	Anheuser-Busch Inc.	USA (Missouri)	$12,262
12	Groupe Danone	France	$12,184
13	Kirin Brewery Co. Ltd.	Japan	$11,287
14	Asahi Breweries Ltd.	Japan	$11,050
15	Tyson Foods	USA (Arkansas)	$10,751

The Green Revolution

During and shortly after World War II, the Rockefeller Foundation based in New York City initiated attempts to stimulate agricultural productivity in developing countries, beginning in Mexico. Led by Norman Borlaug, the application of scientific knowledge to agriculture, including selective breeding of high-yield, high-protein "miracle crops," produced numerous new strains of rice and wheat that grew more rapidly than did earlier varieties.

Miracle crops often yielded two or three harvests per year, compared to one for most traditional breeds, and each crop was two to three times larger; often they were more drought- and disease-resistant as well. Limited in their ability to absorb nitrogen, however, they required nitrogen-based fertilizers to grow properly. In the 1960s these efforts culminated in the green revolution, a sustained effort to introduce these crops in countries with rapidly growing populations such as Mexico, India, the Philippines, Indonesia, Bangladesh, Egypt, and elsewhere. Critics pointed out that such crops often required more water, fertilizer, pesticides, and capital inputs than did indigenous crops, thus increasing costs for small farmers and magnifying social inequalities and increasing their dependence upon the global petroleum industry.

In the long run, however, the results have been indisputable: Asian rice productivity almost doubled, and *green revolution* crops accounted for most of the world's agricultural productivity gains in the 1960s and in the 1970s, and for 80 percent in the 1980s. The program has not been an unqualified success everywhere, however: In Africa such crops have been thwarted by wars, poor soils, and drought.

Recent research has concentrated upon other crops grown in tropical environments. Some of these, such as cassava, soybeans, sorghum, and millet, hold a great deal of promise for the future in Africa.

the long run. The principle of diminishing returns suggests that increased investments in production capacity yield steadily smaller marginal increases in output, although diminishing returns may be offset by continuous technological change and improved productivity.

Because the world supply of arable land is essentially fixed, future gains in supply must occur through productivity increases. Persistent fears that the world cannot feed its people have thus far been countered by technological changes that have overcome diminishing returns. Nevertheless, many areas, notably Africa, have suffered declining food consumption per capita since World War II, largely due to wars that have disrupted farming systems there.

The globalization of agriculture links different production and consumption regions in an integrated industrial complex that spans the world. A network of corporate ties binds farmers, intermediaries, and consumers into an interdependent totality. Global agribusiness is heavily shaped by multinational distribution companies such as Cargill and Archer-Daniels-Midland, which are rarely involved in direct production but can negotiate the complexities of multiple clients and markets, distribution systems, exchange-rate fluctuations, product prices, and different national farming systems. These firms dominate the distribution of cereals, including 80 percent of world wheat trade, in the economically developed countries in Europe, North America, Japan, Australia, and New Zealand, slightly more than one-sixth of the world's population. These countries produce the majority of the world's industrial products and enjoy a high standard of living.

More at:
http://www.agribusinessonline.com/

Tariffs and *quotas* have affected prices and the competitiveness of local farmers in complex ways. Tariffs are a form of domestic *protectionism* in which countries levy a surcharge on imports designed to increase their market price and thus inhibit consumption of imports. Quotas are another form of protectionism in which governments limit the absolute volume of imports in an industry, effectively driving up the market price. Although technically different from a tariff, quotas have much the same impact economically. Tariffs generate additional government revenues and have thus been more popular among countries.

WORLDWIDE ANNUAL PRODUCTION AND ESTIMATED RESERVES OF SELECTED MINERALS, 1998 (MILLIONS OF METRIC TONS)		
METAL	ANNUAL PRODUCTION	ESTIMATED RESERVE BASE
Iron ore	1,020	300,000
Aluminum	22.2	34,000*
Copper	11.9	650
Zinc	7.8	440
Lead	3.08	140
Nickel	1.17	140
Tin	0.216	12
*Bauxite		

MINING AND ENERGY

Minerals and petroleum constitute two of the world's most important nonrenewable natural resources. Metals are indispensable to the production systems of contemporary society. Many of the world's richest mineral deposits have been depleted, and others are insufficiently concentrated to be economically recoverable except at high prices to consumers. This is particularly true of metallic minerals, of which iron is the most important, but also includes lead, aluminum, copper, nickel, tin, bauxite, manganese, and zinc.

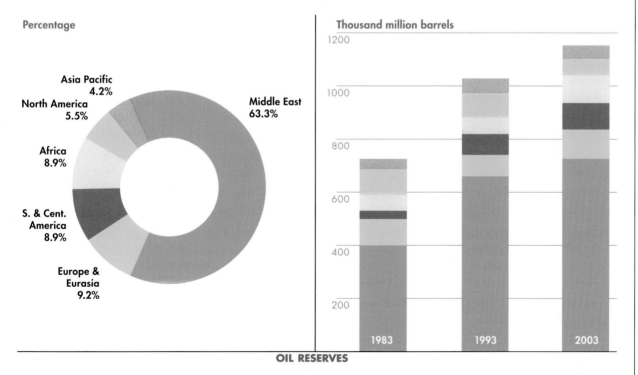

Percentage

Asia Pacific 4.2%
North America 5.5%
Africa 8.9%
S. & Cent. America 8.9%
Europe & Eurasia 9.2%
Middle East 63.3%

Thousand million barrels

1983   1993   2003

OIL RESERVES

Two-thirds of the world's proven oil reserves are found near the Persian Gulf. From 1983 to 2003 oil output from the Middle East nearly doubled .

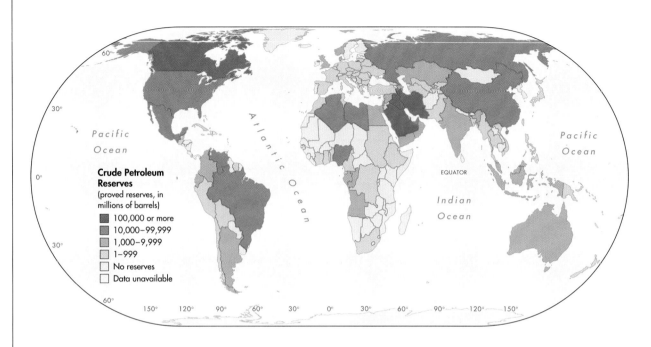

"Proven reserves" are those oil reserves that are recoverable with present technology and prices. "Explored reserves" may include proven, probable, and possible oil sources.

The largest deposits of strategic minerals, those essential to industrial production, are concentrated in five countries: Canada, U.S., Russia, South Africa, and Australia. In contrast, nonmetallic minerals such as potash, clay, and nitrogen are in abundant supply worldwide. To a degree, limited supplies of minerals have been offset by declining consumption per capita because industrialized economies utilize more efficient production methods, find viable substitutes such as plastics and ceramics, and accelerate their reuse and recycling efforts.

In nonindustrial societies, energy needs are satisfied primarily through renewable sources such as wood, charcoal, or dung. Industrialized countries, in contrast, rely on nonrenewable sources, primarily fossil fuels. Indeed, the industrialization of Europe was made possible in part due to extensive deposits of coal that underlay the North European lowlands.

The world's most abundant and commonly used fuel source remains coal, located in vast seams in northern Europe, Russia, China, and North America, especially in Appalachia and the Rocky Mountains. Similarly, natural gas deposits are concentrated primarily in Russia, North America, and the Middle East. U.S. energy production in the 20th century has shifted significantly away from a 19th-century reliance on coal and into petroleum and natural gas. Multinational oil companies such as Exxon, Mobil, Gulf, Texaco, Arco, Phillips, Royal Dutch Shell, and British Petroleum are among the largest firms in the world in terms of sales and exercise considerable influence over the market price.

The geographies of world oil production and consumption are highly uneven. Two-thirds of the world's proven reserves are located in the vicinity of the Persian Gulf,

Deindustrialization

Deindustrialization involves plant closures and contractions in employment industries such as textiles, steel, automobiles, machine tools, shipbuilding, rubber, petrochemicals, and electronics. The process started in Great Britain shortly after the end of World War II because high costs and low productivity levels left Britain increasingly uncompetitive as other countries acquired manufacturing capacity. In the U.S. deindustrialization of the manufacturing belt, starting in the 1970s and later extending to most of the Sunbelt, accelerated the shift from blue-collar to white-collar jobs.

For communities dependent upon industry, deindustrialization is catastrophic because it results in high unemployment, poverty, depressed real estate markets, and out-migration.

including Saudi Arabia, Iran, and Iraq, and North Africa. Secondary reserves are found around the Gulf of Mexico, Alaska, the North Sea, and the South China Sea. The largest petroleum consumers include the U.S.—where about 4 percent of the world's people consume 30 percent of its energy, largely for automobiles—Europe, and Japan.
More at:
http://www.bp.com/subsection.docat egoryld=95&contentld=2006480

The global petroleum industry has been dominated by the geopolitics of the Organization of Petroleum Exporting Countries (OPEC), which includes all the Middle Eastern and North African producers as well as Nigeria, Venezuela, and Indonesia. Oil embargoes imposed on Western countries have raised the price of oil, creating dramatic oil crises with dire consequences for industrialized economies, including recession, inflation, and accelerated

deindustrialization, which occurs when the loss of comparative advantage in manufacturing is brought about by technological displacement and by the rise of foreign competitors.

NUCLEAR ENERGY AS PERCENT OF TOTAL ENERGY SUPPLY, 1996	
France	72.7
Belgium	59.3
Sweden	51.6
Hungary	48.4
South Korea	47.5
Switzerland	40.0
Spain	35.9
Bulgaria	34.0
Finland	33.3
Czech Republic	28.7
Germany	27.6
Japan	23.8
United States	21.7
United Kingdom	20.6
Argentina	19.1
Russia	12.6
South Africa	5.9
Netherlands	4.9
Mexico	3.6
India	1.8
Brazil	0.6

Alternatives to fossil fuels include nuclear energy, hydropower, geothermal sources, and solar and wind energy. Nuclear energy in the form of fission is used primarily in industrialized countries of Europe, although the proportion of total energy derived from this source varies widely. (See Nuclear Energy as Percent of Total Energy Supply table, at left.) In the U.S., nuclear power provided about 20 percent of the energy supply in 1998. The use of nuclear energy is constrained by widespread concerns over the transport and storage of radioactive waste and by the possibility of accidents.

Hydropower is widely used in countries with adequate flows of water. Geothermal energy is rarely used except in countries with large resources of hot groundwater, such as Iceland. Solar and wind energy are inexhaustible and ubiquitous, but economically difficult to obtain in sufficient quantities. (See Renewable and Inexhaustible Resources, page 338.)

Debate continues over the existence of a global-resource supply crisis due to limited supplies of nonrenewable resources, growing population and per capita energy consumption levels, and environmental consequences of high-volume consumption. In the late 20th century, fears of impending shortages were alleviated by a global glut in raw materials, including petroleum and foodstuffs, which severely depressed prices. In many countries, economic development has been constrained by concerns over *environmental degradation*, including pollution and wildlife habitat destruction, and the bio-

physical limits to growth such as soil erosion.

Secondary Sector Activities

Since the mid-1700s, manufacturing has become a vital part of the U.S. and world economies. Unlike primary sector industries, manufacturing involves transforming raw materials into finished goods for consumer use or for further processing, although on occasion the distinction between the primary and secondary sectors is difficult to draw. The evolution of large manufacturing systems was made possible by harnessing the inanimate energy of water power and increasing the use of coal in the 18th and 19th centuries, and petroleum in the 20th century.

THE INDUSTRIAL REVOLUTION

Capitalism as a form of economic and social organization preceded the industrial revolution by several centuries. Beginning in the mid-18th century, the industrial revolution marked a significant qualitative change in the social and spatial organization of society.

Industrialization began in Great Britain and spread to Europe and North America. It was characterized by enormous technological change, including innovations such as the steam engine, cotton gin, railroad, electric generator, telegraph, refrigeration, gasoline engine, the telephone, and the automobile, all of which converted inputs into outputs more efficiently. As a result, productivity levels soared, raising

The Von Thünen Model

Location modeling in economic geography has a long history. One of the first examples was created by a German landowner, Johann Heinrich von Thünen (1783–1850), who analyzed land use patterns on his estates. Von Thünen's model of land use related the economic rent, or maximum potential amount of profit per unit area that a crop could generate, to the market price, production costs, and transportation costs to the market.

Hypothetically, different crops competing for a limited supply of

land form concentric rings of land use centered upon the market, ranging from high-profit, intensive land uses near the market to low-profit, extensive uses which are relatively far from the center. The model is easily illustrated using graphs of different bid–rent lines (reflecting the economic rent each land use generates with distance from the center). This ideal pattern is disrupted by local variations in soil fertility and transportation costs. The model has long been popular among location analysts

for its simplicity and elegance.

In the 20th century neoclassical economic analysis of land markets extended this concept to include consumer demand in the study of intra-urban household and residential location behavior. Although the applicability of the Von Thünen model to the analysis of agricultural land has passed, given that cities today rely on national and global circuits of food production, it does serve as a useful device to illustrate the rent-maximizing behavior of land markets.

the average standard of living as the costs of products declined relative to incomes. Food became cheaper with the industrialization of agriculture, and famine in Europe ended with the exception of the potato famine in Ireland of the late 1840s.

The factory system brought large numbers of workers together for the first time, creating an urbanized workforce and accentuating rural-to-urban migration. In due course, automated assembly lines and use of interchangeable parts accelerated the division of labor in industry, increasing productivity even more through specialization.

Great Britain, as the world's first industrialized country, enjoyed a monopoly over industrialization for several decades late in the 18th century. In the 19th century, industrialization spread to regions of Europe including the lower Seine River in France, the Ruhr Valley in Germany, and the Po River Valley in Italy. By the 1840s North America, starting with New England and later with the manufacturing belt stretching from southern Ontario along the southern shores of the Great Lakes to Milwaukee, became another large concentration of manufacturing, including light industry and shipbuilding on the East Coast and steel, agricultural implements, machine tools, rubber, and auto-mobiles in the Midwest.

After 1868 Japan also industrialized, the only non-Western nation to do so. Industrialization spread to include Russia after Joseph Stalin (1879-1953) became dictator. In the 1960s newly industrializing countries (NICs) included South Korea, China, Mexico, and Brazil.

More at:

http://www.fordham.edu/halsall/mod/modsbook14.html

LOCATIONAL UNDERPINNINGS OF MANUFACTURING

Unlike agriculture, climate and soil type are irrelevant to the geography of manufacturing, but location of raw materials, labor costs, transport costs, and markets are significant. Traditional location theory, developed by German economist Alfred Weber, emphasized the transportation costs of inputs and outputs. Perhaps most important to the location decisions of manufacturing and service firms is the need to agglomerate—or cluster together—a geographic phenomenon that offers advantages such as a common labor pool, shared infrastructure, access to specialized kinds of information, and minimal transport costs of inputs and outputs.

Essential to economic activity, the *infrastructure*—the public works on which companies rely such as sewers, water, and schools as well as the transportation and communications network that allows goods, people, and information to flow across space— is invariably publicly constructed and operated, allowing costs to be socialized but benefits to remain privatized.

Typically, large, capital-intensive sectors—the steel industry, for example—tend to be more *vertically integrated*, that is production processes are located within the confines of an individual corporation, rather than relying upon proximity to suppliers and clients. Smaller, more labor-intensive firms, such as garment producers, tend to require close ties functionally and spatially.

Labor costs are also important in understanding the geography of manufacturing, although this factor varies with the degree of labor- and capital-intensity of the production process. For labor-intensive firms such as the garment industry, the cost of labor is critical; for others, such as petroleum, where labor makes up only 3 percent of total costs, it is unimportant. Labor as a factor of location also involves the skill levels, education, training, experience, and productivity of workers, which vary considerably among industries and locations. Labor markets are often specific to individual types of jobs with unique skill requirements.

INDUSTRY-SPECIFIC ANALYSES OF MANUFACTURING

The geography of manufacturing varies considerably among industrial sectors. Textiles were the leading edge of the industrial revolution in Europe, the U.S., and Japan, and have continued to be so in industrializing countries of East Asia today. Because the total investment capital in this sector is relatively low, start-up costs are not formidable, entry and exit are rapid, and the industry's market structure is

therefore competitive. The historical geography of textile production has thus been typified by a constant search for low labor costs.

In the U.S., the textile industry dominated southern New England throughout the 19th and early 20th centuries, with garment production clustered in New York. By the 1920s and 1930s, however, the industry left New England in favor of the South, particularly North and South Carolina. The industry also has a long presence in East Asia,

beginning with Japan in the late 19th century and continuing with garment production in Taiwan, Hong Kong, southern China, and more recently in Indonesia. In Latin America, garments are produced in Mexico, Guatemala, and Honduras.

In contrast, the iron and steel industry is characterized by high degrees of capital intensity and low degrees of geographical mobility. Few industries are as critical to industrial society as this one, which

Many clothes sold in the U.S. are made abroad, in factories such as this one in Kowloon, Hong Kong. Labor costs are about 50 percent of total costs.

The U.S. Garment and Textile Industry

The textile and garment industry of the United States has long revealed a changing set of locations. In the 19th century, textiles were centered in southern New England in cities such as Manchester, New Hampshire, and Lowell, Massachusetts, where they employed thousands of young Irish women immigrants and women from Quebec.

Garment production was concentrated in New York, particularly the Lower East Side of Manhattan, where it generated jobs for Jewish immigrants from Eastern Europe. In 1900 more than half of the United States garment output originated in New York City.

Between the 1920s and World War II, however, the textile industry relocated to the South, particularly North and South Carolina and Georgia, where labor was cheaper and not as likely to be unionized. Today, most U.S. textile employment remains concentrated in those states, although foreign competition and technological displacement have steadily reduced its size.

Garment production, however, remains largely concentrated in urban areas, particularly in New York and Los Angeles, where large numbers of unskilled immigrant women, primarily from China and Latin America, work in sweatshops, under deplorable, often illegal, conditions.

Long New York's largest form of manufacturing employment, the garment industry has contracted steadily, and today the metropolitan region employs only about 100,000 people in this sector.

Many U.S. clothing retailers, linked to global supply chains, purchase garments produced in East Asia or Central America.
More at:
http://www.itds.treas.gov/Textileprofile.html

has linkages to construction, agriculture, manufacturing, and transportation. With very high start-up costs, the industry has long been dominated by a few companies, such as United States Steel Corporation (now USX), which produced 60 percent of United States steel at the close of the 20th century.

In the 1880s steel became the mainstay of the economy of the U.S. manufacturing belt. During World War II, steel production emerged in California and Maryland in plants that were often publicly subsidized and located near coastal areas to facilitate receiving imported iron ore. By the 1970s in the wake of rising energy costs, old technology and equipment, higher labor costs, and rising imports from Japan and Europe, the U.S. steel industry began a dramatic contraction that devastated many manufacturing belt economies, part of the broader process of deindustrialization or loss of manufacturing jobs in developed countries.

Japan is the largest producer of steel in the world, although other countries such as South Korea and Brazil have also become important producers. In the 1990s the U.S. steel industry experienced a moderate revival as technological changes such as small, highly automated plants restored some degree of competitiveness by relying upon scrap metal inputs, computerization of production, and specialized outputs.
More at:
http://www.steel.org/

The geography of the automobile industry is also significant. Invented in the 1890s by German and French engineers, the automobile in Europe was long produced primarily for the luxury market. Germany remains the largest European automobile producer, although other countries make automobiles, including Sweden, France, Britain, and Italy.

American automobile maker

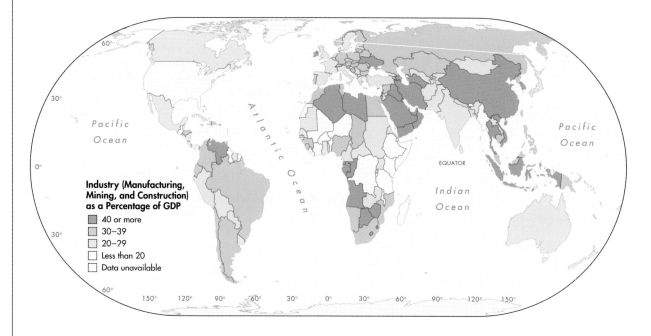

Manufacturing, the production of goods from raw materials, contributes significantly to the economy of the U.S. and other highly industrialized societies. Iron and steel, textiles, lumber, automobiles, electronics, and chemicals are some of the world's most significant manufacturing products.

Henry Ford (1863–1947) perfected the production technique, which made automobiles more affordable for the mass market. He opened an auto factory in 1903 in Dearborn, Michigan, where he had access to a well-developed rail network and parts suppliers such as tire makers in Akron, Ohio. Ford's success was largely predicated upon a finely grained division of labor and standardized job tasks. His method was so widely imitated that the particular form of production ultimately became known as *Fordism*.

Today, automobile assemblers put together roughly 15,000 parts produced by separate firms in the textiles, metals, rubber, glass, chemicals, and electronics industries. And ancillary services such as engineering, finance, advertising, and legal became increasingly important to car manufacturing.

As the automobile gained popularity, particularly after World War I, it dramatically reshaped the social and spatial nature of American cities, including the waves of suburbanization that followed World War II. By the 1980s only three U.S. firms— General Motors, Ford, and Chrysler—were producing cars, and then Mercedes-Benz bought Chrysler in 1998.

Today, foreign imports constitute roughly one-third of U.S. automobile purchases, although the globalization of the industry and location of foreign, particularly Japanese, assembly plants in the U.S. has made national origins notoriously difficult to ascertain.

The production and assembly of parts has spread to many countries, including assembly plants in Mexico, Spain, and Britain. Japan remains the largest automobile producer in the world (28 percent), and provides much of the capital and expertise for the growing productive capacity in South Korea and Southeast Asia.

The electronics sector is among

the most important and rapidly growing industries in the world today. Unlike heavy industry such as steel or automobiles, transport costs are insignificant to electronics production, while labor skills, agglomerative economies, and military spending are critical. The mass production of electronic goods can be traced to the development of the radio industry in the northeastern U.S. in the 1920s.

World War II delivered a major boost to electronics with the invention of radar, the computer, and the vacuum tube. In 1947 Americans William Shockley, John Bardeen, and Walter Brattain of Bell Laboratories invented the transistor, which led to the formation of the Silicon Valley complex south of San Francisco, the largest agglomeration of electronics production in the world. Other major U.S. centers include southern California; Boston's Massachusetts Route 128 (now I-95); Austin, Texas; Phoenix, Arizona; and North Carolina's Research Triangle of Raleigh, Durham, and Chapel Hill.

The semiconductor revolution of

Henry Ford and Fordism

American automobile maker Henry Ford (1863–1947) developed a form of economic organization characterized by the vertically integrated production of homogeneous goods, automated assembly lines, mass markets, and large economies of scale. Adopted by many industries in the early and mid-20th century, Fordism was accompanied by political innovations such as the Keynsian welfare state and internationally by expanding trade and reduced protectionism. Today Fordist production systems have been largely displaced by systems that integrate computerized technology such as just-in-time delivery systems and niche marketing.

More at:
http://www.willamette.edu/~fthompso/MgmtCon/Fordism_&_Postfordism.html

Car manufacturing technology has evolved from conveyor-belt assembly lines of 1913 to today's electronic automation, using computers and robots.

the 1970s and 1980s, which allowed for the digitization of information, revolutionized the electronics industry, dramatically lowering production costs, greatly enhancing computer speed and memory, and triggering numerous applications in other sectors. Electronics has often been labeled the key industry of the late 20th century. As the industry steadily internationalized, low-wage, low-skilled assembly functions decentralized to selected developing countries, including Mexico but particularly South Korea and Malaysia. Japan is the largest producer of electronics in the world, particularly in consumer goods, although China has acquired a growing capacity as well. The U.S. remains very competitive in this industry, dominating the high-end research part of the industry as well as software production.

Tertiary Sector Activities

The tertiary economic sector is synonymous with services, which encompass a diversity of occupations and industries, ranging from lawyers to plumbers. Services also include the movement of goods and people. Eighty percent of the labor force in most industrialized countries works in services.

REASONS FOR GROWTH OF SERVICES

Services employment increased steadily in developed countries throughout the 20th century despite low rates of population growth and significant manufacturing job losses. While services and manufacturing are intimately intertwined, services exhibit growth and

Government Programs and Policies Affecting Population Redistribution

Services involve the production of intangible outputs, and thus stand in contrast to manufacturing, the product of which can be "dropped on one's foot." What, for example, is the output of a lawyer? a social worker? a doctor? It is impossible to measure these outputs accurately and quantitatively, yet they are real nonetheless. To complicate matters, many services generate both tangible and intangible outputs. Consider a fast-food franchise: The output is assuredly tangible, yet it is considered a service; the same is true for a computer software firm, in which the output is stored on disks.

Services encompass so many different industries that it is misleading to speak of the service sector, as if the industries were all the same. However, a broad consensus exists as to the major components of services:

♦ Finance, insurance, and real estate, including commercial, savings, and investment banking, insurance of all types, and commercial and residential real estate.

♦ Business services subsume legal services, advertising, engineering and architecture, public relations, accounting, computer services, and some types of consulting. Financial and business services are often labeled producer services, those that primarily sell their output to other firms rather than to households.

♦ Transportation and communications include telecommunications, trucking, shipping, railroads, airlines, and local transportation such as taxis, and buses.

♦ Wholesale and retail trade firms are the intermediaries between producers and consumers; closely affiliated are eating and drinking establishments, personal services such as beauticians, income tax consultants, veterinarians, and repair and maintenance services.

♦ Entertainment related industries, including film, television and radio, publishing, and lodging, are also an important part of services. Also included is tourism, the world's largest industry in employment terms.

♦ Government at the national, state, and local levels includes state bureaucracies, the armed forces, and all those who provide public services.

♦ Nonprofit agencies include charities, churches, museums, membership organizations, and private, nonprofit health care agencies.

locational dynamics that are somewhat different from those in manufacturing, although both constitute commodity production in varying forms. The reasons for the increase in services employment throughout the world may be:

a) Rising per capita incomes, particularly in the industrialized world, have contributed markedly to rising services employment. The demand for many services is income-elastic, that is, increases in real (post-inflationary) personal income tend to generate proportionately larger increases in the demand for many kinds of services rather than a demand for manufactured goods.

Services with particularly high income-elasticities include entertainment, health care, and transportation. U.S. households, for example, spend slightly more on services (about 51 percent of disposable income) than they do on durable goods such as cars and refrigerators and nondurable goods such as food and clothing combined.

An important reason contributing to this growth is the increasing value of time that accompanies rising incomes—especially with two income earners per family. As the value of time climbs relative to other commodities, consumers generally minimize the time needed to accomplish ordinary tasks. This phenom-

enon also explains the demand for washing machines, dishwashers, and automobiles, and it is especially important for the growth of personal and retail services, such as fast-food restaurants.

Similarly, the growth of repair services reflects increasingly sophisticated technologies—automobiles or televisions, for example—and limited quantities of free time. Thus, the increasing value of time has led to more household functions being accomplished outside of the home: What was once taken care of at home is now a commodity to be purchased.

b) Rising levels of demand for health and educational services compose an important part of the broader growth of the service economy. The provision and consumption of health care has increased steadily, in large part because of the changing demographic composition of the populations of industrialized nations. The most rapidly growing age groups in the industrialized West today are the middle-aged and the elderly, precisely those demographic segments that require relatively high per capita levels of medical care. Consequently, medical services as a proportion of *gross national product* (GNP), the sum total of the value of goods and services produced by a country in one year, have increased steadily

throughout Europe, North America, and Japan, often leading to political conflicts about how to contain the associated costs. Medical services in the U.S. composed 13 percent of the GNP in 1998.

Similarly, a changing labor market and increasing demand for literacy, mathematical literacy, and computer skills at the workplace have driven the increasing demand for educational services at all levels, a process reflected in higher enrollments in universities, degrees from which have become prerequisites to obtain middle-class jobs. Thus, today almost 66 percent of secondary-school graduates in the U.S. start a college or university education, and the proportion of the labor force with a college degree has risen to almost 30 percent.

c) The growth of services reflects the rising proportion of nondirect production workers, including firms in manufacturing. Most corporations devote considerable resources to dealing with highly segmented and rapidly changing markets and legal environments, including specialized clients, complex tax codes, environmental and labor restrictions, international competition, sophisticated financial systems, and real estate purchases. To do so, they require administrative bureaucracies to process information and make strategic decisions: clerical workers process paperwork, sales people and researchers study market demand and create new products, and legions of advertisers, public relations experts, accountants, lawyers, and financial experts assist in a complicated decision-making environment.

d) The increasing size and role of the public sector further underpins growth of the service sector. The state is a major actor, subsidizing firms and framing the broader legal and institutional context of economic activity. Government contributes to the growth of services in two ways: In the United States, for example, public sector employment has increased steadily due to rising demands for the services that it provides, ranging from defense to local libraries. Today, the federal government is the largest single employer in the U.S., employing more than 2.5 million people, and

federal employment is dwarfed by state, county, and municipal government employment, which totals more than 15 million. A second way in which government contributes to the growth of services is through laws, rules, restrictions, and regulations, contributing to the increase in specialists such as tax attorneys, accountants, consultants, and others who assist firms in negotiating with the legal environment.

e) Rising levels of service exports within and among nations stimulates service sector growth. Many cities, regions, and nations derive a substantial portion of their aggregate revenues from the sale of services to clients located elsewhere in the same country or overseas. For example, New York banks make loans to clients around the U.S. and abroad. Services are extensively traded on a global basis, composing roughly 20

percent of international trade. Internationally, the U.S. is a net exporter of services, which is one reason services employment has expanded domestically; services compose roughly one-third of total U.S. export revenues. These sales overseas take many forms, including tourism, fees and royalties, sales of business services, and repatriated profits from bank loans.

LABOR MARKET AND GEOGRAPHIC STRUCTURE OF SERVICES

Location of consumer services essentially is dictated by the location of the client base, typically segmented demographically by age, sex, ethnicity, and purchasing power. Thus, the suburbanization of consumer services—mainly through the shopping mall—has occurred in tandem with the movement of the middle class to the periphery of cities. The geography

Changing Labor Markets

The emergence of services has changed labor markets in the Western world in several important respects. First, services tend to employ far larger numbers of women than did manufacturing, which was predominantly a male domain. Women compose 45 percent of the U.S. labor force. Many women, however, work in low-paying occupations such as

retail trade or daycare, leading to concerns of equality.

Second, services tend to be less unionized than manufacturing; deindustrialization and rising services employment have lowered the share of the labor force in unions from 45 percent in 1948 to 14 percent in 1999. Third, services on the whole tend to pay less than manufacturing, leading to fears that the

transition to a service-based economy may depress household incomes. On average, clerical jobs, for example, pay 60 percent of what manufacturing pays, and retail trade only 50 percent as much. Some services, of course, consist of well-paying professional jobs, but the more rapid growth has been in part-time, low-skilled, and low-paying jobs.

Central Place Theory

Central place theory is an interpretation of city-systems first articulated by German geographer Walter Christaller in 1933, which centers upon consumer demand, including the maximum distance consumers will travel for a given item (range) and the minimum market size necessary to sustain distributors of different goods (threshold). A hierarchy of central places arises to distribute a hierarchy of goods and services from places with varying degrees of specialization. Influential in economic and urban geography, central place theory posits that urban *hinterlands* will form nested hexagons.

of retail trade is a complex topic in its own right. It involves not only traditional marketing variables such as the size of the client base, purchasing power, and transportation networks, all of which change through time, but also the design, location, and symbolic meanings of shopping malls and strip-mall developments, which have come to dominate the American retail landscape. This geography is continually reworked by the development of franchises, the homogenization of consumer tastes, changing incomes and prices, demographic shifts, and advertising, which plays a critical role in the social construction of demand.

The location of *producer services* is somewhat more complex than that of retailing. Producer services tend to be concentrated in metropolitan areas, particularly the skyscrapers of a downtown, although the most rapid rates of growth have occurred on the urban periphery. In general, high value-added service functions such as headquarters and administrative functions tend to cluster in metropolitan areas, whereas relatively low-wage, low value-added functions such as data entry have largely dispersed to the urban periphery and beyond.

Financial services are a particularly important part of the national and global economic landscape. These include commercial, savings, and investment banks, and insurance companies of various types. They provide credit, allowing the costs of production and purchasing goods and services to be separated over space and time, and linking borrowers and savers with widely varying needs. Because banking is so critical to national economies, it is highly regulated throughout the world. Banks play important roles in determining national money supplies, which in turn heavily affect interest, inflation, and currency exchange rates. In the U.S., banking has been shaped by the Federal Reserve System, which forms the equivalent of a national bank.
More at:

http://www.census.gov/econ/www/

Quaternary Sector Activities

As tertiary sector activities have become increasingly complex and sophisticated, a new economic sector has emerged that includes jobs related to knowledge creation, manipulation, and distribution. The quaternary sector encompasses high-end services such as research and development, technology and communications, education and medicine, and financial management, among others.

Financial services were particularly affected by the new telecomunications technologies—especially fiber optics—that came into play in the 1980s. Because they are very information-intensive, financial services have been at the forefront of the construction of extensive worldwide leased and private communication networks. Electronic funds transfer networks form the nervous system of the international economy, enabling banks to move capital around at a moment's notice, make money by playing on interest-rate differentials, take

Two hundred miles of freeways reveal Houston's reliance upon the automobile for primary transportation. A major industrial, commercial, and financial hub, Houston is the center of the U.S. aerospace industry, the national petroleum industry, and one of the country's busiest ports.

advantage of favorable exchange rates, and avoid political unrest.

Subject to digitization, information and capital are two sides of the same coin. In the *securities markets*, telecommunications systems facilitate the linking of stock and bond dealers through computerized trading programs. Trade on the New York Stock Exchange, the world's largest, rose from 10 million shares per day in the 1960s to more than 1 billion per day in the 1990s, and brokers buy and sell to foreign clients with as much ease as they sell to clients next door.

The ascendancy of electronic money has shifted the function of finance from investing to transacting. Traveling at the speed of light, global money performs an electronic dance around the world's networks in astonishing volumes. The world's currency markets, for example, trade roughly $800 billion every day, dwarfing the $25 billion that changes hands daily to cover global trade in goods and services. The boundaries of countries have little significance in this context: It is much easier, say, to move a billion dollars from London

to New York than a truckload of oranges from Florida to Georgia.

These circumstances have also favored the growth of *offshore banking,* financial activities in deregulated sites that seek to lure foreign accounts with liberal tax and regulatory legislation. Offshore banking has become important to many small countries including the Bahamas in the Caribbean, Luxembourg, San Marino, and Liechtenstein in Europe, Cyprus and Bahrain in the Middle East, and Vanuatu in the South Pacific. Thus, as the technological barriers to moving money around internationally have fallen, legal and regulatory ones have increased in importance, and financial firms have found the topography of regulation to be of the utmost significance in choosing locations.

TELECOMMUNICATIONS

As multiple economic activities have expanded over ever-larger distances, including the worldwide spaces of the global economy, the need and means to transmit information have grown accordingly. As a result, contemporary economic landscapes are closely tied to the deployment of telecommunications systems.

Geography is the study of how societies are spread over Earth's surface, and a vital part of that phenomenon is how people come to know and feel about space and time.

Space and time are in fact social constructions and every society develops different ways of dealing with and perceiving them. For example, urbanites accustomed to the hustle and bustle of city life tend to view time and distance in markedly different terms from those who live in the relative quiet of rural areas.

Telecommunications have been critical to the ongoing reconfiguration of time and space for more than 150 years, accelerating the flow of information across distance and bringing places closer to one another in relative space through the process of *time-space compression*. Time-space compression is the notion that distance can be measured in terms of the time required to cross it via transportation and communications. Steady improvements in the velocity of transportation and communications have reduced the time as well as the cost required to interact among places.

Despite the proliferation of new technologies, the telephone remains by far the most commonly used form of telecommunications. Faced with mounting competition, telephone companies have steadily upgraded their copper cable systems to include fiber-optic lines, which allow large quantities of data to be transmitted rapidly, securely, and virtually error free.

With the digitization of information in the late 20th century, telecommunications steadily merged with computers to form integrated networks, most spectacularly through the Internet. Incontestably, the Internet is the largest electronic network on the planet, connecting an estimated 100 million people in more than 100 countries.

The Internet emerged on a global scale via its integration with existing telephone, fiber-optic and satellite systems. Popular access systems in the U.S. allow any individual with a microcomputer and modem to plug into cyberspace, the world of electronic computerized spaces encompassed by the Internet and related technologies such as the World Wide Web. Cyberspace may exist in an office, a sailboat, or virtually anywhere.

As millions of new users log onto

Securities Markets

Securities markets, one of a series of financial markets and institutions involved in buying and selling equities (stocks and bonds), foreign currency exchange, and investment management (takeovers, buyouts, and mergers) in the U.S. are legally differentiated from banking by the Glass-Steagall Act of 1933, which prohibits commercial banks from trading stocks. Deregulation of finance, globalization of markets, and new telecommunication systems have restructured the industry —long dominated by London, New York, and Tokyo—including screen-based trading on the National Association of Securities Dealers Automated Quotation system (NASDAQ) stock market.

the Internet each year, cyberspace has expanded rapidly in size and in use and importance, including e-mail and electronic commerce. While popular mythology holds that cyberspace exists "everywhere," in fact access is uneven, both socially and geographically. Within the U.S., the heaviest users tend to be white, middle-class males. Inequalities in access to the Internet internationally reflect the long-standing divide between developed and developing countries.

In fact, telecommunications are generally a poor substitute for face-to-face meetings, the medium through which most corporate interaction occurs, particularly when the information involved is proprietary and nonstandardized in nature. For this reason, a century of technological change, from the telephone to fiber optics, has left most high-wage, white collar, administrative command and control functions clustered in downtown areas, despite their high rents.

In contrast, telecommunications is ideally suited for the transmission of routine, standardized forms of data, facilitating the dispersal of functions involved with their processing. Whereas the costs of communications have decreased, other factors have risen in importance,

including local regulations, the cost and skills of the labor force, and infrastructure investments. Economic space, in short, will not evaporate because of the telecommunications revolution.

Back offices, an example of dispersed function to low-wage regions, are the part of large service corporations that involve unskilled clerical functions, typically filled by women, such as data entry and processing of medical, insurance, or billing records.

Historically back offices were located next to headquarters activities in downtown areas to ensure close management supervision and rapid turnaround of information. Today back offices have become increasingly detached through the deployment of telecommunications systems and currently employ about 250,000 people in the U.S., frequently operating on a 24-hour basis.

Recently, back offices have also begun to relocate on a much broader, continental scale, making them increasingly footloose, meaning they are not linked to any particular place. Many financial and insurance firms and airlines have moved their back offices from New York, San Francisco, and Los Angeles to communities in the Midwest

and South, which have lower wages and lower rents.

Internationally, this trend has taken the form of the offshore office, which generates cost savings for U.S. firms by tapping labor pools in developing countries. Several New York-based life insurance companies, for example, have relocated back office facilities to Ireland, shipping in documents by express mail services and exporting the digitized records back via satellite or one of the numerous fiber-optic lines that connect New York and London.

Likewise, the Caribbean, particularly English-speaking countries such as Jamaica and Barbados, has become an important locus for American back offices. Such trends indicate that telecommunications may accelerate exporting many low-wage, low value-added information-based jobs from the United States.

SPECIALIZATION AND TRADE IN THE WORLD ECONOMY

Understanding conomic geography means seeing how regions and nations are linked by goods and services, business, and investment. Since the expansion of capitalism in the 16th and 17th centuries, the economic structure of local places has been increasingly tied to their role

The creation of the Internet links the world in an unprecedented way, and is rapidly changing the face of international trade and communication. As millions of new users log on to the Internet each year, cyberspace has expanded in size, use, and importance, including email and electronic commerce. Still, inequalities in access to the Internet around the globe reflect the divide between developed and developing countries.

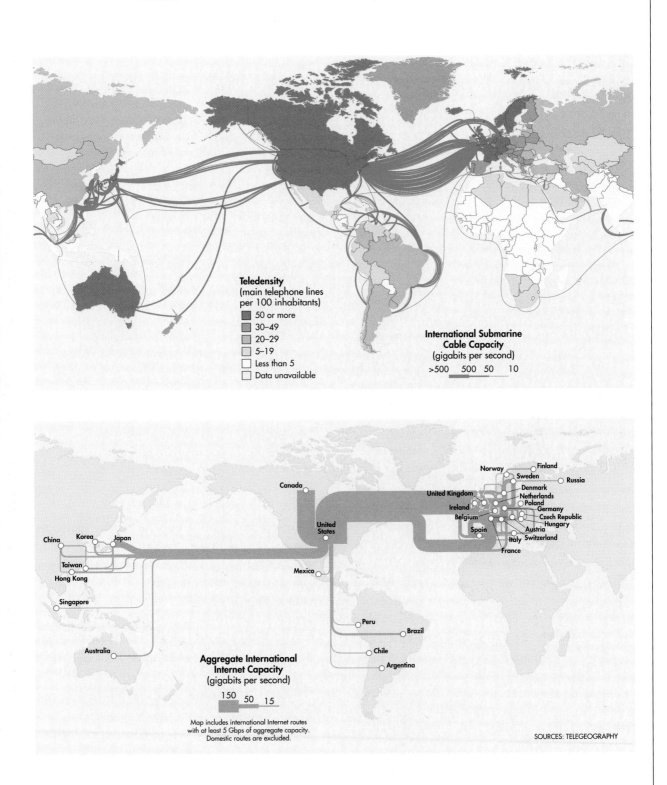

Teledensity
(main telephone lines
per 100 inhabitants)
- 50 or more
- 30–49
- 20–29
- 5–19
- Less than 5
- Data unavailable

**International Submarine
Cable Capacity**
(gigabits per second)
>500 500 50 10

Canada

United
States

Mexico

China
Korea
Japan
Taiwan
Hong Kong

Singapore

Australia

Peru
Brazil
Chile
Argentina

Norway Finland
Sweden Russia
United Kingdom Denmark
Ireland Netherlands Poland
Belgium Germany
Czech Republic
Hungary
Spain Austria Switzerland
Italy
France

**Aggregate International
Internet Capacity**
(gigabits per second)

150 50 15

Map includes international Internet routes
with at least 5 Gbps of aggregate capacity.
Domestic routes are excluded.

SOURCES: TELEGEOGRAPHY

A Brief History of the Internet

The Internet can be traced to 1969, when the U.S. Department of Defense founded ARPANET (Advanced Research Projects Administration Net) at Menlo Park, California, a series of electronically connected computers with high-capacity transmission lines designed to withstand a nuclear onslaught.

The durability and high quality of much of today's network owes its existence to its military origins. In 1984, ARPANET was expanded and opened to the scientific community when it was taken over by the National Science Foundation.

The foundation incorporated ARPANET into its system, NSFNET, and other Internet backbones that linked supercomputers in New York, Washington, Chicago, Dallas, San Francisco, and Los Angeles.

Networking allowed previously independent personal computers, then rapidly declining in price, to be connected through modems and copper cable, and, later, fiber-optic lines of great capacity. Early uses of the Internet, such as e-mail, rapidly were complemented by graphical interfaces made possible by hyper text markup language (HTML), which allowed the creation of the World Wide Web in 1989.

When NSFNET was decomissioned in 1995 and the Internet was privatized, several telecommunications companies provided the backbone, including MCI and Sprint. The corporate presence has grown rapidly, with electronic advertising and shopping. In 1998 40 percent of U.S. households owned personal computers, although only 12 percent were connected to the Internet. The Internet has grown rapidly on a global scale – with more than 600 million users worldwide – through the integration of existing telephone, fiber-optic and satellite systems, a task made possible by the technological innovation of packet switching, in which individual messages may be decomposed, the constituent parts transmitted by various technologies, and then reassembled, virtually instantaneously, at the destination.

The notion that "telecommunications will render geography meaningless" is naive. This view holds that electronic communications are a viable substitute for virtually all activities, allowing everyone to work at home via telecommuting, dispersing economic functions, and spelling the obsolescence of cities. This argument fails in the face of persistent growth of dense, urbanized places.

in the global economy; this trend was particularly true in the 20th century and, no doubt, will become even more relevant in the 21st century.

Industrial Location

A central theme in understanding where and why industries locate where they do is *comparative advantage,* which is the ability of a place to produce an output more cheaply, efficiently, or profitably than other places. The theory of comparative advantage explains the spatial division of labor and the specialization of places in the production of some types of goods and services and not others. When linked to trade, comparative advantage induces a specialization of production that rewards efficient producers and punishes inefficient ones.

Comparative advantage is readily evident across the globe: Developing countries frequently produce and sell primary sector materials (foodstuffs and mineral ores) and purchase manufactured goods. For example, OPEC countries produce oil; South Africa exports gold and diamonds. Within the United States, Appalachia, an economically depressed region, has long been a coal-producing region, the Southeast a producer of cotton and tobacco, and the agricultural Midwest a world-renowned wheat, corn, and dairy producing region.

Individual cities often specialize too. New York, for example, is the country's leading producer of financial and producer services, and during the heyday of the manufacturing belt, Pittsburgh was the world's leading steel-making city; Akron, Ohio, the rubber-making capital of the world; and Memphis, Tennessee the leading cotton seed oil processing center.

The theory of comparative advantage holds that specialization and trade allow resources—land and labor, for example—to be reallocated toward a region's or country's most efficient or profitable uses. To realize comparative advantage, trade networks must thus be well developed. Specialization improves the efficiency of production, lowering costs to consumers and raising standards of living. Specialization may entail costs for inefficient regional producers who, facing decline at the hands of imports from other, more competitive regions, call for protectionism.

Protectionism is government policy designed to protect domestic producers from foreign competition by limiting imports through tariffs, quotas, and nontariff barriers. Arguments for protectionism center around job protection in threatened industries, infant industries just initiating production, and national defense; critics, including most economists, argue that protectionism rewards inefficiency and removes incentives to become competitive internationally. For this rea-

A vista of sleek skyscrapers and cranes reveals Singapore's prominence. Its container port vies with Hong Kong's as number one in the world.

son, economists typically favor unrestricted trade.

To realize a comparative advantage and become part of the global system, places must forfeit their economic independence and become reliant upon a system of sales and purchases of goods and services with other places. Exports do not necessarily mean foreign sales of goods and services, although they might, but rather sales outside the local area of locally produced outputs. Thus, states and cities in the U.S. "*export*" to one another all the time. Further, exports are not synonymous with manufactured or agricultural goods. Cities frequently export services, including financial services, advertising, legal, and medical services, entertainment and tourism, or government services. Because they generate revenues, and thus sustain firms and generate jobs, exports are often viewed as the engine that drives both local and national economic growth.

More at:

http://ita.doc.gov/td/industry/otea/

Comparative advantages do not remain fixed in one place indefinitely. Regions may see their comparative advantages and export bases decline as other regions gain advantage when capital flows across space. Some regions are abandoned and new ones created. For example, the decline of the textile industry in New England in the 1930s and 1940s was matched by its corresponding growth in parts of the South, particularly the Carolinas and Georgia. The rise of the East Asian newly industrializing entities such as Hong Kong and Singapore, as well as Mexico in Latin America, coincided with the deindustrialization of much of the manufacturing belt in the U.S. In this way, the flow of capital across space

Kondratieff Waves

Named for Russian economist Nikolai Kondratieff who first identified them in the 1920s, *Kondratieff waves* reflect the instability of capitalist economies over time. Rather than smooth, continuous changes in output, prices, and employment, capitalism is typified by business cycles, swings, and oscillations of varying amplitudes and durations. Kondratieff identified long-term cycles in which commodity prices and output change in episodes lasting roughly 50 years. He linked these to clusters of innovations and innovative industries. While there is disagreement as to the precise dates that demarcate these cycles, most social scientists agree that the history of capitalism has seen roughly four waves: The first cycle, approximately 1770 to 1820, was linked to the textile industry; the second—about 1820 to 1880—reflected the introduction of steam power to steamships and railroads; the third—about 1880 to 1930—in which electricity and the modern steel industries were critical; and the fourth—about 1930 to the 1970s—in which the automobile, petrochemicals, and aerospace industries were significant. Many social scientists conclude that the 1970s marked the end of the Fordist Kondratieff wave and the beginning of a fifth, post-Fordist one in which electronics and producer services and telecommunications are highly important.

Geographically, Kondratieff waves help to explain changes in the comparative advantage of individual places as their competitiveness rises or falls with the introduction of new products, industries, and production technologies. More recent interpretations of business cycles have added other, shorter term cycles to these long-term ones. These new cycles are of roughly 15 to 20 years' duration—less than half the original. They are linked to aggregate patterns in the formation and depreciation of investment capital.

More at:

http://faculty.washington.edu/modelski/IPEKWAVE.html

reproduces uneven development, as some regions gain while others lose.

One of the more intriguing theoretical issues in economic geography today is the interrelationship between changing cycles of capital investment and the construction of regional landscapes. Regional landscapes may be thought of as consisting of numerous layers of investment, each corresponding to a different regional comparative advantage with its associated export base, labor markets, infrastructure, and landscape, that developed over time, much as geological strata are deposited successively one atop the other. For example, New England's textile industry was followed by the rise in the electronics industry, which nestled within the vestiges of textile production, sometimes literally in old factories such as those in Lowell, Massachusetts, creating a landscape that reflects the superimposition of one activity on top of the other.

As specialization became more pronounced and as transport costs and protectionist barriers declined throughout the 20th century, world trade increased in magnitude and importance. Since World War II, world trade has grown more rapidly than total world output, reflecting the increased integration of national economies. Today, roughly 25 percent of the world's total out-

TOP 15 PARTNERS IN TOTAL U.S. TRADE—2003	
COUNTRY	COMBINED EXPORTS AND IMPORTS (US$ MILLIONS)
Canada	393,936
Mexico	235,531
China	180,798
Japan	170,093
Germany	96,895
United Kingdom	76,562
Korea, South	61,062
Taiwan	49,088
France	46,289
Malaysia	36,358
Italy	36,007
Ireland	33,539
Singapore	31,734
Netherlands	31,675
Brazil	29,102

put is traded among countries. Foreign trade is equivalent to roughly 24 percent of the U.S. economy, including exports (11 percent) and imports (13 percent), a difference manifested in the nation's persistent, large trade deficits in recent decades—often more than 100 billion dollars annually.

The majority of international trade is among industrialized countries, particularly the three major centers of Europe, North America, and Japan, all of which have the requisite purchasing power and productive capacities. Fluctuations in the volume and type of trade reflect many variables, including the global supply, demand, and prices of goods, the locations of production, currency exchange rates, levels of protectionism, and transport costs, all of which change regularly. More at:

http://www.wto.org/english/res_e/statis_e/statis_e.htm

The Multinational Corporation

Increasing global economic integration since World War II has largely been accomplished through a handful of large firms that conduct business in more than one country, that is, *multinational corporations* (MNCs), also called transnational corporations. Despite the stereotype of MNCs as global in scope, most operate within only a handful of countries. Most originate in the industrialized nations—the U.S., Japan, Britain, France, and Germany—and concentrate their investments in other industrialized countries, generating foreign direct investment, with mixed costs and benefits to the host country. *Foreign direct investments* (FDI) are tangible investments in productive capacity, such as buildings and equipment owned by firms from one country but located in another. Most FDI originates from industrialized countries and is located in other industrialized countries.

MNCs dominate global trade, finance and investment, research and development, technology transfer, and the commodity chains that permeate the world economy. In this capacity, they are the most important force in the acceleration of globalization, although the extent of their reach varies widely around the world.

MNCs are, by definition, not bound to a particular country, although all are headquartered in individual countries. In the 1990s, total MNC employment worldwide was estimated to be roughly 65 million, of which 43 million, or 66 percent, was located in developed countries. Many MNCs have a gross output larger than that of small countries.

MNCs invest overseas for reasons that center around maximizing profit, including the quest for low-cost labor, resources, and markets. Frequently, they enter into joint ventures and strategic alliances with one another or with local firms. MNCs usually have operations in relatively few countries, generally no more than two or three; only a few of the largest MNCs are truly global in scope.
More at:
http://www.itcilo.it/actrav/actrav-english/telearn/global/ilo/multi-nat/multinat.htm

Foreign direct investment by MNCs may take the form of direct investments, affiliates and subsidiaries, or acquisitions of existing firms and facilities. American firms lead the world in FDI, although their share of the world total

has gradually declined. Roughly 50 percent of FDI by U.S. firms is located in Europe, 20 percent is in Canada, and only 3 percent is in Japan, although the most rapid rate of growth is in East Asia, which, since the shift to export-promotion in the 1960s, has eagerly courted foreign capital.

Conversely, many foreign firms invest in the U.S., including Japanese, British, Dutch, Canadian, and German companies, to gain access to the large American market and to escape possible protectionist measures and exchange rate fluctuations. In 2003 total FDI in the United States amounted to almost $830 billion, and this was primarily in manufacturing.
More at:
http://www.bea.doc.gov/bea-home.html

In developing countries, MNCs are major sources of investment capital and the primary drivers behind the industrialization of selected countries, most notably the East Asian newly industrializing countries (NICs) and Mexico. One major difference between the industrialization of the developed and less developed countries is the domination of foreign capital in the latter.

The relative benefits and costs of MNC presence have been widely debated. The benefits include enhanced pools of investment cap-

ital, often in short supply in developing countries; greater employment opportunities, including job training and human capital formation (skills and experience); technical and managerial expertise; access to more productive machinery and equipment, technology transfer; superior distribution, marketing, maintenance, and sales networks; and improved balance of payments as exports increase in volume and value.

Disadvantages include potential disruptions of local producers and subsequent distortions of local, indigenous economic relations; the incentives that most countries offer to attract foreign capital; and, possible decline in national sovereignty. For small, relatively weak countries, large foreign corporations can be intimidating. Often, linkages with local producers, such as subcontracting, are minimal. The jobs created may not exceed the jobs displaced by foreign competition.

If MNCs remit the bulk of their profits to their home base, the host country stands to gain little. Many developing countries therefore regulate MNC activity and require profits to be reinvested locally. Because the net benefits of MNC presence are often unclear, tensions arise between an MNC's drive to maximize profits and the host country's needs for economic development.

Maquiladoras

In the 1980s, given the global glut of petroleum, the industry upon which Mexico had staked its economic future in the 1970s, that country undertook a systematic expansion in the diversity of its economy.

During this period, many countries opened varying forms of export-processing platforms—small, tax-free, often subsidized centers designed to attract foreign capital, stimulate jobs, and increase foreign revenues. These broadly corresponded to the shift from import-substitution to export-promotion globally. A maquiladora is a branch plant (mostly U.S. or Japanese) corporation in Mexico, which assembles goods or electronics components and relies on local women as the workforce. Located mainly in northern Mexico maquiladoras were fueled in large part by American corporations seeking low-wage labor.

Many U.S. firms paired their operations in Mexico with cities along the U.S.–Mexican border in California and Texas.

With the low price of petroleum in the 1990s, the maquiladoras have become the largest source of foreign revenues for Mexico. Despite occasionally deplorable standards of living, the region, long one of the poorer parts of Mexico, has gained in income, and the northern states have become the wealthiest part of the country.

Mexico's entrance into NAFTA with the U.S. and Canada in 1994 will have uncertain effects, because flows of capital (but not labor or goods) were already relatively unrestricted.

Over the long run more rapid rates of growth on both sides of the border are possible.
More at:
http://www.umich.edu/~snre492/Jones/maquiladora.htm

Regional Trade Blocs

Because there is widespread recognition that protectionist barriers to trade are detrimental to national economies, most governments have reduced tariffs and quotas on imports. Many countries, for example, have simple, bilateral agreements with trading partners in which trade barriers are minimized or eradicated on a product-by-product basis.

Since World War II, the primary vehicle for serving this purpose on a global basis has been the General Agreement on Trade and Tariffs (GATT), which, through a series of negotiations, systematically lowered tariff rates worldwide, contributing to the post-World War II global economic boom. GATT members originally were almost exclusively developed countries; today, most countries in the world are members.

In 1995 the GATT metamorphosed into the World Trade Organization (WTO), a permanent organization in Geneva, Switzerland, that also regulates trade disputes. The WTO regulates trade in services, but has yet to include important nontariff barriers, such as export restraints, inspection requirements, health and safety standards, and import licensing, which inhibit imports.
More at: http://www.wto.org/

In addition to these broad global agreements, many countries have joined *regional trading blocs*, associations of countries designed to reduce protectionism and enhance economic intercourse among member states. Countries hope that such blocs will enhance their competitiveness internationally, lower the costs of imports, and ultimately lead to gains in national income and purchasing power.

The most well developed regional trading bloc is the European Union (EU) headquartered in Brussels. Founded in 1957 by six member countries (Italy, France, West Germany, Belgium, the Netherlands, and Luxembourg), the

European Economic Community (EEC), as it was originally known, expanded to include most of Western Europe.

The EEC became an important factor in the Europe's attempts to recover from the 1970s oil crisis and slow economic growth, particularly in the face of intense Japanese and U.S. competition.

In 1992 the EEC launched an ambitious plan that extended far beyond simple elimination of trade barriers among its members to include free movement of factor inputs, including capital and labor, land, raw materials, and energy. Corporations from member states may now invest anywhere within the EEC without restriction, and workers may seek employment without restraint.

In addition, the EEC, which became the European Union in 1995, harmonized many production and trade regulations and has moved steadily, if unevenly, toward a common currency, the euro, launched in early 1999, effectively binding diverse countries into a single economy. In May 2004, the EU added ten more countries, bringing its total membership to 25 countries, with two more scheduled for membership in 2007.

With over 450 million people, the EU is the largest single market in the world. The long run conse-quences are likely to be accelerated economic growth rates as specialization intensifies and firms achieve economies of scale. Within Europe, the Mediterranean states may benefit more than their northern counterparts as labor moves north and capital flows south. Whether this trade regime will succeed in equalizing development within Europe is still to be determined.

More at:

http://europa.eu.int/index_en.htm

Compared to the EU, the North American Free Trade Agreement (NAFTA) is more modest. NAFTA's origins lie in the 1988 U.S.-Canada Free Trade Agreement, which gradually eliminated trade restrictions between the world's two largest trading partners.

In 1994 NAFTA was expanded to include Mexico, the first time a developing country was included in the same trade bloc as developed ones. However, whereas the EU allows free movements of labor, NAFTA does not, largely due to U.S. fears of unrestricted flows of Mexican labor northward.

More at:

http://www.mac.doc.gov/nafta/

As critics of NAFTA have emphasized, freedom of capital movements, however, is guaranteed, accelerating the growth of the maquiladora assembly plants located in Mexico that utilize Mex-ico's cheap labor. Maquiladoras are branch plants of foreign (mostly U.S. or Japanese) corporations located in Mexico. They are typically automobile or electronics assembly plants employing young women laborers. The possibility exists that NAFTA may be expanded to include other parts of Latin America, perhaps creating a free trade zone extending from Alaska to Tierra del Fuego.

More at:

http://www.mexconnect.com/business/mex2000maquiladora2.html

Several other trade blocs are found across the world. The Association of Southeast Asian Nations (ASEAN) includes most countries of this rapidly growing region. In Latin America, the Andean Common Market aims to enhance investment and growth, while the Caribbean Community (CARICOM) promotes cooperation among island countries. In West Africa, French-speaking countries formed the Union Douaniere et Economique de l'Afrique Centrale, the UDEAC (Central African Customs and Economic Union) with a central bank; in southern Africa, the Southern Africa Development Coordination Conference (SADCC) integrates national economies.

More at:

http://www.worldbank.org/research/trade/trade_blocs_abs.htm

European Union
- Member country
- Candidate country
- Other country
- *1995* Year of admission

ICELAND

Atlantic Ocean

NORWAY

SWEDEN
1995

FINLAND
1995

ESTONIA
2004

LATVIA
2004

LITHUANIA
2004

RUSSIA

DENMARK
1973

IRELAND
1973

UNITED
KINGDOM
1973

BELARUS

NETHERLANDS
1952

BELGIUM
1952

LUX.
1952

GERMANY
1952

POLAND
2004

UKRAINE

LIECH.

CZECH REP.
2004

SLOVAKIA
2004

MOLDOVA

SWITZ.

AUSTRIA
1995

HUNGARY
2004

FRANCE
1952

SLOVENIA
2004

CROATIA

BOSN. &
HERZG.

SERB. &
MONT.

ROMANIA

Black Sea

ANDORRA

BULGARIA

PORTUGAL
1986

SPAIN
1986

ITALY
1952

ALBANIA

MACEDONIA

GREECE
1981

TURKEY

MALTA
2004

CYPRUS
2004

Mediterranean Sea

AFRICA

Acting as a trade bloc, the European Union is a major player in the global economy. Even before the addition of 10 new members in 2004, the EU accounted for almost twice the comdbined trade of the U.S. and Canada plus Asia. Today, the three regions together account for more than 80 percent of all world exports and imports.

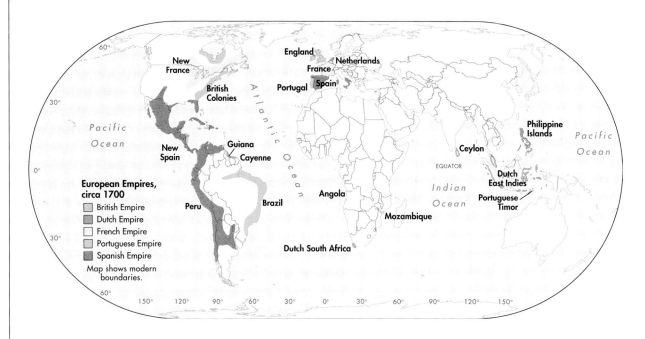

European Empires, circa 1700
- British Empire
- Dutch Empire
- French Empire
- Portuguese Empire
- Spanish Empire

Map shows modern boundaries.

Colonialism has long been the subject of intense controversy. While European civilization may have brought certain economic, technical, and educational benefits, at the same time colonialism disrupted entire societies and destroyed cultures.

DEVELOPING COUNTRIES' ECONOMIC PROBLEMS

An important concern of economic geography is the origins and nature of poverty in developing countries and the possibilities of its alleviation through economic development. The division of Earth into a relatively small group of wealthy countries comprising one-fifth of the world's population and a large number of relatively poor countries is perhaps the defining characteristic of the world's economic geography today. This section explores this important issue by reviewing colonialism and neocolonialism, examining the differences between developed and undeveloped countries, summarizing the major economic problems that are confronting developing countries, and reviewing three major ways in which these important issues have been analyzed conceptually.

Colonialism

Colonialism is the economic and political system by which some countries dominate others. Although colonialism is not an exclusively European phenomenon—for example, the Japanese colonialism in Korea and Taiwan—it was typified by Europe's conquest of the Americas, Africa, and Asia and the subsequent formation of a world economy that disproportionately benefited Europe.

In 1500, Europeans politically and economically dominated 9 percent of the world's land surface; by 1914 they dominated 85 percent, including the Americas, Africa, Asia, Australia, and the islands of the Pacific. Only Japan escaped completely, managing to become the only non-European country to rival the West on its own terms.

The incentives for Europeans to colonize were the resources, including labor, the colonies offered, as well as the markets they provided, which were essential to expanding capitalist enterprises. These economic considerations were overlain with other motivations, such as religious conversions. The means by which Europe accomplished this feat are not well understood, but include technological superiority—better ships, guns, horses, and after the industrial revolution, inanimate energy sources—and the inadvertent introduction of diseases—particularly in the Americas, where 98 percent of the native population succumbed to smallpox and to measles within a century of the Columbian encounter.

Two major waves of colonialism may be identified, which correspond to Europe's expansion before and during the industrial revolution. The size of these empires varied considerably among the colonizers as well as over time. (See European Empires map, page 282.)

The Spanish and Portuguese established the first great empires. Under the terms of the 1494 Treaty of Tordesillas, Spain claimed much of what is now the southwestern U.S. and Florida, as well as Guam, the Philippines, Cuba, Puerto Rico, Mexico, and Central America, and the western parts of South America. Portugal took what are now Brazil, Guinea-Bissau, Angola, Mozambique, Goa, Timor, and Macau. The Spanish empire provided vast quantities of silver and

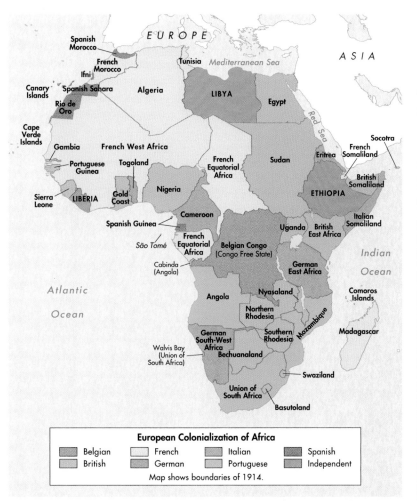

In 1884 the European powers, already colonizing Africa, met in Berlin to draw these boundaries of their possessions. Decolonization here and around the globe took place after World War II.

gold for its rulers, while the Portuguese empire centered mainly on plantation agriculture. The Spanish empire ended with the Spanish-American War in 1898; remnants of the Portuguese empire lingered into the 20th century.

France also conquered vast domains in Quebec, West and Central Africa, Madagascar, Indochina,

some islands in the Caribbean, as well as portions of the Arab world, including what are now Syria, Lebanon, and Algeria. France's empire effectively ended with their defeat in Vietnam in 1954 and the independence of Algeria in 1962.

In the 18th century the British claimed that "the sun never sets on the British Empire," which

The Berlin Conference, 1884: Europe's Division of Africa

In 1884 the European powers, then busily colonizing Africa, met in Berlin to draw the boundaries of their possessions in an attempt to minimize rancor among them.

The Berlin Conference essentially established the boundaries of the current states of Africa, which were created with no consideration to the inhabitants of the places they demarcated.

Thus, Great Britain claimed West African colonies such as Ghana and Nigeria, East African ones such as Kenya, and in Southern Africa, Botswana and South Africa.

The French took a large swath of the states in West Africa stretching from Mauritania to Chad, as well as Central African regions such as Gabon and Brazzaville (Republic of Congo).

Belgium took over almost all of the Congo River Basin, which became the private reserve of King Leopold II (1835–1909), under whose rule more than half the people died.

Portugal retained its long-standing colonies in Guinea-Bissau, Angola, and Mozambique, leaving in 1975.

Italy, having invaded Ethiopia and Eritrea, briefly claimed Libya and part of Somalia.

Finally, Germany, a late colonial power, took control of Togo, Namibia, German West Africa, Rwanda, Tanzania, and a section of Mozambique, losing these as a result of its defeat during World War I.

The impacts of these borders long after the colonies became independent is difficult to exaggerate. Most African states are collections of varying cultures, with radically different languages, religions, and economic practices. Civil war, tribal conflict, and secessionism have wracked the continent.

For example, in the 1960s, the Igbo people of Nigeria, which has more than a hundred tribes, attempted to form the state of Biafra; a government-imposed ban on food imports left a million dead of starvation. The Sudan has been caught in a murderous civil war between the Arab north and Christian and animist south that has killed another million. Hutu and Tutsi in Rwanda and Burundi, lumped together in the same states, have repeatedly engaged in genocide. Congo consists of more than 200 tribal groups, and has degenerated into civil war and anarchy.

Such conflicts not only brought death and misery to tens of millions of people but also deterred foreign and domestic investment and economic progress.

The poverty of Africa is thus in part political in root, a reflection of the particular way in which its landscapes were divided administratively in the late 19th century.

stretched over one-quarter of Earth, including, at various times, most of North America, parts of the Caribbean, Guyana, large parts of Africa (Ghana, Sierra Leone, Nigeria, Kenya, Sudan, Rhodesia, Southern Africa), portions of the Arab world (Egypt, Iraq, Palestine, the Persian Gulf, Yemen), southern Asia (Afghanistan to what is now Bangladesh), Burma (now Myanmar), Malaya (now Malaysia), Hong Kong, Australia, New Zealand, and many islands of the Pacific.

The Dutch were successful in New Amsterdam (later New York), Suriname and Curaçao, South Africa, and above all, Indonesia. The Belgians were confined to the Congo. Germany's short-lived excursions overseas saw it take what are now Togo, Rwanda and Burundi, Namibia, and portions of Tanzania and New Guinea, only to lose it all after World War I. Finally, Italy made brief conquests in Libya, Somalia, and Ethiopia.

Although the broad contours were similar, colonialism took different forms in different regions. In the Americas, disease (particularly smallpox and measles) and genocide effectively depopulated two continents, resulting in the deaths of 50 to 80 million Native Americans. From Africa, the slave trade brutally shipped 15 million people to the labor-short New World, particularly to South America and the Caribbean.

Africa's colonial boundaries, drawn during the Berlin Conference of 1884, collapsed roughly a thousand tribes into 50 states. In Africa, states generally consist of disparate ethnic, linguistic, or religious groups lumped together with little obvious rationale other than European economic advantage, a factor that has greatly aggravated contemporary civil wars and secessionist movements. (See European Empires and European Colonization maps, pages 282, 283.)

Colonialism had profound economic, political, and social effects. In large part, production systems in colonies became centered exclusively around exports of primary sector goods, particularly minerals and foodstuffs.

The plantation system—including cotton and tobacco in North America, sugar and coffee in Latin America, cotton in Egypt, peanuts, tea, and cocoa in Africa, tea in India and Ceylon (now Sri Lanka), and rubber and coffee in Southeast Asia—saw *cash crops,* crops grown for sale for profit, displace subsistence agriculture in many regions. Silver mines in Mexico and Bolivia, copper in Chile and southern Africa, gold and diamonds in South Africa, and tin in Borneo exemplified a different form of resource extraction. Even Canada was colonized primarily around exports of lumber, fish, furs, and wheat. This restructuring has had long-term consequences for many developing countries that find their exports confined to relatively cheap commodities that generate few foreign revenues, resulting in poor terms of trade.

Politically, colonialism accentuated the power of small groups of local people who prospered by cooperating with the colonizers. The assistance of this elite in governance, taxation, military rule, and administration was essential. Frequently, colonizers employed a particular ethnic minority to help in ruling their own people. For the bulk of the population living in rural areas, colonialism offered few benefits and generated unequal patterns of land ownership, taxation, job opportunities, and the disruption of indigenous agriculture.

Colonialism's social polarization was mirrored in its dual geographies. Port cities, for example, became the primary centers of economic activity, including Buenos Aires, Lima, Lagos, Cape Town, Calcutta, Rangoon, Singapore, Jakarta, Saigon, and Hong Kong, often displacing indigenous landlocked capitals (such as Cuzco, Timbuktu, Mandalay, Delhi, and Jogjakarta). Railroads from these port centers offered access to the wealth of the interiors, allowing minerals as well as plantation products to be exported easily. As a result, rural-to-urban migration in the colonized regions increased quite dramatically.

Europe's colonial empires effectively came to an end in the aftermath of World War II, when many nationalist and independence movements succeeded in establishing nominally independent states. Often led by Western-educated intellectuals such as Ho Chi Minh (1890–1969) in Vietnam and Kwame Nkrumah (1909–72) in Ghana, these movements were frequently violent, although Mahatma Gandhi's (1869-1948) program in India was a notable exception.

From the 1940s through the 1980s the process of decolonization witnessed the emergence of dozens of independent states around the world. Today those states number nearly 200. (See Dates States Became Independent map, page 313.) However, the formal end

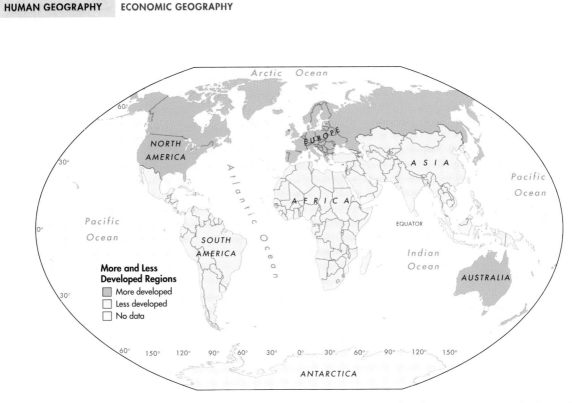

The geography of the more and less developed regions of the world reveals the long-term economic effect of European empires on subordinate colonies.

of colonial political rule did not automatically translate into de facto economic independence. Despite the promise of independence, most colonies had little preparation, scarce investment capital (which was mostly owned by multinational corporations), no management experience, inadequate infrastructures, poor education systems, and could offer only cheap labor and raw materials.

Competitively, then, most newly independent countries were disadvantaged in the world market from the outset. These economic problems were compounded by political instability, including unrest and civil wars, which complicates attempts to attract capital. Thus, formal colonialism is replaced by *neocolonialism,* that is political independence limited by lingering economic dependence. Although the U.S. is not usually labeled a formal colonial power in the same sense as European countries were, the hegemony of American MNCs after World War II has made neocolonialism largely, but not exclusively, an American phenomenon.

Development Problems in Developing Countries

The majority of the world's people live in developing countries, and most of them are poor. Although definitions and perceptions of poverty vary over time and are specific to different social and historical contexts, developing countries face a series of interlinked problems that severely hamper their opportunities for economic progress and improvement in life chances for billions of people. The severity and prevalence of these problems vary widely across the world, and some of them may also be found within portions of the developed world, such as the inner cities of the U.S.

♦ Unequal land distribution is a major political and economic issue in societies in which the bulk of the population lives in rural areas where land is frequently the major source of wealth and income. Often, a small, wealthy elite may own the bulk of arable land, as in Latin America, where the legacy of the colonial Spanish land grant system is still evident.

Much of this land is used to produce crops for export on the international market, leaving large numbers of the rural poor landless, able only to sell their labor at below-subsistence wages. Struggles and conflicts over land tenure are a major cause of social disruption, such as in parts of Central America and southern Africa. Conflicts over

North and South

The long-term consequences of colonialism are difficult to overemphasize in their significance for the contemporary world economy. A widely used term to describe the schism between the world's have and have nots is the global North and South. North, in the context of the global North-South divide, is the economically developed countries (or First World) of Europe, Japan, and North America, including roughly one-fifth of the world's population. South, in terms of the global North-South distinction, is essentially the Third World countries of Latin America, Africa, and Asia (excluding Japan), and includes roughly four-fifths of the world's population.

Such broad terms as First or Third World, or North and South, conceal as much as they reveal. The Third World, for example, consists of an enormously diverse collection of states and societies in terms of their cultural heritage, languages and religions, ethnic and racial compositions, colonial experience, positions in the global economy, and standards of living. Economically, the Third World includes countries as desperately

poor as Mali and as relatively prosperous as South Korea. Singapore has a higher standard of living than much of Europe, yet is often included in the Third World. Such discrepancies reflect the fact that, ultimately, all geographic labels are biased and somewhat arbitrary.

Whatever terminology one uses, and however one measures it, the division between the world's developed and developing countries is real. Using the flawed labels described above, about 20 percent of the world's 6 billion people live in the developed world and 80 percent live in the developing world; in terms of global output of goods and services, however, 86 percent of the output is generated and consumed by developed countries and only 14 percent by the poorer countries of the South.

The poorest 20 percent of the planet consumes just 1.3 percent, while the richest 20 percent consumes 45 percent of all meat and fish, 58 percent of all energy, 84 percent of all paper, 74 percent of all telephones, and owns 87 percent of all automobiles. A common measure of variations in income is

gross national income in purchasing power parity (GNI ppp), which is a measure of the buying power of a given currency coverted to a standard based on the buying power of U.S. dollars. The GNI ppp varies widely across the surface of Earth, ranging in 2004 from Norway's $36,690 (the U.S. stands at $36,110) to a low of $500 in Sierra Leone, a reflection of extreme poverty and largely unmeasured subsistence production.

Other indexes yield a similar geography, including key measures such as per capita energy consumption, life expectancy, infant mortality rates, and literacy levels, all of which mirror the enormous discrepancies between the global core of wealthy, industrialized countries and the world periphery of poor former colonies. Further, the gap between the world's richest and poorest people has been steadily widening for years; for example, in Africa, the average household consumes 20 percent less than it did 25 years ago.
More at:
http://hdr.undp.org/reports/global/2003/

demands for land reform may cause social disruption and accelerate rural-to-urban migration.

Rapid rates of population growth, which are common in rural areas, add to the constant demand for farmland. This problem is further compounded by agricultural mechanization: In the West, mechanization of farming occurred in the context of labor shortages, but in developing countries it occurs in the context of labor surplus.

◆ Urbanization without industrialization refers to the rapid growth of cities in the developing countries but without the attendant job opportunities that characterized the industrializing cities of Europe, Japan, and North America. Most of the world's largest metropolitan regions are located in the developing world, and the bulk of urban growth in the 21st century will be located there as well.(See World's Largest Urban Agglomerations table, page 303.) The growth of these metropolitan areas is propelled largely by rural-to-urban migration, itself a reflection of problems in rural areas, rather than high natural growth. Within these cities, elevated rates of unemployment, underemployment, and poverty are common.

Frequently, urban economies of developing countries are divided into two sectors: a relatively small,

formal sector includes stable, relatively well-paying jobs with regular working hours—for example in multinational corporations or the government—and often consists of segments of the national economy directly linked to the world-system. The second segment is a larger, *informal sector*, with unstable, unregulated, low-paying jobs—such as day laborers, black-market activities, prostitution and crime, street hawking, and recycling of garbage. The size of the informal sector varies with the level of economic development; in the Asian NICs, for example, it is relatively small, whereas in much of Latin America, Africa, and South Asia, it comprises the majority of employment.

◆ An insufficient supply of housing plagues large numbers of people throughout the developing world. Many urban dwellers construct their own housing out of locally available materials. The result is frequently poorly made, uncomfortable, or unsanitary dwellings that comprise vast slum districts or shantytowns, usually on the outskirts of the urban region, with inadequate electricity, transportation, clean water, or medical care. Densities in such neighborhoods greatly exceed those in Western cities. Whereas in most cities of the developed world the poor are a

minority of the population, in many developing countries—depending upon overall levels of economic status—they comprise the majority.

◆ Most exported goods from developing countries have relatively low value, while imports are higher-valued. Many countries export foodstuffs and minerals on the global market and import relatively expensive manufactured goods, resulting in a negative trade balance. This phenomenon is largely attributable to the colonial global division of labor and the critical role of the plantation and mining systems. Export revenues from raw materials are kept low by the numerous suppliers of such goods, low prices, and low-income *elasticities of demand*. Elasticity of demand is a measure of the degree to which the aggregate demand for a commodity rises or falls with changes in income or price.

The late 20th-century global glut of commodities has severely lowered the prices of most such goods, hampering exporters' ability to generate foreign revenues and purchase badly needed imports, thus creating a cycle in which limited supplies of investment capital inhibit economic diversification and productivity growth.

◆ Foreign debt is a significant shadow that hangs over the prospects of economic progress in many developing countries. The origins of this problem lie in the oil crisis of the 1970s, which halted a generation of development and created the conditions in which many Western banks loaned sizable quantities of funds to the governments of countries in Asia, Africa, and above all, Latin America. (See Foreign Debt Burden map, page 292.)

Such loans often went to finance oil imports or large-scale, poorly managed development projects that yielded little in return due to corruption, mismanagement, or capital flight. For example, Brazil and India both borrowed billions of dollars in the 1980s for the construction of hydroelectric dams, which contributed little to national economic growth.

Countries that staked their economic future on petroleum exports, such as Mexico, were devastated by the collapse in oil prices in the 1980s and 1990s. Constrained by their low export earnings, many countries find the repayment of such loans to be difficult, and in the most dire cases have threatened default.

◆ Inadequate public services and infrastructure are commonly encountered across the global South, although the level and quality of these vary dramatically with degree of economic development and government policy. Much of the infrastructure in developing countries was originally constructed by colonial powers, for example British rail lines in Africa and India and Japanese ports in Taiwan and Korea, although the primary focus was facilitating the colony's role in the colonial economic system. Insufficient maintenance of the infrastructure results in high transportation and communication costs. Inadequate health-care services, including, most critically, polluted-water supplies and sewer systems as well as severe shortages of physicians and pharmaceuticals, contribute to high death rates, particularly in the forms of infant mortality and prevalence of infectious diseases.

◆ Educational systems in many developing countries are insufficiently funded, with poorly paid and trained teachers and large, under equipped classrooms, resulting in low literacy rates and unskilled labor supplies. This problem is particularly true for females, because many families in rural areas are more willing to invest scarce educational resources in their sons than in their daughters; as a result, female literacy rates are considerably lower than male ones around the world. (See Percentage of Population Literate map, page 204.)

◆ Corrupt, inefficient governments with skewed budgetary priorities also plague many developing countries. This problem may be manifested in large, poorly managed public bureaucracies that often act as patronage systems for members of influential tribal groups or individuals. Many governments are repressive, curtailing civil rights, imprisoning dissidents, implementing press censorship, and savagely crushing ethnic, political, labor, religious, or regional opposition movements. Many states have misplaced spending priorities. For example, the average developing country spends ten times as much on military expenditures as agricultural development; India has a nuclear-weapons program but lacks compulsory primary education.

◆ Political instability is a frequent phenomenon in many developing nations, including military coups d'etat, tribal conflicts, civil wars, and secessionist movements. These problems are particularly acute in Africa—Angola, Sudan, and Sierra Leone, for example—where colonial boundaries have exacerbated or created conflicts over land ownership and governance. The consequences of such turbulence include genocide, as in Rwanda, the disruption of agricultural systems, as in Zimbabwe, and hampered efforts to attract foreign investment throughout Africa.

◆ Ecological problems in developing countries are frequently much more severe than in developed countries. Their origins can be attributed to rapidly growing populations and to inadequate legislation to preserve public open spaces and ecosystems; air pollution levels in many cities, which typically have few environmental safeguards, are often hazardous. Deforestation and soil erosion are severe problems in many countries when rural families seek sources of firewood or agricultural land or grazing for herds. These problems accentuate disasters such as mudslides or susceptibility to drought and contribute to the steady erosion of wildlife habitats around the globe. Many residents of developing countries regard ecological protection as a luxury that only wealthy countries can afford. Environmental destruction is, at root, a political phenomenon; to stop it, the structural origins of poverty must be changed.

Theories of Development Among Developing Countries

Explanations of poverty in developing countries, development, and the lack of it, fall primarily into three major schools of thought, including modernization theory, dependency theory, and world-systems theory. Frequent, often heated,

debates over these issues compose a significant part of economic geography today.

MODERNIZATION THEORY

The most frequently encountered approach to Third World development, modernization theory, has its origins in German Max Weber's (1864–1920) theory of capitalist development. Development was portrayed as a "race" in which the West was the leader and the Third World "lagged behind" and must "catch up," that is, follow the historic pathways of the developed world. This view celebrated capitalism as a system whose wealth created "a rising tide [that] lifts all boats" and that its absence was the primary explanation of poverty. Accordingly, the introduction of Western ideas, technology, and economic and political institutions, through foreign aid, trade, and diffusion was held to promote the gradual evolution of developing countries into prosperous, democratic states.

Modernization theory was influential in U.S.-led development initiatives after World War II, such as federal programs administered by the Agency for International Development and the Peace Corps. The most famous expression of modernization theory was American economist Walter Rostow's

work, The Stages of Economic Growth (1960), which compared the development process to that of an airplane taking off. Rostow argued that traditional societies were not simply "primitive" or "backwards" in terms of available technologies and standards of living, but lacked a growth-oriented culture. During the "preconditions to take-off" many urban-centered groups and the government begin to create the conditions that favor growth, often by displacing conservative rural elites. The "take-off" phase, akin to the U.S. in the late 19th century or the East Asian newly industrializing countries after World War II, sees growth become normalized and self-sustaining. In this phase, rapid technological change and capital investments, including foreign investment, create productivity growth rates that exceed population growth, resulting in rising standards of living. The fourth phase, the "drive to maturity," involves the creation of a literate, urbanized labor force and capacity to produce capital goods. Finally, the last phase, "mass consumption," sees prosperity, a large middle class, and living conditions akin to those in the West.

More at:

http://www.mtholyoke.edu/acad/intr el/ipe/rostow.htm

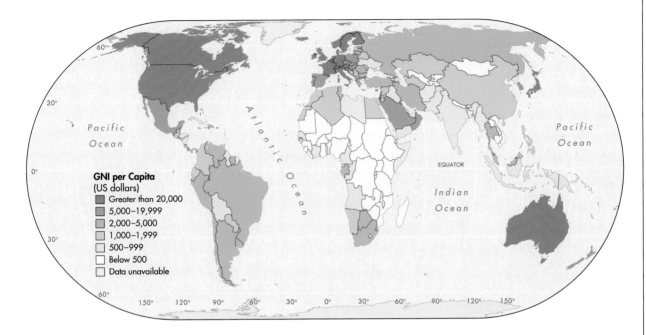

The term Gross National Income (GNI) is used to describe the economic value of a country's output of goods and services in a year. GNI per capita is a measure of potential personal wealth.

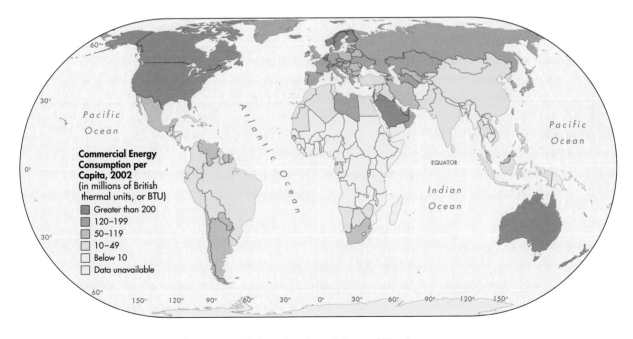

Energy consumption per capita is not only an economic indicator but also an indicator of lifestyle.

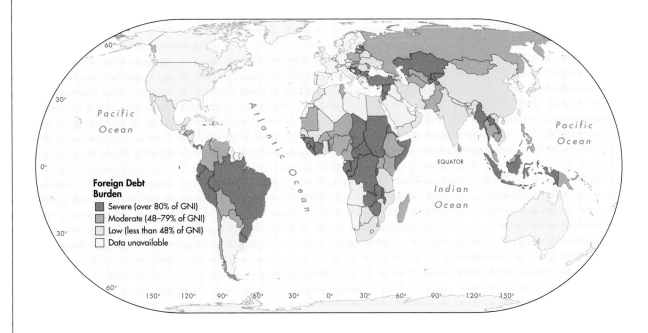

Foreign debt impedes economic progress in much of the developing world. The International Monetary Fund develops programs to help countries in Latin America, Asia, and Africa manage debt crises.

Modernization theory has been roundly criticized by leftists for its obvious ethnocentrism, which denies the validity of non-Western value systems; developing countries are presumed to lack a meaningful history or capacity for development without the introduction of Western institutions, a notion belied by the facts historically. Modernization theory's exclusive focus on the cultural causes of poverty occurs at the expense of external structural features of global capitalism that have historically inhibited economic development, such as colonialism and neocolonialism.

DEPENDENCY THEORY

Drawing upon the intellectual tradition of Marxism developed by German philosopher Karl Marx (1818–1883), dependency theory emphasized the ways in which capitalism simultaneously creates both poverty and wealth through landscapes of uneven development. In this view, poverty is not "natural" and does not simply "happen,"; that is, it is hardly some inherent condition of the Third World. Rather, it is actively created through colonialism and through the mechanics of the contemporary neocolonial world economy.

Dependency theory argued that the *First World* made the Third World poor, that the wealth of the former is a direct result of the poverty of the latter. The Third World is held to be underdeveloped, not simply undeveloped. Thus, the world economy is portrayed as a planetary mechanism for transferring surplus value from developing countries to developed ones. Notions such as comparative advantage and interdependence mask uneven power relations among states that foster exploitation of the weak and poor.

Dependency theory has been criticized on several grounds, including its silence on the role of internal forces that may perpetuate poverty, such as local cultural and political systems. This view erroneously holds that capitalism inevitably generates poverty in developing countries and forecloses possibilities of increases in wealth associated with foreign investment and trade. However, colonial policies varied greatly. By the 1980s, dependency theory had fallen into disuse, and export-promotion strategies designed to maximize foreign revenues and investment had displaced import-substitution ones throughout the developing world.

More at:

http://www.mtholyoke.edu/acad/intrel/depend.htm

WORLD-SYSTEMS THEORY

The reincarnation of dependency theory occurred through the efforts of American sociologist Immanuel Wallerstein, who offered world-systems theory in its place in the 1970s. World-systems theory focuses upon the entire global economy, including the First World. From this perspective, the

Ecological problems confront developing countries. Most temperate forests are now protected by law, but many tropical forests are not. Here in Costa Rica, cleared rain forest abuts Coronado National Park.

global system of states and markets is an indispensable component in the explanation of events within individual areas: Local and national economies and societies can only be understood by reference to their position within the global division of labor.

Wallerstein's original conception viewed the world-system as structured around a core of Western Europe, Japan, the United States, a periphery of the poorer parts of the Third World, and an intermediate, quasi-developed semi-periphery of Eastern Europe, the former Soviet bloc, the East Asian NICs, and Mexico.

This view also holds that capitalism creates wealth and poverty simultaneously through inter-regional and international flows of capital and surplus value. However, unlike dependency theorists, world-systems theory explicitly admits to the possibility of economic development in the developing countries, although global growth is still portrayed as a zero-sum game. Some places may "win," but only if others "lose."

Thus, some peripheral states may rise into the semi-periphery through industrialization, such as the Asian NICs, but such a transition could only occur through the deindustrialization of the global core, as developed countries passed the baton of certain industries to less developed ones. However, world-systems theory has been criticized for its neglect of culture, which can be critical to economic change, for assigning a necessary "role" for the semi-periphery as intermediary between core and periphery, and for conceptually privileging trade over production.

The fastest growing part of the world economy since World War II, East Asia has become an inspiration for other aspiring Third World regions, especially Latin America. The reasons for the NICs' success have been hotly debated. Some observers have suggested that the legacy of 50 years of Japanese colonialism, such as the infrastructures and centralized state bureaucracies, played a role, despite the harsh repression visited on Japan's colonies during World War II; indeed, Japan has long been a model for the NICs and continues to invest actively in East Asia.

Others point to the role of the United States during the Cold War, including development assistance. Yet others stress internal factors such as the Confucian cultural tradition, with its high rates of literacy, continued stress on education, and obedient labor forces, which allow for easily-exploited pools of cheap female labor and which also encourage high national savings rates.

Finally, the NICs clearly were led by efficient, pro-growth governments that encouraged land reform, formed national banks, attracted foreign investment, and promoted exports. In contrast, the nations of Latin America overall have witnessed few attempts at land reform, have had many corrupt and inefficient governments, many contradictory and inefficient development policies persistent import-substitution strategies, numerous strikes, poorly developed credit markets, little foreign investment and relatively little United States foreign assistance.

Economic landscapes are the complex result of influences from the natural environment, from historical legacies, and from contemporary forces that intersect in highly uneven ways over space. Dramatic reductions in the time and costs of transportation and communications have resulted in time-space compression, which alters the relative positions of individual places within a broader grid of economic relations. Economic landscapes are in almost continual flux, simultaneously changing at a very slow rate as well as changing dramatically and almost literally at the speed of light.

More at:

http://www.fordham.edu/halsall/mod/wallerstein.html

URBAN GEOGRAPHY

URBANIZATION

Almost half the world's population lives in cities. Between 1960 and 2003, the number of city dwellers tripled from 1 billion to 3 billion, while the world's total population only doubled from just over 3 bil-lion to 6.4 billion. Analysts predict that there will be 5 billion urban dwellers by 2030, and 80 percent of those urbanites will live in devel-oping countries.

Urbanization is not simply a process of demographic growth of villages, towns, and cities: It involves many other social and spa-tial changes. Cities are crucibles of social change, cultural transforma-tion, and economic innovation. Urbanization typically involves, among other things, distinctive ways of life and subcultures and dis-tinctive patterns of individual behavior and social interaction. Contemporary urbanization is a remarkable geographical phenom-enon, involving some of the most important processes shaping the world's landscapes. More and more

An aerial view of Central Park in New York City reveals successful city planning: the ultimate urbanscape nestles against an emerald rural setting.

of the world's economic, social, cultural, and political processes are played out within and between the world's villages, towns, and cities.

CITIES IN HISTORY

The earliest impulses toward urbanization developed independently in the various hearth areas of the first agricultural revolution. (See Cultural Geography, page 222; Economic Geography: Agriculture, page 249.) The first region of independent urban evolution was in the Middle East, in the valleys of the Tigris and Euphrates Rivers (in Mesopotamia) and in the Nile Valley, beginning around 3500 B.C.

Together, these intensively cultivated river valleys formed the Fertile Crescent. By 2500 B.C. towns and cities had appeared in the Indus Valley of Asia and by 1800 B.C. they were established in northern China. Other areas of independent urbanization include Central America and southwestern North America (from around 600 B.C.) and Andean America (from around A.D. 800). Meanwhile, the original Middle Eastern hearth continued to foster successive generations of city-based empires, including those of Athens, Rome, and Byzantium.

The urbanized economies of these empires were a precarious phenomenon. In Europe the cities introduced by the Greeks and reestablished by the Romans almost collapsed during the early Medieval Age, from about A.D. 476 to 1000. Yet it was from this beginning that an elaborate system of towns and cities developed, the largest centers eventually growing into what have today become the centers of a global world economy. The feudal system, in which lords leased lands to subjects in exchange for military assistance, services, and loyalty, was displaced by a money economy predicated on trade, which provided the foundations for a new phase of urbanization.

URBAN POPULATIONS SELECTED REGIONS 1975–2030												
REGION	TOTAL POPULATION (MILLIONS)			URBAN POPULATION (MILLIONS)			PERCENT URBAN			RATE OF URBANIZATION (%)		
	1975	2000	2030	1975	2000	2030	1975	2000	2030	1975	2000	2030
AFRICA	408	796	1398	103	295	748	25.3	37.1	53.5	2.12	1.54	1.22
ASIA	2398	3823	4887	575	1367	2664	24.0	37.1	54.5	1.47	1.75	1.28
EUROPE	676	728	685	46	529	545	66.0	72.7	79.6	1.02	0.38	0.30
LATIN AMERICA/ CARIBBEAN	322	520	711	197	393	602	61.2	75.5	76.8	1.52	0.84	0.38
NORTH AMERICA	243	316	408	180	250	354	73.8	79.1	86.9	0.58	0.28	0.31
AUSTRALIA AND OCEANIA	22	31	41	15	23	31	71.7	72.7	74.9	0.67	.006	0.10

The Spread of Urbanization

Aggressive overseas colonization made Europeans the leaders, persuaders, and shapers of the rest of the world's economies and societies. Between 1520 and 1580, the Spanish and Portuguese established colonial city systems in Latin America. The Spanish established colonial towns such as Bogotá (Colombia), Lima (Peru), and Quito (Ecuador) mainly as administrative and military centers from which they could occupy and exploit their claims in the New World. Spanish conquistadors founded towns on easily defensible sites such as hilltops and situated them strategically in relation to the populations they governed. Portuguese colonists, in contrast, founded their cities with commercial rather than administrative considerations in mind. Their strategy was to site towns to facilitate collection and export of products from mines and plantations. They thus sought coastal sites in Brazil with good harbors such as Recife, or inland sites along navigable rivers such as Belém on the Pará and Manaus on the Río Negro, a tributary of the Amazon.

Within Europe, Renaissance reorganization—from about the beginning of the 14th century to the 17th century—saw the centralization of political power and the formation of nation-states, the beginnings of industrialization, and the funneling of plunder and produce from distant colonies. In this new political and economic context, the port cities of the North Sea and Atlantic coasts enjoyed a decisive locational advantage. By 1700 London had grown to 500,000, while Lisbon and Amsterdam had each grown to about 175,000.

Merchant Capitalism

Between the 15th and 17th centuries, the cities of Europe and the world's economy were transformed. In Europe merchant capitalism, in contrast to industrial capitalism, increased in scale and sophistication. Merchant capitalism is a form of economic organization in which capital accumulation is based on trade in primary products such as agricultural, fish and forest products, minerals, and handicrafts. Industrial capitalism is a form of organization in which capital accumulation is based on manufacturing processes.

City Planning and Urban Design

Most ancient Greek and Roman settlements were laid out on grid systems, within which the siting of key buildings and the relationship of neighborhoods to one another were carefully thought out. In ancient China cities were laid out with strict regard to Taoist ideas about the natural order of the universe, with different quarters representing the four seasons of the year, and the placement of major streets and the interior layout of buildings designed to be in harmony with cosmic energy. This kind of mystical interpretation of the disposition and alignment of prominent landscape features and sacred sites, is geomancy; its application in design is called feng shui. (See Feng Shui sidebar, page 237.)

It was during the Renaissance and baroque periods in Europe (between the 15th and 17th centuries), that Western ideas of city planning and design were first applied by rich and powerful rulers who used urban design to create extravagant symbols of wealth, power, and destiny. At about the same time, dramatic advances in military ordnance, particularly cannon and artillery, brought a surge of planned urban redevelopment that featured impressive fortifications, geometrically shaped redoubts, or strongholds, and an extensive *glacis militaire*, or clear zones-of-fire.

Inside new walls, cities were recast according to a new aesthetic of grand design: Geometrical plans, streetscapes, gardens emphasizing views with dramatic perspectives, and palaces deliberately designed to show off the power and the glory of state and church. Cities of this era include Charleville (France), Copenhagen (Denmark), Karlsruhe (Germany), Nancy (France), and Philippeville (Belgium).

Colonialism and Gateway Cities

The most important aspect of urbanization during the 17th century was the establishment of *gateway cities*, cities that because of their physical situation served as links between one country or region and others. These include, Boston, Charleston, Savannah, Georgetown, Recife, and Rio de Janeiro in the Americas; Luanda and Cape Town in Africa; Aden in Yemen; Masqat, Goa, Cochin, and Colombo around the Indian Ocean; and Malacca, Makassar (Ujungpandang), Manila, and Macau in the Far East.

Protected by fortifications and European naval power, they began as trading posts and colonial administrative centers. Before long they developed their own manufacturing industries, along with more extensive commercial and financial services, in order to supply the pioneers' needs. (See European Empires map, page 282.)

As colonies were developed and trading networks expanded, some of these ports grew rapidly, acting as gateways for colonial expansion into continental interiors. Rio de Janeiro grew on the basis of gold

Brazil's largest city at more than ten million, Rio de Janeiro began as an outpost for Portuguese explorers in the 1500s. It grew to become Brazil's capital, a major port, and a cultural center.

mining in the Brazilian interior, São Paulo on the basis of coffee, and Buenos Aires, Argentina, on mutton, wool, and cereals. Accra, now capital of Ghana, founded by the British and Dutch in the 17th century, grew on the basis of cocoa, and Calcutta, India, founded by the British in 1690, grew on the basis of jute, cotton, and textiles. As they became major population centers, these cities also became important markets for imported European manufactures, adding even more to their functions as gateways for international transport and trade.

Industrial Cities

Between the late 1700s and the end of the 1800s a distinctive phase of economic development took root in Europe and North America: Industrialization blossomed in towns and cities, driven by competition among small family businesses and with few constraints or controls imposed by governments or public authorities. Very quickly, it became evident that industrial economies could only be organized effectively with large pools of labor, transportation networks, physical infrastructure of factories, warehouses, stores, and offices, and the consumer markets provided by cities. As industrialization spread throughout Europe in the first half of the 19th century and then to other parts of the world, urbanization increased at an ever faster pace. The higher wages and greater variety of opportunities in urban labor markets attracted migrants from rural areas.

In Europe the demographic transition caused a rapid growth in population as death rates fell dramatically. This growth in population provided a massive increase in the labor supply throughout the 19th century, further boosting the rate of urbanization, not only in Europe but also in Australia, Canada, New Zealand, South Africa, and the United States. Emigration carried industrialization

and urbanization to the frontiers of the world economy.

The model city of 19th-century European industrialization was Manchester, England, which grew from a small town of 15,000 in 1750 to a town of 70,000 in 1801, a city of 500,000 in 1861, and a metropolis of 2.3 million by 1911. Cities like Manchester, England, were engines of economic growth, and their prosperity attracted migrants who made for rapid population growth. In this urbanization process, there was a close and positive relationship between rural

and urban development. The appropriation of new land for agriculture, together with mechanization, resulted in increased agricultural productivity. This extra productivity prompted rural-urban migration, enabling surplus rural labor to work in the growing manufacturing sectors in both towns and cities. At the same time, it provided the additional produce needed to feed growing urban populations. The whole process was reinforced by the capacity of urban labor forces to produce agricultural tools, machinery, fertilizer,

Urban Systems

One of the most important ways in which geographers conceptualize urbanization processes is through the attributes and dynamics of urban systems. An urban system, or city system, is an interdependent set of urban settlements within a given region. Thus, for example, there is a German urban system, European urban system, and a global urban system. Every town and city is part of one of the interlocking urban systems that link local-, regional-, national- and international-scale human geographies in a complex web of economic interdependence. These urban systems organize space through hierarchies of cities of different sizes and functions. Many of these hierarchical urban

systems have common attributes and features, particularly relative size and spacing of individual towns and cities.

Urban systems tend to exhibit clear functional differences within hierarchies of settlements of different sizes. The geographical division of labor resulting from processes of economic development means that many medium size and larger size cities perform quite specialized economic functions and so acquire quite distinctive characters, such as the steel-producing cities of Sheffield, England, and Pittsburgh, Pennsylvania. Some towns and cities, of course, do evolve as general-purpose urban centers, providing an evenly balanced range of

functions for their own particular sphere of influence.

Within the American urban system, for example, the top tier of cities consists of centers of global importance (including Chicago, New York, and Los Angeles) that provide sophisticated functions to an international marketplace. A second tier consists of general-purpose cities with diverse functions but only regional importance, such as Atlanta, Miami, and Boston, while third and fourth tiers consist of more specialized centers of subregional and local importance, such as Charlotte, North Carolina, Memphis, Tennessee, and Richmond, Virginia. At the hierarchy's base are local service centers.

and other products that made for still greater increases in agricultural productivity.

In this self-sustaining process of urbanization, the personal and corporate incomes generated by industrialization allowed for higher potential tax yields, which could be used to improve public utilities, roads, schools, health services, recreational amenities, and other components of the infrastructure. These investments improved the efficiency and attractiveness of cities for further rounds of private investment in industry. The whole process was one of *cumulative causation*, in which an upward spiral of advantages accumulated as a result of a city's development of *external economies* and *localization economies*. External economies are cost savings and other benefits that result from circumstances beyond a firm's own organization and methods of production—in particular, savings and benefits that accrue to producers from associating with similar producers in geographic settings that encompass the specialized business services that they need. For example, it is a financial advantage for a wire-making factory to locate near a steel mill, not just to save the cost of transporting steel for wire and to save on the cost of reheating the steel but also to have

Lights twinkle along the boulevards and foundation of the Eiffel Tower in Paris. Some 2,000 years old, Paris was redesigned in the 1800s under Baron Georges Haussmann for Emperor Napoleon III. A grand example of city planning, Paris is a symbol of power, authority, and national pride.

access to the specialized marketing, transportation, and engineering services associated with the steel industry. Localization economies are cost savings that accrue to particular industries as a result of clustering at a specific location, for example being able to share a pool of labor with special skills or experience, or joining together to create a marketing organization or research institute.

Systems of Cities

By the late 18th century, the industrial revolution and European *imperialism* had created unprecedented concentrations of humanity that were intimately linked into networks and hierarchies of towns and cities that geographers describe as urban systems, which are an interdependent set of urban settlements within a specified region. (See Urban Systems sidebar, page 299.) Within industrializing nation states such as the United States, Great Britain, France, and Germany, tight-knit networks of towns and cities evolved as production platforms. *Production platforms* are geographic settings in which specialized, interrelated manufacturing activities are bound together by the creation and the exploitation of external economies. These settings in turn became the bases and foundations of thriving national economies.

As European nation states sought to establish economic and political control over continental interiors, colonial cities such as Calcutta, Saigon (Ho Chi Minh City), Hong Kong, Jakarta, Lagos (Nigeria), Manila, and Singapore were established or reinforced as centers of administration, political control, and commerce.

Cities and Modernization

As societies and economies became more complex with the transition from *merchant capitalism* to industrial capitalism, national rulers and city leaders looked to city planning not just as a means of symbolizing new seats of power and authority but also as a means of imposing order, safety, and efficiency. In Europe between 1853 and 1870, Baron Georges Haussmann (1809–1891) demolished large sections of old Paris to make way for broad, new tree-lined avenues, with numerous public open spaces and monuments. His design set the precedent for other capital cities, including Berlin (Germany), Brussels (Belgium), Vienna (Austria).

In the United States the late 19th-century *City Beautiful movement* drew heavily on Haussmann's ideas and its associated Beaux Arts designs. The objective of the City Beautiful movement was to remake cities to reflect the higher values of society by implementing where possible neoclassical architecture, grandiose street plans, parks, and inspirational monuments and statues. American architect Daniel Burnham (1846–1912) developed a plan for Chicago in 1909 that exploited urban design and planning as an uplifting and civilizing influence, emphasizing civic pride and power.

European imperial powers imposed similar designs on their colonial capitals and administrative centers, including the French in Casablanca (Morocco) and Saigon (now called Ho Chi Minh City in Vietnam); the British in New Delhi (India), Pretoria (South Africa), and Rangoon (Burma—now called Yangon in Myanmar); and the Germans in Windhoek (Namibia).

The cultural response to pressures of industrialization and urbanization was the *modern movement*, which incorporated the idea that buildings and cities should be designed and run like machines. Equally important to Modernists was that urban design should not reflect prevailing social and political values but, rather, help to create a new moral and social order. This led to the idea of imposing order and creating safety and efficiency through building codes, planning regulations, and exclusionary land use *zoning*.

Cities and Advanced Capitalism

After World War II, an important transformation in the nature of capitalism took place within developed countries: There was a shift away from industrial production toward services, particularly sophisticated business and financial services, as the basis for profitability. This was another step in the evolution of capitalism: from merchant capitalism to industrial capitalism to advanced capitalism. The decline in manufacturing jobs in developed countries, combined with the ability of huge multinational corporations to outmaneuver both governments and labor unions, contributed to a destabilization of urban and economic geographies throughout much of the world.

There are some 40,000 multinational corporations, which control about 180,000 foreign subsidiaries and account for more than six trillion dollars in worldwide sales. These multinational corporations have been central to a phase of global geographical restructuring that has been under way for the last 25 years or so. Firms of all sizes have had to adjust their operations by restructuring their activities and reorganizing and redeploying their resources among different countries, regions, and cities.

As a result, contemporary cities must be understood within the con-

Zoning

Exclusionary land use zoning has been a central tool of modern western city planning since the 1900s. Zoning is based on the idea that order and predictability of land uses within a city makes for efficiency, promotes investment by minimizing uncertainty, and lessens conflict between non-compatible land users. In 1924 the U.S. Department of Commerce drafted a model zoning law. Exclusionary zoning in the village of Euclid, Ohio, was tested by Euclid v. Ambler. The case went to the U.S. Supreme Court, which recognized, in 1926, single-purpose, land use zoning as a legitimate application of a local government's powers to police health, safety, and welfare issues. Specifically, it allowed local governments to protect the character of neighborhoods with single-family dwellings against the threat of lower property values and nuisances associated with apartment housing, commercial activities, and industrial land uses.

In practice, zoning in the United States, as in other developed countries, has also been motivated by the desire to exclude social, racial, or ethnic groups deemed undesirable. By the end of the 1970s, single-use exclusionary zoning came to be seen by critics as rigid and inflexible, and one of the reasons that caused city centers to lose their vitality. The cities had been planned to death. By then, zoning and other tools of western city planning had been exported to developing countries. Western-style land use planning was unable to cope with the pressure of urban growth in less developed countries, and it frequently forced land and property prices up, thus reducing most households' access to shelter. Nevertheless, zoning remains an important tool of urban planning. More at:

http://www.abanet.org/rppt/cmtes/rp/c1/cases/Village_of_Euclid_v._Ambler_Realty_Co.pdf

text of constantly changing networks of economic interdependence created by multinational corporate strategies. At the top of the global urban hierarchy are world cities in which a disproportionate part of the world's most important business is conducted. These cities—London, New York, and Tokyo—are, effectively, the control centers for the networked world economy.

WORLD CITIES IN A WORLD ECONOMY

Ever since the advent of merchant capitalism in the world economy in the 15th century, certain cities, known as world cities, have played key roles in organizing space beyond their own national boundaries. In earlier phases of urbanization, these roles involved organization of trade and execution of colonial, imperial, and geopolitical strategies. The world cities of the 17th century were London, Amsterdam, Antwerp, Genoa, Lisbon, and Venice. By the 18th century they included Paris, Rome, and Vienna, while Antwerp and Genoa had become less influential. Berlin, Chicago, Manchester, New York, and St. Petersburg became world cities in the 19th century, and Venice became less influential. Today, with the globalization of the economy, roles of world cities are less about deployment of imperial power and orchestration of trade and more about multinational corporate organization, international banking and finance, supranational government, and the work of international agencies.

London, New York, and Tokyo are at the peak of the current modern global *urban system*. The second tier of world cities, including Brussels, Chicago, Frankfurt, Los Angeles, Paris, Singapore, Washington, D.C., and Zürich, exercise influence over large regions of the world economy; the third tier includes important international

WORLD'S LARGEST URBAN AGGLOMERATIONS BY POPULATION (MILLIONS) 2003			
URBAN AREA	1975	2003	2015 (projected)
Tokyo	26.6	35.0	36.2
Mexico City	10.7	18.7	20.6
New York	15.9	18.3	19.7
São Paulo	9.6	17.9	20.0
Mumbai (Bombay)	7.3	17.4	22.6
Delhi	4.4	14.1	20.9
Kolkata (Calcutta)	7.9	13.8	16.8
Buenos Aires	9.1	13.0	14.6
Shanghai	11.4	12.8	12.7
Jakarta	.8	12.3	17.5
Los Angeles	8.9	12.0	12.9
Dhaka	2.2	11.6	17.9
Osaka-Kobe	9.8	11.2	11.4
Rio de Janeiro	7.6	11.2	12.4
Karachi	4.0	11.1	16.2
Beijing	8.5	10.8	11.1
Cairo	6.4	10.8	13.1
Moscow	7.6	10.5	10.9
Metro Manila	5.0	10.4	12.6
Lagos	1.9	10.1	17.0

in the globalized world economy, are able not only to generate powerful spirals of local economic development but also to act as pivotal points in the reorganization of global space. They are control centers for the flow of information, cultural products, and finance that, collectively, sustain the economic and cultural globalization of the world.

World cities also provide an interface between the global and the local. They contain the economic, cultural, and institutional apparatuses that channel national and regional resources into the global economy and that also transmit the impulses of globalization back to national and regional centers.

CONTEMPORARY PATTERNS OF URBANIZATION

There are thousands of cities functioning both as local centers of industry and commerce and as interdependent nodes in national, regional, and subregional urban systems. The most comprehensive source of statistics is the United Nations, whose data suggest that almost half of the world's population is now urban. A 1996 United Nations document, "Global Report on Human Settlements: An Urbanizing World," stated that the rate of urbanization of the world's population is accelerating significantly as a result of the global shift to tech-

cities with more limited or more specialized international functions, including Amsterdam, Houston, Madrid, Mexico City, Miami, San Francisco, Seoul, Sydney, Toronto, and Vancouver. The fourth tier is those cities with national importance and with some multinational functions, such as Barcelona, Boston, Dallas, Manchester, Montreal, Munich, and Philadelphia.

These cities have infrastructures essential for the delivery of services to clients whose activities are international in scope, such as specialized office space, financial exchanges, communications networks, and airports. They have also

established a comparative advantage in two major areas: the availability of specialized firms and expert professionals and the availability of high-order cultural amenities for both high-paid workers and their out-of-town business visitors.

Above all, these cities have established themselves as centers of authority—with a critical mass of people-in-the-know about market conditions, trends, and innovations—people who can gain one another's trust through frequent face-to-face contact, not just in business settings but also in the informal settings of clubs and office bars.

They have become places that,

World Cities—Global Power Brokers

World cities are the settings for most of the leading global markets for commodities, commodity futures, investment capital, foreign exchange, equities, and bonds; for clusters of specialized, high-order business services such as advertising, design, and market research; for concentrations of corporate headquarters and national and international headquarters of trade and professional associations; for the most powerful and internationally influential media organizations, news and information services, and culture industries; and for most of the leading nongovernmental organizations (NGOs) and intergovernmental organizations (IGOs) that are international in scope. Geneva, Switzerland, is base for such international NGOs as the World Council of Churches and the World Business Council for Sustainable Development, and Paris is headquarters for the Organization for Economic Cooperation and Development (OECD), an example of an international IGO.

There is synergy in the various functional components of world cities. New York, for example, attracts multinational corporations because it is a center of culture and communications, and it attracts specialized business services because it is a center of corporate headquarters and of global markets. Corporate headquarters and specialized legal, financial, and business services cluster in New York because of mutual cost savings. At the same time, different world cities, such as London or Tokyo, fulfill specialized roles within the world system, making for differences in the nature of their world-city functions, as well as differences in their degree of importance as world cities.

nological, industrial, and service-based economies. It concluded that few countries are able to handle the consequent urban population crush, which is causing problems on an unprecedented scale: Ten million people die annually in densely populated urban areas from conditions produced by substandard housing and poor sanitation. About 500 million people, worldwide, are either homeless or living in unfit housing that is life threatening.

Rates of Urban Growth

In many parts of the world, urban growth is taking place at such a pace and under such chaotic conditions that it is impossible for geographers and demographers to do more than provide informed estimates of just how urbanized the world has become. National demographic statistics need to be analyzed cautiously because countries employ quite diverse definitions of what constitutes a town, a city, or a metropolitan region.

In 1975 there were 174 urban agglomerations of 1–5 million, and 17 urban areas of 5–10 million. Today there are approximately 350 metropolitan areas of 1–5 million and 40 metropolitan areas of 5–10 million. Looking ahead, population projections for 2015 suggest that there will be almost 500 metropolitan areas with a population of 1–5 million, and at least 39 with 5–10 million.

More at:

http://www.un.org/esa/population/publications/wup2003/WUP2003Report.pdf

The most important aspect of world urbanization, from a geographical perspective, is the striking difference in trends and projections between the world's developed and developing regions. In 1950, 20 of the world's 30 largest metropolitan areas were located in developed countries—11 in Europe and 6 in North America. By 1980 the situation reversed—with 19 of the largest 30 metropolitan areas were located in less developed regions. By 2015 all but 6 of the 30 largest metropolitan areas are expected to be located in developing regions.

Asia provides some of the most dramatic examples of this trend. From a region of villages, Asia is fast becoming a region of cities and towns. Its urban population rose more than sixfold between 1950 and 2003, to almost 1.5 billion people. Already, Asia has more than 48 percent of the world's urban population and 16 of the 30 largest cities in the world.

By 2030 more than half of Asia's 4.9 billion projected population will be living in urban areas. Nowhere is the trend toward rapid urbanization more pronounced than in China, where for decades the communist government imposed strict controls on where people were allowed to live. The government feared the liberating effects that cities might have on rural migrants. By tying jobs, school admission, and even the right to buy food to the places where people were registered to live, the government made it almost impossible for rural residents to migrate to towns or cities. As a result, more than 70 percent of China's billion people still lived in the countryside in 1985. The Chinese government, having recognized that towns and cities can be engines of economic growth, has not only relaxed residency laws but also drawn up plans to establish more than 430 new cities. Between 1980 and 2000 the number of people living in cities in China more than doubled—from 196 million, or about 20 percent of China's total population in 1980, to 456 million, or about 35 percent of China's total population in 2000.

Levels of Urbanization

In the world's industrialized regions, levels of urbanization, the percentage of the total population living in places designated as "urban" by a country's national census authorities, are high and have been high for some time. According to their government definitions, Belgium, the Netherlands, and the United Kingdom, report more than 90 percent of their populations are urbanized; the populations of Australia, Canada, Denmark, France, Germany, Japan, New Zealand, Spain, Sweden, and the United States are all more than 75 percent urbanized. Levels of urbanization are also very high in many of the world's newly-industrialized countries. Brazil, Hong Kong, Mexico, Taiwan, Singapore, and South Korea, for example, are all at least 75 percent urbanized. Almost all developing countries, meanwhile, are experiencing high rates of urbanization, with forecast growth of unprecedented speed and unmatched size. Karachi, Pakistan, a metropolis of 1.1 million in 1950, is estimated to reach 11.1 million in 2003, and 16.2 million in 2015. Likewise, it is estimated that Cairo, Egypt, will grow from 2.1 million to 10.8 million between 1950 and 2003 and may reach 13.1 million in 2015. Mumbai (Bombay), India's second largest city, Mexico City, and São Paulo are all projected to have populations in excess of 20 million by 2015.

Many of the large cities in developing regions are growing at annual rates of between 4 and 7 percent. Metropolitan areas such as Mexico City and São Paulo are adding half a million persons to their population each year, or nearly 10,000 every week, even after accounting for deaths and out-migrants. It took London 190 years to grow from half a million to ten million, and New York 140 years. By contrast, Mexico City, São Paulo, Buenos Aires, Calcutta, Rio de Janeiro, and Mumbai all took less than 75 years to grow from half a million each to ten million inhabitants each. One consequence of such unprecedented growth is that these urban systems have become more volatile. They exhibit characteristics of greater *centrality*—the functional dominance of cities in terms of economic, political, and cultural activity within an urban system—than would be expected in comparison to the rank-size relationships that characterized the cities of the developed world.

Rural-Urban Migration in Developing Regions

Urban growth processes in the world's developing regions have been entirely different from those in developed regions. In contrast to the self-sustaining urban growth of the world's industrial regions, the urbanization of developing regions generally has been a consequence of demographic growth that has preceded economic development. The unprecedented rates of urban growth in developing regions have been driven by rural push—overpopulation and the lack of increase in employment opportunities in rural areas—rather than the pull of prospective jobs in towns and cities. This can be observed in many Latin American nations.

In much of the developing world, fast-growing rural populations, a result of the onset of the demographic transition, face an apparently hopeless future of drudgery and poverty. In the past, emigration provided a safety valve, but most of the more affluent countries have now put up barriers to immigration. The

Rank-Size Relationships and Primacy

Functional interdependency among places within urban systems tends to result in a distinctive relationship between the population size of cities and their ranking from largest to smallest within a particular urban system. This relationship is known as the *rank-size rule*, which describes a statistical regularity in the city-size distributions in countries and regions. The relationship is such that the nth largest city in a country or region is 1/n the size of the largest city in that country or region. In some urban systems, however, the top of the rank-size distribution is distorted as a result of the disproportionate size of the largest—and sometimes also the second-largest—city. In Argentina, for example, Buenos Aires is more than ten times the size of Rosario, the second-largest city. In the United Kingdom, London is more than three times the size of Birm-

ingham. In France, Paris is nearly eight times the size of Marseilles. In Brazil, São Paulo is nearly five times the size of Belo Horizonte. This condition is known as *primacy*, a situation in which the population of the largest city in an urban system is disproportionately larger than the second- and third-largest cities in that system.

In a developing country, primacy is usually a consequence of the primate city's early role as a gateway city. Primacy in industrialized countries is usually a consequence of primate cities' roles as imperial capitals and centers of administration, politics, and trade for a much wider area than their own domestic system.

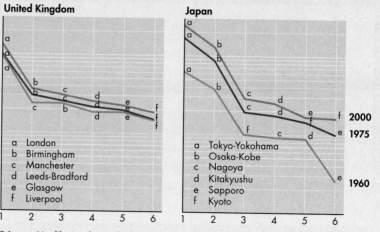

Primate cities like London and Tokyo-Yokohama far surpass in size a nation's other big cities.

Centrality

When cities' economic, political, and cultural functions are disproportionate to their population size, the condition is known as centrality, the functional dominance of cities within an urban system. Cities that account for a disproportionately high share of economic, political, and cultural activity have a high-degree of centrality within their urban systems. Bangkok, for instance, with 9 percent of the Thai population, accounts for approximately 38 percent of the country's overall gross domestic product (GDP), more than 85 percent of the country's GDP in banking, insurance, and real estate, and 75 percent of its manufacturing. Very often, it is primate cities that exhibit centrality, but cities do not necessarily have to be primate in order to be functionally dominant within their urban systems.

only option for the growing numbers of impoverished rural residents has been to move to relatively nearby towns and cities, where at least there is hope of employment and the prospect of access to schools, health clinics, a safe water supply, and the kinds of public facilities and services often unavailable in rural regions. The lure of cities has, meanwhile, been intensified by images of modern amenities and consumer goods beamed into rural areas through satellite TV. Overall, the cities of developing regions have absorbed four out of five of the 1.2 billion city dwellers who have been added to the world's population since 1970.

Rural migrants have poured into cities out of desperation and hope, rather than being drawn by jobs and opportunities. Because these migration streams have been composed disproportionately of teenagers and young adults, an important additional component of urban growth has followed: exceptionally high rates of natural population increase. In most developing countries the rate of natural increase of urban populations exceeds the rate of net in-migration. On average, about 60 percent of current urban population growth in developing countries is attributable to natural increase.

One striking result of the combination of high rates of rural-urban migration and high rates of natural increase in urban populations has been the emergence of megacities, cities with populations of ten million or more. In 1960 New York and Tokyo were the only cities with ten million or more inhabitants. By 2003 there were twenty, including: Beijing, Cairo, Calcutta, Jakarta, Lagos, Manila, Mexico City, São Paulo, Shanghai, and Dhaka. Most of these megacities provide important intermediate roles between the metropolises of the developed world and the provincial towns and villages of large regions of the less developed world. They not only link local and provincial economies with the global economy but also provide a point of contact between the traditional and the modern and between formal and informal economic sectors.

The informal sector is composed of economic activities that take place beyond official record and are not subject to formalized systems of regulation or remuneration. In addition to domestic labor, these activities include illegal activities such as drug peddling and prostitution, as well as legal activities such as casual labor in construction crews, domestic piecework, street trading, scavenging, and providing personal services such as shining shoes or writing letters.

Overurbanization

A consequence of urban population growth in developing countries has been that many cities have grown more rapidly than the jobs and housing they can sustain. This is called *overurbanization*, and it produces

Mexico City is one of the world's largest metropolitan areas. It currently boasts almost 19 million inhabitants and exhibits strong growth. Like many developing countries, Mexico is highly urbanized, with almost 75 percent of its inhabitants living in cities.

instant slums, characterized by shacks set on unpaved streets, often with open sewers and no basic utilities. Shelters are constructed of any material that comes to hand, such as planks, cardboard, tar paper, thatch, mud, and corrugated iron. Such is the pressure of in-migration that many instant slums are squatter settlements, built illegally on land that is neither owned nor rented by its occupants.

Squatter settlements are not necessarily slums, but many of them are. In Chile squatter settlements are called *callampas*, mushroom cities; in Turkey they are called *gecekondu,* meaning they were built after dusk and before dawn. In India they are called *bustees*; in Peru *barriadas*; in Brazil *favelas*, and in Argentina simply *villas miserias*, villages of misery. They typically accommodate well over one-third of the population and sometimes as much as three-quarters of the population of major cities in less developed countries. A 1996 United Nations report on human settlements estimated that 600 million people worldwide were living under health- and life-threatening situations in cities, with some 300 million people living in extreme poverty. The same report showed that, in some developing countries, including Bangladesh, El Salvador, Gambia, Guatemala, Haiti, and Honduras, more than 50 percent of the urban population lives below their respective country's national poverty line.

Collectively, it is these slums and squatter settlements that have to absorb the unprecedented rates of urbanization in the megacities of the less developed regions. Many neighborhoods are able to develop self-help networks and organizations that form the basis of community amid dauntingly poor and crowded cities. Nevertheless, over-urbanization is causing acute problems throughout the less-developed world.

CITIES AS MIRRORS OF CHANGE: THE INTERNAL STRUCTURE OF CITIES

These broad processes of urbanization have not only produced systems of towns and cities of different sizes and with different economic functions but also have imprinted themselves into the physical fabric and cultural settings within individual cities. Cities acquire a distinctiveness that comes in large part from their physical characteristics: the layout of streets, the presence of monumental and symbolic structures, and

the building types and architectural styles that fill out the built environment. The resulting urban landscapes reflect a city's history, its physical environment, and its people's social and cultural values.

Geographers are interested in urban landscapes because they can be read as multilayered texts that show how, when, and why cities have developed, how they are changing, and how people's values and intentions take expression in urban form. The built environment, whether it is planned or unplanned, is what gives expression, meaning, and identity to the various forces involved in urbanization—it becomes a biography of urban change. At the same time, the built environment provides people with cues and contexts for behavior, with landmarks for orientation, and with symbols that reinforce collective values such as civic pride and a

sense of identity.

The distinctiveness of cities also stems in part from patterns of land use and the functional organization of economic and social sub-areas in cities. The resulting patterns of neighborhoods and districts are partly a product of the economic, political, and technological conditions at the time of the city's growth, and partly a product of regional cultural values. The most striking contrasts in such patterns are to be found between the cities of developing countries and those of developed countries. The structure and urban landscapes of some cities in developing countries retain strong elements of traditional structure, such as Islamic towns and cities, which are organized around a central mosque, a citadel, and a main covered bazaar, or souk. Colonial towns may still have a fort, a cantonment, and civil lines, planned

Uncontrolled Growth

A 1995 UNICEF report, "The Progress of Nations," blamed "uncontrollable urbanization" in less developed countries for the widespread creation of "danger zones," where increasing numbers of children are forced to become beggars, prostitutes, and laborers before they reach their teens. Pointing out that urban popula-

tions are growing at twice the general population growth rate, the report concluded that too many people are being squeezed into cities that do not have the jobs, housing, or schools to accommodate them. As a consequence, the family and community structures that support children are being destroyed, with the result that

more and more young children have to work, whether shining shoes, guiding cars into parking spaces, chasing other street kids away from patrons at an outdoor café, working as domestic help, making fireworks, or selling drugs. More at: http://www.unchs.org/Istanbul+5/statereport.htm

residential developments for officials and their families. In most cities in less developed regions, once-distinctive patterns of spatial organization and land use have all but disappeared as a result of congestion and overcrowding caused by over-urbanization, so that the dominant feature of their internal structure is the juxtaposition in geographic space of the formal and informal sectors of an economy, based on a core of modern commerce, retailing, service, and industry, and its associated residential areas, contrasted with an informal economy that is manifest in street markets and extensive areas of makeshift, shanty housing.

Urban Land Use Patterns

In the more developed regions of the world, urban structure has, from the onset of the industrial revolution, been strongly influenced by successive transportation technologies. In the mid-19th century, railways established a hub-and-spoke framework for urban development.

At the end of the 19th century, streetcars triggered the first significant wave of suburbanization. Since the 1950s automobiles and trucks have allowed cities to sprawl ever outward, in a network of retailing, services, and industry . Meanwhile, household competitions for the best and most accessible sites meant that the urban social patterns tended to develop into distinctive sectors and zones of neighborhoods, all organized around the city's commercial and industrial core.

The typical U.S. city has long been structured around the *central business district* (CBD), which is then surrounded by a *zone in transition*— a transitional area of mixed industrial, commercial, warehousing, and residential land uses, which, in turn is surrounded with tiers of successively more recent suburbs, then nodes of secondary business districts and strips of commercial development and industrial districts. Poorer households, unable to afford the recurrent costs of long journeys to work, tend to trade off amounts and qualities of living space for accessibility to jobs. As a result, poorer households end up in high-density conditions, at relatively expensive locations near low-wage jobs.

Cities experiencing high rates of in-migration have tended to become structured into a series of concen-

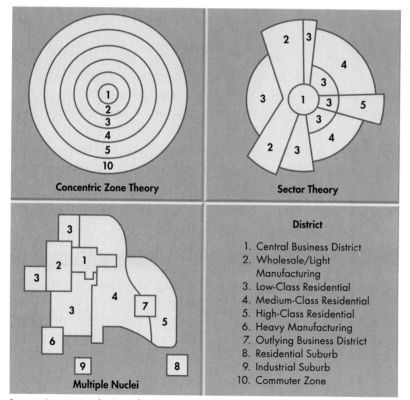

Concentric Zone Theory

Sector Theory

Multiple Nuclei

District

1. Central Business District
2. Wholesale/Light Manufacturing
3. Low-Class Residential
4. Medium-Class Residential
5. High-Class Residential
6. Heavy Manufacturing
7. Outlying Business District
8. Residential Suburb
9. Industrial Suburb
10. Commuter Zone

From an inner core, a business district, a city sprawls outward via rails and roads. The concentric structured city usually has in-migration, with circles of ethnicity. The sector city has many neighborhoods all clustering close to the core. A multiple nuclei city has more than one business district.

Edge Cities

In the United States and some other developed countries, the decentralization of retailing and offices from central cities has reached the point where some suburban commercial centers have grown so large that they compete directly with the CBD for highly specialized functions such as mortgage banking, corporate legal offices, accounting services, publishing, luxury hotels, and exclusive clubs. The success of these suburban commercial centers has been paralleled by the development of private, master-planned residential subdivisions, new towns, and office parks and business campuses. With distinctive local labor markets and commuting patterns, they have collectively come to be known as edge cities. In northern Virginia, Tysons Corner, bounded by I-66, I-495 , (the Washington Beltway), and Virginia 267, (the toll road to Dulles International Airport), is the archetypal example. Administratively, it is still rural, an unincorporated area within Fairfax County, but with almost 40,000 residents and almost 100,000 jobs. One of the ten largest concentrations of office and retailing space in the United States, it incorporates more than 25 million square feet of office space, several million square feet of retail space, ten major department stores, almost 4,000 hotel rooms, and parking for more than 90,000 cars. Yet, while this edge city incorporates the largest retail concentration on the East Coast with the exception of Manhattan, it has little of the apparatus of urban governance or civic affairs. It has a branch of Tiffany's but no public open space; an exclusive business club but no public forum; dozens of sportswear stores but no public recreation centers, swimming pools, or bicycle paths. It is a place of great affluence but at the same time a focus of intense concern over traffic congestion, inflated land values, service provision, and land use conflicts.
More at:
http://www.brookings.edu/es/urban/publications/langmiami.pdf

tric zones of neighborhoods of different ethnicity, demographic composition, and social status. Distinctive neighborhoods emerge over time through these processes of *residential mobility* when similar kinds of households go through similar search patterns and then make similar decisions about where to live.

Recently, however, the internal structure of cities in developed regions has been reorganized as a result of the economic transformation to postindustrial economies. Traditional manufacturing and related activities have closed or been moved out of central cities, leaving decaying neighborhoods and a growing, residual population of elderly and marginalized people. However, new, postindustrial activities such as business services have begun to cluster in redeveloped central business districts and in *edge cities*— nodes of commercial and residential development located on metropolitan fringes typically near major highway intersections. (See Edge Cities, above.)

Metropolitan growth, in a few cases, has become so complex and extensive that hundred-mile cities have begun to emerge, with half a dozen or more major commercial and industrial centers forming the nuclei of a series of interdependent urban realms. The archetype is megalopolis, such as the urbanized region extending from Washington, D.C., through Baltimore, Philadelphia, Newark, and New York to Boston. Cities such as Mumbai, Hong Kong, Lagos, São Paulo, and Shanghai represent the dominant future urban forms in a world characterized by accelerating population growth and a global economy.

POLITICAL GEOGRAPHY

GEOGRAPHY IN DECISION-MAKING

Political geography examines the spatial and environmental contexts of political decisions made by entities such as governments, multinational corporations, and rebel organizations. Political decision-makers form policies and use power to implement those policies; these decisions inevitably have a geo-graphic basis and geographic consequences. Annexation of a rural area outside of a city, redrawing boundaries for electoral districts, and occupation of territories by ethnic groups such as Kurds and Basques all involve political decisions that change the geography of an area. Political geographers analyze political activities at local, regional, national, and global or international scales.

Political decisions and their geographic contexts change constantly because of changes in the interests of populations and constituencies, in the groups making decisions, in the authority and power they wield, and in the boundaries of political units. In- and out-migration and aging change the demographic composition of regions and the priorities of their populations. Elections, uprisings, and coups may change the composition and priorities of those in power and even change political boundaries.

All political problems involve a

Policy-makers, such as those in London's Parliament, use geographic knowledge of cultures, territories, and boundaries to make and implement laws.

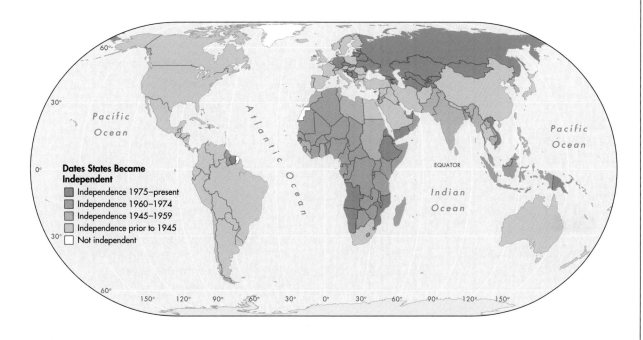

Dates States Became Independent
- Independence 1975–present
- Independence 1960–1974
- Independence 1945–1959
- Independence prior to 1945
- Not independent

Political power is focused at the state or national level. Most of the world's major powers achieved independence prior to 1945. Independence has been more recent for many countries in historically colonized Africa and parts of Asia.

scale dimension in which the consequences of political decisions impinge unequally on groups within or outside the political region. A government enforces political decisions within a physical territory, often one whose spatial limits were established by previous agreements or decisions. In some cases, enforcement of the decisions is opposed by internal regions and groups, who may also oppose the entire decision-making process. Calls for autonomy by Québecois in Canada, Basques in Spain, and Chechens in Russia are examples of ethnic groups who oppose the national decision-making process. In other cases, political

decisions affect populations in areas outside the region who also object to the decision or to the decision-making process. For example, political decisions by Germany on immigration have caused problems with Turkey, and political decisions on pollution policies by countries bordering the Mediterranean Sea affect people throughout the region regardless of boundaries.

Political geography, like politics itself, is shaped by the use of power by one group to affect the behavior of other groups and achieve specified goals. Power is the capacity to modify events and actions by others through the use of persua-

sion, purchase, barter, or coercion. Official power or authority is possessed by governments—recognized administrators of political units—and used to enforce compliance with laws and regulations. The federal government of the United States can impose laws regulating air pollution on all industries and can also influence manufacturing in Japan by imposing tariffs. Rebel organizations, an example of unofficial bodies, such as the Irish Republican Army in Ulster and the Zapatistas in Mexico, have power to affect behavior within their homelands.

When political goals are not

achieved by a government or other political entity or when the underlying policy is opposed by those affected, conflicts over decision-making may arise. Political conflicts between a national government and regional interests can threaten national unity. The United States, for example, formed after a civil war with Great Britain over taxation and representation.

POLITICAL ORGANIZATION OF TERRITORY

States

Political power exists at a variety of spatial levels, but it has been concentrated at the state—or the national—level for centuries. State is used interchangeably with country in political science, but is also used in reference to a constituent unit within a state, such as Missouri in the United States or New South Wales in Australia. *Nation* refers to an ethnic group that shares a common sense of identity and is committed to a political area or homeland.

The English comprise a nation, and England together with Scotland, Wales, and Northern Ireland are combined to form a state, the United Kingdom.

The defining moment in establishing the concept of modern states occurred at the Peace of Westphalia in 1648, which ended the Thirty Years' War. Combatants agreed on mutual recognition and on establishing boundaries for their principalities and duchies. This treaty recognized more than 300 princes as sovereigns, giving them the political authority to control both their internal and external affairs.

A state comes into existence when other established states recognize it and when it possesses basic geographic characteristics. Recognition involves an established state acknowledging that the new state's government has specific authority to determine its internal affairs and external relations.

More at:

http://www.state.gov/s/inr/states/

Geographically, a state must possess territory and a permanent population. An organized economy and transportation and communication systems are also important in the formation of a state. No minimum threshold of territory or of population is now required for a state, but a state must be able to defend its territory from outside threats. Such a defense is called *territoriality*, although the term can be applied on a scale ranging from gangs protecting their turfs to armed forces defending their country from invasion.

During the latter half of the 20th century, a tremendous number of new states were created, more than 40 in Africa alone, as independence movements by former colonies were successful. Fewer than 60 states existed prior to World War II, of which only 3 were in Africa: South Africa, Egypt, and Liberia. During the 1990s, as the Soviet Union and Yugoslavia dissolved, 19 new states were recognized by the world community and have since joined the United Nations, bringing the membership to a total of 191 countries. There remain, however, a number of would-be states.

More at:

http://www.un.org/members/index.html

Every state has institutions through which governmental power promotes and protects the interests of its citizens or, more specifically, the constituencies supporting the government. State institutions such as commerce departments, welfare agencies, and police forces can be viewed as "crisis-managers," which defuse potential societal conflicts and challenges to the state. Governments enact laws in reaction to events and situations that alarm the public or threaten public health, safety, or welfare. Political geographers are increasingly interested in how institutions are used to secure the support of society and to protect political and economic systems.

International Law

International law, the body of rules that has evolved to guide interstate relations, requires recognition before such international relations as trade or exchange of ambassadors may occur. Dutch scholar, Hugo Grotius (1583–1645), published De jure belli ac Pacis, the Law of War and Peace, in 1625, which became the basis for subsequent studies of international law. In 1778 France's decision to recognize and support the United States in its war against Great Britain was critical for the survival of the new state. More than a century and a half later, in 1948, the United States recognized Israel the day after the state of Israel was declared, and the U.S. could therefore, under international law, provide arms, materiel, and other resources to enable Israel to defend itself from attacks by its neighbors.

Capitals and Core Areas

National governments of states are physically located in capital cities. Political decisions are made by national institutions such as legislatures to ensure the stability of the political system by enacting and implementing policies and resolving conflicts. In the United States, the Capitol Building, where members of Congress work and the Supreme Court Building, office for the justices of the judicial branch, are located in Washington, D.C. In Canada the Parliament buildings and the Supreme Court building are located in Ottawa. Well-known national government buildings, such as the Kremlin in Moscow and Whitehall in London, are commonly used when referring to the governments of Russia and the United Kingdom.

Capitals are symbols of the states as well as centers of power, and therefore the selection of a capital city is a major political decision. A number of capitals, especially in Europe, are located within traditional core areas of the states where population has historically been concentrated. The capital cities of Athens, Paris, and London are state capitals that have been dominant cities within the core areas of Greece, France, and the United Kingdom, respectively, for centuries. Other capitals are located in capital territories, which are not part of a constituent state. This reduces competition between their cities. Canberra, Australia's capital city, for example, is located in the Australian Capital Territory midway between the country's two largest cities, Sydney and Melbourne, and it replaced Melbourne as the country's capital.

A number of national capitals have been relocated for symbolic reasons, and some relocations have been made for economic reasons. Moscow was the traditional capital of Russia, until Tsar Peter the Great moved it to the newly created city of St. Petersburg in 1712 to symbolize Russia's interest in Western Europe. The capital was returned to Moscow, again for symbolic reasons, in 1918 following the Bolshevik Revolution. Israel moved its capital from the seaport of Tel Aviv to the traditional Jewish capital of Jerusalem in 1980, despite resistance by Palestinians and surrounding Muslim populations for whom Jerusalem is also a sacred city. Other relocations of capitals have been made to shift the focus of economic development within the state: Brazil moved its capital from Rio de Janeiro to Brasília in 1960 for this reason, and Nigeria built an interior capital in Abuja to shift the focus of development from Lagos.

Territorial Characteristics

Size, shape, and location of state territories relative to terrestrial features and to other states can affect the states' internal administration, the natural resources available to

In a space shuttle view, Chile sweeps the length of South America, with the Andes mountain range as its defining feature and boundary. A nation's size, shape, and location affect its internal administration, the natural resources available to its people, and its external security.

states, and external security. The United States, for example, has the fourth largest territory of any country, a fragmented shape—since Hawaii and Alaska are separated from the main territory—two ocean coastlines, and it shares land borders with Canada and Mexico, two non-threatening states. In contrast, Belarus is one of the smaller states, with an area just a bit less than that of Kansas. It has a compact shape, no coastline, and is bordered by five states—Lithuania, Latvia, Poland, Ukraine, and Russia. Compared to the United States, the territorial

characteristics of Belarus provide it with fewer resources, more external challenges, but relatively easier internal administration.

Size of territory has been traditionally associated with military and economic power, because greater area often meant more population and more abundant natural resources. Large countries such as China and the United States possess abundant natural resources and large populations, and both are world powers. But neither resources nor habitable environments are uniformly distributed over Earth, so size alone can be a mis-

leading indicator of power. Canada, despite possessing the second largest territory after Russia, the largest state, had a population of only 32 million in 2004 and is not a military power; yet based on total GNP, Canada's economy ranks as one of the highest in the world.

States with extremely small territories often have small populations and limited natural resources and economic diversity. As a result, microstates are often dependent on one or a very few industries. Tuvalu, an island country in the South Pacific, relies on copra; Malta, in the Mediterranean Sea, relies on tourism and shipping; and Brunei, a fragmented state on the island of Borneo, relies on petroleum. Vatican City, in Rome, is a state despite its name. It is also the smallest state, with a territory of 0.2 square miles and a population of fewer than a thousand. In addition, it is economically dependent on the activities of the Roman Catholic Church.

A state's global location, especially with respect to ocean shorelines, is extremely important to the economic development of states. Landlocked states, of which there are 15 in Africa alone, lack direct access to oceans and therefore lack shipping, fishing, and other coastal opportunities. Many landlocked states, such as Niger, Bolivia, and Nepal, are among the poorest coun-

Territorial Morphology

Territorial shape can affect the number of bordering states—and therefore the number of potential conflicts—as well as internal administration, and ocean access. Political geographers traditionally categorize states by their shape: compact, elongated, fragmented, perforated, and prorupt. Compact states, such as Cambodia, Poland, and Uruguay, are roughly spherical, which facilitates internal control even if the capital is not centrally located, as in the case of Montevideo on Uruguay's coast. Topographic features such as mountains or an ethnically diverse population can interfere with administration, as in the case of Nigeria, which contains about 250 tribal groups and is divided in general between a Muslim north and a Christian and animist south.

Elongated states such as Chile, Italy, and Laos have narrow, linear shapes. Internal diversity is more likely to occur in elongated states, as seen in the economic division in Italy between the wealthy north and the poor south; and in the ethnic division in Laos between the northern Kha and southern mountain tribes. Fragmented states such as Angola, Indonesia, and Oman have territories separated by bodies of water or another state. Fragmentation can lead to regional identities and reduce linkages between territories, thus exacerbating administrative problems. Such problems were responsible for the breakup of the fragmented state of Pakistan, which was separated geographically into East and West Pakistan by India. East Pakistan became Bangladesh in 1971.

There are only a few states that are perforated, that is they totally surround another state. Italy surrounds San Marino, and South Africa surrounds Lesotho; the larger, surrounding states influence trade and transport systems of the smaller states. Prorupt states have extensions—proruptions—reaching outward from their main territories. Most proruptions provide access to resources, though some form buffers between neighboring states. The Caprivi Strip is Namibia's proruption and provides access to the Zambezi River, and the Shaba province of the Democratic Republic of Congo provides access to mineral resources such as cobalt and zinc. Afghanistan's proruption was designed by the British in the 1880s to serve as a buffer between Russia's expanding territory and the United Kingdom's expanding influence in the Indian subcontinent. Thailand's proruption extends down the Malay Peninsula.

tries; Switzerland and Austria are exceptions. Russia, despite its huge size and long coastlines, has no ice-free ports in winter and therefore faces difficulties in shipping.

The proximity of more powerful or hostile bordering states is another important, though changing, aspect of global location. Canada's foreign relations with the United States are generally very friendly, but many of the states bordering Germany, Russia, and China have at one time been threatened by these three powerful states. States may also face economic or political threats when bordering states are far poorer or are attempting to export their political system through revolution. The United States has coped with illegal Mexican immigration for several decades, and Laos was threatened by guerrilla activity from Vietnam during the 1970s.

Intrastate Political Organization

Political decision-making within every state includes the subdivision of the national territory into smaller political units administered by subnational governments. Not all problems and issues are national in scale and so policies and decisions addressing these issues may be more effectively formulated and implemented at the local or regional scale. Political disputes can arise when a

Divided Powers

Some unitary states have found it expedient to grant greater power, even autonomy, to regional territories and governments to increase national stability. In this process of devolution the national governments technically retain ultimate power, but in fact a hybrid form of government has been created. Devolution of authority has occurred in Brazil, Belgium, and Spain, for example; and Scotland and Wales within the United Kingdom were granted greater power in 1998.

In federal states, such as the United States and Australia, decision-making power is divided between a national government and subnational governments, and a constitution is the contract specifying what authority is held by each level of government. National governments in federal states ordinarily are responsible for decisions involving national defense, foreign relations, and foreign trade. Subnational governments usually make decisions concerning such regional concerns as health, welfare, and public safety that apply only within their individual territories, although these can also be dealt with at the national level. Subnational units have numerous designations, including provinces (Canada), cantons (Switzerland), and states (the United States and India.)

Within the United States, the national government and state governments have each established new political units and governing agencies with limited powers to address problems at intermediate spatial scales. The national government also has authority over outlying territories such as Guam, American Samoa, and the U.S. Virgin Islands, all of which are unincorporated territories.

problem involves a territory larger than that administered by a subnational government or when a political decision is enforced over a territory and population that disagrees with that decision.

Various types of state organizations exist, differentiated primarily by the degree to which power is centralized in the national government. In unitary states, all political power derives from the national government; the majority of states have unitary systems. In *federal states*, political power is shared by both the national government and a number of subnational governments; fewer than 25 states can be classified as federal.

Unitary states, which include Paraguay, Libya, Japan, and Sweden, vary in how zealously the central governments protect their prerogatives. Central governments in unitary states may be elected, hereditary royalty, or controlled by military dictators. The most centralized unitary states are usually run by dictators who use their centralized power to counter internal diversity and challenges to their authority. In less centralized systems, assertions of centralized power are not needed to prevent regional challenges, and local governments are allowed more leeway in decision-making.

Subnational administrative units exist in unitary states, but the powers of their governments are delegated from and controlled by the national government. France, for example, is divided into 22 administrative regions, which are subdivided into 96 *departements*. The departements are progressively subdivided into *arrondissements, cantons*, and *communes*. There are about 3,600 communes. Local issues are dealt with at the commune level. The larger administrative regions were established to address economic development and environmental issues.

All 50 states have established county subdivisions—called parishes in Louisiana and boroughs in Alaska—as well as municipalities and towns whose governments have varying degrees of authority. County governments have decreased in power as urban populations have outpaced those of rural areas, but most still have responsibilities in law enforcement, highway construction, and other concerns.

The number of municipalities is growing in the United States—nearly 20,000 have been incorporated as legal entities—and their governmental powers vary according to the state in which they exist. Cities often create special-purpose districts, such as school and fire districts, within their municipal limits to manage regional services better, much as counties have used irrigation districts and soil conservation districts to improve the provision of those services.

Electoral Geography

Electoral geography is the analysis of the spatial patterns of voting and representation in political systems. In nearly all representative governments, elected officials represent a political territory—called an electoral district in the United States and a riding in Canada—and are voted into office by residents of that territory. One notable exception is Israel, whose nationally elected officials represent the votes for particular parties rather than for individual representatives; this pro-

portional representation ensures that a wide range of viewpoints is represented in the Knesset, the Israeli legislature, but makes it difficult to achieve majority approval on controversial issues.

Analyses of voting behavior in the United States have become increasingly important to political parties seeking to have their candidates elected and to special interest groups seeking to have an issue approved or refuted. In recent years the voting patterns of growing ethnic groups have become of increasing interest. The rapid growth of the Hispanic population of the southern United States, of whom the majority vote Democratic, may increase their representation in Congress and cause both major political parties to revise their stands on immigration, welfare, and other issues of concern to Hispanic voters.

Studies of representation focus on voters' characteristics in electoral districts and changes in the areas, boundaries, and demographics of those districts. Shifts in population mean shifts in representation, political power, and federal monies. In the United States and Canada, there are constitutional requirements to redraw voting districts following every ten-year census to ensure their relatively equal populations. This process is called reapportionment, which can result

in adding or subtracting the number of representatives from states and provinces as well as redistricting, or redrawing district boundaries within states and provinces. Voting tendencies of residents in electoral districts can be determined fairly accurately, and therefore incumbents who run for reelection may resist changes in district boundaries that will threaten their chances in subsequent elections.

The U.S. Constitution requires Congress to reapportion the number of representatives to the House of Representatives (currently 435) among the states after each ten-year census, and then requires states to redistrict to ensure the equal populations of districts within the states. Arizona, Florida, Georgia, and Texas had the greatest increases in population between the 1990 and 2000 censuses and gained two representatives each; New York and Pennsylvania each lost two representatives; many other northeastern states also lost a number of representatives.

More at:

http://www.census.gov/population/www/censusdata/apportionment.html

Redistricting the boundaries of the electoral districts can become an exercise in *gerrymandering*, which provides unfair advantage to one segment of voters over others.

Gerrymandering, named after Governor Elbridge Gerry (1744–1814) of Massachusetts, who was known to support unfair redistricting, has often been used by members of major political parties to ensure that their party remains in power. Electoral districts can be redrawn to divide the voting power of the competing party among several districts or to concentrate it into one district.

Even attempts at righting past voting wrongs, such as trying to ensure minority representation from a state, can be determined to be gerrymandering. Following a 1982 amendment to the 1965 Voting Rights Act, several states created majority-minority districts to ensure that minority ethnic groups would have a majority vote in some congressional districts. North Carolina's 1st and 12th congressional districts were drawn to ensure African-American majorities, but the United States Supreme Court ruled in *Shaw v. Reno* (1993) and *Miller v. Johnson* (1996) that both of these districts used race unfairly as the basis of redistricting, which was considered to be unconstitutional gerrymandering.

REGIONALISM AND POLITICAL FRAGMENTATION
Nations and Nation-States

Distinct societies committed to their homelands, territories with which they identify a shared history and culture, form nations. A nation is usually an ethnic group, a people sharing the same language, religion, history, and icons—symbols of their distinctiveness. Flags, anthems, heroes, and heroines are icons. Maps, too, can be icons: Argentine maps have shown the Islas Malvinas (Falkland Islands) as Argentine territory, not British.

The identification of a nation with its homeland is often translated into proprietary behavior or defensive territorial behavior. Territoriality occurs at all spatial scales, including an individual defending his or her personal space, but it becomes political behavior and affects political geography when groups or nations act or seek to act as decision-makers for their homeland. Separatist actions taken by the Moros on Mindanao challenged the government of the Philippines much as actions taken by Corsicans challenge the French government. A peace was achieved in 1996 in the Philippines, but the Corsicans have not yet reached an accord with the French national government.

Ideally, the homelands of nations should coincide with the territories of states and form *nation-states*, thereby reducing potential conflicts, but few states approach this ideal. Island states such as Iceland and Japan, which can more easily control immigration, are close to the ideal of a nation-state. Japan is the classic example and has a population that is more than 99 percent ethnic Japanese, although Ainu, an indigenous, aboriginal people, inhabit the northern island of Hokkaido and the number of Koreans in Japan is growing. In Western Europe, Denmark is nearly a nation-state, but Greenlanders (Inuit) and a growing population of Turkish laborers prevent a perfect match between nation and state.

Occasionally the territory of one state, occupied by members of its national group, is located within the boundaries of a second state, forming an exclave of the first state and an enclave within the second. Nagorno-Karabakh, an exclave of Armenia, is an enclave within Azerbaijan and was the scene of fierce fighting between the two states in the early 1990s. Typically, territory within one state includes part of the homeland of a nation dominating a neighboring state; the Albanian population of Kosovo in Yugoslavia and the Hungarian population of Transylvania in Romania are examples. Attempts by a state to incorporate outlying territories containing its nationals is called irredentism and has led to wars in the past and continues to threaten some areas. German incursions into Czechoslovakia and Poland in 1939

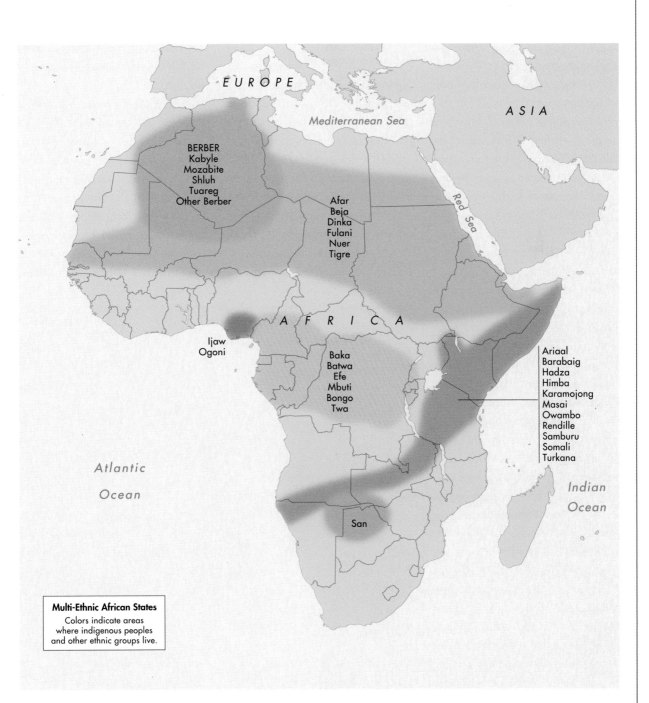

EUROPE

ASIA

Mediterranean Sea

Red Sea

BERBER
Kabyle
Mozabite
Shluh
Tuareg
Other Berber

Afar
Beja
Dinka
Fulani
Nuer
Tigre

A F R I C A

Ijaw
Ogoni

Baka
Batwa
Efe
Mbuti
Bongo
Twa

Ariaal
Barabaig
Hadza
Himba
Karamojong
Masai
Owambo
Rendille
Samburu
Somali
Turkana

Atlantic

Ocean

San

Indian

Ocean

Multi-Ethnic African States
Colors indicate areas
where indigenous peoples
and other ethnic groups live.

African state boundaries, drawn principally by Europeans who sought to maximize profits from natural resources, ignored the boundaries of ethnic nations and homelands. The resulting political turmoil continues to this day.

are historical examples of irredentism as are current efforts by Somalia to capture the Ogaden region of Ethiopia.

Most states are best described as multi-ethnic, which can create political problems, especially when ethnic groups occupy spatially discrete homelands within the state. In the United States, few ethnic groups are so geographically concentrated that they present a separatist challenge to the federal government. Native American tribes, however, do have limited decision-making authority in governing their reservations, although these territories are rarely their original homelands.

In other states, especially those in Africa, there may be dozens of nations and homelands within state boundaries. Most of the boundaries in Africa were established by European states to divide the resources of the continent among themselves under the 1884 Treaty of Berlin. (See The Berlin Conference sidebar, page 284.) Recent political disruptions and civil wars in Nigeria, the Democratic Republic of Congo, Sierra Leone, and other African states reflect the problems that can result when political boundaries are designed to separate areas on the basis of conquest, resources, and economic development rather than on ethnicity.

Nationalism and Separatist Movements

Nationalism, sometimes specified as ethnic nationalism, is the concept that nations deserve the right to self-determination, as either autonomous, self-governing regions or as sovereign states. Ethnic nationalism instigated the breakup of the overseas empires of European states during the 20th century. In 1990 and 1991, ethnic nationalism contributed to the independence of the 15 republics of the Union of Soviet Socialist Republics and the breakup of Yugoslavia.

There are dozens of stateless nations, and many actively seek political power from the governments of states in which their territories lie. Many of these nations, such as Baluchistan in southern Pakistan and Iran, are little known outside their region. Other nations, such as Kurdistan, have achieved much more publicity. Kurdistan,

This map shows the ethnic groups of the former Yugoslavia. In 1990 and 1991, ethnic nationalism, or the concept that nations deserve the right to self-determination, contributed to the breakup of Yugoslavia, as well as to the independence of the 15 republics of the U.S.S.R.

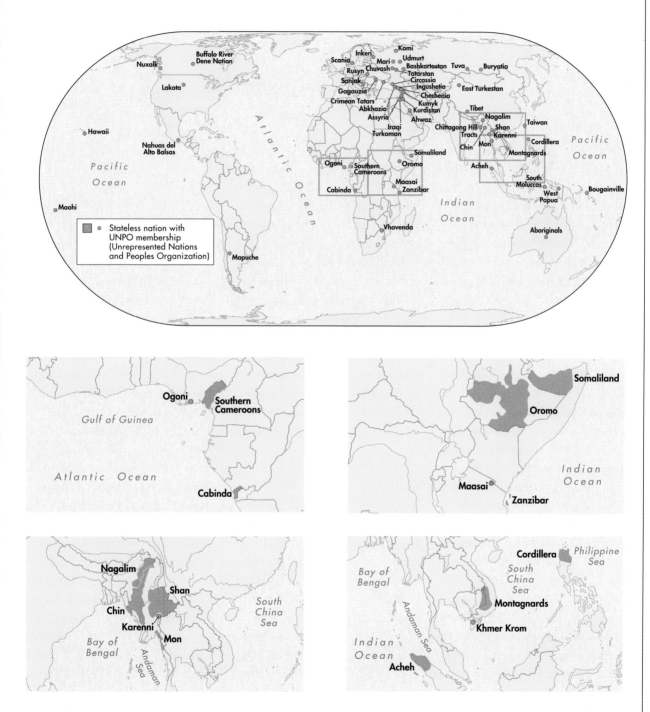

Dozens of stateless nations exist in the world today. Many seek political power from the states within which their territories lie. This may be achieved through nonviolence or extended conflict.

Nationalism in Canada

The struggle for independence by French-Canadians in Quebec has been nonviolent, except for a few terrorist acts by the Quebec Liberation Front (FLQ) in the late 1960s. In 1976 a political party, the Parti Québecois, was elected on the platform of Quebec independence and protection of the French language and culture; it continues to seek the approval of Quebec voters to separate from Canada. On another political front in Canada, the Inuit, a native group in northern Canada, was granted the right to self-government and in 1999 the territory of Nunavat was created from the Northwest Territories.

like many other homelands that existed prior to the formation of modern states, pre-dates state boundaries and overlaps several independent states—Iraq, Turkey, Syria, and Iran.

More at:

http://www.krg.org/

The *devolution* of power from a state government to a homeland may take forms ranging from limited authority—over local issues, for example—to statehood. The process of devolution may be either violent or non-violent. In the case of Scotland, a regional legislature with limited powers of taxation was approved in 1997 by both Scottish voters and the national government of the United Kingdom. Eritrea, on the other hand, won independence and sovereignty from Ethiopia in 1993, but only following a 31-year struggle.

In many cases, the political struggle of a nation for power becomes an armed struggle because the state governments involved try to preserve their territorial control. Guerrilla warfare and terrorism may be used by nationalist movements because states are better funded, manned, and armed; although in some cases outside powers may supply the nationalist movement with funding, weapons, and other resources. In recent years the Basque nation in Spain, the Chechens in Russia, the Corsicans in France, and numerous other groups have been engaged in armed struggles to gain political power.

BOUNDARIES: SPATIAL LIMITS ON POWER

Frontiers and Boundaries

The boundaries of a state or any political unit define its jurisdiction and are the spatial limits of its ability to enforce compliance with political decisions. Laws, whether on taxation or criminal activities, can be enforced only within the spatial jurisdiction of the governing body. As a result, international boundaries remain an important part of the international system.

Boundaries, also called borders, are both inclusive and exclusive. Governments use them as limits on the flow of goods, people, money, and communications both into and out of the state. Relationships with bordering states are managed along international boundaries, but political decisions on how an international boundary functions change with governments and situations. During the last century the United States has alternately excluded Mexicans and then allowed Mexican citizens into this country, depending upon economic conditions and on political sentiment. Mexicans were excluded during the Great Depression, but during the 1940s they were encouraged to migrate under the Emergency Farm Labor Program, also called the Bracero Program, which lasted until 1964.

Frontiers, undeveloped regions separating political units, are now almost nonexistent. Expanding economies, growing populations, improving weaponry, and increasing claims on land have caused boundaries to replace frontiers throughout the world.

Boundaries between states are understood to extend above the surface, beneath it, and, for those with coastlines, offshore. Airspace under the jurisdiction of a state currently extends to the limit of powered flight, so that overflights by satellites—at least for now—are not considered invasions of airspace. The downward extent of boundaries is theoretically to Earth's center and includes all resources, although fluid resources such as petroleum, which can flow between states, may cause disputes. (See Boundary Disputes, page 327.)

Formal establishment of a boundary occurs in stages, not all of which necessarily take place. First, the boundary is defined by a treaty between bordering states that specifies the boundary's location. This written description is followed by delimitation, which means that the boundary is drawn on a map, and its location is recognized by the bordering states. Next, boundaries are demarcated, that is they are marked on the landscape by walls, posts, clearings, or other features. The United States, for example, has constructed walls along part of the border between California and Mexico's Baja California Norte to control the flow of people across that segment of their boundary. Finally, boundaries are administered by officials to fulfill such governmental functions as customs, health, and immigration checks.

Boundaries also separate subnational political units. Within the United States, boundaries separate states and limit the enforcement of state laws. For example, states have different income taxes, except for Alaska, Florida, Nevada, South Dakota, Texas, Washington, and Wyoming, which have no state taxes. City limits are boundaries for city taxes, city services, and city laws—such as land-use planning ordinances—which are not enforced or provided beyond city limits. (See Zoning sidebar, page 302.) Annexation of adjacent rural areas expands city boundaries to include additional areas, thereby enlarging both the tax base and the urban service area. While some cities have doubled their areas through annexation, some rural populations fight annexation to remain beyond city laws and taxes.

United Nations Convention on the Law of the Sea

The international consensus for offshore boundaries of states is the result of three United Nations Conferences on the Law of the Sea held in 1958, 1960, and from 1973 to 1982, known as UNCLOS I, II, and III. (See Alliances and Organizations, page 329.) Territorial seas, over which states have virtually the same authority as over land, extend 12 nautical miles offshore. States must, however, allow "innocent passage" of foreign ships through these waters. Each state's exclusive economic zone (EEZ), over which the state controls resource development, is the water and seabed within 200 nautical miles of shore. Problems with locating EEZ boundaries are common because many coastal states—such as Denmark, Germany, and Sweden— are not separated by 400 nautical miles of ocean, and therefore lines halfway between the nearest shorelines of the adjacent states must be determined. Beyond EEZs lie the high seas, open to all states for transport and resource development, although there is an international movement to have seabed resources under these waters benefit all states, including landlocked ones. Open access to high-sea fisheries has been limited by specific treaties or organizations, such as the International Whaling Commission, but not by general international laws.

More at: http://www.unclos.com/

Types of Boundaries

Boundaries can be classified, based on the surface features they follow, as physical, geometric, or ethnic. Each classification will have advantages and disadvantages in managing relations with bordering states. Physical boundaries follow natural features in the physical environment, such as rivers, mountains, and lakes. Rivers are easily recognized as boundaries, but questions of proportionate water use by the two states may come up, and rivers may also change location because of meandering. The United States and Mexico negotiated the boundary along the Rio Grande, called Río Bravo in Mexico, numerous times in the 20th century to reflect changes in the location of the river's channel.

Geometric boundaries are usually straight lines, often lines of latitude or longitude, and only rarely follow other geometric shapes. The ease of establishing geometric boundaries is countered by the problems caused by dividing natural resources such as water and dividing ethnic populations and homelands. The 49th parallel dividing western Canada and the United States created problems with water development because it divided major watersheds, such as that of the Columbia River. The linear boundary between Ethiopia and Somalia resulted in the inclusion of a large Somali population in Ethiopia, which has created a major boundary dispute between the two states.

Ethnic boundaries, sometimes called cultural boundaries, attempt to separate different ethnic groups to establish national identities and prevent irredentist claims. However, not only are ethnic settlements not grouped into discrete regions, but settlements often contain residents of different ethnicity. The difficulty in accurately defining ethnic boundaries became evident in the partitioning of the former Yugoslavia: Ethnic Bosnians in particular were dispersed throughout Bosnia and Herzegovina, and Croatian and Serbian settlements and families also were found in most parts of the new state.

Boundaries can also be classified according to the timing of their establishment and location relative to territorial development as antecedent, consequent, superimposed, and relict boundaries. Antecedent boundaries are those established prior to substantial development of the territory, and therefore state governments can influence the population and types of activities in the areas. For example the boundary on Borneo between Malaysia and Indonesia predates development of Borneo's resources.

Consequent boundaries are imposed subsequent to settlement within a territory and attempt to separate different ethnic groups. The boundaries between Ireland and Northern Ireland and between Pakistan and India were an attempt to divide rival religious groups, but neither boundary was perfectly matched to the populations. Despite the migration of millions of Muslims and Hindus between Pakistan and India, problems remain today from these imperfect divisions.

Superimposed boundaries are also subsequent boundaries but are not concerned with patterns of ethnic settlements. The boundaries of most modern African states are the result of European colonial powers such as France, Portugal, the United Kingdom, and Spain dividing Africa to benefit themselves without concern for the homelands of hundreds of African nations. A continuing legacy of these superimposed boundaries has been ethnic conflict. In Rwanda and Burundi, for example, the rivalry between Hutus and Tutsis resulted in hundreds of thousands of deaths in 1994; around one million died in the late 1960s as the Ibo nation attempted to separate from the rest of Nigeria.

Relict boundaries are inactive and represent earlier political units, nevertheless their impacts are often evident on landscapes. The reunification of Germany in 1990 erased the international boundary between West Germany and East Germany, but differences in industry and development between the two Germanies have not yet disappeared and will persist for decades because of the high costs of implementing policies to help the east reach parity with the west.

More at:
http://www.siue.edu/GEOGRAPHY/ONLINE/German_border/border/001.htm

Boundary Disputes

Boundaries impose spatial limits on the power of states and other political units, yet the political, economic, and social interests of governments and their constituencies extend across boundaries. As a result, boundary disputes are extremely common between states and between subnational units. Claims to territory, resources, and populations across boundaries, and disagreements over boundary management, such as limiting imports or migrants, cause disputes and can cause wars. Ecuador's claims to territory now in northern Peru, Libya's claims to northern Chad, and the claims of Brunei, China, Malaysia, the Philippines, Taiwan, and Vietnam to the Spratly Islands are among the dozens of boundary disputes in the world.

Boundary disputes are rarely straightforward. An argument in the early 1900s between the United States and Canada over the position of the boundary between Alaska's Panhandle and British Columbia focused on the interpretation of 1824 treaties among Russia, the United States, and the United Kingdom. A judgement by an international tribunal in 1903 gave the United States control over coastal areas and therefore over the ports of Dyea and Skagway serving miners arriving for the Yukon gold rush.

Territorial disputes arise when land is claimed by two or more states, each of which has historical claims to the territory. In Latin America alone there are nearly a dozen territorial disputes, many of which can be traced back to boundaries imprecisely defined in the 1800s. The 1982 Falkland Islands War between Argentina and the United Kingdom was the result of competing claims to these islands, dating back to 1833.

Resource disputes involve natural resource development. The invasion of Kuwait by Iraq in 1990 was triggered by Kuwait pumping oil from the shared Rumaila oil field without a prior agreement with Iraq; several other factors were also involved, including Iraq's historic claims to Kuwaiti territory. Many

recent resource disputes have arisen over the offshore extent of EEZs and authority over good fishing grounds. Canada and the United States disputed their offshore boundary and fishing rights in the Gulf of Maine until 1984, when the two nations finally agreed to accept a judgment by the International Court of Justice.

Other disputes revolve around policies one state enforces on transboundary movements, including immigration, imports, and smuggling, that result in inconveniences to another state. North Korea has barred South Koreans from visiting family members in the north since the two states divided in 1948. The United States has complained for

Lebensraum

Friedrich Rätzel (1844–1904) envisioned states as being similar to organisms requiring additional living space or lebensraum to remain healthy. Karl Haushofer (1869–1946), a German geographer and army officer, expanded Rätzel's work to justify German expansion in the first half of the 20th century. Haushofer specifically defended Germany's need to acquire parts of Poland and Czechoslovakia and coined the term geopolitik to justify Nazi aggression against its Slavic neighbors.

Flags fly outside United Nations Headquarters in New York. In this supranational organization, many states give limited power to a central authority.

years about the number of illegal immigrants and drug smugglers regularly crossing the United States-Mexican border.

IMPERIALISM, MULTISTATE ORGANIZATIONS, AND THE WORLD ORDER

State interests such as security and trade cannot be fully contained within state jurisdictions, so states have acted to exert their influence beyond their boundaries. Over the last five centuries, powerful states—especially European states—conquered lands around the globe and

established colonies to improve their political, economic, and cultural influence.

During the 20th century most of these empires collapsed, and the colonies became independent thus changing the political and economic relationships between states and creating incentives for them to act jointly. During the last decades of the 20th century, the end of the Cold War and increases in biological weapons and terrorism created new threats to state security, and the growth of multinational corporations and expansion of the

global economy generated new benefits from cooperative actions. The result has been the formation of a new political and economic order, a new world order.

Imperialism

Imperialism, the policy of dominating colonies or other states and maintaining those relationships to increase state power, was embraced by European states in the 15th century. European imperialism at one time claimed almost the entire non-European world. The territories of North and South

America, Australia, most of Africa, and much of Asia were claimed as colonies by European states. The phrase "The sun never sets on the British Empire" referred to the global extent of British colonies on all continents except Antarctica, as well as on island in the Pacific and Atlantic Oceans.

Russia's empire, unlike those of other Western European states, extended overland rather than overseas, which enabled Russian, and after 1917, Soviet forces, to maintain their presence and retain control in what are now central Asian states. Japan and the U.S. also had empires. Japan's empire in eastern and southeastern Asia was established between 1875 and World War II, and that of the United States, although rarely referred to as an empire, included Cuba, the Philippines, and other islands in both the Caribbean and the Pacific.

Colonies were controlled politically and economically by the imperial power. Often, colonies provided strategic locations for their rulers: Cape Town, Malta, and Singapore provided the British with critical locations for supplying and repairing ships. Economically, colonies provided raw materials for the industries of the imperial states; not only did this remove much of the wealth from the colony, but it established strong production and trading relationships that are still present today. Côte d'Ivoire, a former French colony, has France as its major trading partner; Ghana, a former British colony, has the United Kingdom as its major trading partner; and so on.

Empires began to break apart in the 19th century and continued to dissolve at the end of the 20th century, although some colonies had gained independence earlier. As recently as 1960 there were only four independent states in Africa: South Africa, Egypt, Ethiopia, and Liberia; all other territories were colonies. Now, however, the majority of colonies are recognized as states, as evidenced by their admissions to the United Nations, which began with only 51 members at its inception in 1945, but had grown to 191 by 2004. (See Alliances and Organizations, below.)

Alliances and Organizations

States have numerous incentives to cooperate with each other—to promote economic growth and military security, for example—but cooperation necessarily involves shared decision-making power and some loss of individual initiative. To preserve state powers, most alliances are established as international organizations that do not create a powerful central authority, but rely instead on collaborative decision-making. The British Commonwealth and the Organization of African States (OAU), for example, are international organizations. Supranational organizations have also been created, in which three or more member states delegate limited powers to a central authority. Although states lose some sovereignty, the organization has increased power to implement policies promoting joint interests. The United Nations and the European Union (EU) are supranational organizations.

As empires dissolved and international conflict and exchange increased in the 20th century,

Neocolonialism

One enduring legacy of imperialism is a global division between less developed former colonies in Africa, Latin America, and southern Asia, and the more developed colonizing states of Europe, the U.S., and Japan. Neocolonialism, the economic dependence of former colonies on their former colonizers, is a modern form of imperialism that compromises the political sovereignty of the former colonies. Since the economies of former colonies remain dependent on supplying raw materials for the developed states, their financial policies are necessarily constrained.

dozens of organizations formed as both new and old states sought to protect and promote their national interests. Most of these organizations postdate World War II. In some instances the formation of one organization spurred the formation of another as the non-members grouped together to be economically or militarily competitive. Three years after Belgium, the Netherlands, Luxembourg, France, West Germany, and Italy formed the European Economic Community (EEC) in 1957, the European Free Trade Association (EFTA) was formed by the United Kingdom, Switzerland, Norway, Sweden, Denmark, Austria, and Portugal.

Organizations of states can be classified as primarily economic, military, or political in nature. Economic organizations are based on the premise that combining markets, resources, and production will create economies of scale, inspire greater productivity, and increase the organization's role in the international economy. The Central American Common Market (CACM), North American Free Trade Agreement (NAFTA), Economic Community of West African States (ECOWAS), Organization of Petroleum Exporting Countries (OPEC), and Asia-Pacific Economic Cooperative Group (APEC) are examples of economic groups.

The European Union (EU) has special status as an economic organization because of its commitment to social and political as well as economic integration. Building on earlier organizations, the EU is the latest name for the organization that started in 1957, when it was called the EEC or Common Market. By 1999, 15 European states were members; in May 2004, 10 more countries joined the EU. Almost no restrictions or movements across members' boundaries remain, and the euro, a common currency, is now the official currency of 11 EU members. Plans for a common defense policy and coordinated social policies are underway.

More at:
http://europa.eu.int/index_en.htm

Military organizations provide regional security by committing members to a joint defense against an attack on any other member. Two such organizations, the North Atlantic Treaty Organization (NATO) and the Warsaw Pact, symbolized the Cold War because the United States dominated NATO and the Soviet Union dominated the Warsaw Pact. After the Soviet Union disintegrated in 1991, the Warsaw Pact also dissolved. NATO remains, but in a changed role. NATO now has 26 member states, including some of the former Warsaw Pact states.

More at:
http://www.nato.int/

Other organizations are designed to promote social, cultural, economic, and other links among members and are called political organizations. Most political organizations have a limited number of members because the shared concerns are limited regionally or topically; a few organizations, however, seek global participation. The United Nations (UN), founded in 1945 after the end of World War II and based in New York City, is the most important global organization and is the successor to the now defunct League of Nations formed after the end of World War I. Except for a very few states, most notably Switzerland, all states on Earth are members of the UN.

The purpose of the UN is to promote international peace, security, and cooperation among states, as well as to protect human rights. Two chambers are key to achieving these goals: the General Assembly and the Security Council. All members of the UN have a vote and the right to speak in the General Assembly, making it the most inclusive parliament in the world. While debates and speeches cannot resolve major issues, representatives from all states—and from some stateless nations awaiting recognition—are present and able to contact each other. The Security Council, with a membership of 15, has fewer

SELECTED INTERNATIONAL ORGANIZATIONS (2004 MEMBERSHIP)	
Asia-Pacific Economic Cooperation (APEC)	Australia, Brunei Darussalam, Canada, Chile, People's Republic of China, Hong Kong (China), Indonesia, Japan, Republic of Korea, Malaysia, Mexico, New Zealand, Papua New Guinea, Peru, Philippines, Russia, Singapore, Chinese Taipei, Thailand, United States, Vietnam
Economic Community of West African States (ECOWAS)	Benin, Burkina Faso, Cape Verde, Côte d'Ivoire, Gambia, Ghana, Guinea, Guinea-Bissau, Liberia, Mali, Niger, Nigeria, Senegal, Sierra Leone, Togo
European Union (EU)	Austria, Belgium, Cyprus, Czech Republic, Denmark, Estonia, Finland, France, Germany, Greece, Hungary, Ireland, Italy, Latvia, Lithuania, Luxembourg, Malta, Netherlands, Poland, Portugal, Slovakia, Slovenia, Spain, Sweden, United Kingdom. Candidate Countries: Bulgaria, Croatia, Romania, Turkey
North Atlantic Treaty Organization (NATO)	Belgium, Bulgaria, Canada, Czech Republic, Denmark, Estonia, France, Germany, Greece, Hungary, Iceland, Italy, Latvia, Lithuania, Luxembourg, Netherlands, Norway, Poland, Portugal, Romania, Slovakia, Slovenia, Spain, Turkey, United Kingdom, United States
Organization of American States (OAS)	Antigua and Barbuda, Argentina, Bahamas, Barbados, Belize, Bolivia, Brazil, Canada, Chile, Colombia, Costa Rica, Cuba*, Dominica, Dominican Republic, Ecuador, El Salvador, Grenada, Guatemala, Guyana, Haiti, Honduras, Jamaica, Mexico, Nicaragua, Panama, Paraguay, Peru, Saint Kitts and Nevis, Saint Lucia, Saint Vincent and the Grenadines, Suriname, Trinidad and Tobago, United States, Uruguay, Venezuela (*Cuba is excluded from participation)
Organization of Petroleum Exporting Countries (OPEC)	Algeria, Indonesia, Iran, Iraq, Kuwait, Libya, Nigeria, Qatar, Saudi Arabia, United Arab Emirates, Venezuela. Associate Member: Gabon

members but more power than the General Assembly. Specifically, the council has the authority to investigate threats to international peace and to send UN troops on peacekeeping operations throughout the world. In 1949 and 1950 this meant involvement in conflicts in both Kashmir and Korea, and in 1999 UN troops were in 16 locations, including Sierra Leone, Haiti, Tajikistan, and Cyprus.
More at:
http://www.un.org/english

Other agencies with global concerns, including the Food and Agriculture Organization (FAO), the International Monetary Fund (IMF), and the International Bank for Reconstruction and Development (World Bank), work in tandem with the UN. The World Bank and IMF have become especially important because their decisions on loans for major projects and currency stabilization affect not just the states requesting assistance but, through

financial linkages, the health of the global economic system.
More at:
http://www.worldbank.org/ and http://www.imf.org/

Geopolitics and the New World Order

The global system of power relationships whereby more powerful states dominate and influence the actions of less powerful states, establishes a world order. Geopolitics is the study of international power relations from a spatial perspective. At the beginning of the 20th century, the United Kingdom dominated much of the world, and at the end of the 20th century the United States held the dominant role.

Geographers have long attempted to model global power relationships and Englishman Sir Halford Mackinder (1861–1947) developed one of the earliest geopolitical theories, the *heartland theory*, in 1904. Mackinder argued that the relative location and hostile environment of central Russia—the heartland—made it nearly impregnable to invasion by other states and was the key to world dominance. Eastern Europe, however, was the entryway to the heartland, and therefore the strategy of would-be global powers should be control over Eastern Europe. Some writers see the application of this philosophy in the Nazi invasions of Eastern Europe that began World War II.

Mackinder's theory oversimplified the influences of environment and distance, but it influenced numerous other writers. A competing view of world power relations was developed during World War II by Nicholas Spykman (1893–1943), a Dutch-American. Spykman's rimland theory argued that accessibility, productivity, and the history of relations among states meant that the coastal rimland of

Eurasia, not its central part, was the key to world dominance. Preventing the rimland from uniting, Spykman argued, would also prevent any power from achieving world dominance; the policy of the United States during the Cold War to contain the spread of communism to rimland states reflects the viewpoint of this theory.

During the 1960s, American geographer Saul Cohen argued that there were four geostrategic regions on Earth, the *spheres of influence* of the United States, Maritime Europe, the Soviet Union, and China. During revisions of this theory over the next 30 years, other regions were noted, including South Asia, areas prone to political instability (shatterbelts) such as the Middle East and Southeast Asia, and regions linking the different spheres (gateway regions) such as Eastern Europe. In his later revisions, Cohen modified the divisions and focused on cooperation among states in an increasingly interdependent world.

In 1991, even as the Union of Soviet Socialist Republics was beginning to break up, the United States and the USSR agreed on the need to confront Iraq over its invasion of Kuwait and a *New World Order* began. The inflexible, bipolar aspects of the Cold War were over, and optimists believed increasing economic and political linkages

between states and nations and the emergence of supranational organizations would create a balance of powers: Multinational actions would replace unilateral decisions and actions. The growing interconnectedness within the European Union and the United Nation's coalition against Iraq in 1991 were cited as evidence of these changes.

Late in the 20th century new threats to a peaceful world order emerged as terrorism. Terrorist attacks against symbols of authority in Europe and the U. S. created an atmosphere of unease. Contemporary terrorism has two major motivations: challenging the global role of the United States and asserting national self-determination. A worldwide network of terrorist cells, loosely affiliated with al Qaeda, has targeted buildings and activities that represent Western economic, political and cultural activity, culminating in the attacks of September 11, 2001 in New York City, suburban Washington, D.C., and rural Pennsylvania. A more traditional form of terrorism is ethno-national terrorist activity within states. The drive for national separation or self-determination is behind terrorist activities in, for example, Kashmir, Chechnya, and West Africa.

More at:

http://www.fas.org/irp/threat/terror.htm

ENVIRONMENT AND SOCIETY

NETWORKS AND INTERACTIONS

Geographers study interactions between human societies and natural environments, how societal—and individual—actions affect environmental processes, and how environmental processes affect societies. The flora and fauna in different regions and their physical milieus, including climate, soil, and hydrology comprise Earth's natural environments. Each component of the natural environment is linked to the others by flows of energy, water, and nutrients. For example, vegetation removes carbon dioxide from the atmosphere and nutrients from the soil while providing energy to grazing animals and oxygen and water vapor to the atmosphere. A system of interrelated processes is at work in each environment and creates resources such as forests, clean water, and fertile soils, but these processes can also contribute to hazards such as floods, tornadoes, and locust plagues.

Humans, more than any other species, can change natural environments. As a society inhabits a region and develops its resources, the society unavoidably interferes with environmental processes by displacing animal and plant species, interrupting cycles of nutrients, and modifying water flows. The society may drain swamps to reduce disease, build dams to control flooding, construct power plants to

Coal-burning plants such as this one can range far in their polluting power. In the 1970s, plants in Ohio contributed to acid rain in New England.

provide energy, and apply pesticides to increase agricultural yields. All of these actions may benefit humans, but at the expense of other species. Drained swamps support far fewer species of plants and animals; dams prevent the migration of salmon and the distribution of silt, but cannot prevent the largest floods; power plants pollute air and water; and pesticides kill far more creatures than the pests for which they were intended.

Societies cannot always predict the effects of their actions on the environment nor can they control many environmental processes. During the 1970s and 1980s, coal-burning power plants in the Ohio Valley contributed to acid rain in New England that killed fish, amphibians, and trees. Attempts to increase rainfall during droughts by intervening in atmospheric

Resource-Management Strategies

There are four major categories of resource-management strategies:

◆ Exploitation strategies focus on the immediate market value that will be realized by developing the resource.

◆ Utilitarian, or conservation, strategies emphasize ensuring long-term productivity by conserving resources for the future.

◆ Ecosystem strategies, another form of conservation, attempt to balance commercial and noncommercial values by developing resources without seriously affecting environmental processes.

◆ Preservation strategies concentrate on managing resources to preserve noncommercial values of the environment such as species preservation and biodiversity.

Economic, political, and cultural considerations all play a role in resource-management strategies. Natural resources are nearly always limited in quality, quantity, and location, which means that development involves selecting among different, and often competing, management strategies. Economic concerns are compelling

in many cases because resource development creates jobs, provides marketable products, and improves an economy—at least in the short run. Resources such as clean air, biological diversity, and wilderness, for which there is no market mechanism, do not compete well with commercial resources that provide measurable financial benefits.

Economic and political influences on resource-management strategies can be driven externally as well as internally. Countries are linked into a *world system* of trading patterns in which natural resources are often produced by less developed (peripheral) countries and traded to developed (core) countries. An unequal relationship exists between these two groups of countries: Developed countries have greater power than less developed countries because their economies are larger and more diversified. Less developed countries can be trapped into resource-based economies, which are vulnerable to changes in demand by the developed countries. Honduras, for example,

exports mostly bananas and coffee, Uganda exports coffee, and Libya exports mostly petroleum. The value of these exports rises and falls depending on the demands by other countries.

Natural resources and resource-management strategies change over time as economics, politics, and culture change. When the price of steel drops, iron deposits that cannot be mined at a profit are no longer developed. Changes in a political administration can cause changes in resource decisions. The definition of wetlands, lands saturated with water in which a majority of endangered species live, was changed in 1991 by the administration of President George H. W. Bush to include only half the areas that had previously been under federal protection. Cultural change, as seen in public opinion polls, was influential in slowing the development of nuclear power in United States and Europe following accidents at Pennsylvania's Three Mile Island reactor in 1979 and at Ukraine's Chernobyl site in 1986.

processes have met with limited and mixed success. Earthquakes, hurricanes, and other natural hazards threaten societies around the globe, and we are unable to control the environmental processes behind these events.

RESOURCES AND RESOURCE MANAGEMENT

All societies rely on energy and materials extracted from the environment to maintain and improve their quality of life. *Resources* are substances, qualities, or organisms that have use and value to a society: Resources that are taken from the environment are natural resources. Societies often perceive resources differently because of differences in cultural values, levels of technology, and economic concerns. Iron ores, fertile soils, coal deposits, and groundwater are considered resources by most societies. Clean air, diverse wildlife, wetlands, and beautiful views are also valued and may therefore be considered resources by every society. A society makes choices when developing a resource that may prevent development of another. For example, iron ore is necessary for making steel, but mining the ore, processing it, and transporting it to factories scars the land and pollutes air and water unless specific preventive or restorative actions are taken.

Decisions on which natural resources should—and should not—be developed, to what degree, in what manner, and for whom are resource-management issues. Resource-management decisions vary tremendously in their impacts on society and the environment as different weights are assigned to resources by corporations, federal governments, and other groups based on immediate financial gain, long-term productivity, and environmental protection.

All resource decisions occur within and are influenced by political systems. In the United States, elected officials are ultimately responsible for establishing the laws regulating resource development. The U.S. Forest Service manages some 191.8 million acres of national forests and grasslands; private loggers buy the rights to harvest timber, but Congress legislates the rules for multiple use, including recreation and wildlife protection and timber harvesting, and sustainable yield.

More at:

http://www.fs.fed.us/

The cultural values of a society affect *resource management* because management choices reflect the values of the decision-makers. During the 1960s in the United States, for example, the environmental movement gained strength. Congress and various administrations reacted to this shift in public values by passing laws such as the Clean Air Act (1965), the Endangered Species Act (1966, with amendments in 1973), and the National Environmental Policy Act (1969). Cultural values of different populations can also come into conflict and affect resource management. In 1986 the International Whaling Commission banned commercial whaling, but Japan—where eating whale meat is part of the cultural heritage—and a few other countries continued taking whales.

Nonrenewable Resources

Mineral resources, such as coal and copper, that form from extremely slow geologic processes, are considered nonrenewable, that is the amounts available will not appreciably increase for millions of years. When a mineral resource has been discovered, measured, and determined to be of commercial value, it is called a proven reserve. As yet undiscovered, but inferred from past discoveries, deposits are potential reserves.

Nonrenewable mineral resources can be divided into three main groups based on their uses and physical properties: metals, fossil fuels, and industrial minerals. Metal reserves occur as ores, rocks composed primarily of a valuable mineral that can be mined and processed for commercial gain. The spatial distribution of metal resources on Earth reflects past geologic processes, processes often caused by plate-tectonic activity. (See Plate Tectonics, page 104.) As a result, some countries have larger reserves of more types of minerals than other countries. South Africa, Russia, and Canada all have substantial proven reserves of gold, nickel, chromite, and zinc. South Africa produces more than 25 percent of the world's gold, Russia 25 percent of the world's nickel, and Canada 17 percent of the world's zinc. The United States has substantial deposits of lead, copper, silver, and some other metals but imports all manganese, which is used in making steel, and 99 percent of bauxite, which is used to make aluminum.

The amount and distribution of metal reserves change over time because of discoveries of new deposits, exhaustion of old deposits, changes in societal demand, and improvements in mining and processing. Future development of metal resources will be affected by a combination of factors, including the changing value of metals, size and location of proven reserves, ease of recycling metal products, availability of substitutable materials, and political concerns. Certain

metals such as niobium and cobalt are labeled strategic resources because of their use in military, space, and energy programs. Such metals will retain their high value, and development of deposits of these metals will remain a national priority. Other metals, which are widespread and abundant, should remain relatively inexpensive.

Nonrenewable mineral fuels, particularly fossil fuels, are the major source of energy used globally today, and provide the energy to run machinery ranging from cars and combines to computers and canneries. The demand for energy is rising as populations increase and industrialization spreads. Between 1970 and the present, the total amount of energy used around the globe almost doubled to more than 150 quadrillion Btu (British thermal unit).

More at:

http://www.bp.com/

Three fossil fuels—petroleum, coal, and natural gas—account for more than 80 percent of global energy use, and uranium, the other major mineral fuel, accounts for another 5 percent. Renewable energy sources such as hydropower, wind, and solar energy provide the remaining 15 percent. Petroleum, also called oil or crude oil, is a mixture of liquid hydrocarbons derived from microscopic marine organisms

trapped in seafloor sediments hundreds of millions of years ago. Natural gas, mostly methane, usually forms along with petroleum from the same organisms. Coal forms from swamp vegetation, and coal beds show the extent of ancient swamps. Deposits of uranium are created by igneous activity.

Unlike metal resources, mineral fuels are fully consumed when used and cannot be recycled. The amount of reserves is therefore extremely important for determining management strategies. At current rates of consumption and current technology, worldwide proven petroleum reserves should last until about 2040, coal reserves until 2200, natural gas reserves until 2060, and uranium until the year 3000 or later.

International trade in mineral resources—in fossil fuels, in particular—is a critical component of the world economy. More than half of Earth's petroleum reserves are located in the Middle East, and the major oil consumers are developed countries that must import petroleum. Coal deposits are more widespread—the U.S. alone contains 23 percent of the global reserves. The global exchange of coal is less extensive than of oil because of its widespread distribution and because of its greater weight per unit output of energy. Natural gas, the cleanest

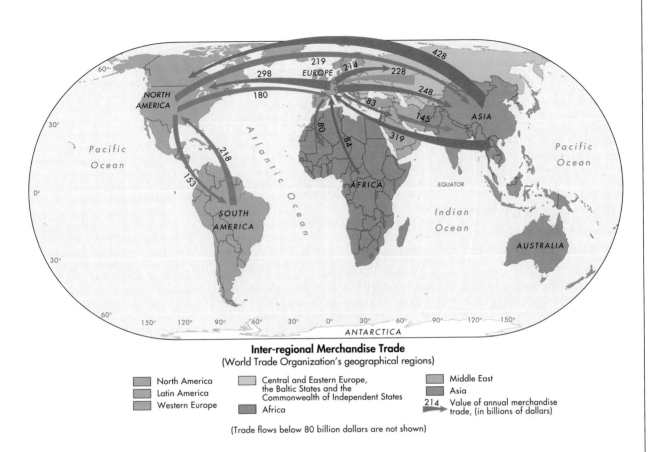

Inter-regional Merchandise Trade
(World Trade Organization's geographical regions)

- North America
- Latin America
- Western Europe
- Central and Eastern Europe, the Baltic States and the Commonwealth of Independent States
- Africa
- Middle East
- Asia
- 214   Value of annual merchandise trade, (in billions of dollars)

(Trade flows below 80 billion dollars are not shown)

Petroleum is, along with natural gas, the main source of energy for the world. However, as the century progresses, limited resources, geopolitics, and the imbalance between supply and demand of oil will necessitate a conversion to other renewable sources of energy.

PETROLEUM RESERVES AND LEADING PRODUCERS AND CONSUMERS, 2003		
LARGEST PROVEN RESERVES (THOUSAND MILLION BARRELS)	**% TOTAL PRODUCTION**	**% TOTAL CONSUMPTION**
Saudi Arabia—262.7	Saudi Arabia—12.8%	U.S.A.—25.1%
Iran—130.7	Russia—11.4%	China—7.6%
Iraq—115.0	U.S.A.—9.2%	Japan—6.8%
U.A.E.—97.8	Iran—5.1%	Germany—3.4%
Kuwait—96.5	Mexico—5.1%	Russia—3.4%
Venezuela—78.0	China—4.6%	India—3.1%

burning fossil fuel, is concentrated in Russia, which contains 40 percent of Earth's reserves and exports natural gas, mostly via pipelines, to European countries. Uranium deposits are found on all continents, but extensive data on reserves are not available because of security considerations.

In the future, as the reduction of petroleum reserves drives energy costs upward, the use of alternative sources of fossil fuels should increase, as should the use of renewable energy sources. Oil shales and tar sands are rock and sediment deposits containing fossil fuels and are alternatives to petroleum and natural gas. Today's high production costs make these deposits noncompetitive with traditional fossil fuels at present, but they may become competitive in the future. In addition, increased conservation of fuel resources should result from increased technological efficiency.

Industrial minerals, minerals used in the construction and chemical industries, such as sand, gravel, and limestone, are nonrenewable, but are for the most part plentiful. The weight of these materials increases transport costs so that mines and quarries are usually developed near areas of high, local demand. Salt, sulfur, phosphate, and other minerals are important to the chemical industry. Salt and sulfur deposits are

United States Strategic Petroleum Reserves

United States dependency on imported oil and its vulnerability to interrupted imports became evident during the 1973 oil embargo by members of the Organization of Petroleum Exporting Countries (OPEC), which decreased supplies, thereby increasing prices. In reaction, the United States government decided in 1975 to stockpile one billion barrels (1 barrel = 42 gallons) of oil by pumping it into salt domes in Louisiana. These strategic reserves started filling in 1977, but by the mid-1990s a leak formed in one of the salt domes and only 561 million barrels had been stored by December 1998.

substantial and are widely distributed around the world; however, only Morocco, South Africa, and the United States can claim major phosphate deposits.

Renewable and Inexhaustible Resources

Resources that are not limited by slow rates of geological formation are classified as either renewable or inexhaustible. *Renewable resources* such as fisheries, forests, and soils, are regenerated by either biologic reproduction or by environmental

processes. However, mismanagement of these resources can cause depletion and destruction. *Inexhaustible resources* such as solar, wind, water, and tidal energy are generated continuously, and their production is not reduced or hampered through mismanagement.

A variety of plant and animal species are valued as renewable resources for food and materials; and ecosystems, the interacting community of plants and animals in a region, are increasingly viewed as a resource in their own right. Harvesting an individual species in an ecosystem, such as salmon or redwoods, raises the issue of how to establish the level of *sustained yield*, the amount that can be harvested annually without depletion.

A major source of energy in many less-developed countries has been biomass, that is wood and biologic wastes (including field debris and animal dung) that can be burned as fuel or be processed into ethanol or gaseous fuels. In less developed countries such as Uganda and Haiti, biomass can supply more than 90 percent of local fuel demand, often in the form of wood. Globally, biomass provides about 15 percent of total energy consumption.

Attempts to establish sustained yields for fishing stocks are not easily achieved because of incomplete knowledge of all factors affecting

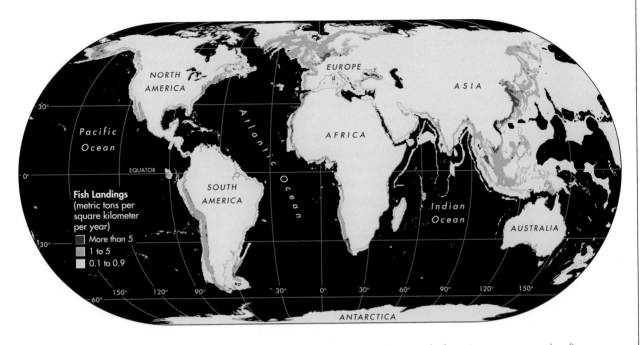

Fisheries are considered to be renewable resources because they can be regenerated in fairly short periods of time. Proper management benefits many people with food and other products. Mismanagement, such as overfishing or pollution, can seriously deplete or even destroy these valuable resources.

fish populations and because of difficulties in limiting the amount of fish harvested. Managing ocean fisheries has been a major problem in the past because all coastal countries have access to the major fishing grounds around the world. The oceans were considered a commons, with free access to all, and with no authority to limit the amount of fish harvested. As a result, numerous fish populations have been depleted. The 200-nautical mile exclusive economic zone (EEZ) each country claims off its shores has not resolved problems with overfishing because many fish species migrate to the open seas where fishing is not regulated.

Fertile soils, clean air, and clean water are also renewable resources that can be perpetuated indefinitely if not abused. While soil, air, and water are always available, each can be degraded in the process of developing other resources, such as coal mining; this degradation is an unintentional side effect of the development process that is often omitted in considering costs and benefits of development. Soil formation and fertility are primarily the results of the

TWENTY COUNTRIES LEADING 80% OF THE WORLD'S MARINE CATCH			
1. China	6. Russian Fed.	11. India	16. Spain
2. Peru	7. Thailand	12. Iceland	17. Taiwan, Prov China
3. Japan	8. Indonesia	13. Philippines	18. Canada
4. Chile	9. Korea, Rep.	14. Korea, DPR	19. Mexico
5. USA	10. Norway	15. Denmark	20. Vietnam

climate in a region; soils retain their fertility if protected from erosion and overuse. Air, like water, has the ability to remove particulates, chemicals, and other substances through settling and chemical changes, if the amounts are not excessive. Water, too, cleanses itself if protected from excessive additions of substances.

Inexhaustible resources are sources of energy that derive directly or indirectly from the sun, Earth's interior, or gravity. As a rule these are considered clean forms of energy because they do not pollute the air or water while producing energy,

Deforestation in Nepal

Overharvesting of wood in forests near towns and villages in Nepal has resulted in deforestation and depletion of local resources. In the 1980s more than 300 square miles were deforested each year, an annual loss rate of 3 percent. The side effects include increased soil erosion and resulting loss of crops and property, loss of wildlife, and increased riverbed silt buildup and flooding downstream in India and Bangladesh.

although hydropower can cause other environmental problems. Each of these energy resources—solar, wind, hydropower, geothermal, fission, ocean thermal, and tidal—faces economic, technologic, or other difficulties, and so, despite high hopes by conservationists, they have not easily replaced fossil fuels as major sources of energy.

Solar energy uses radiation from the sun for heating or to generate electricity; wind power and hydropower both depend indirectly on the sun's radiation to create winds or the rainfall that accumulates

Soil fertility is governed by the amount of organic material, such as dead and decaying plant and animal debris, it contains. A renewable resource, soils retain their fertility if protected from erosion and overuse.

Three Gorges Dam

In 2009 one of the largest hydro-electric and flood control dams ever constructed, the Three Gorges Dam on China's Chang Jiang (Yangtze River), is to be completed. Major problems will be resolved by this dam. It should control floods that have plagued China for centuries and reduce the number of coal-powered generating plants that create air pollution problems. At right, water gushes from the existing dam in Yichang after a 2004 flood. Some 1.2 million people will have to be resettled as the reservoir behind the new dam fills. It will extend 400 miles, cover 1.7 million acres of land, and modify seasonal variations in flow levels of the river.

behind dams. Solar heating is economically competitive with energy from fossil fuels in many regions, but producing electricity in large quantities using photovoltaic cells is not. Wind power has been tapped for centuries—in the Netherlands to pump water out of polders behind dikes, and in the United States to grind grain—but improved storage facilities are needed to make wind-powered electricity competitive with fossil fuels because the timing of winds and of power demand do not always match. Hydroelectricity generated by water in reservoirs flowing through turbines is extremely

important in a number of countries, such as Ghana, Laos, Canada, Austria, and other places where precipitation is high and rugged terrain provides good dam sites. Dams do, however, trap sediment and eventually lose their storage capacity and can disrupt local environments.

Inexhaustible energy is also derived from rising and falling tides and from tapping Earth's geothermal energy. Energy from nuclear fusion holds perhaps the greatest potential of all inexhaustible energy sources. At this stage, however, many problems remain, and environmental issues involving radioac-

tive wastes, leaks, and disasters are of great concern.

The development of natural resources is essential for modern societies. Over time populations have expanded numerically and spatially, technology has improved, and standards of living have risen. These factors have resulted in greater demand for resources and therefore have increased development activities in environments around the world. The effects of these increased activities include species extinctions, water and air pollution, and other forms of environmental degradation.

Wilderness

Wilderness areas are a relatively new type of resource, valued for their solitude, biological diversity, and scenery. The United States Wilderness Act of 1964 defined a wilderness as "an area where Earth and its community of life are untrammeled by humans, where humans themselves are visitors who do not remain." Unlike most other renewable resources, wilderness areas specifically need preservation, where no exploitation of individual resources occurs, rather than conservation, where resources such as rangelands are managed to yield benefits both to humans and the environment.
More at:
http://www.wilderness.org/index.cfm

HUMAN DEGRADATION OF THE ENVIRONMENT

Whenever natural resources are developed for use by societies, the environment is changed, and when these changes interfere with existing biological and environmental processes environmental degradation results. Different levels and scales of degradation can occur. Soil and bedrock are removed and displaced by mining, species are harvested by fishing and forestry, and water is dammed for flood control; all these actions benefit human societies, but all affect local species and flows of water, energy, and nutrients. Some amount of degradation is unavoidable if a society wishes to survive and retain its standard of living, but resource-management strategies need to consider and address levels and locations of degradation.

Atmosphere

Air pollution occurs when human activities add substances or energy—such as radiation—to the atmosphere in high enough concentrations or levels to harm organisms and cause other undesirable changes in the environment. Specific levels of unsafe concentrations of *pollutants* have been established by the United States Environmental Protection Agency (EPA)—sulfur dioxide, nitrogen oxides, particulates, carbon monoxide, and ozone—and these have been integrated into a Pollutant Standards Index (PSI).
More at:
http://www.scorecard.org/env-releases/def/cap_psi.html

Air pollution is a side effect of resource use, and much pollution results from the use of fossil fuels and industrial processes. An unhealthy concentration of pollutants is usually caused by an increase in the production of pollutants combined with stable meteorological conditions such as *temperature inversions*. When an atmospheric inversion occurs, pollutants are unable to disperse and therefore accumulate; dangerous inversions are common in cities such as Mexico City, Santiago, and Los Angeles.

Tragedy of the Commons

In 1968 biologist Garrett Hardin compared Earth's global ecosystem to a commons, a public grazing land traditionally open to all residents of a village or town. Where there is no supervision or limitations on a commons, villagers may graze as many animals as possible to take as much advantage as possible of the land. What is individually rational, however, is collectively disastrous because overgrazing ultimately destroys the resource for all. Today's global commons, such as ocean fisheries, are also subject to collective disasters because individuals may act as though there is no limit to the use of the resource.
More at:
http://dieoff.org/page95.htm

The sources of pollutants vary between regions and countries depending on their levels of industrialization and technology. In the United States about 55 percent of atmospheric pollutants are from vehicles, 21 percent from power plants and energy production, and 16 percent from industrial activities; incinerators and miscellaneous activities produce the rest. The concentration of population and industry in the northeastern states and other regions is mirrored in the concentration of pollution sources. Winds, however, can transport pollutants to other areas. Acid precipitation afflicting eastern Canada from the 1960s to the 1990s was caused by pollutants from power plants in the Ohio Valley; and haze near the Arctic Circle is the result of pollutants transported from industrial and urbanized regions in Europe and Russia.

Human health is affected by long-term, chronic exposure to air pollutants and by short-term, acute episodes of pollutants. Chronic exposure, especially in metropolitan areas, can cause bronchitis and emphysema, and it is implicated in some lung cancers. Children are at greater risk than adults because their respiratory systems are smaller and immune systems less developed. Air pollution is a hazard to other species and to global

Barry Commoner's "Rules of Ecology" (1971)

- Everything is linked to everything else: Changes in air or water eventually affect groundwater, land, and ecosystems.

- Everything must go somewhere: Incineration of garbage changes material in landfills into atmospheric pollutants because material cannot disappear.

- Nature knows best: The environment is a complex, interdependent, self-regulating system that will not be improved by human modification.

- There is no such thing as a free lunch: All benefits obtained from the environment have their associ-ated costs, to humans or toother species.

atmospheric conditions, too. Increased levels of carbon dioxide and other greenhouse gases are slowly but measurably warming Earth's climate, and chlorofluorocarbons (CFCs) and halons released into the atmosphere from appliances such as refrigerators and air conditioners are thinning the protective ozone layer 12 miles above Earth's surface. (See Atmosphere, page 342.)

More at:
http://www.epa.gov/air/

Control of air pollution is primarily through restrictive legislation. Local, state, national, and international laws have been established to fight air pollution problems at different spatial scales. Eighteen major U.S. cities had laws promoting carpools by the mid-1990s. The federal Clean Air Act of 1970 was revised substantially in 1990, and 150 countries signed the Framework Convention on Climatic Change at the 1992 Earth Summit in Rio de Janeiro. The 1970 Clean Air Act was passed to limit pollutants from power plants, vehicles, and factories; the 1990 amendment introduced a new market approach to reduce pollution by establishing pollution allowances to major electric utilities and then allowing bidding on air pollution permits. When utility companies purchase pollution permits they can increase the amount of air pollutants they can release without breaking federal regulations while the seller must reduce its pollution output. The total amount of air pollution remains the same, but regional differences result.

More at:
http://www.epa.gov/oar/oaqps/peg_caa/pegcaain.html

Air pollution control strategies can target capturing pollutants after

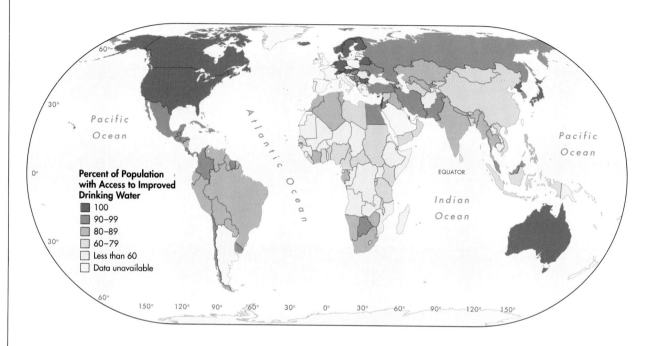

As world population grows, so does the demand for fresh, unpolluted water. Obtained from either surface or underground sources, water is a valuable but often limited renewable resource.

producing them, or reducing the amount and type being produced. Air pollution control raises several important issues, including the trade-off between environmental protection and jobs. The costs of lowering pollution can result in higher prices for goods or reduce profits, which can create concern among manufacturers. In the late 1990s, however, Harris public opinion polls showed 80 percent of Americans were very or somewhat willing to pay higher taxes to protect the environment.

Water

Societal demands for greater amounts of unpolluted water are increasing around the world. Water is a renewable resource, which moves perpetually through the hydrologic cycle. (See hydrologic cycle diagram, page 65.) It has the capacity to cleanse itself of pollutants, but the increasing demands and uses of water have made it a valuable and limited commodity in many parts of the world.

There are distinctions between demand for water and need for water and between withdrawal of water and consumption of water. Demand for water is the amount of water at a given quality that will be used at a given price; demand is higher when the price is lower, and water is usually under priced in the United States because of government subsidies for storage and treatment. Need for water, on the other hand, is the amount of water necessary for survival, and the amount that would be used at any price. Demand has increased far faster than need because water contributes to development and more comfortable lifestyles.

Withdrawal of water refers to the amount of water removed from the environment, either from surface water or groundwater; consumption is that portion of withdrawn water not returned directly to its source but to another part of the hydrologic cycle, often by evaporation and transpiration. Water that is consumed by drinking or irrigation is unavailable to downstream users, and this can cause economic problems. The Colorado River in the southwestern United States has an average annual flow of 13 million acre-feet, but only 1.5 million acre-feet reach Mexico downstream because the

majority of the water is used for irrigation or used in municipalities such as Los Angeles and Phoenix.

Development of water resources almost inevitably results in changing the seasonal variations in flow levels and the water quality. Groundwater flow is far more consistent throughout the year because the low permeability of most rocks prevents groundwater from reacting quickly to seasonal changes in input.

River flow levels are primarily controlled by dams that store water to meet societal demands. Streams may be channelized and confined by levees to control their flows. The benefits of dams include flood control, hydropower, recreation, and a secure water supply; and the effects on the environment include changes in the amounts and locations of erosion and deposition by streams, interruption of fish migra-

Between 1961 and 1971, the U.S. Army Corps of Engineers attempted to straighten Florida's Kissimmee River to form a flood-control canal. As a result of the historic decision to reverse that project, the river is now being restored to its original channel, regaining its curves, marshes, and wildlife.

tions, flooding of land covered by the reservoir, and draw-downs of reservoir water for power generation that affect shoreline ecosystems. Salmon populations of the entire Columbia and Snake River systems were destroyed by construction of the Grand Coulee, Bonneville, and other dams.

Groundwater supplies, unlike surface water, cannot be regulated by physical structures, and excessive withdrawal of groundwater affects future groundwater availability. In areas with dry climates, groundwater can be pumped out of aquifers far faster than environmental processes replace it. (See Groundwater, page 143.) As a result, the upper level of groundwater, the water table, drops, making further withdrawals more difficult and eventually uneconomic. In Texas overuse of the Ogallala aquifer has caused the water table in places to drop more than a hundred feet since the 1930s. Surface subsidence can also be triggered by groundwater removal.

Water quality is degraded by the introduction and concentration of substances and energy—especially heat—into water bodies. Water pol-

In 1989 the Exxon Valdez oil spill of 11 million gallons spelled ecological disaster in Prince William Sound, Alaska. Here, the Exxon Baton Rouge (on the right) pumps remaining oil from undamaged cargo tanks.

lution occurs when concentrations of these substances are high enough to interfere with environmental processes or to make use of the water unsafe. In less developed countries where rivers may also serve as sewers, water pollution can be extreme.

More at:

http://water.usgs.gov/nawqa/

Water Pollutants

Water pollutants are generated by a wide range of activities. In the United States, agriculture, industry, mining, and municipalities generate the most pollutants. Agricultural activites contribute fertilizers, biocides, and excessive amounts of animal wastes to water bodies and are linked to two specific pollution problems: nitrate pollution and *eutrophication*. A source of nitrate pollution is nitrates associated with fertilizers that can accumulate in groundwater supplies in agricultural areas. Agricultural areas in the United States, Israel, and eastern England have reported high levels of nitrates in groundwater.

Eutrophication, a loss of dissolved oxygen (DO), occurs when large amounts of sewage, fertilizers, or animal wastes enter a stationary water body. In addition, nutrients from these pollutants cause the population of phytoplankton to bloom, which in turn

causes the water to become cloudy and can kill sea grasses that produce DO. Although bacteria and other microorganisms survive without DO, fish die and eutrophication creates what are called dead lakes.

Industrial wastes include heavy metals such as mercury, synthetic chemicals such as PCBs (now banned in the United States), and other substances such as oil. Accumulation of these pollutants in water supplies creates health hazards for humans and for other species. Sudden large releases of industrial pollutants can be devastating to the environment. For example, accidental releases of mercury used in gold mining have sterilized sections of streams in the Amazon Basin. Oil tanker accidents, such as the 1989 grounding of the Exxon Valdez in Alaska's Prince William Sound, are responsible for the deaths of sea mammals, birds, and fish.

Residential and commercial activities in municipalities add detergents, salt, and chemicals to water, but sewage is a major water pollutant in the United States and throughout the world. Currently, only about half of the population of the United States lives in cities where sewage is treated to meet federal water quality standards. Worldwide, more than 1.7 billion people return their sewage to the environ-

ment untreated, and waterborne diseases associated with sewage, such as cholera, typhoid, and dysentery, are linked to the deaths of perhaps 25 million people each year.

Water can cleanse itself of pollutants if the amounts of pollution are not too great. Pollutants are removed from water, especially moving water, by gravity settling, by chemical changes or decomposition, and by dilution of the concentration to non-harmful levels. Stationary water sources, such as lakes and groundwater supplies in aquifers, are far slower in cleansing themselves than streams because their flow rates are slower.

Some water pollution can be treated after it occurs; preventive strategies are necessary to control other types of water pollution. Clean-up controls work with *point source pollution*, where pollutants are emitted from a specific and limited area, such as a sewage pipe or factory. *Nonpoint source pollution*, which includes runoff from streets, mining areas, and agricultural fields, is more difficult to control.

Federal legislation to control water pollution in the United States was first passed in 1972 with the Federal Water Pollution Control Act, called the Clean Water Act (CWA) since it was amended in 1977. Under this legislation, Congress funded more sewage treat-

Water Quality in the United States
- More serious water quality problems
- Less serious water quality problems
- Better water quality
- Data unavailable

Although U.S. waterways have become cleaner because of pollution controls, extensive problems remain in lakes, streams, reservoirs, and estuaries.

ment facilities, required a minimum level of secondary sewage treatment, and established minimum quality standards for rivers and lakes. Water pollution levels have been drastically reduced since 1972 in a number of cases, but further protection of water resources is necessary to meet CWA standards.
More at:
http://www.epa.gov/region5/water/cwa.htm

Land and Soil
Earth's land and soil are degraded by both the development of mineral resources and by discarding material waste products of society.

The distribution of valuable metals within larger ore bodies necessitates the removal of large amounts of rock, which, after processing to remove the metals, is returned to the environment. Surface mining, which includes strip mining, hydraulic mining, and open-pit mining, scars Earth's surface, eliminates native vegetation and wildlife, and changes surface and subsurface drainage. Excavations may range from small quarries of less than an acre to the Bingham

Canyon copper mine near Salt Lake City, which is now more than half-a-mile deep and about two-and-a-half-miles across.

Land reclamation, where mined land is recontoured, resoiled, and revegetated, alleviates much of the degradation caused by mining, but the costs involved can be substantial. In the United States, the 1977 Surface Mining Control and Reclamation Act was enacted that requires coal-mining companies to reclaim strip-mined land.
More at:
http://www.osmre.gov/smcra.htm

Land degradation also occurs when fluid resources such as petroleum, natural gas, and groundwater are pumped from the ground. Compaction of sediments can cause surface subsidence, which endangers buildings as well as future subsurface fluid supplies.

Soil loss and pollution due to agricultural practices occur throughout the world. In the United States about one third of all farmland is suffering from some soil erosion. The worst erosion, which involves loss of more than 75 percent of the topsoil and the presence of gullying, results from poor management of thick, fertile soils, and is found mostly in Iowa and the Mississippi Valley south of Illinois. Worldwide, soil loss is most pronounced in countries such as El Salvador and Colombia, where steep slopes and high annual rainfall magnify effects of poor soil management. Population pressures will increase agriculture and erosion.

Mismanagement of soils can also cause salinization and desertification, which involve long-term loss of land use. Salinization of soils—when chloride and sulphate salts accumulate on or near the soil surface as soil water evaporates—is most common in dry areas with poorly drained soils. Desertification, the transition of a semiarid grassland region to arid desert conditions, was once thought to be solely climatic in nature, but overgrazing and changes in vegetation contribute to desertification by changing soil water availability.

More at:

http://www.fao.org/desertification/

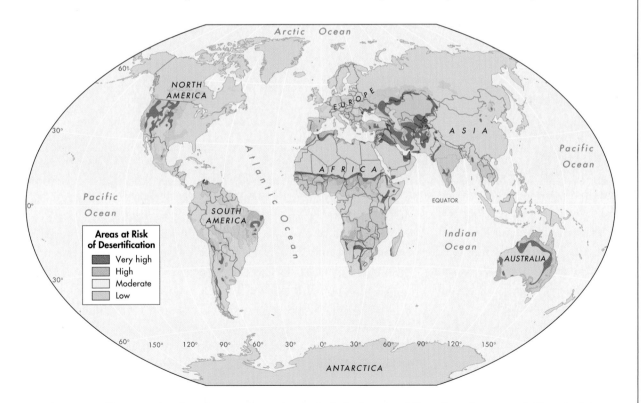

Desertification caused by overgrazing, the replacement of natural vegetation by food crops, and climate fluctuation turns semiarid grasslands into desert. Regions bordering tropical deserts are most vulnerable to this process.

In the People's Republic of China, a grid made from straw prevents drifting sand from burying railroad lines in the Tengger Desert.

Disposal of municipal and industrial waste degrades soil, water, and air through the volume of garbage involved and through the toxicity of some of the waste. In the U.S. Americans on average produce about 1,500 pounds of garbage a year, about twice as much as the average Japanese and four times as much as the average Pak istani or Indian. About one third of the volume of garbage in the United States consists of packaging—paper, plastic, and plastic foam—and another third consists of yard and food wastes and wood. While most of these substances are considered biodegradable, able to be decomposed by organisms in the environment, the dry conditions within landfills do not allow this to occur rapidly.

In the United States the majority of garbage, currently around 67 percent, is deposited in landfills, about 10 percent is incinerated, and 23 percent is recycled. There are significant differences between states in the disposal of garbage because of political cultures, population pressures, and other factors. Landfills can account for as much as 95 percent of waste disposal in Arizona or as little as 20 percent in Maine.

Sanitary landfills, where garbage is buried under a layer of soil, have replaced open dumps, but they still have a number of problems. Precipitation that percolates down through the garbage can dissolve or leach chemicals from the waste and become leachate; leachate then pollutes the groundwater unless downward drainage is controlled by an underlying and impermeable layer of clay, plastic, or other material. A problem for municipalities and industries seeking to dispose of garbage cheaply is the closing of landfills as they fill up. Between 1970 and 2000, the number of landfills in the United States has dropped from around 18,000 to about 3,500. Solid waste disposal is a worldwide problem; more than half the countries represented at the 1992 Earth Summit in Rio de Janeiro listed waste disposal as a major environmental problem and concern.

Few new landfills have opened in the U.S. in recent years, and most residents oppose opening them in their area. Landfills are often located in areas where residents are poor and have little political power or other economic opportunities; the location of many landfills in or near minority neighborhoods has led to charges of environmental *racism*. In recent years, some countries have paid others to allow dumping of solid waste. The United States and Japan have exported solid wastes to African and Latin American countries, and Canada was exporting almost 500,000 tons of garbage a year to the United States in the early 1990s.

Incineration of garbage can be used to reduce the volume by as much as 75 percent and generate electricity from the heat, but burning creates air pollution and ash that must still be disposed of. Ash from incinerators must now be tested for toxicity and, if levels are high enough, must be disposed of in licensed, hazardous waste landfills.

Recycling saves landfill space and energy used in processing and transporting raw materials. During the last decades, recycling has increased in the United States—especially for newspapers, glass, and metals—from about 7 percent of the total volume in 1970 to more than 20 percent in the mid-1990s. Greater public involvement is needed to expand recycling in the future to overcome the inconveniences involved, and to generate markets for recycled material. State and city governments are working to encourage recycling.

More at:

http://www.earth911.org/

Toxic, or hazardous waste, which poses a substantial threat to human health or the environment, is a common component of industrial pollutants. The Environmental Protection

Hazardous waste can be extremely harmful to people and the environment. Here, workers collect samples of toxic waste from open containers.

Agency classifies more than 400 substances as hazardous wastes, including caustics, explosives, flammables, and poisons. Nuclear waste is a special category of hazardous waste. Leachate from hazardous waste landfills has polluted groundwater and has been responsible for health problems. The enactment of the Superfund Act in 1980 was in response to hazardous waste pollution, and it was used to clean up nearly 200 sites by the mid-1990s. Some experts believe that there are between 2,000 and 10,000 sites that still need extensive environmental cleanup.

More at:

http://www.epa.gov/superfund/

Nuclear waste disposal, particularly high-level radioactive waste, which will continue to be radioactive and a health hazard for more than 10,000 years, is problematic; low-level radioative wastes are relatively safe after a decade or so. No totally satisfactory disposal system has yet been found for high-level waste such as plutonium. An extensive search by the United States Department of Energy for an isolated and secure burial place proposed Yucca Mountain, Nevada, in 1987. However, political resistance is delaying development of this prospective site, and a site near Carlsbad, New Mexico, is currently being used.

Vegetation and Wildlife

Developing natural resources unavoidably affects native vegetation and wildlife and has led to regional losses and to global extinction of species. A change in any component of an ecosystem—soils, water, or atmosphere—changes the flow of energy or nutrients and therefore changes both the food web and the species most able to survive in the changed environment.

In the United States the Endangered Species Act (ESA), enacted in 1966 and amended in 1973, protects endangered species, those in immediate risk of extinction, from hunting and loss of habitat. The Department of the Interior's Fish and Wildlife Service is responsible for determining if species are endangered or threatened, at risk but not immediate risk. Placing a new species on the endangered list threatens resource development in regions where its habitat is found.

In 2002, nearly 1,000 species in the United States were considered endangered and 275 species threatened, with many more species awaiting full evaluation. Public awareness of endangered species is greatest for large mammals and birds, but insects, clams, and crustaceans are also endangered, and endangered plants outnumber endangered animals by a considerable margin.

More at:

http://endangered.fws.gov/wildlife.html#Species

Worldwide, human activities contribute to about a thousand species becoming extinct each year, though estimates vary widely. Certain ecosystems, especially rain forests and wetlands, have especially high diversity, and their development threatens the loss of more species than development of rangeland ecosystems. The loss of any species affects an ecosystem, but when a keystone species, one whose activities affect large numbers of others, is lost, the effects are greater.

The major causes of extinction are loss of habitat, degradation of

Tigers

Only about 5,000 to 7,200 tigers, the world's largest cat, still live in the wild, about 5 percent of the number in 1900. The demand for pelts, body parts for medicinal purposes, a continuing loss of habitat, and a shrinking gene pool are making extinction in the next ten years increasingly likely. Although tigers once roamed much of Asia, their spatial range is steadily shrinking, and the Bali, Caspian, and Javan tiger subspecies are already extinct.

More at:

http://www.5tigers.org/

habitat, and overhunting. Deforestation, clearing for agriculture, and urbanization all cause the loss of habitat. As human populations grow around the world, increasing amounts of habitat are lost. This problem is compounded by the fragmentation of those ecosystems involved: Ecological islands, small acreages of habitat surrounded by developed land, are unable to effectivly support the diversity of life of larger areas.

Toxics, such as pesticides, and pollutants cause degradation of habitats, resulting in unintentional deaths of wildlife. Many pesticides, such as DDT, pass through the food web and accumulate in predators at the top of the food chain by biomagnification. The decline of bald eagles and peregrine falcons in the United States prior to the banning of DDT in 1972 was partially the result of the eagles' and falcons' consumption of prey that had eaten insects and worms containing DDT.

Overhunting has caused a number of birds and mammals to become extinct or endangered. Large mammals, such as blue whales, cheetahs, and black rhi-

A clear-cut hillside in the Olympic Mountains of Washington State reflects the controversy surrounding forest management. Valued resources of recreation, clean water, wilderness, and biodiversity, forests must be managed carefully to permit long-term, sustainable use.

noceroses, seem especially vulnerable to overhunting, and their extinctions would have unknown ripple effects in their ecosystems.

The value of species and ecosystems diversity to societies is not always commercial but can be philosophical or ethical in nature. Loss of species also means the loss of gene pools and potential resources. Plants provide a large number of medicines. The loss of rain forest and of wetland is particularly harmful to the search— the so-called chemical prospecting—for new medicines.

More at:

http://www.redlist.org/

The introduction of an alien species, also called an exotic species, into a new environment changes the ecosystem and often results in the loss of native species. Every species is part of the food web within an ecosystem, and an exotic species is a competitor for food and, too often, a new predator. Intentional introduction of alien species may provide initial benefits to humans in an area, but unanticipated problems often counteract these benefits. Kudzu, an Asian vine that prevents soil erosion by stabilizing slopes, was introduced into the southeastern U.S. early in the 20th century and has spread over the countryside, covering trees and buildings.

Zebra Mussels

Although more than a hundred exotic species now live in the Great Lakes, the zebra mussel is often singled out as a classic example of problems introduced species may cause. Zebra mussels settle on hard surfaces where there is flowing water; therefore, they accumulate on and block intake pipes for water supplies and power stations. The damage caused by the mussels may reach into the billions of dollars within the next decade, especially if the species spreads farther into the Mississippi, Hudson, and St. Lawrence River systems.

More at:

http://www.cptr.ua.edu/kudzu/

Unintentional introduction of alien species, especially insects, is almost unavoidable because of increasing exchange between regions. The Asian gypsy moth, the Mediterranean fruit fly, and the Asian cockroach have all been accidentally introduced into the United States. New species can exterminate native species: In Hawaii, for example, cannibal snails are eliminating Oahu tree snails, and koa trees are being choked by the invasions of banana poka vines and wild blackberries.

More at:

http://www.invasivespecies.gov/

NATURAL HAZARDS

Natural environments are increasingly vulnerable to degradation by human actions, but it is not a one-way relationship. Humans are part of the environment and are therefore unavoidably affected by environmental processes. When local, global, or regional environmental processes such as earthquakes, tornadoes, or lightning threaten harm to humans or damage to developments, they are considered natural hazards: Blizzards or volcanic eruptions in unpopulated areas are natural events, but not natural hazards.

All societies face risk from natural hazards, both to property and to persons because no location is immune to hazardous events. Control over the environmental processes causing hazardous events is neither feasible nor possible, so future losses from hazards are inevitable but can be reduced by adopting more effective hazard-mitigation strategies.

Vulnerability to Hazards

Natural hazards are natural events that threaten the human system and range from avalanches to fog to plagues of locusts; hazards are often extreme events, such as large earthquakes rather than minor temblors or droughts rather than dry summers. In an average year, about 250,000 people die from natural hazards—more than 200,000 of these in less developed countries—and about 50 billion dollars is spent annually, mostly on recovery efforts, but also on prediction, prevention, and insurance premiums.

Natural hazards include meteorological, geological, and biological events; and although hazardous events, especially extreme events, are by definition uncommon, they are not abnormal. The environmental processes operating on Earth will generate high magnitude events, such as Category Five hurricanes and earthquakes with energy levels of 8.0 on the Moment Magnitude Scale, when conditions cause sufficient energy to accumulate. Fortunately, the frequency of extreme events is inversely related to their magnitude; small earthquakes, for example, far outnumber large earthquakes.

As humanity has spread over Earth, people of necessity settled in locations where extreme events will occur because natural resources are often found alongside natural hazards. The San Francisco Bay area of California has both earthquakes as well as a mild climate. The midwestern United States has both tornadoes and fertile soils. Some locations experience greater numbers or more dangerous hazards than others. The coastline of Bangladesh, for example, is a low-lying delta that experiences cyclones (hurricanes), floods, and an occasional tsunami. Residents evaluate how hazardous a place is against the benefits of living there and the opportunities to relocate.
More at:
http://earthobservatory.nasa.gov/NaturalHazards/

Worldwide, the risk of natural hazards is rising despite efforts by individual communities, countries, and international organizations to protect people. Two key factors underlie the rise in risk: growing populations in dangerous locations and increasing concentration of wealth in those locations. In the United States, for example, the growth of population and income in Florida and along the rest of the Atlantic coastline increases the potential for disaster because of the hurricane hazard.

If a natural hazard occurs and causes sufficient damage to humans then a natural disaster is said to have occurred. In the United States, the term disaster is commonly restricted to those events in which more than a hundred people die or more than a million dollars in damage occurs. On average the federal government declares 15 states of emergency each year in response to requests by state governments. As a rule, financial losses are most common in developed countries and loss of life is greatest in less-developed countries because of the better warning systems and stronger structures in developed countries. Many of the deadliest disasters in human history have occurred in China where huge populations live on floodplains and in seismically active areas.

Hazard Adjustments

Adjusting to hazards requires individuals and societies to assess the risk of hazards at a location and then to choose among the available strategies that address those risks. In an ideal world, the statistical risk of each hazard would be known, the costs and benefits of each stategy measured and evaluated, and the final decision would consequently be made rationally and objectively. None of these conditions exists in real life, and so management decisions can often be, or at least appear to be, inadequate.

The major factors affecting the choice of hazard adjustment are the perception of hazard risk and the limitations on choices economically, politically, and socially. Perception, the mental image held of both hazard risk and possible responses, is critical because people react to what they believe rather than to what is. The perceptions among individuals vary widely because of differences in knowledge base, which is associated with age, frequency of hazard experience, income and investments to be protected, and personal values and emotions. Individuals may be fatalistic about hazards or they may be activists, risk-seekers or risk-avoiders, or optimizers who seek satisfactory rather than optimum answers. In earthquake-prone regions, people may decide to buy insurance, build safer houses, relocate, or deny that the risk exists; each of these options will be considered sound by the individual selecting it.

Perception of risk is a major problem because data are often unreliable, anecdotal experiences and recent events are over-emphasized, and statistics misunderstood. Flood risk, to take one example, is based on records of flooding, and the records themselves may be incomplete and cover too short a time period. In addition, expected flood heights are probabilities rather than assertions of frequency: A hundred-year flood does not occur every hundredth year; rather it has a one percent chance of occurring each year. Residents were surprised when the Patuxent River between Baltimore, Maryland, and Washington, D.C., flooded in 1971, an event expected less than once a century. Amazingly, a comparable flood occurred the following year.

MAJOR NATURAL DISASTERS SINCE 1900			
	COUNTRY	YEAR	ESTIMATED DEATH TOLL
10 WORST EARTHQUAKES	China	1976	242,000
	China	1927	200,000
	China	1920	180,000
	Indonesia-Sri Lanka-Thailand	2004	150,000
	Japan	1923	143,000
	Former Soviet Union	1948	110,000
	Italy	1908	75,000
	China	1932	70,000
	Peru	1970	66,794
	Pakistan	1935	60,000
10 WORST FLOODS	China	1931	3,700,000
	China	1959	2,000,000
	China	1939	500,000
	China	1935	142,000
	China	1911	100,000
	China	1949	57,000
	Guatemala	1949	40,000
	China	1954	30,000
	Venezuela	1999	30,000
	Bangladesh	1974	28,700
10 WORST DROUGHTS	China	1928	3,000,000
	India	1942	1,500,000
	India	1900	1,250,000
	Former Soviet Union	1921	1,200,000
	China	1920	500,000
	India	1965	500,000
	India	1966	500,000
	India	1967	500,000
	Ethiopia	1984	300,000
	Ethiopia	1974	200,000

Collective hazard adjustments are limited by the perceptions of the decision-makers on risks, possible adjustments, and on available resources and political support. Decision-makers are also influenced by the advice of technical experts and the public; studies show that officials' perceptions of hazards are often based on what officials believe is the public perception. Finances are also a limitation, especially for hazards that are not seen as imminent threats. When seismologists of the United States Geological Survey informed California officials in 1976 about the possibility of a major earthquake, it was several years before actual evacuation plans were completed and requirements established for building and construction safety.

The range of possible hazard adjustments, from which individuals and societies can choose, focuses on accepting, reducing, or avoiding future losses. Technical and technological responses involve construction and engineering, while social responses involve regulating activities and the locations at which they occur. Incidental adjustments, such as improved construction, communication, and transportation, are made, not necessarily to reduce negative impacts of hazards, but they have a positive impact nevertheless. Purposeful adjustments,

such as purchasing hazard insurance and building tornado-warning systems, are made intentionally to cope with the hazard.

People make a conscious choice in selecting their purposeful adjustment to a hazard and, in addition to their perceptions, are affected by the magnitude, duration, and amount of forewarning of the hazard. Earthquakes, floods, and droughts may all cause comparable amounts of death and damage, but the duration of each hazard is so different that different types of adjustment should be expected. Acceptance of losses, for example, may be a realistic reaction in response to an earthquake. Reduction of loss, by modifying the event or preventing its negative effects can be another way of adjusting. Building dams, for example, may be an appropriate way to deal with the hazards of floods. Choosing to change the location of land use or to move people could also be required for survival in drought-prone areas, especially in developing countries.

Hazard adjustments are made not only by individuals and communities but also at national and international levels. Following the devastating 9.0 magnitude earthquake and subsequent tsunami in Southeast and South Asia in December 2004, the international com-

munity pledged hundreds of millions of dollars in aid. Humanitarian concerns are part of the reason for assistance from those outside the area of impact, but the increasing economic and political linkages between countries can cause losses from a disaster to be felt nationally and internationally, as well as regionally or locally. Loss of productivity, jobs, and investments add an economic dimension to the personal tragedies that disasters bring. More at:
http://temp.water.usgs.gov/tsunami/

AWARENESS FOR THE FUTURE

Humans are both dependent upon and interdependent with their environments. Consider the striking image of Earth taken by the Voyager I spacecraft in 1977. Enlarged, it reveals oceans and clouds—a blue planet. It presents Earth as a whole, as a closed system in terms of the amount of matter. Since that image was made, humans have come to appreciate their impact on this planet. We are responsible for global warming, deforestation, drained wetlands, desertification, and far more. These are the human imprint. And they call the questions: What has been done, why has it been done, and most important, what can be done now and in the future to ensure the well-being of every resident here?

SOURCES OF FURTHER INFORMATION

Barber, Benjamin. *Jihad vs. McWorld*. New York: Ballantine Books, 1995.

Berry, Edgar Conkling, and D. Michael Ray. *The Global Economy in Transition*. Upper Saddle River, N.J.: Prentice Hall, 1997.

Black, Jeremy. *Maps and Politics*. Chicago: University of Chicago, 1998.

Boutros-Ghali, Boutros. *Unvanquished: A U.S.-U.N. Saga*. New York: Random House, 1999.

Brown, Lester R., Gary Gardner, and Brian Halweil. *Beyond Malthus: Nineteen Dimensions of the Population Challenge*. New York: W.W. Norton, 1999.

Brunn, S., J. Williams, and D. Zeigler. *Cities of the World*. New York: Rowman & Littlefield Publishers, 2003

Cadwallader, M. T. *Urban Geography: An Analytical Perspective*. Englewood Cliffs, N.J.: Prentice Hall, 1996.

Castells, Manuel. *The Rise of the Network Society*. Blackwell: Oxford, 1996.

de Souza, Anthony and Frederick Stutz. *The World Economy*. New York: Macmillan, 1994.

Dicken, Peter. Global Shift: *The Internationalization of Economic Activity*. New York: Guilford Press, 1998.

Fadiman, Anne. *The Spirit Catches You and You Fall Down*. Farrar, Strauss, & Giroux, 1997.

Gebhard, Arlene, Carl Haub and Mary M. Kent. "World Population: Beyond Six Billion?" Population Bulletin 54 (1). Washington, D.C.: Population Reference Bureau.

Hall, P. *Cities in Civilization*. New York: Pantheon, 1998.

Hall, Ray and Paul White, editors. *Europe's Population: Toward the Next Century*. London: University College London Press, 1996.

Harr, Jonathan. *A Civil Action*. New York: Vintage Books, 1995.

Hugill, Peter. *World Trade since 1431: Geography, Technology, and Capitalism*. Baltimore: Johns Hopkins University Press, 1993.

Johnson, Ron, Peter Taylor and Michael Watts. *Geographies of Global Change*. Oxford: Blackwell, 1995.

Knox, Paul. Urbanization: *An Introduction to Urban Geography*. Englewood Cliffs, N.J.: Prentice Hall, 1994.

Knox, Paul and S. Marston. *Human Geography: Places and Regions in Global Context*.

Englewood Cliffs, N. J.: Prentice Hall, 2003.

McCrum, Robert et al. *The Story of English*. New York: Penguin USA, 1993.

McPhee, John. *The Control of Nature*. New York: Farrar, Straus & Giroux, 1990.

Malecki, Edward. *Technology and Economic Development*. Essex, U.K.: Longman, 1991.

Martin, Philip and Elizabeth Midgley. "Immigration to the United States," Population Bulletin 54 (2). Washington, D.C.: Population Reference Bureau, 1999.

Martin, Philip and Jonas Widgen. "International Migration: A Global Challenge," Population Bulletin 51 (1). Washington, D.C.: Population Reference Bureau, 1996.

Monmonier, Mark. *Cartographies of Danger: Mapping Hazards in America*. Chicago: University of Chicago, 1997.

Moynihan, Daniel P. *Pandemonium: Ethnicity in International Politics*. New York: Oxford University Press, 1993.

Peterson, William. *Population*. New York: Macmillan Publishing Co., 1975.

Plane, David A. and Peter A. Rogarson. *The Geographical Analysis of Population: With Applications to Planning and Business*. New York: John Wiley & Sons, 1994.

Robinson, Marilynne. *Mother Country*. New York: Farrar, Strauss, & Giroux, 1989.

Sassen, Saskia. *Cities in a World Economy*. Thousand Oaks, Calif.: Pine Forge Press, 1994.

Schlesinger, Arthur M., Jr. *The Disuniting of America*. New York: W.W. Norton & Co., 1998.

Shipler, David. *Arab and Jew: Wounded Spirits in a Promised Land*. New York: Penguin USA, 1987.

So, Alvin. S*ocial Change and Development: Modernization, Dependency, and World-System Theories*. Newbury Park, Thousand Oaks, Calif.: Sage Publications, 1990.

Stegner, Wallace. *American West as Living Space*. Ann Arbor: University of Michigan Press, 1987.

Suzuki, David and Amanda Mconnell. *The Sacred Balance: Rediscovering Our Place in Nature*. Amherst, N.Y.: Prometheus Books, 1998.

United Nations Centre for Human Settlements, *An Urbanizing World: Global Report on Human Settlements 1996*. New York: Oxford University Press, 1996.

Zelinsky, Wilbur. *The Cultural Geography of the United States*. Upper Saddle River, N.J: Prentice Hall, 1992.

PLACES

The number of nations has soared from 65 in 1946 to 192 today—with 28 new countries added to the world map since 1990. The first country to gain independnce in the 21st century is Timor-Leste. The following pages provide a guide to the world's countries—from Afghanistan to Zimbabwe.

READING THE WORLD'S NATIONS

THE 192 INDEPENDENT COUNTRIES of the world that follow are those counted in 2005 by the National Geographic Society, whose cartographic policy is to recognize de facto countries. The section on maps of the world and its regions begins on page 485.

Countries vary widely in their ability and resources for collecting and tracking statistical information. Even when gathered efficiently and accurately, the data require revision as soon as they are published; human geography issues, by their nature, are in a constant state of flux. Consideration was given to ensure comparability from country to country.

The **country name** listed is used by the National Geographic Society, which consults the U.S. Board on Geographic Names, country embassies, and country governments. The **official name**, which follows the country name and is listed underneath it, is the conventional long form of the name in English translation.

The **continent** name references on which of six continents the country is located. Antarctica contains no independent nation, so it does not appear in the continent entries.

Arca accounts for the sum of all land and inland water delimited by international boundaries, intranational boundaries, or coastlines. It is shown in square kilometers and square miles.

Population figures are 2003–2004 estimates from the Population Reference Bureau's World Population Data Sheet. Population numbers refer to the total population of the country.

Capital gives the seat of government followed by the city's population. The population is that of the entire capital city. In some cases, (as in Bolivia, for example) population figures are provided for both the administrative and the constitutional capital.

Under **Religion**, the most widely practiced faith appears first. Subseqently, the names of other widely practiced faiths are listed. "Traditional" or "indigenous" connotes beliefs of important local sects, such as the Maya in Guatemala.

Language refers to the country's official language. If a country has an official language, it is listed first. Often, a country lists more than one official language. Otherwise, both religion and language are ordered by rank.

Literacy generally indicates the percentage of the population above the age of 15 who can read and write. There are no universal standards of literacy, so these U.S. Census Bureau estimates are based on the most common definition available for that nation, which makes comparisons among countries difficult.

Life expectancy at birth refers to the average number of years an infant born in 2003 can be expected to live if current mortality trends remain constant in the future.

Currency gives the country's official medium of exchange. In Europe, for example, the euro has become the official currency for many of the member nations of the European Union. However, some, such as the United Kingdom, (which still uses the British pound) have not adopted the euro as their official currency.

GDP per capita is derived by dividing the gross domestic product (**GDP**), which is the value of all final goods and services produced within a country in a given year, by midyear population estimates. GDP-per-capita numbers for independent nations use the purchasing power parity (**PPP**) conversion factor designed to equalize the purchasing powers of different currencies. Because it is an average, it hides extremes of poverty and wealth and does not account for factors that also affect quality of life, such as environmental degradation, educational opportunites, and health care.

Economy provides a concise list of the main sectors of industry, the main products of agriculture, and the leading exports. Agriculture, because of the structured nature of the text, serves as an umbrella term not only for crops but also for livestock, products, and fish.

Abbreviations used:

EU	European Union
GDP	gross domestic product
N.A.	data not available
pop.	population
PPP	purchasing power parity
sq km	square kilometer (multiply by 0.39 to get square miles)
sq mi	square mile (multiply by 2.60 to get square kilometers)
IND	industry
AGR	agriculture
EXP	exports
U.K.	United Kingdom
UN	United Nations
U.S.	United States

**COLOR KEY
TO THE NATIONS
LOCATOR GLOBES**

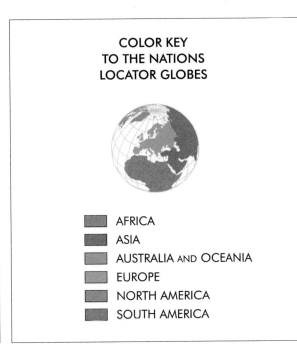

- AFRICA
- ASIA
- AUSTRALIA AND OCEANIA
- EUROPE
- NORTH AMERICA
- SOUTH AMERICA

COUNTRY BY COUNTRY

Afghanistan
TRANSITIONAL ISLAMIC STATE OF AFGHANISTAN

CONTINENT	Asia
AREA	652,090 sq km
	(251,773 sq mi)
POPULATION	28,717,000
CAPITAL	Kabul 2,956,000
RELIGION	Sunni and Shiite Muslim
LANGUAGE	Pashtu, Afghan Persian
	(Dari), Uzbek, Turkmen,
	30 minor languages
LITERACY	36%
LIFE EXPECTANCY	46
CURRENCY	afghani
GDP PER CAPITA	$700
ECONOMY	IND: small-scale produc-
	tion of textiles, soap, furni-
	ture, shoes. AGR: opium,
	wheat, fruits, nuts, wool.
	EXP: opium, fruits and
	nuts, handwoven carpets,
	wool, cotton.

Since Alexander the Great, invading armies and peaceful migrations have brought in diverse peoples to this Central Asian crossroads. As a result, Afghanistan is a country of ethnic minorities: Pashtun (38 percent), Tajik (25 percent), Hazara (19 percent), and Uzbek (6 percent). The towering Hindu Kush range dominates and divides Afghanistan. The northern plains and valleys are home to Tajiks and Uzbeks. Pashtuns inhabit the desert-dominated southern plateaus. Hazara live in the central highlands. Kabul, south of the Hindu Kush, is linked by narrow passes to the northern plains.

In 1989 the nine-year Soviet occupation ended, and Muslim rebels toppled the communist regime in 1992, after which rival groups vied for power. From among the various factions arose the Taliban ("students of religion"), a militant Islamic movement. The Taliban seized Kabul in 1996 and imposed Islamic punishments, including amputation and stoning, and banned women from working. In 2001 the Taliban destroyed giant Buddha statues at Bamian in defiance of international efforts to save them. Three weeks after the September 11 attacks on New York and Washington, D.C., the U.S. and Britain bombed terrorist camps in Afghanistan; by November 2001 Kabul fell to anti-Taliban forces.

After decades of war, Afghanistan is rebuilding its economy, which is mostly agricultural, and attempting to stabilize its government. The government faces problems with health care, security, and opium.

Albania
REPUBLIC OF ALBANIA

CONTINENT	Europe
AREA	28,748 sq km
	(11,100 sq mi)
POPULATION	3,135,000
CAPITAL	Tirana 367,000
RELIGION	Muslim, Albanian
	Orthodox, Roman Catholic
LANGUAGE	Albanian, Greek
LITERACY	87%
LIFE EXPECTANCY	74
CURRENCY	lek
GDP PER CAPITA	$4,400
ECONOMY	IND: food processing,
	textiles and clothing,
	lumber, oil. AGR: wheat,
	corn, potatoes, vegeta-
	bles, meat. EXP: textiles
	and footwear, asphalt,
	metals and metallic ores,
	crude oil.

Albania lies along the Adriatic Sea in southeastern Europe. The narrow coastal plain rises to mountains that are almost 2,000 meters (6,500 feet) high, which cover most of the country. These mountains are rich in mineral resources such as chrome, iron, nickel, and copper; however, mining requires investment that Albania lacks. It is one of the poorest countries in Europe (and the only one with a Muslim majority). It suffered from more than 40 years of communist rule, which ended in 1991. War in neighboring Kosovo brought 480,000 ethnic Albanian refugees into Albania in 1999—straining the country's resources. The largely agricultural economy is growing thanks to remittances from Albanian workers abroad—mostly in Greece and Italy.

Algeria
PEOPLE'S DEMOCRATIC REPUBLIC OF ALGERIA

CONTINENT	Africa
AREA	2,381,741 sq km (919,595 sq mi)
POPULATION	31,746,000
CAPITAL	Algiers 3,060,000
RELIGION	Sunni Muslim
LANGUAGE	Arabic, French, Berber dialects
LITERACY	70%
LIFE EXPECTANCY	70
CURRENCY	Algerian dinar
GDP PER CAPITA	$5,400
ECONOMY	IND: petroleum, natural gas, light industries, mining. AGR: wheat, barley, oats, grapes, sheep. EXP: petroleum, natural gas, petroleum products.

Algeria, in northwest Africa on the Mediterranean coast, is the second largest country in Africa after Sudan. The Sahara covers more than four-fifths of its territory, where the inhabitants are concentrated in oases surrounded by desert. More than 90 percent of Algerians live along the Mediterranean coastlands on only 12 percent of the country's land. The Atlas Mountains cross Algeria east to west along the Mediterranean coast, with the north-facing slopes receiving good winter rainfall; the southern slopes, southern ranges, and interior plateaus get little rain.

Since antiquity Algeria has enticed settlers—Phoenicians, Romans, Arabs, Turks—and, in the 19th century, French farmers. The French organized Algeria into departments and disenfranchised native Arabs and Berbers. In 1954 Algerians rebelled; the war that followed took a million lives before independence in 1962, and a million French colonists left. Socialist-style, military-dominated governments pinned their hopes on huge oil and natural gas reserves in the Algerian Sahara. But low petroleum prices, a high birthrate, and austere policies produced a dismal economic picture. Algerians demanded democratization, and many migrated to France. Since 1991 Algerian politics has been dominated by violence between the military and Islamic militants. The government is also challenged by unrest from the Berber-speaking minority in the mountainous northeast.

American Samoa (U.S.)
TERRITORY OF THE UNITED STATES

CONTINENT	Australia/Oceania
AREA	233 sq km (90 sq mi)
POPULATION	62,000
CAPITAL	Pago Pago 4,300
RELIGION	Christian Congregationalist
LANGUAGE	Samoan, English

American Samoa is a group of seven, mostly volcanic, islands in the South Pacific. It is the southernmost territory of the United States and is strongly linked to the U.S. by trade. Canned tuna is the main export.

Andorra
PRINCIPALITY OF ANDORRA

CONTINENT	Europe
AREA	468 sq km (181 sq mi)
POPULATION	67,000
CAPITAL	Andorra la Vella 21,000
RELIGION	Roman Catholic
LANGUAGE	Catalan, French, Castilian (Spanish)
LITERACY	100%
LIFE EXPECTANCY	NA
CURRENCY	euro
GDP PER CAPITA	$19,000
ECONOMY	IND: tourism, cattle raising, timber, banking. AGR: rye, wheat, barley, oats, sheep. EXP: tobacco products, furniture.

Tiny, landlocked Andorra sits almost hidden in the high Pyrenees between France and Spain. A co-principality since the 13th century, mountainous Andorra has two princes as heads of state: France's president and Spain's Bishop of La Seu d'Urgell (a historic town just south of Andorra). The country adopted a democratic constitution in 1993, creating a parliament and limiting the power of the co-princes. The economy is based on tax-free shopping, tourism, and international banking.

Angola
REPUBLIC OF ANGOLA

CONTINENT	Africa
AREA	1,246,700 sq km (481,354 sq mi)
POPULATION	13,087,000
CAPITAL	Luanda 2,623,000
RELIGION	Indigenous beliefs, Roman Catholic, Protestant
LANGUAGE	Portuguese, Bantu
LITERACY	42%
LIFE EXPECTANCY	40
CURRENCY	kwanza
GDP PER CAPITA	$1,700
ECONOMY	IND: petroleum, diamonds, iron ore, phosphates, feldspar, bauxite, uranium, gold. AGR: bananas, sugarcane, coffee, sisal, livestock, forest products, fish. EXP: crude oil, diamonds, refined petroleum products, gas.

Angola lies on the Atlantic coast of southwestern Africa; its small (but oil-rich) northern province, Cabinda, is separated from the rest of the country by a small part of the Democratic Republic of the Congo and the Congo River. Angola's narrow coastal plain, where most of the people live, rises to a high interior plateau with rain forests in the north and dry savanna in the south.

A Portuguese colony since the 15th century, Angola was the source for millions of slaves sent to Brazil (also a Por-

tuguese colony) and other places across the Atlantic. It won independence in 1975 after 14 years of guerrilla war. Civil war then broke out between major ethnic groups. The Mbundu, about a quarter of the population and based in the north, took the capital, Luanda. The Ovimbundu, with over a third of the population, and others set up a rival government in the interior city of Huambo. The civil war brought more bloodshed—and economic collapse—as 300,000 Portuguese fled. After 16 years of civil war a peace agreement made elections possible in 1991. But the elections were disputed, and peace was shattered 18 months later by renewed conflict. The 27-year-long civil war ended in 2002 with hopes of a lasting peace. The war ravaged the country's political and social institutions and littered the country with land mines. Angola is one of the world's poorest countries, with life expectancy (40 years) among the lowest in Africa.

Anguilla (U.K.)
BRITISH OVERSEAS TERRITORY

CONTINENT	North America
AREA	96 sq km (37 sq mi)
POPULATION	12,000
CAPITAL	The Valley 1,000
RELIGION	Anglican, Methodist
LANGUAGE	English

The northern Caribbean island of Anguilla was settled by British and Irish settlers in 1650. Today this small, flat, beach-lined island is a self-governing British territory. Tourism and banking are important to the economy.

Antigua and Barbuda
ANTIGUA AND BARBUDA

CONTINENT	North America
AREA	442 sq km (171 sq mi)
POPULATION	74,000
CAPITAL	St. John's 28,000
RELIGION	Anglican, other Protestant, Roman Catholic
LANGUAGE	English, local dialects
LITERACY	89%
LIFE EXPECTANCY	71
CURRENCY	East Caribbean dollar
GDP PER CAPITA	$11,000
ECONOMY	IND: tourism, construction, light manufacturing. AGR: cotton, fruits, vegetables, bananas, livestock. EXP: petroleum products, manufactures, machinery and transport equipment, food and live animals.

Antigua and Barbuda, small islands in the eastern Caribbean, were colonized by the English in 1632. Sugar plantations ruled the economy, and African slaves were brought in as laborers. Independent of Britain since 1981, this three-island nation remains within the Commonwealth. Antigua, one of the first Caribbean islands to promote tourism, in the early 1960s, is the wealthiest. Barbuda seeks to balance resort development with protection of its varied wildlife. Tiny Redonda is uninhabited.

Argentina
ARGENTINE REPUBLIC

CONTINENT	South America
AREA	2,780,400 sq km
	(1,073,518 sq mi)
POPULATION	36,925,000
CAPITAL	Buenos Aires 13,349,000
RELIGION	Roman Catholic
LANGUAGE	Spanish, English, Italian,
	German, French
LITERACY	97%
LIFE EXPECTANCY	74
CURRENCY	Argentine peso
GDP PER CAPITA	$10,500
ECONOMY	IND: food processing,
	motor vehicles, consumer
	durables, textiles. AGR:
	sunflower seeds, lemons,
	soybeans, livestock. EXP:
	edible oils, fuels and
	energy, cereals, feed,
	motor vehicles.

Argentina, meaning "land of silver," is a rich and vast land—second largest (after Brazil) in South America and eighth largest in the world. Its heartland is a broad grassy plain known as the Pampas (pronounced PAHM-pahs). Here Argentina's gaucho, like the U.S. cowboy, has galloped into the country's folklore.

The Spanish first arrived around 1516, and Argentina gained independence in 1816. The small native population died from European diseases, and today's population is over 95 percent European. For Spanish, Italian, German, and other immigrants in the late 19th century, Argentina held great promise. Today the literacy and urbanization rates are high, the birthrate and the infant mortality rate are low, and most Argentines consider themselves middle class.

The recent past has been tumultuous. Some 30,000 people disappeared—the *Desaparecidos*—in the "dirty war" during the military junta's 1976–1983 rule. In April 1982 Argentine forces invaded the British-held Falkland Islands, which Argentina calls the Islas Malvinas. Defeat by Britain during the 1982 Falkland Islands war loosened the military dictatorship's stranglehold on democracy.

Since then much has been won: greater freedom of the press, tolerance of opposition, and increased foreign investment. However, a deep recession caused economic collapse at the end of 2001—followed by fragile economic growth. Unemployment plagues the economy, even as the nation enjoys the continent's highest per capita income.

The Andes mark Argentina's western edge, forming the boundary with Chile. The highest peak in the Western Hemisphere, Aconcagua, dominates the Andes at 6,960 meters (22,834 feet). From the Andes, gently rolling plains extend eastward toward the sea. Much of the Pampas, including a rich agricultural section, occupies this region. Northeast Argentina features rain forests and Iguazú Falls. These spectacular falls, on Argentina's border with Brazil, drop along a 2.7-kilometer (1.6-mile) front in a horseshoe shape. South of the Pampas, dry and windswept Patagonia stretches to the southernmost tip of South America with the world's southernmost city, Ushuaia.

Armenia
REPUBLIC OF ARMENIA

CONTINENT	Asia
AREA	29,743 sq km
	(11,484 sq mi)
POPULATION	3,220,000
CAPITAL	Yerevan 1,079,000
RELIGION	Armenian Apostolic
LANGUAGE	Armenian, Russian
LITERACY	99%
LIFE EXPECTANCY	72
CURRENCY	dram
GDP PER CAPITA	$3,600
ECONOMY	IND: metal-cutting
	machine tools, forging-
	pressing machines,
	electric motors, tires.
	AGR: fruit (especially
	grapes), vegetables, live-
	stock. EXP: diamonds,
	mineral products, food-
	stuffs, energy.

Smallest of the former Soviet republics, Armenia lies landlocked and earthquake ridden in rugged mountains. In A.D. 301, Armenia became the first Christian nation; today it is almost surrounded by Islamic nations. During World War I the Ottoman Turks brutally forced out Armenians, causing a diaspora to foreign havens. Armenia gained independence in 1918, but succumbed to a Red Army invasion in 1920. In 1988 a devastating earthquake killed 25,000 people, and conflict erupted with Azerbaijan over Nagorno-Karabakh (a region of 140,000 ethnic

Armenians). Armenia won independence in 1991; by 1994 Armenians had defeated Azeri forces and had control of Nagorno-Karabakh—but the dispute remains unresolved.

Aruba (Netherlands)
OVERSEAS REGION OF THE NETHERLANDS

CONTINENT	North America
AREA	193 sq km (75 sq mi)
POPULATION	95,000
CAPITAL	Oranjestad 29,000
RELIGION	Roman Catholic, Protestant
LANGUAGE	Dutch, Papiamento, English, Spanish

The Caribbean island of Aruba, just north of Venezuela, was part of the Netherlands Antilles until 1986—when it gained full autonomy within the Dutch realm. Tourists enjoy miles of white sand beaches fringed with palm trees.

Australia
COMMONWEALTH OF AUSTRALIA

CONTINENT	Australia/Oceania
AREA	7,692,024 sq km (2,969,906 sq mi)
POPULATION	19,917,000
CAPITAL	Canberra 373,000
RELIGION	Protestant, Roman Catholic
LANGUAGE	English, native languages
LITERACY	100%
LIFE EXPECTANCY	80
CURRENCY	Australian dollar
GDP PER CAPITA	$26,900
ECONOMY	IND: mining, industrial and transportation equipment, food processing, chemicals, steel. AGR: wheat, barley, sugarcane, fruits, cattle. EXP: coal, gold, meat, wool, alumina, iron ore.

An island continent located between the Indian and Pacific Oceans, Australia combines a wide variety of landscapes. The highest mountains are part of the Great Dividing Range that line the east coast from Cape York Peninsula south to the state of Victoria. Most people reside along the southeast coast, in cities like Melbourne or Sydney, because winds from the southeast release rain there—leaving the interior beyond the mountains arid or semiarid. West of the Great Dividing Range the landscape consists mostly of plains and plateaus; the Macdonnell Ranges near the country's center are an exception. The Great Artesian Basin provides underground water for a region that would otherwise be desert. Vegetation ranges from rain forests in the far north to steppes and deserts in the vast interior (which Australians call the Outback). There are more than 130 species of marsupials, such as kangaroos, koalas, and wombats. The Murray-Darling River Basin, covering about 14 percent of the continent, helps sustain wheat and wool industries.

Founded in 1788 as a British convict colony, Australia was a place of banishment until gold strikes in 1851 opened floodgates of immigration. Independence came in 1901, with a constitution adapted in part from that of the United States. Immigration has been key to Australia's development since 1788; from 1945 through 2000 almost six million immigrants arrived. Aborigines number 410,000, and the government is making efforts to settle aboriginal land rights. Australia has one of the world's highest living standards with 85 percent living in urban areas.

EXTERNAL TERRITORIES

Ashmore and Cartier Islands

Christmas Island

Cocos (Keeling) Islands

Coral Sea Islands

Heard and McDonald Islands

Norfolk Island

Austria
REPUBLIC OF AUSTRIA

CONTINENT	Europe
AREA	83,858 sq km (32,378 sq mi)
POPULATION	8,157,000
CAPITAL	Vienna 2,179,000
RELIGION	Roman Catholic, Protestant
LANGUAGE	German
LITERACY	98%
LIFE EXPECTANCY	79

CURRENCY	euro
GDP PER CAPITA	$27,900
ECONOMY	IND: construction, machinery, vehicles and parts, food. AGR: grains, potatoes, sugar beets, wine, dairy products, lumber. EXP: machinery and equipment, motor vehicles and parts, paper, metal goods, chemicals, iron ore, oil, timber.

Bordering eight countries in Europe's center, Austria is mountainous in the south and west. Fertile lowlands in the east are part of the Danube River basin. Accepted in 1995 as a member of the European Union (EU), Austria has increased its competitiveness by privatizing industries and reducing subsidies. Manufacturing, powered by hydroelectricity, drives the nation's export trade; Austria also profits from iron ore, oil, and timber. Austria is one of the most forested countries in Europe with almost half its territory covered in forest—and forested area is increasing steadily thanks to Austria's "green lung" projects. In 2002 the euro replaced the Austrian schilling; the EU common currency benefits trade and the Austrian economy.

Natural grandeur lures visitors to Tirol and the Hohe Tauern National Park—the largest protected natural area in Central Europe. Seat of the former Habsburg empire, Vienna is a world center of the arts, the site of many splendid palaces, and the headquarters for many international organizations. Tourists can visit the houses of Ludwig van Beethoven, Wolfgang Amadeus Mozart, or Johann Strauss. Salzburg, Mozart's birthplace, celebrates his 250th birthday in 2006.

Azerbaijan
REPUBLIC OF AZERBAIJAN

CONTINENT	Asia
AREA	86,600 sq km (33,436 sq mi)
POPULATION	8,233,000
CAPITAL	Baku 1,816,000
RELIGION	Muslim, Russian Orthodox, Armenian Orthodox
LANGUAGE	Azerbaijani, Russian, Armenian
LITERACY	97%
LIFE EXPECTANCY	72
CURRENCY	Azerbaijani manat
GDP PER CAPITA	$3,700
ECONOMY	IND: petroleum and natural gas, petroleum products, oil field equipment, steel. AGR: cotton, grain, rice, grapes, cattle. EXP: oil and gas, machinery, cotton, foodstuffs.

South of Russia, Azerbaijan is on the west coast of the Caspian Sea; the Caucasus Mountains define the northwestern border of this republic. South and west of Baku, the oil-rich capital, there are extensive lowlands, often below sea level. To the west, separated from the main part of the country by Armenia, is the autonomous region of Naxçıvan with about 300,000 people.

From the 10th to the 12th century, Turkic tribes migrated into the area. By the 1400s the Shiite sect of Islam linked the people more closely to Persians than to their Turkic kin. Russia dominated the region in the 19th century, which also saw the emergence of a national consciousness and the development of an oil industry. Azerbaijan briefly gained independence in 1918—terminated by the Red Army in 1920.

When the Soviet Union collapsed in 1991, Azerbaijan was one of the first to declare independence; but independence escalated the war over Nagorno-Karabakh, a predominately ethnic Armenian region within Azerbaijan. A cease-fire was reached in 1994 after some 30,000 deaths with at least 900,000 Azeris and 300,000 Armenians displaced by the conflict. Ten years later, Armenian forces still control Nagorno-Karabakh—and about 16 percent of Azerbaijan's territory remains occupied.

Bahamas
COMMONWEALTH OF THE BAHAMAS

CONTINENT	North America
AREA	13,939 sq km (5,382 sq mi)
POPULATION	311,000
CAPITAL	Nassau 222,000
RELIGION	Baptist, Anglican, Roman Catholic
LANGUAGE	English, Creole
LITERACY	96%

LIFE EXPECTANCY 72

CURRENCY Bahamian dollar

GDP PER CAPITA $15,300

ECONOMY IND: Tourism, banking, e-commerce, cement, oil refining and transshipment. AGR: citrus, vegetables, poultry. EXP: fish and crawfish, rum, salt, chemicals, fruit and vegetables.

The Bahamas, 700 islands and 2,400 cays, dot the Atlantic Ocean from Florida almost to Haiti. Only 30 of the islands are inhabited. When Christopher Columbus first set foot in the New World on San Salvador in 1492, the Arawak Indians were the only inhabitants there. Today, about 85 percent of Bahamians are of African heritage. New Providence, one of the smallest of the major islands, is home to almost 70 percent of the population. The Bahamas takes in more than three billion dollars annually from nearly four million tourists. International banking and investment management augment the economy, with more than 400 banking institutions from 36 countries.

Bahrain
KINGDOM OF BAHRAIN

CONTINENT Asia

AREA 717 sq km (277 sq mi)

POPULATION 678,000

CAPITAL Manama 139,000

RELIGION Shiite and Sunni Muslim

LANGUAGE Arabic, English, Farsi, Urdu

LITERACY 89%

LIFE EXPECTANCY 74

CURRENCY Bahraini dinar

GDP PER CAPITA $15,100

ECONOMY IND: petroleum processing and refining, aluminum smelting, offshore banking, ship repairing, tourism. AGR: fruit, vegetables, poultry, shrimp. EXP: petroleum and petroleum products, aluminum, textiles.

Bahrain consists of 33 islands in the Persian Gulf (Arabian Gulf). The islands are mostly desert, and most of the population lives in or near Manama, the capital. Since the 1930s the oil industry has replaced pearl diving, and Bahrain has become a financial and communications hub. It is connected to Saudi Arabia by the 26-kilometer (16 mile) King Fahd Causeway. Since independence from Britain in 1971, there has been conflict between the ruling Sunni tribe and the Shiite majority. A new constitution in 2002 provided for an elected parliament and gave women the right to vote and stand as candidates.

Bangladesh
PEOPLE'S REPUBLIC OF BANGLADESH

CONTINENT Asia

AREA 147,570 sq km (56,977 sq mi)

POPULATION 146,733,000

CAPITAL Dhaka 12,560,000

RELIGION Muslim, Hindu

LANGUAGE Bangla (Bengali), English

LITERACY 43%

LIFE EXPECTANCY 59

CURRENCY taka

GDP PER CAPITA $1,800

ECONOMY IND: cotton textiles, jute, garments, tea processing. AGR: rice, jute, tea, wheat, beef. EXP: garments, jute and jute goods, leather, frozen fish and seafood.

Bangladesh, meaning "Bengal nation," is a low-lying country formed by the alluvial plain of the Ganges-Brahmaputra river system—the largest delta in the world. The rivers' annual floods bring silt to renew farmland fertility, often creating new islands in the delta that are quickly claimed as farmland. Much of the land is barely above sea level, with the exception of hills east and south of Chittagong. The monsoon winds come in summer (June to September) and bring heavy rainfall and cyclones. Bangladesh is one of the poorest countries on Earth, and most people are subsistence farmers.

Supported by India, East Pakistan became Bangladesh in 1971 after a war of independence against Pakistan. Bangladesh has the third largest Muslim population in the world after Indonesia and Pakistan. With more than 1,000 people per square kilometer (2,600 per square mile), the country is one of the most crowded on Earth. Rich soils yield three rice harvests a year, but major cyclones cause storm surges that smash into the delta, sweeping people, livestock, and crops from the lowlands. Deforestation in upper watersheds of the Ganges and Brahmaputra Rivers worsens flooding downstream. The government protects the Sundarbans mangrove forest—one of the largest in the world and home to threatened species like the Bengal tiger.

Barbados
BARBADOS

CONTINENT	North America
AREA	430 sq km (166 sq mi)
POPULATION	253,000
CAPITAL	Bridgetown 140,000
RELIGION	Protestant, Roman Catholic
LANGUAGE	English
LITERACY	97%
LIFE EXPECTANCY	73
CURRENCY	Barbadian dollar
GDP PER CAPITA	$15,000
ECONOMY	IND: tourism, sugar, light manufacturing, component assembly for export. AGR: sugarcane, vegetables, cotton. EXP: sugar and molasses, rum, other foods and beverages, chemicals.

Barbados is the most easterly of the Caribbean islands and first in line for seasonal hurricanes. The west coast has white sandy beaches and calm water, but the east coast faces the turbulent Atlantic. Settled by the British in 1627, it won independence in 1966 but retains a strong British flavor. With more than 625 people per square kilometer, Barbados is one of the world's most densely populated nations. It has a stable democracy and a relatively prosperous economy, based largely on tourism and sugar. The grapefruit originates from Barbados.

Belarus
REPUBLIC OF BELARUS

CONTINENT	Europe
AREA	207,595 sq km (80,153 sq mi)
POPULATION	9,873,000
CAPITAL	Minsk 1,705,000
RELIGION	Eastern Orthodox, Roman Catholic, Protestant, Jewish, Muslim
LANGUAGE	Belarusian, Russian
LITERACY	100%
LIFE EXPECTANCY	69
CURRENCY	Belarusian ruble
GDP PER CAPITA	$8,700
ECONOMY	IND: metal-cutting machine tools, tractors, trucks, earthmovers. AGR: grain, potatoes, vegetables, sugarbeets, beef. EXP: machinery and equipment, mineral products, chemicals, metals, textiles.

Belarus, meaning "White Russia," is in Eastern Europe and consists of flat lowlands separated by low hills and uplands. Forests cover a third of this republic, and the Pinsk Marshes occupy much of the south. Settled by a Slavic people, Belarus was dominated by Kiev during the 13th century, by Lithuania and Poland into the 18th century, and by Russia after 1772. The region suffered grievously during World War II, losing more than two million people. Postwar years saw heavy industrial development, centered at Minsk. The 1986 nuclear disaster at Chornobyl (Chernobyl), just south of Belarus in Ukraine, contaminated one third of Belarus—70 percent of the radiation fell on its territory. Belarusians continue to suffer from high incidences of cancer and birth defects, and about 25 percent of the land is considered uninhabitable.

With independence in 1991 came economic decline. The government continues to stifle democracy and oppose privatization of money-losing state enterprises. Belarus remains heavily dependent on Russia—especially to meet its energy needs. Minsk is the administrative headquarters for the Commonwealth of Independent

States, and Belarus uses this organization to seek greater economic and political integration with Russia.

Belgium
KINGDOM OF BELGIUM

CONTINENT Europe
AREA 30,528 sq km
(11,787 sq mi)
POPULATION 10,380,000
CAPITAL Brussels 998,000
RELIGION Roman Catholic,
Protestant
LANGUAGE Dutch, French, German
LITERACY 98%
LIFE EXPECTANCY 78
CURRENCY euro
GDP PER CAPITA $29,200
ECONOMY IND: engineering and
metal products,motor
vehicle assembly,
processed food and
beverages, chemicals.
AGR: sugar beets, fresh
vegetables, fruits, grain,
beef. EXP: machinery
and equipment, chemi-
cals, diamonds, metals
and metal products.

Belgium, a small country in Western Europe, is generally flat except for the hilly and forested southeast (Ardennes) region. After centuries of invasion and occupation, this crossroads of Europe now stands at center stage in the newly unifying continent. Belgium is greatly

attached to democratic values, and in 1994 a revised constitution made it a federal state, giving political representation to its Dutch, French, and German cultures. There are three regions in the federal state: Flanders in the north (with 5.8 million people) where the official language is Dutch; Wallonia in the south (with 3.3 million people) where French is the official language; and Brussels-Capital region (950,000 people) where both languages are used. The smaller German-speaking community (about 70,000) is in Wallonia. The regions and language communities enjoy autonomy in cultural and economic matters.

Belgium is bolstered by a strong economy and can compete in the new single-currency European marketplace. It is the world's largest producer of azaleas. Antwerp, Belgium's second largest city, is the diamond capital of the world. Major tourism attractions include the medieval city of Brugge and the town of Bastogne, in the Ardennes, for Battle of the Bulge sites. Brussels is the headquarters of the European Union and the North Atlantic Treaty Organization.

Belize
BELIZE

CONTINENT North America
AREA 22,965 sq km
(8,867 sq mi)
POPULATION 271,000
CAPITAL Belmopan 9,000
RELIGION Roman Catholic,
Protestant
LANGUAGE English, Spanish, Mayan,
Garifuna, Creole

LITERACY 94%
LIFE EXPECTANCY 67
CURRENCY Belizean dollar
GDP PER CAPITA $4,900
ECONOMY IND: garment production, food processing,
tourism, construction.
AGR: bananas, coca,
citrus, sugar, fish,
lumber. EXP: sugar,
bananas, citrus,
clothing, fish products.

Belize lies along the Caribbean coast of Central America. The first British settlement came in 1638, and the United Kingdom relinquished its last colony on the mainland of the Americas in 1981. English is the official language, but Spanish is widely spoken. Peace and plentiful land attract refugees from troubled neighboring countries. Tourists flock to see Maya ruins like Altun Ha; wildlife such as jaguars, howler monkeys, and toucans; and the Western Hemisphere's longest coral reef.

Benin
REPUBLIC OF BENIN

CONTINENT Africa
AREA 112,622 sq km
(43,484 sq mi)
POPULATION 7,041,000
CAPITAL Porto-Novo 238,000
RELIGION Indigenous beliefs,
Christian, Muslim

LANGUAGE French, Fon, Yoruba, tribal languages
LITERACY 41%
LIFE EXPECTANCY 51
CURRENCY CFA franc
GDP PER CAPITA $1,100
ECONOMY IND: textiles, food processing, chemical production, construction materials. AGR: cotton, corn, cassava (tapioca), yams, livestock. EXP: cotton, crude oil, palm products, cacao.

Countless slaves were shipped to the New World from this West African state during the 18th and 19th centuries. After independence from France in 1960, the country—then called Dahomey—was plagued by many coups; in 1975 a Marxist military government renamed the nation Benin. About 42 African ethnic groups live in Benin, with most people living in the south. In 1989 the government renounced Marxism—since then it has held numerous free elections. The agricultural economy is based largely on cotton.

Bermuda (U.K.)
BRITISH OVERSEAS TERRITORY

CONTINENT North America
AREA 53 sq km (21 sq mi)
POPULATION 64,000
CAPITAL Hamilton 1,000

RELIGION Protestant, Anglican, Roman Catholic
LANGUAGE English, Portuguese

In the western Atlantic, the eight main islands of Bermuda form a chain about 35 kilometers (22 miles) long, interconnected by bridges. The climate is warm, and tourists are drawn to the pink sand beaches, golf courses, and shops.

Bhutan
KINGDOM OF BHUTAN

CONTINENT Asia
AREA 46,500 sq km (17,954 sq mi)
POPULATION 945,000
CAPITAL Thimphu 35,000
RELIGION Lamaistic Buddhist, Hindu
LANGUAGE Dzonkha, Tibetan and Nepali dialects
LITERACY 42%
LIFE EXPECTANCY 66
CURRENCY ngultrum, Indian rupee
GDP PER CAPITA $1,300
ECONOMY IND: cement, wood products, processed fruits, alcoholic beverages. AGR: rice, corn, root crops, citrus, dairy products. EXP: electricity (to India), cardamom, gypsum, timber, handicrafts.

Bhutan is a tiny, remote, and impoverished country between two powerful neighbors, India and China. Violent storms coming off the Himalaya gave the country its name, meaning "Land of the Thunder Dragon." This conservative Buddhist kingdom high in the Himalaya had no paved roads until the 1960s, was off-limits to foreigners until 1974, and launched television service only in 1999. Fertile valleys (less than 10 percent of the land) feed all the Bhutanese. Bhutan's ancient Buddhist culture and mountain scenery make it attractive for tourists, whose numbers are limited by the government.

Bolivia
REPUBLIC OF BOLIVIA

CONTINENT South America
AREA 1,098,581 sq km (424,164 sq mi)
POPULATION 8,595,000
CAPITAL La Paz (administrative capital) 1,477,000; Sucre (constitutional capital) 212,000
RELIGION Roman Catholic
LANGUAGE Spanish, Quechua, Aymara
LITERACY 87%
LIFE EXPECTANCY 63
CURRENCY boliviano
GDP PER CAPITA $2,500
ECONOMY IND: mining, smelting, petroleum, food and beverages. AGR: soybeans, coffee, coca, cotton, timber. EXP: soybeans, natural gas, zinc, gold, wood.

Bolivia—named for Simon Bolívar, liberator of much of South America— is poor, mountainous, and landlocked. Over 60 percent of Bolivia's people are Indian, mostly Quechua or Aymara; the rest are European and mixed. Many are subsistence farmers on the Altiplano (pronounced ahl-tee-PHAH-noh). Here La Paz, with 1.5 million people, sprawls amid snowy peaks near Lake Titicaca. The waters of Lake Titicaca help warm the air, otherwise La Paz, the world's highest capital city at 3,600 meters (11,800 feet), would not be livable. Bolivia has a second capital at Sucre, named after its first president, where the supreme court resides.

In 1987 Bolivia made the world's first debt-for-nature swap with an international conservation organization for the 135,000-hectare Beni Biosphere Reserve—a portion of Bolivia's foreign debt was purchased to support the reserve. Bolivia continues to conserve its environment with the 1995 creation of the 1,895,750-hectare Madidi National Park. Madidi includes everything from Andean glaciers to rain forests; it helps Indians, like the local Quechua, develop ecotourism, which includes watching some 1,000 bird species, tracking tapirs, or white-water rafting.

Large natural gas deposits in the Santa Cruz area and expansion of soybean cultivation help the economy. But a historic boundary dispute with Chile and cocaine from the Cochabamba area plague the national government.

Bosnia and Herzegovina
BOSNIA AND HERZEGOVINA

CONTINENT	Europe
AREA	51,129 sq km
	(19,741 sq mi)
POPULATION	3,892,000
CAPITAL	Sarajevo 579,000
RELIGION	Muslim, Orthodox, Roman Catholic
LANGUAGE	Croatian, Serbian, Bosnian
LITERACY	NA
LIFE EXPECTANCY	72
CURRENCY	marka
GDP PER CAPITA	$1,900
ECONOMY	IND: steel, coal, iron ore, lead, zinc, manganese, bauxite, vehicle assembly. AGR: wheat, corn, fruits, vegetables, livestock. EXP: metals, clothing, wood products.

In mountainous southeastern Europe, Bosnia's Muslims, or Bosniacs, trace their ancestry to Christian Slavs who converted to Islam under the Ottomans for tax and landholding advantages. Yugoslavia recognized Bosniacs as a separate people in 1969. Muslim Slavs and Roman Catholic Croats voted in early 1992 for independence from Yugoslavia; most Eastern Orthodox Serbs were fiercely opposed. In the ensuing 1992-95 civil war, some 250,000 people died. The Dayton Peace Accord ended the war and partitioned the country into a Muslim-Croat region and a Serbian region (Serbian Republic). High unemployment and ethnic tensions continue to hamper the country.

Botswana
REPUBLIC OF BOTSWANA

CONTINENT	Africa
AREA	581,730 sq km
	(224,607 sq mi)
POPULATION	1,573,000
CAPITAL	Gaborone 199,000
RELIGION	Indigenous beliefs, Christian
LANGUAGE	English, Setswana
LITERACY	80%
LIFE EXPECTANCY	37
CURRENCY	pula
GDP PER CAPITA	$8,500
ECONOMY	IND: diamonds, copper, nickel, salt, soda ash, potash, livestock processing, textiles. AGR: sorghum, maize, millet, livestock. EXP: diamonds, copper, nickel, soda ash, meat.

A landlocked country in southern Africa, Botswana enjoys a mild climate in the east; the Kalahari Desert dominates the west and south. The Okavango Delta and Chobe National Park in the north are areas of outstanding natural beauty, rich in animal life. Elephants at Chobe are the largest in body size of all living elephants; they number about 120,000. Stable and prosperous, Botswana has blossomed since independence from Britain in 1966. It is Africa's longest continuous democracy and one of the world's biggest diamond producers.

Brazil
FEDERATIVE REPUBLIC OF BRAZIL

CONTINENT	South America
AREA	8,547,403 sq km
	(3,300,169 sq mi)
POPULATION	176,464,000
CAPITAL	Brasília 3,099,000
RELIGION	Roman Catholic
LANGUAGE	Portuguese, Spanish,
	English, French
LITERACY	86%
LIFE EXPECTANCY	69
CURRENCY	real
GDP PER CAPITA	$7,600
ECONOMY	IND: textiles, shoes,
	chemicals, cement,
	lumber, iron ore.
	AGR: coffee, soybeans,
	wheat, rice, beef. EXP:
	transport equipment,
	iron ore, soybeans,
	footwear, coffee.

Brazil is the giant of South America with nearly half of the continent's area and people; worldwide it ranks fifth in both area and population, which is as diverse as it is large. About 54 percent (95 million) are mainly of European origin, descendants of immigrants from Portugal, Italy, Spain, Germany and Eastern Europe. More than 45 percent (80 million) are black or of mixed-race, a legacy of the African slave trade. Less than 1 percent (700,000) are from indigenous groups, mostly Indians in the Amazon region; smaller numbers of Japanese, other Asians, and Arabs live in the larger Brazilian cities.

The motto "Ordem e Progresso"—(Order and Progress) appears on Brazil's flag. Political progress continues after years of military dictatorship gave way to civilian rule in 1985. Recent censuses reveal social progress, with lower infant mortality rates and higher literacy rates. Brazil's growing urbanization rate helps economic development (some 80 percent of Brazilians live in urban areas), but creates serious social and environmental problems in cities.

São Paulo, with some 18 million people, is Brazil's largest city—and the world's fifth largest metropolis. It is the leading industrial producer and financial center, but problems with pollution, overcrowding, and poverty abound. The Southeast region of Brazil includes São Paulo, Belo Horizonte, and Rio de Janeiro—the economic hub of Brazil containing more than 40 percent of the country's population. South of São Paulo is a rich agricultural region with European-style standards of living, where German and Italian are still spoken alongside Portuguese.

Itaipú, the largest hydroelectric dam in the world, provides electricity to power-hungry São Paulo.

Brazil's second most populous region is the Northeast region, from Maranhão in the north down to Bahia (the most African of Brazilian states). The architecture of cities like Recife and Salvador (Portuguese colonial capital, 1549–1763) shows an earlier age of plantation wealth, but today this is a poor region subject to devastating droughts. Millions have left here for jobs in the Southeast. However, tourism has begun to boom due to sunny weather, samba music, and soft sand beaches.

The North, dominated by the Amazon, is the largest region with the fewest people. The government is making progress in conserving the tropical rain forest and protecting the indigenous people. Tumucumaque National Park, created in 2002, is the world's largest tropical forest park.

British Virgin Islands (U.K.)
BRITISH OVERSEAS TERRITORY

CONTINENT	North America
AREA	153 sq km (59 sq mi)
POPULATION	21,000
CAPITAL	Road Town 12,000
RELIGION	Methodist, Anglican
LANGUAGE	English

The British Virgin Islands, in the northern Caribbean, consist of 16 inhabited islands. Lush vegetation, beaches, coral reefs, and yachting marinas make the islands a natural tourist attraction.

Brunei
NEGARA BRUNEI DARUSSALAM

CONTINENT Asia
AREA 5,765 sq km
(2,226 sq mi)
POPULATION 360,000
CAPITAL Bandar Seri Begawan
61,000
RELIGION Muslim, Buddhist,
Christian, indigenous
beliefs
LANGUAGE Malay, English, Chinese
LITERACY 92%
LIFE EXPECTANCY 76
CURRENCY Bruneian dollar
GDP PER CAPITA $18,600
ECONOMY IND: petroleum,
petroleum refining,
liquefied natural gas,
construction. AGR:
rice, vegetables, fruits,
chickens. EXP: crude
oil, natural gas, refined
products.

A small country, Brunei is located on the island of Borneo in Southeast Asia. Consisting of just two jungle enclaves—inhabited by Malays, Chinese, and indigenous tribes—Brunei won its independence in 1984 after almost a century as a British protectorate. Rich in oil and natural gas, the people of this Muslim sultanate enjoy high subsidies and generous health care. Brunei has an absolute monarchy and a hereditary nobility, with the sultan awarding titles to commoners.

Malay is the official language, but English is widely understood and used in business.

Bulgaria
REPUBLIC OF BULGARIA

CONTINENT Europe
AREA 110,994 sq km
(42,855 sq mi)
POPULATION 7,519,000
CAPITAL Sofia 1,076,000
RELIGION Bulgarian Orthodox,
Muslim
LANGUAGE Bulgarian
LITERACY 99%
LIFE EXPECTANCY 72
CURRENCY lev
GDP PER CAPITA $6,500
ECONOMY IND: electricity, gas
and water, food, bever-
ages, and tobacco.
AGR: vegetables, fruits,
tobacco, livestock. EXP:
clothing, footwear, iron
and steel, machinery
and equipment.

Bulgaria, in southeastern Europe, is dominated by rugged mountains, except for the Danube lowland in the north that it shares with Romania. Rich farmland in the Danube Valley, 130 kilometers (80 miles) of sandy beaches on the Black Sea, and mountainous terrain characterize one of Eastern Europe's least densely populated nations. Most of the population

is urban; about 83 percent are Orthodox Christians, and some 12 percent are Muslim—the Rhodope Mountains, along the border with Greece, are home to many Muslims, including an ethnic Turkish minority.

Bulgarians have a tradition of gratitude towards Russians, who in 1878 helped end 500 years of Ottoman Empire rule. After World War II, communists gained control, and agriculture led the economy until the 1950s, when Russians from the Soviet Union helped finance steel mills, chemical plants, and machine shops. In 1989 communist rule ended and democratic change began.

Tourists flock to Black Sea resorts and to Rila National Park, Bulgaria's largest. Kazanluk, a town in central Bulgaria and heart of the famous Valley of the Roses, exports rose oil—a precious ingredient in world perfume production. As economic conditions improve, Bulgaria joined NATO in 2004 and plans to become a member of the European Union by 2007.

Burkina Faso
BURKINA FASO

CONTINENT Africa
AREA 274,200 sq km
(105,869 sq mi)
POPULATION 13,228,000
CAPITAL Ouagadougou 821,000
RELIGION Muslim, indigenous
beliefs, Christian
LANGUAGE French, native African
languages
LITERACY 27%
LIFE EXPECTANCY 45
CURRENCY CFA franc

GDP PER CAPITA	$1,100
ECONOMY	IND: cotton lint, beverages, agricultural processing, soap. AGR: cotton, peanuts, shea nuts, sesame, livestock. EXP: cotton, livestock, gold.

Half the population of Burkina Faso, or "land of the honest people," claims descent from the Mossi warriors who ruled over one of the most powerful empires in West Africa from the 11th to the 19th century. The landlocked country, desert in the north and savanna in the center and south, is home to 63 ethnic groups. Formerly known as Upper Volta, the French colony gained independence in 1960. Its mostly agricultural economy has been hurt by droughts and political instability. Parks protect the largest elephant population in West Africa and other wildlife.

Burundi
REPUBLIC OF BURUNDI

CONTINENT	Africa
AREA	27,834 sq km (10,747 sq mi)
POPULATION	6,096,000
CAPITAL	Bujumbura 378,000
RELIGION	Roman Catholic, indigenous beliefs, Muslim, Protestant
LANGUAGE	Kirundi, French, Swahili

LITERACY	52%
LIFE EXPECTANCY	43
CURRENCY	Burundi franc
GDP PER CAPITA	$500
ECONOMY	IND: light consumer goods (blankets, shoes, soap), assembly of imported components. AGR: coffee, cotton, tea, corn, beef. EXP: coffee, tea, sugar, cotton, hides.

Small, poor, densely populated, and landlocked, Burundi lies just south of the Equator in central Africa. From the capital, Bujumbura, on Lake Tanganyika, a great escarpment rises to fertile highlands. Agriculture employs 90 percent of the people, with most being subsistence farmers. Since independence in 1962, Burundi has been plagued by ethnic conflict between the majority Hutus and the Tutsis, who tend to dominate the government and army—but are only 14 percent of the population. A 2003 cease-fire and new government offer hope for peace.

Cambodia
KINGDOM OF CAMBODIA

CONTINENT	Asia
AREA	181,035 sq km (69,898 sq mi)
POPULATION	12,558,000
CAPITAL	Phnom Penh 1,157,000
RELIGION	Theravada Buddhist

LANGUAGE	Khmer, French, English
LITERACY	70%
LIFE EXPECTANCY	56
CURRENCY	riel
GDP PER CAPITA	$1,600
ECONOMY	IND: tourism, garments, rice milling, fishing, wood and wood products. AGR: rice, rubber, corn, vegetables. EXP: timber, garments, rubber, rice, fish.

A mostly flat and forested land, Cambodia is a small, compact country. But for more than 500 years, Angkor (in northwestern Cambodia) was the capital of the Khmer Empire, which controlled mainland Southeast Asia from the 9th to the 13th century. Thailand and Vietnam encroached upon the kingdom until 1863, when France made Cambodia a protectorate. Independence came in 1953.

The Vietnam War spilled into Cambodia, igniting conflict, and in 1970 a pro-Western military government overthrew longtime ruler Prince Norodom Sihanouk. Five years later Pol Pot's Khmer Rouge guerrillas began brutally enforcing radical communism, killing some two million Cambodians. After intense border clashes, Vietnam invaded and occupied Cambodia from 1978 to 1989, with up to 200,000 troops. In 1991 three rebel groups and the Phnom Penh government signed a UN-sponsored peace accord. Returned from exile in 1993, Sihanouk became king, leading the new constitutional monarchy.

Coming into the 21st century, Cambodia enjoys relative stability; but subsistence farming employs 75 percent of the workforce and many live in poverty. Cambodians hope that tourism focused on Angkor Wat, meaning "capital monastery," will bring prosperity; it is the largest temple at Angkor—its image is on Cambodia's flag.

Cameroon
REPUBLIC OF CAMEROON

CONTINENT	Africa
AREA	475,442 sq km
	(183,569 sq mi)
POPULATION	15,746,000
CAPITAL	Yaoundé 1,616,000
RELIGION	Indigenous beliefs,
	Christian, Muslim
LANGUAGE	French, English, 24 major
	African language groups
LITERACY	79%
LIFE EXPECTANCY	48
CURRENCY	CFA franc
GDP PER CAPITA	$1,700
ECONOMY	IND: petroleum produc-
	tion and refining, food
	processing, light con-
	sumer goods, textiles.
	AGR: coffee, cacao,
	cotton, rubber, livestock,
	timber. EXP: crude oil
	and petroleum products,
	lumber, cacao,
	aluminum.

Cameroon, in West Africa, is a mixture of desert plains in the north, mountains in the central regions, and tropical rain forest in the south. Along its western border with Nigeria are mountains, which include the volcanic Cameroon Mountain—the highest point in West Africa at 4,100 meters (13,451 feet). Some 250 ethnic groups speaking about 270 languages and dialects make it a remarkably diverse country. The Republic of Cameroon is a union of two former United Nations trust territories—French Cameroun, which became independent in 1960, and southern British Cameroons, which joined it after a 1961 UN-sponsored referendum. English and French are both official languages, but there is growing tension between the country's five million English speakers and the government.

While oil earnings have helped fund industrial expansion, fluctuations in prices of export commodities have forced austerity. After the legalization of opposition parties in 1990, the government adopted IMF and World Bank programs to increase business investment and to foster efficiency in agriculture and trade. Recent elections have been marred by irregularities. In 1994 and 1996 Cameroon fought border wars with Nigeria over a disputed oil-rich coastal area. The International Court of Justice (ICJ) ruled on the disputed territory in 2002—largely in Cameroon's favor. The UN is work-

ing with both Cameroon and Nigeria to implement the ICJ ruling. The Chad-Cameroon oil pipeline, completed in 2003, brings new business to Cameroon's port of Kribi.

Canada
CANADA

CONTINENT	North America
AREA	9,984,670 sq km
	(3,855,101 sq mi)
POPULATION	31,630,000
CAPITAL	Ottawa 1,093,000
RELIGION	Roman Catholic,
	Protestant
LANGUAGE	English, French
LITERACY	97%
LIFE EXPECTANCY	79
CURRENCY	Canadian dollar
GDP PER CAPITA	$29,300
ECONOMY	IND: transportation
	equipment, chemicals,
	processed and un-
	processed minerals, food
	products. AGR: wheat,
	barley, oilseed, tobacco,
	dairy products, forest
	products, fish. EXP:
	motor vehicles and parts,
	industrial machinery, air-
	craft, telecommunications
	equipment, chemicals,
	timber, crude petroleum.

The second largest country in area after Russia, Canada has coastlines on the Atlantic, Arctic, and Pacific Oceans, giving it the longest coastline of any country. In area, Canada is slightly larger than the United States, but has only 11 percent as many people. It is one of the least densely inhabited and most prosperous countries. A vast region of swamps, lakes, and ancient rock, known as the Canadian Shield, radiates out from Hudson Bay to cover half of the country; it is agriculturally poor with few people but rich in mineral deposits and forests. The shield stretches from the Arctic to the Great Lakes and Labrador, cutting the country in half and contributing to a division between easterners and westerners. The Canadian Shield and rugged western mountains experience subarctic climates, resulting in a near empty north—an estimated 75 percent of Canadians live within 161 kilometers (100 miles) of the U.S. border.

France pioneered settlement, but Britain gained control in 1763. In 1867 the British North America Act united English-speaking Upper Canada (Ontario) and French-speaking Lower Canada (Quebec) with Nova Scotia and New Brunswick in a self-governing confederation—with independence in 1931. Canada is a multicultural society dependent on immigration for growth. Some 28 percent are of British descent, 23 percent claim French descent (concentrated in Quebec), 2 percent are aboriginal peoples—other minorities include Italians, Germans, Ukrainians, and Chinese. Canada's population is highly urbanized, with most people living in four areas: southern Ontario, Montreal region, Vancouver city and southern Vancouver Island, and the Calgary-Edmonton corridor. The urban economy has a large manufacturing base, and the North American Free Trade Agreement (NAFTA) has brought an economic boom—about 80 percent of Canada's trade is with the U.S.

Cape Verde
REPUBLIC OF CAPE VERDE

CONTINENT	Africa
AREA	4,036 sq km (1,558 sq mi)
POPULATION	474,000
CAPITAL	Praia 107,000
RELIGION	Roman Catholic, Protestant
LANGUAGE	Portuguese, Crioulo (Kriolu)
LITERACY	77%
LIFE EXPECTANCY	69
CURRENCY	Cape Verdean escudo
GDP PER CAPITA	$1,400
ECONOMY	IND: food and beverages, fish processing, shoes and garments, salt mining. AGR: bananas, corn, beans, sweet potatoes, fish. EXP: fuel, shoes, garments, fish, hides.

Off the coast of West Africa, Cape Verde consists of ten volcanic islands and five islets. The islands were uninhabited until discovered by the Portuguese in 1456; African slaves were brought here to work on plantations. Independence from Portugal came in 1975. African culture is most evident on the island of Sao Tiago—where half the population lives. Cape Verde enjoys a stable democratic system. Water shortages hinder agriculture, but tourism is a growing industry.

Cayman Islands (U.K.)
BRITISH OVERSEAS TERRITORY

CONTINENT	North America
AREA	262 sq km (101 sq mi)
POPULATION	42,000
CAPITAL	George Town 24,000
RELIGION	Protestant, Roman Catholic
LANGUAGE	English

The three Cayman Islands are northwest of Jamaica in the Caribbean. About 94 percent of the population live on the largest island, Grand Cayman. Offshore banking and tourism are key contributors to the economy.

Central African Republic
CENTRAL AFRICAN REPUBLIC

CONTINENT	Africa
AREA	622,984 sq km (240,535 sq mi)
POPULATION	3,684,000
CAPITAL	Bangui 698,000
RELIGION	Indigenous beliefs, Protestant, Roman Catholic, Muslim
LANGUAGE	French, Sangho, Arabic, tribal languages
LITERACY	51%
LIFE EXPECTANCY	43
CURRENCY	CFA franc

GDP PER CAPITA	$1,200
ECONOMY	IND: diamond mining, logging, brewing, textiles. AGR: cotton, coffee, tobacco, manioc (tapioca), timber. EXP: diamonds, timber, cotton, coffee, tobacco.

Deep in the heart of the continent, the Central African Republic was part of French Equatorial Africa before independence in 1960. Most of the country is savanna plateau, with rain forests in the south. The economy, moribund by 1979 when self-proclaimed Emperor Bokassa's 14-year reign of terror ended, remains in poor condition. It is one of the world's least developed countries, with most of the population engaged in subsistence farming. Timber and uncut diamonds are sources of export revenue. Political instability continued with a coup and rebellions in 2003.

Chad
REPUBLIC OF CHAD

CONTINENT	Africa
AREA	1,284,000 sq km (495,755 sq mi)
POPULATION	9,253,000
CAPITAL	N'Djamena 797,000
RELIGION	Muslim, Christian, animist
LANGUAGE	French, Arabic, Sara, Sango, more than 120 different languages and dialects

LITERACY	48%
LIFE EXPECTANCY	49
CURRENCY	CFA franc
GDP PER CAPITA	$1,000
ECONOMY	IND: oil, cotton textiles, meatpacking, beer brewing, natron (sodium carbonate). AGR: cotton, sorghum, millet, peanuts, cattle. EXP: cotton, cattle, gum arabic.

A landlocked nation in north central Africa, Chad consists of fertile lowlands in the south but is arid in the center and largely desert in the north. Since independence from France in 1960, Chad has suffered instability stemming mostly from tension between the African-Christian south and the Arab-Muslim north and east. A border dispute with Libya over the Aozou Strip went to the United Nation's International Court of Justice for arbitration; in 1994 the court ruled in favor of Chad. One of Africa's poorest countries, the start of large-scale oil production in 2004 helps the economy.

Channel Islands (U.K.)
BRITISH CROWN DEPENDENCIES

Guernsey

CONTINENT	Europe
AREA	78 sq km (30 sq mi)
POPULATION	65,000
CAPITAL	Saint Peter Port 16,500
RELIGION	Anglican, Roman Catholic
LANGUAGE	English, French, Norman French dialect

Chile
REPUBLIC OF CHILE

CONTINENT	South America
AREA	756,096 sq km (291,930 sq mi)
POPULATION	15,774,000
CAPITAL	Santiago 5,623,000
RELIGION	Roman Catholic, Protestant
LANGUAGE	Spanish
LITERACY	96%
LIFE EXPECTANCY	76
CURRENCY	Chilean peso
GDP PER CAPITA	$10,100
ECONOMY	IND: copper, other minerals, foodstuffs, fish processing, iron and steel. AGR: wheat, corn, grapes, beans, beef, fish, timber. EXP: copper, fish, fruits, paper and pulp, chemicals.

Chile extends like a ribbon down the west coast of South America for over 4,000 kilometers (2,485 miles)—but averages 150 kilometers (93 miles) wide. From the mineral-rich Atacama Desert to haunting Torres del Paine National Park and stormy Cape Horn, "Chile," wrote Nobel Prize winner Pablo Neruda, "was invented by a poet." This elongated country, wedged between the deepest ocean and the longest mountain chain, straddles a tectonically unstable region. Mountains cover 80 percent of Chile.

Most Chileans are of European or mixed European and indigenous ancestry—only about 5 percent are indigenous (mostly Mapuche). Chile is highly urbanized, with 40 percent of the population living in the Santiago area. Chileans once enjoyed Latin America's longest tradition of political stability and civil liberty. But in 1973 a bloody coup overthrew Salvador Allende's elected Marxist government and ushered in 16 years of dictatorship under Gen. Augusto Pinochet. Democracy was restored in 1989.

Privatization of industries and increased agricultural exports have boosted the economy. The Chuquicamata and Escondida copper mines, in the arid Atacama, rank as the world's largest. Tourism is a major business; a popular attraction is Easter Island, 3,700 kilometers (2,299 miles) west of Chile. Here one thousand *moai* (giant figures carved in stone) fascinate visitors to this small Polynesian island.

China
PEOPLE'S REPUBLIC OF CHINA

CONTINENT	Asia
AREA	9,596,960 sq km
	(3,705,405 sq mi)
POPULATION	1,288,679,000
CAPITAL	Beijing 10,849,000
RELIGION	Taoist, Buddhist, Muslim
LANGUAGE	Chinese (Mandarin),
	Cantonese, other dialects
	and minority languages

LITERACY	86%
LIFE EXPECTANCY	71
CURRENCY	Yuan, also referred to as
	the Renminbi
GDP PER CAPITA	$4,700
ECONOMY	IND: iron and steel, coal,
	machine building, armaments, textiles and
	apparel, petroleum,
	cement. AGR: rice, wheat,
	potatoes, sorghum, pork,
	fish. EXP: machinery and
	equipment, textiles and
	clothing, footwear, toys
	and sporting goods, mineral fuels.

China is the world's most populous country with about 1.3 billion people— 20 percent of the Earth's population. Occupying most of East Asia, it is the fourth largest country in area (after Russia, Canada, and the U.S.). China's geography is highly diverse, with hills, plains, and river deltas in the east and deserts, high plateaus, and mountains in the west. Climate is equally varied, ranging from tropical in the south (Hainan) to subarctic in northeastern China (Manchuria). China's geography causes an uneven population distribution; 94 percent live in the eastern third of the country. Shandong province, with its mild coastal climate, has 91 million people, but Tibet, with its harsh mountain plateau climate, has only 2.6 million. The coastal regions are the most economically developed—acting

as a magnet for an estimated 90 million Chinese migrants from the poor rural interior.

China has perhaps the world's longest continuous civilization; for more than 40 centuries its people created a culture with strong philosophies, traditions, and values. The start of the Han dynasty 2,200 years ago marked the rise of military power that created an empire—one that provided a golden age in art, politics, and technology. Ethnic Chinese still refer to themselves as the "People of Han," and Han Chinese constitute 92 percent of the country's population. Successive dynasties developed a system of bureaucratic control that gave agrarian-based China an advantage over rivals. China remains a predominantly rural society, with only 39 percent living in urban areas.

The first half of the 20th century saw the fall of the last Chinese emperor, Japanese invasion, World War II, and civil war between Chinese Communist and Nationalist forces—ending with the retreat of the Nationalists to Taiwan. The People's Republic of China from 1949 to 1976 imposed state control on the economy. Since 1979, China has reformed its economy and allowed competition, and today has the world's highest rate of growth. Rapid industrial development has increased pollution—with China having seven of the world's ten most polluted cities. The largest producer and consumer of coal, the country is turning away from coal toward clean hydroelectric resources, such as the Three Gorges Dam. Politically it still maintains strict control over its people. Chinese rule over Tibet remains controversial, fighting with Muslim separatists in Xinjiang continues, and political issues with Taiwan remain unresolved. China regained Hong Kong from Britain in 1997 and Macau from Portugal in 1999. In 2003 China became only the third nation (after Russia and the U.S.) to launch a manned spaceflight—with plans to reach the moon by 2010.

Hong Kong
SPECIAL ADMINISTRATIVE REGION OF CHINA

CONTINENT Asia
AREA 1,092 sq km
(422 sq mi)
POPULATION 7,182,000
RELIGION Local religions,
Christian
LANGUAGE Chinese (Cantonese),
English

Macau
SPECIAL ADMINISTRATIVE REGION OF CHINA

CONTINENT Asia
AREA 25 sq km (10 sq mi)
POPULATION 450,000
RELIGION Buddhist, Roman
Catholic
LANGUAGE Portuguese, Chinese
(Cantonese)

Taiwan

CONTINENT Asia
AREA 35,980 sq km
(13,891 sq mi)
POPULATION 22,568,000
CAPITAL Taipei 2,550,000

RELIGION Mixture of Buddhist,
Confucian, and Taoist;
Christian
LANGUAGE Mandarin Chinese,
Taiwanese

Taiwan, located about 130 kilometers (80 miles) off the China coast, has a fertile plain along the west coast that rises to one of the highest mountain ranges in Asia. About 84 percent of the people consider themselves native Taiwanese (descendents of Chinese who migrated to the island by the 19th century).

In 1945, the Nationalist Chinese started administering Taiwan, which had been ruled by Japan for 50 years. Seen as liberators at first, the Nationalists imposed an authoritarian government favoring mainlanders, and resentment among natives grew. More than two million Chinese fled here from the mainland after Communist forces defeated the Nationalists in 1949. Formally known as the Republic of China, Taiwan made the transition from an authoritarian state to a multiparty democracy in the early 1990s. Chen Shui-bian, of the Democratic Progressive Party, won the presidential election in 2000, ending the Nationalist Party's 55-year monopoly.

Politically most nations and the UN acknowledge the position of the People's Republic of China that Taiwan is one of 23 provinces of China; however, most countries have commercial relations with it—including China. Taiwan is one of the world's largest suppliers of computer technology, and its investment in China alone is estimated at 70 billion dollars. The new Taipei 101 building, considered the world's tallest, reflects the island's economic prosperity.

Christmas Island (Australia)
AUSTRALIAN EXTERNAL TERRITORY

CONTINENT Australia/Oceania
AREA 135 sq km (52 sq mi)
POPULATION 2,770
CAPITAL The Settlement 1,500
RELIGION Buddhist, Muslim,
Christian
LANGUAGE English, Chinese, Malay

Cocos (Keeling) Islands (Australia)
AUSTRALIAN EXTERNAL TERRITORY

CONTINENT Australia/Oceania
AREA 14 sq km (5 sq mi)
POPULATION 630
CAPITAL West Island 200
RELIGION Sunni Muslim
LANGUAGE Malay (Cocos dialect),
English

Colombia
REPUBLIC OF COLOMBIA

CONTINENT South America
AREA 1,141,748 sq km
(440,831 sq mi)
POPULATION 44,172,000
CAPITAL Bogotá 7,594,000
RELIGION Roman Catholic
LANGUAGE Spanish
LITERACY 93%
LIFE EXPECTANCY 71
CURRENCY Colombian peso

GDP PER CAPITA $6,100

ECONOMY IND: textiles, food processing, oil, clothing and footwear. AGR: coffee, cut flowers, bananas, rice, forest products, shrimp. EXP: petroleum, coffee, coal, apparel, bananas, cut flowers.

Colombia is the only South American country with coastlines on both the Pacific Ocean and Caribbean Sea. Three mighty north-south Andean cordilleras separate the western coastal lowlands from the almost empty eastern jungles, with 54 percent of Colombia's land but only 3 percent of the people. Most Colombians are of mixed ethnicity; about 20 percent claim European descent. Native Indians, about one percent of the population, live in the eastern jungles.

The Andes contribute to the concentration of Colombia's people into separate clusters. Some live in the Caribbean lowlands in cities like Barranquilla and Cartagena; some live in isolated mountain valleys in cities like Cali and Medellín. Bogotá, the capital and largest city, is in a remote mountain basin at 2,500 meters (8,200 feet).

Colombia has had a turbulent history. Civil war (1899–1902) claimed 100,000 lives, and La Violencia (1948–1957) cost 300,000 more. In the 1980s, as the government

worked with the U.S. to disrupt the lucrative illicit drug trade, violence came from cocaine traffickers, who targeted judges, newspaper editors, and community officials. The drug cartels continue to be a disruptive force into the 21st century, in spite of efforts to arrest the more powerful leaders.

Farmers raise world-renowned coffee on the Andean slopes. Colombia sells much of the world's emeralds and considerable amounts of gold, silver, and platinum and has the continent's highest coal production— most from the Guajira Peninsula. However, oil development suffers from sabotage by guerrilla groups, and large parts of Colombia are beyond government control.

Comoros
UNION OF THE COMOROS

CONTINENT	Africa
AREA	1,862 sq km (719 sq mi)
POPULATION	633,000
CAPITAL	Moroni 53,000
RELIGION	Sunni Muslim, Roman Catholic
LANGUAGE	Arabic, French, Shikomoro
LITERACY	57%
LIFE EXPECTANCY	56
CURRENCY	Comoran franc
GDP PER CAPITA	$700
ECONOMY	IND: tourism, perfume distillation. AGR: vanilla, cloves, perfume essences, copra. EXP: vanilla, ylang-ylang, cloves, perfume oil, copra.

The Comoros are a group of volcanic islands in the Mozambique Channel between northern Madagascar and Africa. The people share African-Arab origins. In 1975 three of the so-called perfume islands voted for independence from France; the fourth, Mayotte, elected to remain a dependency. Some 18 coups, or attempted coups, since independence have created great instability. In 1997 the islands of Anjouan and Mohéli declared independence, but a new federal constitution in 2001 brought the islands back together. Most inhabitants make their living from subsistence agriculture or fishing; exports include vanilla and essences used in the manufacture of perfumes.

Congo
REPUBLIC OF THE CONGO

CONTINENT	Africa
AREA	342,000 sq km (132,047 sq mi)
POPULATION	3,723,000
CAPITAL	Brazzaville 1,080,000
RELIGION	Christian, Animist, Muslim
LANGUAGE	French, Lingala, Monokutuba
LITERACY	84%
LIFE EXPECTANCY	50
CURRENCY	CFA franc
GDP PER CAPITA	$900
ECONOMY	IND: petroleum extraction, cement, lum-

ber, brewing, sugar, palm oil, soap. AGR: cassava (tapioca), sugar, rice, corn, forest products. EXP: petroleum, lumber, plywood, sugar, cacao.

Astride the Equator in west central Africa, Congo has a small population that is concentrated in the country's southwest, with a virtually uninhabited jungle in the north. Most people live between Brazzaville, the capital, and Pointe-Noire, Congo's port city and focus for the oil industry. Since it achieved independence in 1960, political turmoil has hampered the country. After almost three decades of Marxist rule, Congo adopted a multiparty, democratic system in 1992. Conflict in the late 1990s derailed democracy, but a new constitution in 2002 brings the promise of stability.

Congo, Dem. Rep. of the
DEMOCRATIC REPUBLIC OF THE CONGO

CONTINENT	Africa
AREA	2,344,885 sq km (905,365 sq mi)
POPULATION	56,625,000
CAPITAL	Kinshasa 5,717,000
RELIGION	Roman Catholic, Protestant, Kimbanguist, Muslim, traditional
LANGUAGE	French, Lingala, Kingwana, Kikongo, Tshiluba
LITERACY	66%
LIFE EXPECTANCY	48
CURRENCY	Congolese franc
GDP PER CAPITA	$600
ECONOMY	IND: mining (diamonds, copper, zinc), mineral processing, consumer products. AGR: coffee, sugar, palm oil, rubber; wood products. EXP: diamonds, copper, crude oil, coffee, cobalt.

Straddling the Equator, the Democratic Republic of the Congo is the third largest country in Africa (after Sudan and Algeria). The mighty Congo River flows north and then south through a land rich in minerals, fertile farmlands, and rain forests. The country has a tiny coast on the Atlantic Ocean, just enough to accommodate the mouth of the Congo River. The forested Congo River basin occupies 60 percent of the nation's area, creating a central region that is a communication barrier between the capital, Kinshasa, in the west, the mountainous east, and the southern mineral-rich highlands. As many as 250 ethnic groups speaking some 700 local languages and dialects endure one of the world's lowest living standards. War, government corruption, neglected public services, and depressed copper and coffee markets are contributing factors.

In 1960 the Belgian Congo became independent as the Democratic Republic of the Congo. Gen. Joseph Mobutu came to power in a coup in 1965; he changed his name to Mobutu Sese Seko and the country's name to the Republic of Zaire. Mobutu's corruption-ridden government continued in power until 1997 when rebel forces led by Laurent Kabila—supported by Rwanda and Uganda—took Kinshasa and changed the country's name back to the Democratic Republic of the Congo. A rift between Kabila and his former allies caused a new rebellion in 1998, backed by Rwanda and Uganda. What became known as "Africa's world war" started as Zimbabwe, Angola, Namibia, and Chad sent troops to support Kabila. The war claimed some three million lives, with all sides plundering the country's natural resources—especially diamonds from south-central Congo. A UN-supported peace agreement and the formation of a transitional government in 2003 seemed to signal an end to the five-year conflict. The government works to unify the country.

Cook Islands (New Zealand)
SELF-GOVERNING IN FREE ASSOCIATION WITH NEW ZEALAND

CONTINENT	Australia/Oceania
AREA	240 sq km (93 sq mi)
POPULATION	18,000
CAPITAL	Avarua 13,000
RELIGION	Christian
LANGUAGE	English, Maori

Costa Rica
REPUBLIC OF COSTA RICA

CONTINENT	North America
AREA	51,100 sq km
	(19,730 sq mi)
POPULATION	4,171,000
CAPITAL	San José 1,085,000
RELIGION	Roman Catholic,
	Evangelical
LANGUAGE	Spanish, English
LITERACY	96%
LIFE EXPECTANCY	79
CURRENCY	Costa Rican colon
GDP PER CAPITA	$8,300
ECONOMY	IND: microprocessors,
	food processing, textiles
	and clothing, construc-
	tion materials. AGR:
	coffee, pineapples,
	bananas, sugar; beef,
	timber. EXP: coffee,
	bananas, sugar, pine-
	apples, textiles.

Located in Central America, Costa Rica has coastlines on the Caribbean Sea and Pacific Ocean. The tropical coastal plains rise to mountains, active volcanoes, and a temperate central plateau where most people live (San José, the capital, is here). The only country in Central America with no standing army, it enjoys continuing stability after a century of almost un-interrupted democratic government. Tourism, which has overtaken bananas as Costa Rica's leading foreign ex-change earner, bolsters the economy.

A quarter of the land has protected status; the beauty of rain forest pre-serves draws more and more visitors.

Côte d'Ivoire (Ivory Coast)
REPUBLIC OF CÔTE D'IVOIRE

CONTINENT	Africa
AREA	322,462 sq km
	(124,503 sq mi)
POPULATION	16,962,000
CAPITAL	Abidjan (administrative
	capital) 3,516,000;
	Yamoussoukro (legislative
	capital) 416,000
RELIGION	Christian, Muslim,
	indigenous beliefs
LANGUAGE	French, Dioula, 60 native
	dialects
LITERACY	51%
LIFE EXPECTANCY	43
CURRENCY	CFA franc
GDP PER CAPITA	$1,400
ECONOMY	IND: foodstuffs, bev-
	erages, wood products,
	oil refining, truck and
	bus assembly. AGR: cof-
	fee, cacao, bananas,
	palm kernels, timber.
	EXP: cacao, coffee, tim-
	ber, petroleum, cotton.

Formerly known as the Ivory Coast, this West Africa country officially uses its French name, Côte d'Ivoire (pronounced kot dee-VWAHR). The geography ranges from coastal beaches and forests in the south to a savanna plateau in the north. Muslims, who make up 40 percent of the population, live mostly in the north, while Chris-tians (35 percent) mostly inhabit the south. The country has 60 ethnic groups; the Baoule, inhabiting the central region, is the largest group. The dominance of the Baoule in running the country since independence is a major issue with other groups.

This former French colony gained independence in 1960, but only held its first multiparty elections in 1990. Felix Houphouet-Boigny (a Baoule) was president from 1960 until his death in 1993. His home village, Yamoussoukro, was designated the capital in 1983 (most government functions remain in Abidjan). Yamous-soukro was also the site of an extrav-agant replica of Rome's St. Peter's Basilica. Once a model of stability, the country in 1999 started slipping into the kind of tribal strife that has plagued many African countries. Vio-lence between northerners and south-erners in 2000 destroyed churches and mosques and killed dozens of people. In 2002 a failed coup turned into a rebellion that split the country in two. Rebel forces gained control of a large area in the north, including the major cities of Korhogo, Bouaké, and Man. The political crisis, combined with low commodity prices, caused personal income to decrease, while unemployment and urban migration increased. Dependent on such com-modities as cacao and coffee, which are subject to drastic price swings, this once prosperous nation faces ongoing economic problems.

Croatia
REPUBLIC OF CROATIA

CONTINENT	Europe
AREA	56,542 sq km
	(21,831 sq mi)
POPULATION	4,287,000
CAPITAL	Zagreb 688,000
RELIGION	Roman Catholic, Orthodox
LANGUAGE	Croatian
LITERACY	99%
LIFE EXPECTANCY	74
CURRENCY	kuna
GDP PER CAPITA	$9,800
ECONOMY	IND: chemicals and plastics, machine tools, fabricated metal, electronics. AGR: wheat, corn, sugar beets, sunflower seed, livestock. EXP: transport equipment, textiles, chemicals, foodstuffs, fuels.

A crescent-shaped country in southeast Europe, Croatia extends from the fertile plains of the Danube to the mountainous coast of the Adriatic Sea. In the Adriatic, Croatia has 1,185 islands—many are major tourist areas. The 1991–95 civil war between Croats and Serbs caused massive damage to cities and industries. War halted the tourist trade and cut industrial output, including a lucrative ship-building business. Since the war, Croatia has progressed politically and economically; it applied for European Union membership in 2003.

Cuba
REPUBLIC OF CUBA

CONTINENT	North America
AREA	110,860 sq km
	(42,803 sq mi)
POPULATION	11,279,000
CAPITAL	Havana 2,189,000
RELIGION	Roman Catholic, Protestant, Jehovah's Witness, Jewish, Santeria
LANGUAGE	Spanish
LITERACY	97%
LIFE EXPECTANCY	76
CURRENCY	Cuban peso
GDP PER CAPITA	$2,700
ECONOMY	IND: sugar, petroleum, tobacco, chemicals. AGR: sugar, tobacco, citrus, coffee; livestock. EXP: sugar, nickel, tobacco, fish, medical products.

The Republic of Cuba encompasses more than 4,000 islands and cays; the main island is the largest in the West Indies. Since its discovery five centuries ago by Christopher Columbus, strategic location and agricultural wealth have made Cuba a coveted prize. By the mid-1800s sugar plantations were satisfying a third of world demand.

Heavy U.S. involvement after Cuba's independence from Spain in 1902 was supplanted by an association with the Soviet Union following Fidel Castro's 1959 revolution. In 1961 Castro established the first communist state in the Western Hemisphere. Since then more than a million Cubans have moved to the U.S. Cuba's centrally planned economy initially brought dramatic gains in education, health care, and social welfare. The economy crashed after Cuba was abandoned by the former U.S.S.R. and Eastern European trading partners in the early 1990s, and food and energy were tightly rationed.

Despite the U.S. trade embargo, Cuba's economy is improving due to Canadian, European, and Latin American investments—especially in tourism. The natural beauty of Cuba brings millions of tourists, with many coming to see the thousands of species of plants and animals that live nowhere else on Earth. Protected natural areas take up nearly 22 percent of Cuban territory, providing habitats for crocodiles, flamingos, orchids, and more. Politically the government still restricts human freedoms, although religious rights saw some progress in 1998 with Christmas, December 25, being reinstated as a national holiday. In southeastern Cuba, the U.S. maintains a presence at Guantanamo Bay; the 1934 treaty between Cuba and the U.S. grants a perpetual lease for the naval base—voided only by mutual consent or by U.S. abandonment of the base.

Cyprus
REPUBLIC OF CYPRUS

CONTINENT	Europe
AREA	9,251 sq km
	(3,572 sq mi)
POPULATION	934,000

CAPITAL Nicosia 205,000

RELIGION Greek Orthodox, Muslim

LANGUAGE Greek, Turkish, English

LITERACY 98%

LIFE EXPECTANCY 77

CURRENCY Cypriot pound, Turkish lira

GDP PER CAPITA $15,000

ECONOMY IND: food, beverages, textiles, chemicals, metal products. AGR: potatoes, citrus, vegetables, barley. EXP: Greek Cypriot area: citrus, potatoes, pharmaceuticals, cement; Turkish Cypriot area: citrus, potatoes, textiles.

The island of Cyprus, in the eastern Mediterranean, was divided in 1974 when Turkish troops invaded to stop Greek military plans for enosis (union) with Greece. Tensions between the Greek Cypriot majority and the Turkish Cypriot minority had been high since independence from Britain in 1960. Fighting in 1974 displaced more than a third of the population as some 180,000 Greek Cypriots fled south and 45,000 Turkish Cypriots went to the northern Turkish-occupied area (37 percent of the island). The UN patrols the dividing line and works to settle ethnic enmities. However, it failed to reunify the island before May 2004—when Cyprus joined the European Union. The northern area, known as the Turkish Republic of Northern Cyprus, is not recognized by the UN so only the southern Republic of Cyprus could join the EU.

Czech Republic
CZECH REPUBLIC

CONTINENT Europe

AREA 78,866 sq km (30,450 sq mi)

POPULATION 10,177,000

CAPITAL Prague 1,170,000

RELIGION Roman Catholic, Protestant, atheist

LANGUAGE Czech

LITERACY 100%

LIFE EXPECTANCY 75

CURRENCY Czech koruna

GDP PER CAPITA $15,300

ECONOMY IND: metallurgy, machinery and equipment, motor vehicles, glass, armaments. AGR: wheat, potatoes, sugar beets, hops, pigs. EXP: machinery and transport equipment, intermediate manufactures, chemicals, raw materials, fuel.

Breaking a nearly 75-year union with the Slovak Republic in 1993, this independent country in Central Europe consists of the regions of Bohemia and Moravia—once part of the Great Moravian Empire formed by Slav tribes in the early ninth century. The Bohemian kingdom arose here during the tenth century, its 600-year reign a highlight of Czech history. Bohemia is a plateau surrounded by mountains, and Moravia, to the east, is mostly hills and lowlands. Austria's Habsburgs took control of both regions at the start of the 16th century.

With the end of the Austro-Hungarian Empire in 1918, the Czechs and Slovaks came together to create Czechoslovakia.

Separated again in 1939, Czech lands were annexed by the Nazis during World War II, while Slovakia became a puppet state of the Germans. Communists took charge of a reunited Czechoslovakia in 1948, crushing an attempt at liberalization in 1968, only to be forced out in 1989.

After its break with the Slovak Republic, the Czech nation rapidly privatized state-owned businesses. State ownership of businesses was at about 97 percent under communism—today it is less than 20 percent. The country is also reducing its dependence on highly polluting brown coal as an energy source, turning more toward nuclear energy. Tourism is a rapidly developing sector, and millions come to Prague to visit castles, palaces, and spas. Although the political and financial crises of 1997 eroded somewhat the country's stability and prosperity, the Czech Republic succeeded in becoming a NATO member in 1999 and a European Union member in 2004.

Denmark
KINGDOM OF DENMARK

CONTINENT	Europe
AREA	43,098 sq km
	(16,640 sq mi)
POPULATION	5,395,000
CAPITAL	Copenhagen 1,066,000
RELIGION	Evangelical Lutheran
LANGUAGE	Danish, Faroese,
	Greenlandic
LITERACY	100%
LIFE EXPECTANCY	77
CURRENCY	Danish krone
GDP PER CAPITA	$28,900
ECONOMY	IND: food processing,
	machinery and equip-
	ment, textiles and cloth-
	ing, chemical products.
	AGR: barley, wheat,
	potatoes, sugar beets,
	pork, fish. EXP: machin-
	ery and instruments, meat
	and meat products, dairy
	products, fish, chemicals.

Located in northern Europe, Denmark consists of the mainland of Jutland and 406 islands. Fertile farmland covers 64 percent of the country, which is among the flattest in the world. A stepping-stone between the European mainland and the Scandinavian peninsula, Denmark has been integral to NATO defense since 1949. Membership in the European Union gives ready access to markets for its pork and dairy products. However, Danes rejected adopting the euro currency in 2000.

Denmark has earned more from manufacturing than from agriculture since the 1960s. Devastation of lobster colonies by industrial pollution has prompted imposition of some of the world's strictest environmental standards. More and more Danes are using alternative energy sources—wind power, solar energy, and geothermal heat—for environmental and economic reasons. Wind is an increasingly important source of energy in Denmark, and windmills are an important export.

With its palaces and gardens, Copenhagen hosts more visitors than any other Nordic city. Tivoli, founded in 1843, is a world-famous amusement park in downtown Copenhagen. Also popular is Legoland, near Vejle on the Jutland peninsula, with famous features created with Lego blocks— like Mount Rushmore. English, a required subject in the public school system, is widely spoken in Denmark. The Kingdom of Denmark, a constitutional monarchy, includes the self governing territories of the Faroe Islands, and Greenland, the world's largest island.

OVERSEAS REGIONS

Faroe Islands

Greenland

Djibouti
REPUBLIC OF DJIBOUTI

CONTINENT	Africa
AREA	23,200 sq km
	(8,958 sq mi)
POPULATION	658,000
CAPITAL	Djibouti 502,000
RELIGION	Muslim, Christian
LANGUAGE	French, Arabic, Somali,
	Afar
LITERACY	68%
LIFE EXPECTANCY	43
CURRENCY	Djiboutian franc
GDP PER CAPITA	$1,300
ECONOMY	IND: construction,
	agricultural processing.
	AGR: fruits, vegetables,
	goats. EXP: reexports,
	hides and skins, coffee
	(in transit).

Gateway for Red Sea shipping, Djibouti lies in northeast Africa. A French territory until 1977, France's naval base and garrison generate about half of the country's income. The capital of this resource-poor nation profits as a regional banking center with a free port and modern air facilities. Terminus of the railway from Addis Ababa, it handles much of Ethiopia's trade. A civil war in the early 1990s ended with a power-sharing agreement between the two main ethnic groups, the Issa of Somali origin and the Afar of Ethiopian origin.

Dominica
COMMONWEALTH OF DOMINICA

CONTINENT	North America
AREA	751 sq km (290 sq mi)
POPULATION	70,000

CAPITAL	Roseau 27,000
RELIGION	Roman Catholic, Protestant
LANGUAGE	English, French patois
LITERACY	94%
LIFE EXPECTANCY	73
CURRENCY	East Caribbean dollar
GDP PER CAPITA	$5,400
ECONOMY	IND: soap, coconut oil, tourism, copra. AGR: bananas, citrus, mangoes, root crops. EXP: bananas, soap, bay oil, vegetables, grapefruit.

Mountainous, densely forested, with waterfalls and exotic birds, much of Dominica is protected as national wilderness. Volcanic activity provides boiling pools, geysers, and black-sand beaches. Most Dominicans are descendants of African slaves brought in by colonial planters. Independent from Britain since 1978, Dominica remains poor and dependent on banana exports. Governments, including that of Mary Eugenia Charles, the first female prime minister in the West Indies, have sought to broaden the economic base with tourism and light industry. Home to 3,000 Carib Indians, Dominica is the last bastion of this once populous Caribbean tribe.

Dominican Republic
DOMINICAN REPUBLIC

CONTINENT	North America
AREA	48,442 sq km (18,704 sq mi)
POPULATION	8,716,000
CAPITAL	Santo Domingo 1,865,000
RELIGION	Roman Catholic
LANGUAGE	Spanish
LITERACY	85%
LIFE EXPECTANCY	69
CURRENCY	Dominican peso
GDP PER CAPITA	$6,300
ECONOMY	IND: tourism, sugar processing, ferronickel and gold mining, textiles. AGR: sugarcane, coffee, cotton, cacao, cattle. EXP: ferronickel, sugar, gold, silver, coffee.

Occupying the eastern two-thirds of Hispaniola, the Dominican Republic is the second largest country, after Cuba, in the West Indies. This mountainous land includes Pico Duarte—the highest point in the Caribbean. Colonized in 1493 by Spaniards, it offered the first chartered university, hospital, cathedral, and monastery in the Americas. Santo Domingo, founded in 1496, is the oldest European settlement in the Western Hemisphere. The nation became independent in 1844, but endured political instability and repressive governments. Today it is a democracy, economically dependent on agriculture and tourism.

Ecuador
REPUBLIC OF ECUADOR

CONTINENT	South America
AREA	283,560 sq km (109,483 sq mi)
POPULATION	12,558,000
CAPITAL	Quito 1,451,000
RELIGION	Roman Catholic
LANGUAGE	Spanish, Quechua
LITERACY	93%
LIFE EXPECTANCY	71
CURRENCY	US dollar
GDP PER CAPITA	$3,200
ECONOMY	IND: petroleum, food processing, textiles, metal work. AGR: bananas, coffee, cacao, rice, cattle, balsa wood; fish. EXP: petroleum, bananas, shrimp, coffee, cacao.

Ecuador's name comes from the Equator, which divides it unequally, putting most of the country in the Southern Hemisphere. It may be the smallest Andean country, but it has four distinct and contrasting regions. The Costa, or coastal plain, grows enough bananas to make the country the world's largest exporter of the fruit. The Sierra, or Andean uplands, offers productive farmland. Oil from the Oriente, jungles east of the Andes, enriches the economy. The Galápagos Islands, volcanic islands 960 kilometers west of Ecuador, bring tourism revenue with its unique reptiles, birds, and plants.

The country is divided ethnically as well as regionally. About 10 percent of the population is of European descent, about a quarter belong to indigenous cultures, and the rest are of mostly mixed ethnicity. Those of Spanish descent often are engaged in administration and land ownership in Quito and the surrounding Andean uplands; this is also where most of the indigenous people live—many are subsistence farmers. As a result, land-tenure reform is an explosive issue. The city of Guayaquil dominates the coastal plain, largely populated by mestizos. Guayaquil—the country's largest city, major port, and leading commercial center—is a rival to Quito. This is the wealthiest part of Ecuador, and complaints that tax revenues are squandered in the capital are common.

Regional and ethnic issues contribute to political instability for Ecuador's democracy.

Egypt
ARAB REPUBLIC OF EGYPT

CONTINENT	Africa
AREA	1,002,000 sq km
	(386,874 sq mi)
POPULATION	72,062,000
CAPITAL	Cairo 11,146,000
RELIGION	Sunni Muslim, Coptic Christian
LANGUAGE	Arabic, English, French
LITERACY	58%
LIFE EXPECTANCY	68
CURRENCY	Egyptian pound
GDP PER CAPITA	$4,000
ECONOMY	IND: textiles, food processing, tourism, chemicals. AGR: cotton, rice, corn, wheat, cattle. EXP: crude oil and petroleum products, cotton, textiles, metal products, chemicals.

A Middle Eastern country in northeast Africa, Egypt is at the center of the Arab world. Egypt controls the Suez Canal, the shortest sea link between the Indian Ocean and the Mediterranean Sea. The country is defined by desert and the Nile, the longest river on Earth. The Nile flows north out of central Africa, cascading over the cataracts (waterfalls) through Upper (southern) Egypt and Lower (northern) Egypt to the Mediterranean Sea—with a mountainous desert to the east, a rolling drier desert to the west, and the vast Sahara to the south.

Ancient civilizations arose along the narrow floodplain of the Nile, protected by the deserts that were natural barriers to invaders. Egyptians take pride in their rich heritage and in their descent from what is considered the first great civilization. Some 4,500 years ago Old Kingdom Egypt possessed enough peace and wealth to cultivate a culture devoted to the afterlife. Some 20,000 to 30,000 citizens mobilized to construct the Great Pyramid at Giza for the pharaoh Khufu; at 147 meters (481 feet) high it was the tallest monument in the world for thousands of years—until the 19th century.

Egypt is Africa's second most populous country after Nigeria, and it has the highest population in the Arab world. About 95 percent of Egyptians live along the Nile—on less than 5 percent of Egypt's territory. The Nile Valley is one of the world's most densely populated areas, containing an average of 1,540 persons per square kilometer (3,820 per square mile). Most Egyptians are Muslim Arabs, but there is a sizeable Coptic Christian population of seven million.

The main sources of foreign currency are remittances from workers abroad, Suez Canal fees, tourism, and oil. The Aswan High Dam, completed in 1971, provides hydroelectricity, as well as a controlled water supply for year-round irrigation and desert reclamation. The new Sheikh Zayed Canal diverts water from Lake Nasser to create new farmland. Harnessing the Nile, however, has reduced silt deposits downstream, increasing erosion and soil salinity.

Gamal Abdel Nasser helped end British control in 1953. In 1970 he was succeeded by Anwar Sadat, who liberalized the economy, distanced Egypt from the Soviets, and pursued peace with Israel. Muslim fundamentalists assassinated Sadat in 1981. President Hosni Mubarak reclaimed Egypt's strength in the Arab world with the return of the Arab League to Cairo. An ally of the West, the government is democratic but authoritarian, seeking to control Islamism and political dissent.

El Salvador
REPUBLIC OF EL SALVADOR

CONTINENT	North America
AREA	21,041 sq km
	(8,124 sq mi)
POPULATION	6,640,000
CAPITAL	San Salvador 1,424,000
RELIGION	Roman Catholic, Evangelical
LANGUAGE	Spanish, Nahua
LITERACY	80%
LIFE EXPECTANCY	70
CURRENCY	US dollar
GDP PER CAPITA	$4,600
ECONOMY	IND: food processing, beverages, petroleum, chemicals. AGR: coffee, sugar, corn, rice, shrimp, beef. EXP: offshore assembly exports, coffee, sugar, shrimp, textiles, chemicals, electricity.

The smallest and most densely populated country in Central America, El Salvador adjoins the Pacific in a narrow coastal plain, backed by a volcanic mountain chain, and a fertile plateau. About 90 percent of Salvadorans are mestizo; 9 percent claim Spanish descent. The rich volcanic soils brought coffee plantations— with a few, rich landowners and a subjugated peasant population. Economic inequality led to the 1980-1992 civil war; many Salvadorans, rich and poor, fled to the United States. El Salvador's democratic

government shows success in adding manufacturing jobs—but faces the challenges of poverty, crime, and natural disasters.

Equatorial Guinea
REPUBLIC OF EQUATORIAL GUINEA

CONTINENT	Africa
AREA	28,051 sq km
	(10,831 sq mi)
POPULATION	510,000
CAPITAL	Malabo 95,000
RELIGION	Roman Catholic, pagan practices
LANGUAGE	Spanish, French, pidgin English, Fang, Bubi, Ibo
LITERACY	86%
LIFE EXPECTANCY	54
CURRENCY	CFA franc
GDP PER CAPITA	$2,700
ECONOMY	IND: petroleum, fishing, sawmilling, natural gas. AGR: coffee, cocoa, rice, yams, livestock, timber. EXP: petroleum, methanol, timber, cacao.

A small country on the west coast of central Africa, Equatorial Guinea comprises the mainland territory of Río Muni (where most people live) and five volcanic islands. The largest island is Bioko on which the country's capital, Malabo, is located. After independence from Spain in 1968,

Equatorial Guinea fell under the rule of Francisco Macías Nguema, who plunged the nation into ruin. He was overthrown and executed in 1979 by his nephew. President Obiang Nguema continues the family dictatorship, and there is wide-spread civil unrest over flawed elections. New oil wealth masks stagnation in the rest of the economy and widespread poverty.

Eritrea
STATE OF ERITREA

CONTINENT	Africa
AREA	121,144 sq km
	(46,774 sq mi)
POPULATION	4,362,000
CAPITAL	Asmara 556,000
RELIGION	Muslim, Coptic Christian, Roman Catholic, Protestant
LANGUAGE	Afar, Arabic, Tigre, Kunama, Tigrinya
LITERACY	59%
LIFE EXPECTANCY	54
CURRENCY	nakfa
GDP PER CAPITA	$700
ECONOMY	IND: food processing, beverages, clothing and textiles. AGR: sorghum, lentils, vegetables, corn, livestock; fish. EXP: livestock, sorghum, textiles, food, small manufactures.

A former Italian colony in northeast Africa, Eritrea joined a UN-administered federation with Ethiopia in 1952, with a guarantee of democratic rights. Ten years later the state was annexed by Ethiopian Emperor Haile Selassie, touching off decades of bitter warfare. In 1993 Eritrea achieved independence from its dominating neighbor. After independence, Eritrea plunged into a 1995 war over Red Sea islands with Yemen and then a more devastating border war with Ethiopia in 1998, causing an estimated 100,000 casualties. A peace agreement in 2000 established a UN-patrolled buffer zone along the Eritrean-Ethiopian border.

Estonia
REPUBLIC OF ESTONIA

CONTINENT Europe
AREA 45,227 sq km
(17,462 sq mi)
POPULATION 1,353,000
CAPITAL Tallinn 391,000
RELIGION Evangelical Lutheran, Russian Orthodox, Eastern Orthodox
LANGUAGE Estonian, Russian, Ukrainian
LITERACY 100%
LIFE EXPECTANCY 71
CURRENCY Estonian kroon
GDP PER CAPITA $11,000
ECONOMY IND: engineering, electronics, wood and wood products, textiles. AGR: potatoes, vegetables, livestock and dairy products, fish. EXP: machinery and equipment, wood and paper, textiles, food products, furniture.

Estonia, smallest in population of the former Soviet republics, is a low-lying land on the Baltic Sea with 1,500 lakes and plenty of forests. Independence blossomed briefly between 1918 and 1940 after centuries of German, Swedish, and Russian rule. During World War II it was invaded first by Russian troops, then Germans, and then Russians again, forcing Estonia into the Soviet Union in 1944. Since independence in 1991, Estonia deals with the legacy of Russian workers brought in during the Soviet years— 26 percent of the population is Russian. As a stable democracy with a market economy, Estonia looks west for trade and security, joining both the European Union and NATO in 2004.

Ethiopia
FEDERAL DEMOCRATIC REPUBLIC OF ETHIOPIA

CONTINENT Africa
AREA 1,133,380 sq km
(437,600 sq mi)
POPULATION 70,677,000
CAPITAL Addis Ababa 2,723,000
RELIGION Muslim, Ethiopian Orthodox, animist
LANGUAGE Amharic, Tigrinya, Orominga, Guaraginga, Somali, Arabic
LITERACY 43%
LIFE EXPECTANCY 42
CURRENCY birr
GDP PER CAPITA $700
ECONOMY IND: food processing, beverages, textiles, chemicals. AGR: cereals, pulses, coffee, oilseed, cattle, hides. EXP: coffee, qat, gold, leather products, live animals, oilseeds.

Ethiopia is a landlocked country in the northeast African region known as the Horn of Africa. The country has a high central plateau, with some mountains reaching more than 4,000 meters (13,000 feet). The Great Rift Valley splits the plateau diagonally. The western highlands get summer rainfall; the lowlands and eastern highlands are hot and dry. Most people reside in the western highlands as does the capital, Addis Ababa—the highest capital city in Africa at 2,400 meters (8,000 feet). The population is almost evenly split between Christians, living in the highlands, and Muslims inhabiting the lowlands. The Oromo, Amhara, and Tigreans are the largest ethnic groups.

Hunger and war plague this nation, whose history spans 2,000 years. During the first millennium A.D. the Ethiopian Orthodox Church held the kingdom's Christianity secure against Islamic holy wars. Emperor Haile Selassie, dethroned in 1974, was the last of the monarchs, all of whom avoided European colonialism, except for Italian occupation from 1936 to 1941.

Most Ethiopians are farmers and herders. But deforestation, drought, and soil degradation have caused crop failures and famine during the past few decades; seven million people face starvation. A high birthrate and refugees from Somalia further strain economic resources. In May 1991, a 30-year civil war between the government and rebel forces aligned with Eritrean nationalists ended with the government's downfall. Under a transitional government, Eritrea became independent in 1993, cutting off Ethiopia's access to the Red Sea. The 1994 constitution divided the newly landlocked country into nine ethnically based regions. A 1998-2000 border war with Eritrea killed tens of thousands and ended with a UN-sponsored agreement to demarcate the ill-defined border.

Falkland Islands (U.K.)
BRITISH OVERSEAS TERRITORY

CONTINENT	South America
AREA	12,173 sq km
	(4,700 sq mi)
POPULATION	3,000
CAPITAL	Stanley 2,000
RELIGION	Anglican, Roman Catholic
LANGUAGE	English

The Falkland Islands are 700 islands in the South Atlantic, with most people living on East Falkland. In 1982 Argentina invaded the Falklands, but a British task force soon retook the islands. The economy is based on fishing and tourism.

Faroe Islands (Denmark)
OVERSEAS REGION OF DENMARK

CONTINENT	Europe
AREA	1,399 sq km (540 sq mi)
POPULATION	48,000
CAPITAL	Tórshavn 18,000
RELIGION	Evangelical Lutheran
LANGUAGE	Faroese, Danish

The Faroes, 18 North Atlantic islands, have been under Danish control since the 14th century and have been self-governing since 1948. Fishing is the main economic activity, and Danish subsidies remain an important source of income.

Fiji Islands
REPUBLIC OF THE FIJI ISLANDS

CONTINENT	Australia/Oceania
AREA	18,376 sq km
	(7,095 sq mi)
POPULATION	865,000
CAPITAL	Suva 210,000
RELIGION	Christian, Hindu, Muslim
LANGUAGE	English, Fijian, Hindustani
LITERACY	94%
LIFE EXPECTANCY	67
CURRENCY	Fijian dollar
GDP PER CAPITA	$5,600
ECONOMY	IND: tourism, sugar, clothing, copra. AGR: sugarcane, coconuts, cassava (tapioca), rice, cattle, fish. EXP: sugar, garments, gold, timber, fish.

The Fiji Islands comprise 333 islands in the South Pacific, with beaches, coral gardens, and rain forests. Most people live on the largest island, Viti Levu, where the capital, Suva, is located. After 96 years as a British colony, Fiji gained independence in 1970. During British rule, indentured servants from India came to work in the sugarcane fields—Indo-Fijians currently constitute 40 percent of the population.

Indo-Fijians are mostly Hindu, while the majority native Fijians are mostly Christian. Tensions between the two communities caused two coups in 1987 and one in 2000. Democracy returned in 2001 and so did a record number of tourists.

Finland
REPUBLIC OF FINLAND

CONTINENT	Europe
AREA	338,145 sq km
	(130,558 sq mi)
POPULATION	5,212,000
CAPITAL	Helsinki 1,075,000
RELIGION	Evangelical Lutheran
LANGUAGE	Finnish, Swedish
LITERACY	100%
LIFE EXPECTANCY	78
CURRENCY	euro
GDP PER CAPITA	$25,800
ECONOMY	IND: metal products, electronics, shipbuilding, pulp and paper, copper refining.

AGR: barley, wheat, sugar beets, potatoes, dairy cattle, fish. EXP: machinery and equipment, chemicals, metals, timber, paper, pulp.

Finland, in northern Europe, is low-lying in the south and center with mountains in the north. One quarter of its territory lies north of the Arctic Circle, and the country experiences long, harsh winters. Most of the population is concentrated in the triangle formed by the cities of Helsinki (the capital), Tampere, and Turku.

Coniferous forests and more than 180,000 lakes grace Finland, which maintains a fleet of icebreakers to keep ports open during the long winters. Despite a short growing season Finland is self-sufficient in meat, grains, and dairy products. For years, the wood and paper industry dominated Finland's exports; but now the metal and engineering aspects of industry have surpassed forest products.

After six centuries of union with Sweden, Finland came under Russian rule in 1809. The Finns declared independence in 1917 but lost territory to the Soviets in World War II. Close economic ties between the two nations existed until the U.S.S.R. was dissolved in 1991. Since then Finland has strengthened links with Western Europe by joining the EU. Helsinki remains a preeminent center for international diplomacy.

France
FRENCH REPUBLIC

CONTINENT	Europe
AREA	543,965 sq km
	(210,026 sq mi)
POPULATION	59,771,000
CAPITAL	Paris 9,854,000
RELIGION	Roman Catholic
LANGUAGE	French
LITERACY	99%
LIFE EXPECTANCY	79
CURRENCY	euro
GDP PER CAPITA	$26,000
ECONOMY	IND: machinery, chemicals, automobiles, metallurgy, aircraft, electronics, textile. AGR: wheat, cereals, sugar beets, potatoes, beef, fish. EXP: machinery and transportation equipment, aircraft, plastics, chemicals, pharmaceuticals, food.

Fertile plains cover two thirds of France, which is the largest country in Western Europe. With more than half the land under cultivation, France leads the European Union in food exports. The mountain ranges are mostly in the south, including the Alps, Pyrenees, and Massif Central. Forests cover 26 percent of France and are a source of environmental and scenic wealth. The north is humid and cool, while the south is dry and warm. Favorable conditions for grape growing in the south make French wines world-renowned—and France the world's largest pro-

ducer. The nation sets a fast pace in telecommunications, biotechnology, and aerospace industries. Sophia Antipolis, a booming high-tech complex on the Riviera, attracts scientists from throughout Europe. Coal and steel industries are concentrated in the northeast near major coalfields.

The government continues to play a large role in directing economic activity. The national road network is the world's densest, and the high-speed train (TGV) runs at speeds of 270 kilometers (167 miles) per hour or more. Both road and rail transport tourists, helping to make France the most visited country on Earth. Nuclear power, which supplies 80 percent of France's electricity, enjoys widespread support, in part because there is virtually no domestic oil. Government policies provide for a 35-hour workweek and five weeks of paid vacation annually.

Paris has long been France's cultural, political, and business epicenter. In the early 19th century Napoleon Bonaparte divided large, traditional provinces into small *departements*, which have since been regrouped into larger, regional units. Low turnout in the 2002 elections was interpreted as voter apathy due to the dominant influence of Paris. Amendments to the constitution, approved in 2003, give more political power to the country's 22 regions and 96 departments.

Heavy losses in both world wars bled France of labor, wealth, and prestige. After World War II, France's colonial subjects, from Algeria to Vietnam, struggled for independence. Immigration from France's former colonies, especially Algeria, contributes to some four million persons of Arab descent living in France today. An independent defense doctrine, launched by President Charles de Gaulle in 1966, has turned the nation into one of the world's largest arms suppliers. France maintains ties with its former colonies through aid, trade, and military pacts. The French have developed modern political ties with former colonies still under French administration. Overseas departments (officially part of France) with their own elected governments are: French Guiana, Guadeloupe, Martinique, and Réunion. Territories with varying degrees of autonomy are: French Polynesia, French Southern and Antarctic Territories, Mayotte, New Caledonia, St.-Pierre and Miquelon, and Wallis and Futuna.

OVERSEAS REGIONS

Clipperton Island

French Guiana

French Polynesia

French Southern and Antarctic Territories

Guadeloupe

Martinique

Mayotte

New Caledonia

Réunion

Saint-Pierre and Miquelon

Wallis and Futuna

French Guiana (France)
FRENCH OVERSEAS DEPARTMENT

CONTINENT	South America
AREA	86,504 sq km
	(33,400 sq mi)
POPULATION	182,000
CAPITAL	Cayenne 56,000
RELIGION	Roman Catholic
LANGUAGE	French

French Guiana, located in northern South America, is 95 percent covered by tropical rain forest. Most Guianans live on the coast. The Guiana Space Center near Kourou is the major launch facility for the European Space Agency.

French Polynesia (France)
FRENCH OVERSEAS TERRITORY

CONTINENT	Australia/Oceania
AREA	4,167 sq km
	(1,608 sq mi)
POPULATION	245,000
CAPITAL	Papeete 126,000
RELIGION	Protestant, Roman Catholic
LANGUAGE	French, Tahitian

French Polynesia consists of five South Pacific archipelagos: Austral, Gambier, Marquesas, Society, and Tuamotu. The 118 islands span an area larger than Europe. The economy is based on tourism.

Gabon
GABONESE REPUBLIC

CONTINENT	Africa
AREA	267,667 sq km
	(103,347 sq mi)
POPULATION	1,327,000
CAPITAL	Libreville 611,000
RELIGION	Christian, indigenous beliefs
LANGUAGE	French, Fang, Myene, Nzebi, Bapounou/ Eschira, Bandjabi
LITERACY	63%
LIFE EXPECTANCY	59
CURRENCY	CFA franc
GDP PER CAPITA	$6,500
ECONOMY	IND: petroleum extraction and refining, manganese and gold mining, chemicals. AGR: cacao, coffee, sugar, palm oil, cattle, okoume (a tropical softwood), fish. EXP: crude oil, timber, manganese, uranium.

Gabon sits on the Equator in western Africa. Oil, timber, and manganese earn this thinly settled republic one of the highest per capita incomes in Africa. However, the income is largely based on oil money going to a few—most live by subsistence farming. France gained control starting in 1839, and Libreville (Free Town), Gabon's capital, got its name when French forces freed slaves there in 1849. With independence in

1960, it functioned mostly as a one-party state until 1991, when a new constitution brought multiparty democracy. In 2002 the country created 13 new national parks—some 11 percent of Gabon's area— to protect its forests and wildlife from logging.

Gambia
REPUBLIC OF THE GAMBIA

CONTINENT	Africa
AREA	11,295 sq km
	(4,361 sq mi)
POPULATION	1,501,000
CAPITAL	Banjul 372,000
RELIGION	Muslim, Christian
LANGUAGE	English, Mandinka,
	Wolof, Fula
LITERACY	40%
LIFE EXPECTANCY	53
CURRENCY	dalasi
GDP PER CAPITA	$1,800
ECONOMY	IND: processing
	peanuts, fish, and hides,
	tourism, beverages,
	agricultural machinery
	assembly. AGR: rice,
	millet, sorghum,
	peanuts, cattle. EXP:
	peanut products,
	fish, cotton lint, palm
	kernels, reexports.

Gambia, in West Africa, is a small, narrow country with an outlet to the Atlantic Ocean. In 1588 Britain purchased from Portugal trading rights to this territory extending along both sides of the Gambia River. Independence came in 1965. After nearly 30 years of democratic rule, Gambia's president was ousted by a military coup in 1994. The constitution was rewritten and approved by national referendum in August 1996, and constitutional rule was reestablished in January 1997. Most people are subsistence farmers; the main export is groundnuts (peanuts).

Gaza Strip
See listing under Israel.

Georgia
REPUBLIC OF GEORGIA

CONTINENT	Asia
AREA	69,700 sq km
	(26,911 sq mi)
POPULATION	4,660,000
CAPITAL	T'bilisi 1,064,000
RELIGION	Georgian Orthodox,
	Muslim, Russian Orthodox
LANGUAGE	Georgian, Russian,
	Armenian, Azeri
LITERACY	99%
LIFE EXPECTANCY	77
CURRENCY	lari
GDP PER CAPITA	$3,200
ECONOMY	IND: steel, aircraft,
	machine tools, electrical
	appliances, mining.
	AGR: citrus, grapes,
	tea, hazelnuts; livestock.
	EXP: scrap metal,
	machinery, chemicals,
	fuel reexports, citrus
	fruits, tea.

Georgia, on the Black Sea, is geographically in Asia—the mountains forming its northern border serve as the Europe-Asia boundary. Rich in farmland and minerals, rugged Georgia is wedged between the Caucasus Mountains and the Lesser Caucasus. Over the centuries it has been an object of rivalry between Persia, Turkey, and Russia, and was annexed by Russia in the 19th century.

After independence came in 1991, ethnic strife caused Georgia to lose control of South Ossetia in 1992 and Abkhazia in 1993. Both the Ossete and Abkhaz regions enjoyed autonomy during Soviet rule, and both allied with Russia to separate from Georgia. The UN works to resolve these separatist disputes.

Germany
FEDERAL REPUBLIC OF GERMANY

CONTINENT	Europe
AREA	357,022 sq km
	(137,847 sq mi)
POPULATION	82,621,000
CAPITAL	Berlin 3,327,000
RELIGION	Protestant, Roman Catholic
LANGUAGE	German
LITERACY	99%
LIFE EXPECTANCY	78
CURRENCY	euro
GDP PER CAPITA	$26,200
ECONOMY	IND: iron, steel, coal,
	cement, chemicals,

machinery, vehicles, machine tools, electronics. AGR: potatoes, wheat, barley, sugar beets, cattle. EXP: machinery, vehicles, chemicals, metals and manufactures, foodstuffs, textiles.

Europe's strongest economic and industrial power, Germany is also the most populous European country outside Russia. Fertile northern plains stretch south from the North and Baltic Seas changing to central highlands and then rising to the rugged Schwarzwald (Black Forest) in the southwest and to the Alps in the far south. Germans are highly urbanized; about 86 percent live in cities and towns. With one of the world's lowest birthrates, Germany is a magnet for foreign workers—some 7.3 million immigrants live here. Some German industry is well known (Daimler Chrysler, Siemens, and Volkswagen); some, like Transrapid (the maglev railway) and Nordex wind turbines represent new environment-friendly technology.

"*Wir sind ein Volk*—We are one people," sang crowds on November 9, 1989, as East Germans breached the Berlin Wall. A year later, just after midnight on October 3, 1990, Germany was reborn. One people, divided since the end of World War II, had one country again. Yet German unity is relatively new. Disparate Germanic principalities did not come together until 1871, when the king of Prussia became kaiser (emperor) of Germany. Defeat in World War I cost Germany its empire and left the nation staggering under heavy reparations. Inflation and unemployment hounded the democratic, but shaky, Weimar Republic. By 1933 a demoralized population had turned to Adolf Hitler. Under Hitler, Germany rearmed and invaded neighboring countries, triggering the Second World War, which killed 55 million people and devastated much of Europe. When Germany surrendered in 1945, it lost eastern lands, like Prussia and Silesia, to the Soviet Union and Poland. The Allies divided the rest of the country, and its capital, Berlin, into four occupation zones. This temporary partition persisted as tensions rose between the U.S.S.R. and other Allied powers. In 1949 the American, French, and British zones formed the Federal Republic of Germany (West Germany), and the Soviet Union established the German Democratic Republic (East Germany). The Berlin Wall went up in 1961 to stop East Germans from fleeing west.

Rejoining two populations after 45 years of separation has been difficult. The economy in eastern Germany remains weak—the population is declining as young people go west for jobs. A bright spot in the east is Berlin as the construction boom continues in Germany's capital and largest city; tourists come to see the innovative architecture, including the Reichstag building with its new glass dome. A founding member of the European Union, Germany stands to gain from increased trade with the 2004 addition of the Czech Republic, Poland, and others to EU membership.

Ghana
REPUBLIC OF GHANA

CONTINENT	Africa
AREA	238,537 sq km (92,100 sq mi)
POPULATION	20,468,000
CAPITAL	Accra 1,847,000
RELIGION	Christian, indigenous beliefs, Muslim
LANGUAGE	English, Akan, Moshi-Dagomba, Ewe, Ga
LITERACY	75%
LIFE EXPECTANCY	57
CURRENCY	cedi
GDP PER CAPITA	$2,000
ECONOMY	IND: mining, lumbering, light manufacturing, aluminum smelting. AGR: cacao, rice, coffee, cassava (tapioca), timber. EXP: gold, cacao, timber, tuna, bauxite.

Ghana, in West Africa, is a land of plains and low plateaus covered by rain forests in the west and Lake Volta in the east—one of the world's largest man-made lakes. The precious metal that once gave its name to the Gold Coast lured Portuguese, Danes, Dutch, Germans, and British. After Ghana's independence from Britain in 1957,

President Kwame Nkrumah emerged as a leading spokesman for Pan-Africanism. A series of military coups brought Jerry Rawlings to power in 1981. Multiparty democracy started with the new 1992 constitution. In December 2000, for the first time in its history, Ghana witnessed the election of an opposition party.

Gibraltar (U.K.)
BRITISH OVERSEAS TERRITORY

CONTINENT Europe
AREA 7 sq km (3 sq mi)
POPULATION 27,000
CAPITAL Gibraltar 27,000
RELIGION Roman Catholic, Church of England
LANGUAGE English, Spanish, Italian, Portuguese

The peninsula that is Gibraltar is in southwest Europe on the southern coast of Spain. British control dates to 1704, though Spain still claims the territory. Tourism is growing, with some seven million visitors a year.

Greece
HELLENIC REPUBLIC

CONTINENT Europe
AREA 131,957 sq km (50,949 sq mi)
POPULATION 10,988,000
CAPITAL Athens 3,238,000
RELIGION Greek Orthodox
LANGUAGE Greek
LITERACY 98%
LIFE EXPECTANCY 78
CURRENCY euro
GDP PER CAPITA $19,100

ECONOMY IND: tourism, food and tobacco processing, textiles, chemicals. AGR: wheat, corn, barley, sugar beets, beef. EXP: food and beverages, manufactured goods, petroleum products, chemicals.

Greece, on the Balkan Peninsula in southeastern Europe, is mostly dry and mountainous, with a large mainland and more than 1,400 islands. The nation where democracy was conceived in the fifth century B.C. has periodically suffered the loss of freedom and welcomed its rebirth. After almost 400 years under Turkish rule, Greece won independence in 1830. Scarred by Nazi occupation during World War II and an ensuing civil war, the nation endured seven years of military dictatorship from 1967 to 1974. The junta fell after a failed Athens-backed coup in Cyprus—which brought the Turkish invasion and occupation of northern Cyprus in 1974. An elected government and new constitution followed. Even though Greece and Turkey are both members of NATO, relations have been tense over Cyprus and Aegean issues. A breakthrough occurred in 1999 when major earthquakes hit Greece and Turkey—both countries and peoples responded generously to the other's need.

Membership in the European Union has helped stimulate industry, agriculture, and shipping. Greece's maritime fleet is the largest in Europe. Recent economic growth, lower inflation, and lower unemployment have helped Greece somewhat overcome its position as one of the poorest of the EU countries in terms of per capita income.

The unique ecosystems of the Prespa Lakes region and the dense woodlands of the Rhodope Mountains have been set aside as international preserves. Athens stepped into the global spotlight as host of the 2004 Summer Olympic Games. Greece's ancient treasures, striking landscapes, and pleasing climate are irresistible, enticing some 12 million visitors a year.

Greenland (Denmark)
OVERSEAS REGION OF DENMARK

CONTINENT North America
AREA 2,166,086 sq km (836,086 sq mi)
POPULATION 58,000
CAPITAL Nuuk (Godthåb) 14,000
RELIGION Evangelical Lutheran
LANGUAGE Greenlandic (East Inuit), Danish, English

An ice sheet that is up to 3 kilometers (1.8 miles) thick covers 85 percent of Greenland—the world's largest island. About 80 percent of the population is Inuit, and the economy is based on fish and fish products.

Grenada
GRENADA

CONTINENT	North America
AREA	344 sq km (133 sq mi)
POPULATION	105,000
CAPITAL	St. George's 33,000
RELIGION	Roman Catholic, Anglican, other Protestant
LANGUAGE	English, French patois
LITERACY	98%
LIFE EXPECTANCY	71
CURRENCY	East Caribbean dollar
GDP PER CAPITA	$5,000
ECONOMY	IND: food and beverages, textiles, light assembly operations, tourism. AGR: bananas, cacao, nutmeg, mace. EXP: bananas, cacao, nutmeg, fruits and vegetables.

Grenada, located in the southeastern Caribbean, consists of the islands of Grenada, Carriacou, and Petite Martinique. Most Grenadians are of African descent. Grenada, the largest and most populous island, is known as The Spice of the Caribbean. Nutmeg replaced sugar as the main crop after the British took the island from France in 1783. Small farms replaced sugar plantations, slavery was abolished, and today the sweet smells of nutmeg and other spices waft on balmy breezes. Independence came in 1974; a military coup in 1983 brought a U.S.-Caribbean force that restored democracy.

Guadeloupe (France)
FRENCH OVERSEAS DEPARTMENT

CONTINENT	North America
AREA	1,705 sq km (658 sq mi)
POPULATION	441,000
CAPITAL	Basse-Terre 12,400
RELIGION	Roman Catholic, Hindu, pagan African
LANGUAGE	French, Creole patois

Located in the eastern Caribbean, Guadeloupe includes the main islands of Basse-Terre and Grande-Terre as well as St. Barthélemy (St. Barts) and the northern half of St. Martin. All are major tourist destinations.

Guam (U.S.)
TERRITORY OF THE UNITED STATES

CONTINENT	Australia/Oceania
AREA	561 sq km (217 sq mi)
POPULATION	164,000
CAPITAL	Hagåtña (Agana) 1,100
RELIGION	Roman Catholic
LANGUAGE	English, Chamorro, Japanese

Lying at the southern end of the Northern Mariana Islands, Guam was a strategic Pacific prize in World War II. Japan captured it in 1941, and the U.S. regained it in 1944. U.S. military bases and tourism are chief employers.

Guatemala
REPUBLIC OF GUATEMALA

CONTINENT	North America
AREA	108,889 sq km (42,042 sq mi)
POPULATION	12,360,000
CAPITAL	Guatemala City 951,000
RELIGION	Roman Catholic, Protestant, indigenous Mayan beliefs
LANGUAGE	Spanish, Amerindian languages
LITERACY	71%
LIFE EXPECTANCY	66
CURRENCY	quetzal, US dollar, others allowed
GDP PER CAPITA	$3,900
ECONOMY	IND: sugar, textiles and clothing, furniture, chemicals. AGR: sugarcane, corn, bananas, coffee, cattle. EXP: coffee, sugar, bananas, fruits and vegetables, cardamom, meat.

Guatemala, meaning "land of trees," is a heavily forested and mountainous nation—and the most populous in Central America. The Pacific coast lowlands in the south rise to the volcanic Sierra Madre and other highlands, then the land descends to the forested northern lowlands, including the narrow Caribbean coast. The highlands, where most Guatemalans live, are temperate in climate compared to the tropical lowlands.

A thousand years ago the remarkable Maya civilization flourished, and its ruins dot the landscape. Today more than half of Guatemalans are descendants of the indigenous Maya peoples; most live in the western highlands and are poor subsistence farmers. By contrast the rest of the population are known as Ladinos (mostly mixed Maya-Spanish ancestry). Ladinos use Spanish and wear Western clothing, while Maya speak some 24 indigenous languages and retain traditional dress and customs. The more urbanized Ladino population dominates commerce, government, and the military. Guatemalan society grew increasingly polarized between a Ladino upper class and Maya lower class when guerrilla groups first formed in 1960 to fight for the poor majority. Warfare between guerrillas and government forces cost 200,000 lives and displaced half a million people. In September 1996 the government and the guerrillas agreed on terms to end the 36-year-long civil war.

The democratic government faces problems of crime, illiteracy, and poverty, but it is making progress in moving the economy away from coffee and agriculture toward manufacturing and tourism. Tikal, in northern Guatemala, may be the premier tourism site, with some 3,000 Maya buildings dating from 600 B.C. to A.D. 900. Tikal's Temple IV is the tallest pre-Columbian structure in the Americas at 65 meters (212 feet).

Guinea
REPUBLIC OF GUINEA

CONTINENT	Africa
AREA	245,857 sq km (94,926 sq mi)
POPULATION	9,030,000
CAPITAL	Conakry 1,366,000
RELIGION	Muslim, Christian, indigenous beliefs
LANGUAGE	French, local languages
LITERACY	36%
LIFE EXPECTANCY	49
CURRENCY	Guinean franc
GDP PER CAPITA	$2,100
ECONOMY	IND: bauxite, gold, diamonds, alumina refining, light manufacturing. AGR: rice, coffee, pineapples, palm kernels, cattle; timber. EXP: bauxite, alumina, gold, diamonds, coffee.

Facing the Atlantic Ocean, Guinea is a West African country with a narrow coastal plain and interior highlands that are forested in the southeast. After independence from France in 1958, repressive socialist rule plunged the country into economic ruin. A 1984 coup brought in a military government until 1990, after which Guinea began the transition to a multiparty democratic system. Liberalized commercial policies, plus diamonds and gold, diversify an economy overly dependent on the bauxite industry.

Guinea-Bissau
REPUBLIC OF GUINEA-BISSAU

CONTINENT	Africa
AREA	36,125 sq km (13,948 sq mi)
POPULATION	1,288,000
CAPITAL	Bissau 336,000
RELIGION	Indigenous beliefs, Muslim, Christian
LANGUAGE	Portuguese, Crioulo, African languages
LITERACY	42%
LIFE EXPECTANCY	45
CURRENCY	CFA franc
GDP PER CAPITA	$700
ECONOMY	IND: agricultural products processing, beer, soft drinks. AGR: rice, corn, beans, cassava (tapioca), timber, fish. EXP: cashew nuts, shrimp, peanuts, palm kernels, sawn lumber.

Guinea-Bissau, on the Atlantic coast of West Africa, has a swampy coast, with forests changing to grasslands in the east. Guerrilla warfare liberated a mix of ethnic groups from Portuguese rule in 1974. In 1994 the country's first multiparty elections were held. An army uprising four years later led to a bloody 1998–99 civil war, which caused severe damage to the nation's infrastructure. Political instability continued with a military coup in 2003. Guinea-Bissau is among the world's least developed countries, with most people engaged in subsistence agriculture and fishing—cashew nuts are the main export crop.

Guyana
CO-OPERATIVE REPUBLIC OF GUYANA

CONTINENT	South America
AREA	214,969 sq km
	(83,000 sq mi)
POPULATION	765,000
CAPITAL	Georgetown 231,000
RELIGION	Christian, Hindu, Muslim
LANGUAGE	English, Amerindian
	dialects, Creole, Hindi,
	Urdu
LITERACY	99%
LIFE EXPECTANCY	63
CURRENCY	Guyanese dollar
GDP PER CAPITA	$3,800
ECONOMY	IND: bauxite, sugar,
	rice milling, timber, tex-
	tiles. AGR: sugar, rice,
	wheat, vegetable oils,
	beef, shrimp. EXP: sugar,
	gold, bauxite/alumina,
	rice, shrimp.

Tropical rain forest shrouds more than 80 percent of this English-speaking former British colony on the north coast of South America. During 150 years of rule, Britain imported Africans and East Indians as laborers, and Guyana forged close trade ties with the Caribbean. Since independence in 1966, Guyanese have supported a parliamentary system. Sugar, bauxite, rice, and gold lead among exports. Guyana's high debt burden to foreign creditors and territorial disputes with Suriname and Venezuela continue to hamper the government.

Haiti
REPUBLIC OF HAITI

CONTINENT	North America
AREA	27,750 sq km
	(10,714 sq mi)
POPULATION	7,528,000
CAPITAL	Port-au-Prince
	1,961,000
RELIGION	Roman Catholic,
	Protestant, Voodoo
LANGUAGE	French, Creole
LITERACY	53%
LIFE EXPECTANCY	51
CURRENCY	gourde
GDP PER CAPITA	$1,400
ECONOMY	IND: sugar refining,
	flour milling, textiles,
	cement. AGR: coffee,
	mangoes, sugarcane,
	rice, wood. EXP:
	manufactures, coffee,
	oils, cocoa.

Haiti, the first Caribbean state to achieve independence, occupies the western third of the island of Hispaniola. Mountainous with a tropical climate, it is the poorest country in the Americas due to decades of violence and instability. There is a huge income gap between the Creole-speaking black majority and the French-speaking mulattos (mixed African and European descent). Mulattos, only 5 percent of the population, control most of the wealth. Haiti became the first black republic in 1804 after a successful slave revolt against the French. As Haiti celebrated 200 years of independence, a rebellion toppled the government in February 2004.

Honduras
REPUBLIC OF HONDURAS

CONTINENT	North America
AREA	112,492 sq km
	(43,433 sq mi)
POPULATION	6,876,000
CAPITAL	Tegucigalpa
	1,007,000
RELIGION	Roman Catholic
LANGUAGE	Spanish, Amerindian
	dialects
LITERACY	76%
LIFE EXPECTANCY	71
CURRENCY	lempira
GDP PER CAPITA	$2,500
ECONOMY	IND: sugar, coffee,
	textiles, clothing.
	AGR: bananas, coffee,
	citrus; beef, timber,
	shrimp. EXP: coffee,
	bananas, shrimp,
	lobster, meat.

Honduras, in Central America, is mountainous and forested—although wide-spread slash-and-burn subsistence farming is destroying many forests. The largely mestizo population speaks Spanish, with English common on the northern coast and Bay Islands. Maya ruins at Copán, which represent the wealth of the past in what today is one of the region's poorest nations, help diversify the economy with tourist revenue. Although agricultural products are plentiful, mostly bananas and coffee, they have failed to enliven the economy of this tenuous democracy. The 2003 U.S.-Central American Free Trade Agreement brings economic hope.

Hong Kong

See listing under China.

Hungary
REPUBLIC OF HUNGARY

CONTINENT	Europe
AREA	93,030 sq km
	(35,919 sq mi)
POPULATION	10,141,000
CAPITAL	Budapest
	1,708,000
RELIGION	Roman Catholic,
	Calvinist, Lutheran
LANGUAGE	Hungarian
LITERACY	99%
LIFE EXPECTANCY	72
CURRENCY	forint
GDP PER CAPITA	$13,300
ECONOMY	IND: mining, metallurgy, construction materials, processed foods, textiles, chemicals. AGR: wheat, corn, sunflower seed,

potatoes, sugar beets; pigs. EXP: machinery and equipment, other manufactures, food products, raw materials.

The Danube River flows north to south through the middle of Hungary, splitting this landlocked central European country almost in half. Hungarians (Magyars) migrated here from Asia more than a thousand years ago and are distinct from the Germanic and Slavic peoples that surround them. Hungary's support for Hungarian minorities in other countries is sometimes criticized as interference by neighboring governments.

Fertile plains lie east of the Danube, with hills to the west and north. Soviet tanks crushed an uprising for democracy in 1956, but Hungary rebounded to become Eastern Europe's first purveyor of "goulash communism," blending personal freedom, prosperity, and a pinch of free enterprise. While other countries in the region suffered shortages, boutiques displaying designer fashions and cafés selling caviar lined Budapest streets.

By the late 1980s reform-minded Hungary had lost faith in communism, shaken by sagging productivity and the highest per capita foreign debt in Eastern Europe. In 1989 the government abolished censorship, dismantled barriers along the Austrian border, and called for privatization of industry, religious freedom, and free elections.

Foreign investment and private companies are flourishing. The economy is strong, with low inflation and falling interest rates. European Union member countries account for more than 60 percent of Hungarian exports. Now a member of NATO, the dream of 1956 has become the reality. Hungary joined the European Union in 2004.

Iceland
REPUBLIC OF ICELAND

CONTINENT	Europe
AREA	103,000 sq km
	(39,769 sq mi)
POPULATION	289,000
CAPITAL	Reykjavík 184,000
RELIGION	Evangelical Lutheran
LANGUAGE	Icelandic, English, Nordic languages, German
LITERACY	100%
LIFE EXPECTANCY	80
CURRENCY	Icelandic krona
GDP PER CAPITA	$30,200
ECONOMY	IND: fish processing, aluminum smelting, ferrosilicon production, geothermal power. AGR: potatoes, green vegetables, chicken, pork, fish. EXP: fish and fish products, animal products, aluminum, diatomite, ferrosilicon.

A volcanic island, Iceland is Europe's westernmost country and home to the world's northernmost capital city, Reykjavík. Although glaciers cover more than a tenth of the island, the Gulf Stream and warm southwesterly winds moderate the climate—most residents occupy the country's southwest. Established in 930, the national assembly, or Althingi, is the world's oldest continuous parliament. Under the Danish crown for more than 500 years, the country became a republic in 1944. Almost all of Iceland's electricity and heating come from hydroelectric power and geothermal water reserves. Explosive geysers, relaxing geothermal spas, glacier-fed waterfalls like Gullfoss (Golden Falls), and whale watching attract more than 270,000 visitors a year.

India
REPUBLIC OF INDIA

CONTINENT Asia

AREA 3,287,270 sq km (1,269,221 sq mi)

POPULATION 1,068,572,000

CAPITAL New Delhi 295,000

RELIGION Hindu, Muslim, Christian, Sikh, Buddhist, Jain, Parsi

LANGUAGE Hindi, English, 14 other official languages

LITERACY 60%

LIFE EXPECTANCY 63

CURRENCY Indian rupee

GDP PER CAPITA $2,600

ECONOMY IND: textiles, chemicals, food processing, steel, transportation equipment, cement, mining.

AGR: rice, wheat, oilseed, cotton, cattle, fish. EXP: textile goods, gems and jewelry, engineering goods, chemicals, leather manufactures.

The South Asian country of India includes a peninsula extending into the Indian Ocean, and it is a land of great contrasts in geography. The barren, snow-capped Himalaya, the world's tallest mountain system, rises along its northern border. South of the Himalaya, the low, fertile Ganges Plain is India's most populous region. The Great Indian Desert lies in the west, but eastern India receives some of the highest rainfall in the world during the monsoon season (June to October). India is second only to China in country population—but India is growing faster (some 16 million a year) and may surpass China by 2030. Although 81 percent of the people are Hindu, India also has 126 million Muslims—one of the world's largest Muslim populations.

Hindu culture evolved out of the mingling of indigenous Dravidian peoples and Aryan-speaking nomads who arrived from Central Asia in 1500 B.C. Islam spread across the subcontinent starting in the eighth century A.D. From the 17th century to the mid-20th century India was the pride of the British Empire. Guided by Mahatma Gandhi, Indians won nationhood in 1947. From British rule they inherited deep poverty but also parliamentary government, the English language, and a far-flung rail system, which helped knit the multiethnic country into a secular democracy—often called the world's largest democracy.

Violence born of separatist yearnings or religious differences gnaws at national unity—and takes lives: Sikh revolts since the 1980s have left thousands dead; in 1984 Prime Minister Indira Gandhi was the victim of assassins' bullets; seven years later her son Rajiv was killed by a bomb. Other problems include a ponderous bureaucracy, illiteracy, a high birthrate, and border disputes with Pakistan and China. A 1948 ceasefire line, known as the line of Control, divides Kashmir between India and Pakistan. India claims that Kashmir legally is part of it, but Pakistan says that the mostly Muslim population should vote on which country to join. Diplomatic talks with China work to resolve border disputes in India's northeast state of Arunachal Pradesh.

The Hindu caste system reflects Indian economic and religious hierarchies. One out of six Indians suffer in the lowest caste as Untouchables. Indians in lower castes can escape serf-like conditions that exist in rigidly-structured rural areas by going to India's chaotic cities. Mumbai (Bombay) is the largest city and is home to "Bollywood"—India's film industry. Bangalore is India's Silicon Valley. India has a burgeoning middle class and has made great strides in engineering and information technology. The country's space program plans to reach the moon by 2007.

Indonesia
REPUBLIC OF INDONESIA

CONTINENT Asia

AREA 1,922,570 sq km
(742,308 sq mi)

POPULATION 220,483,000

CAPITAL Jakarta 13,194,000

RELIGION Muslim, Protestant,
Roman Catholic, Hindu,
Buddhist

LANGUAGE Bahasa Indonesia,
English, Dutch, Javanese,
and other local dialects

LITERACY 89%

LIFE EXPECTANCY 68

CURRENCY Indonesian rupiah

GDP PER CAPITA $3,100

ECONOMY IND: petroleum and natural gas, textiles,
apparel, and footwear;
mining, cement, chemi-
cal fertilizers. AGR: rice,
cassava (tapioca),
peanuts, rubber; poultry.
EXP: oil and gas, electrical appliances, ply-
wood, textiles, rubber.

Indonesia is a vast equatorial archipelago of 17,000 islands extending
5,150 kilometers (3,200 miles) east
to west, between the Indian and Pacific
Oceans in Southeast Asia. The largest
islands are Sumatra, Java, Kalimantan
(Indonesian Borneo), Sulawesi, and the
Indonesian part of New Guinea
(known as Papua or Irian Jaya). Islands
are mountainous with dense rain
forests, and some have active volcanoes. Most of the smaller islands
belong to larger groups, like the
Moluccas (Spice Islands).

Indonesia, the world's fourth
most populous nation, is 87 percent
Muslim—and the largest Islamic country, though it is a secular state. Indonesians are separated by seas and
clustered on islands. The largest cluster is on Java, with some 130 million
inhabitants on an island the size of New
York State (60 percent of the country's
population). Sumatra, much larger
than Java, has only about a third of its
people. Ethnically the country is highly
diverse, with over 580 languages and
dialects—but only 13 have more than
one million speakers.

After independence from the
Netherlands in 1949, the new republic confronted a high birthrate, low
productivity, and illiteracy—areas in
which progress has since been made.
The government used a "transmigra-
tion" policy to address uneven population distribution by relocating
millions of people from Java to other
islands. Unity and stability are improving, although outer areas of the archipelago resent domination by Java. The
Asian financial crisis hit Indonesia
extremely hard. Public unrest, including violent rioting, forced President
Suharto—in office since 1967—to
resign in May 1998. One year later
Indonesia conducted its first democratic elections since 1955.

The democratic government faces
many problems after years of military
dictatorship. Secessionists in the
regions of Papua and Aceh (northwest
tip of Sumatra) have been encouraged
by East Timor's (now Timor-Leste)
1999 success in breaking away after
25 years of Indonesian military occupation. Militant Islamic groups have
become active in recent years, and religious conflict between Muslims and
Christians recently flared in Sulawesi
and the Moluccas. The island of Bali,
a center of Hindu culture, suffered a
terrorist bomb blast in 2002 that killed
over 200 people—mostly tourists.

Export earnings from oil and
natural gas help the economy, and
Indonesia is a member of the Organization of Petroleum Exporting Countries (OPEC). Tourists come to see
the rich diversity of plants and
wildlife—some, like the giant Komodo
dragon and the Javan rhinoceros,
exist nowhere else.

Iran
ISLAMIC REPUBLIC OF IRAN

CONTINENT Asia

AREA 1,648,000 sq km
(636,296 sq mi)

POPULATION 66,582,000

CAPITAL Tehran 7,352,000

RELIGION Shiite and Sunni Muslim

LANGUAGE Persian, Turkic, Kurdish,
various local dialects

LITERACY 79%

LIFE EXPECTANCY 69

CURRENCY Iranian rial

GDP PER CAPITA $6,800

ECONOMY IND: petroleum, petrochemicals, textiles,
cement and other
construction materials.
AGR: wheat, rice,
other grains, sugar
beets, dairy products,
caviar. EXP: petroleum,

carpets, fruits and nuts, iron and steel, chemicals.

Iran is a southwest Asian country of mountains and deserts. Eastern Iran is dominated by a high plateau, with large salt flats and vast sand deserts. The plateau is surrounded by even higher mountains, including the Zagros to the west and the Elburz to the north. Farming and settlement are largely concentrated in the narrow plains or valleys in the west or north, where there is more rainfall. Iran's huge oil reserves lie in the southwest, along the Persian Gulf.

Shah Mohammad Reza Pahlavi, who came to power in 1941, perpetuated a pattern of autocratic rule extending back to Cyrus the Great, whose Persian Empire reached its zenith in the sixth century B.C. Aided by the U.S., the shah initiated social and economic reforms financed by petroleum exports. His opponents reviled Westernization for tainting Iran's Islamic purity and cultural identity. Revolution broke out in 1978. The shah fled, and Ayatollah Khomeini imposed a fundamentalist theocracy, under which an estimated 70,000 critics were executed. The official state religion is the Shiite branch of Islam, practiced by most Iranians.

War with Iraq from 1980 to 1988 cost a million Iranian lives and devastated the economy. Iran confronts political and social transformation as

some promote liberal ideas, while others hold fast to established Islamic traditions. An estimated seven million Iranians have access to the Internet, which has been used to circumvent government censorship. In December 2003 a massive earthquake struck the southeastern city of Bam, killing more than 30,000 people.

Iraq
REPUBLIC OF IRAQ

CONTINENT	Asia
AREA	437,072 sq km
	(168,754 sq mi)
POPULATION	24,205,000
CAPITAL	Baghdad 5,620,000
RELIGION	Shiite and Sunni Muslim
LANGUAGE	Arabic, Kurdish, Assyrian, Armenian
LITERACY	40%
LIFE EXPECTANCY	58
CURRENCY	Iraqi dinar
GDP PER CAPITA	$2,400
ECONOMY	IND: petroleum, chemicals, textiles, construction materials. AGR: wheat, barley, rice, vegetables, cattle. EXP: crude oil.

Iraq occupies the ancient region of Mesopotamia, "land amidst the rivers," a fertile lowland created by the Tigris and Euphrates Rivers. Today these rivers sustain large areas of irrigated farmland and one of the high-

est populations in the Middle East. Beneath the land, Iraq is second only to Saudi Arabia in rich oil reserves. Temperatures range from below freezing in winter to higher than 49°C (120°F) in the summer.

Iraq's diverse population includes some 20 million Arabs consisting of Shiite Muslims (60%), Sunni Muslims (35%), and Christians (3%). Most Shiites live in the southeast, and most Sunnis live in central Iraq. About four million Kurds, a non-Arab Muslim people, live in the mountainous northeast.

Iraq gained independence in 1932 as a monarchy, but a 1958 coup brought a series of military dictatorships. In 1979 Saddam Hussein took control of Iraq; he invaded Iran in 1980 and Kuwait in 1990. Iraq lost both resource-draining wars. Iraqis suffered from high war casualties and from Hussein's persecution of Shiites, Kurds, and others who opposed him.

U.S.-led coalition forces drove the Iraqi occupation army from Kuwait in 1991 and patrolled no-fly zones over Iraq from 1992–2003—protecting Kurds and Shiites from Iraqi warplanes. The Kurdish community, defying the Iraqi army, established its own self-governing region in the early 1990s. Iraq ended cooperation with UN weapons inspectors in 1998, creating concern that Iraq was again developing nuclear or chemical weapons.

Another U.S.-led coalition invaded Iraq on March 20, 2003, reaching Baghdad by April 9, and capturing Hussein on December 14. However, militants continue attacking coalition forces and terrorizing Iraqis working with the new government. Coalition forces administer Iraq until June 30, 2004—the scheduled date for transferring authority to an Iraqi transitional government. Plans call for UN-assisted elections for a national assembly to be held by the end of January 2005.

Ireland
REPUBLIC OF IRELAND

CONTINENT	Europe
AREA	70,273 sq km
	(27,133 sq mi)
POPULATION	3,990,000
CAPITAL	Dublin 1,015,000
RELIGION	Roman Catholic
LANGUAGE	English, Irish
LITERACY	98%
LIFE EXPECTANCY	77
CURRENCY	euro
GDP PER CAPITA	$29,300
ECONOMY	IND: food products, brewing, textiles, clothing, chemicals, pharmaceuticals. AGR: turnips, barley, potatoes, sugar beets, beef. EXP: machinery and equipment, computers, chemicals, pharmaceuticals, live animals.

An island in the North Atlantic, Ireland features coastal mountains in the west and interior agricultural lowlands, with numerous hills, lakes, and

bogs. The Republic of Ireland occupies about 83 percent of the island of Ireland—Northern Ireland, in the northeast, is part of the United Kingdom. Irish, or Irish Gaelic (a Celtic language), is the country's first official language and is taught in schools, but few native speakers remain. Éire (AIR-uh) is the Irish name for the Republic of Ireland. English is the second official language and is more common.

The object of waves of invasion from Europe, the Emerald Isle has been inhabited for 7,000 years. Celtic invaders from Europe came in the sixth century B.C. Tradition holds that, in A.D. 432, St. Patrick began converting the Irish to Christianity. England began seizing land in the 1100s, but many areas remained in Irish hands until the 16th century. In the 19th century Ireland's growing population was becoming ever more dependent on the potato for sustenance. The potato crop could not withstand the large amount of precipitation that fell year after year in the 1840s, causing blight and rotting the harvest. Death and emigration reduced the population from eight to six million by 1856, and it would fall further—the island total today is just 5.7 million residents (four million in the Republic of Ireland).

Eventually, in 1922, the Roman Catholic counties won independence, while mostly Protestant Northern Ireland remained under British control. Since independence, forces for and against uniting the island have claimed thousands of lives. In 1998 a peace agreement was signed by the Northern Ireland parties, Britain, and Ireland—with Ireland giving up its

territorial claim to Northern Ireland. The country's robust growth promotes trade, foreign investment, and industries such as electronics. In the south, the Waterford area enjoys a slightly sunnier climate and is a growing area for business and retirement.

Isle of Man (U.K.)
BRITISH CROWN DEPENDENCY

CONTINENT	Europe
AREA	572 sq km (221 sq mi)
POPULATION	72,000
CAPITAL	Douglas 26,000
RELIGION	Anglican, Roman Catholic, Methodist, Baptist
LANGUAGE	English, Manx Gaelic

Israel
STATE OF ISRAEL

CONTINENT	Asia
AREA	22,145 sq km
	(8,550 sq mi)
POPULATION	6,707,000
CAPITAL	Jerusalem 692,300
RELIGION	Jewish, Muslim, Christian
LANGUAGE	Hebrew, Arabic, English
LITERACY	95%
LIFE EXPECTANCY	79
CURRENCY	new Israeli shekel
GDP PER CAPITA	$19,500
ECONOMY	IND: high-technology projects, wood and paper products, potash and phosphates, food. AGR: citrus, vegetables, cotton, beef. EXP: machinery and

equipment, software,
cut diamonds, agri-
cultural products,
chemicals.

Israel lies on the Mediterranean coast of southwest Asia, with most people living along the coastal plain. The eastern interior is dry and includes the Dead Sea—the lowest point on the Earth's surface. North are the rugged hills of Galilee, and south lies the Negev, a desert plateau. Israel's population is about 81 percent Jewish; most of the rest is Arab. The Israeli-occupied Palestinian territories have some 3.5 million inhabitants—about 11 percent Jewish, 89 percent Palestinian.

Born in battle after the British left Palestine in 1948, Israel has fought six wars with its Arab neighbors. To secure peace, Israel in 1982 ended its 15-year occupation of the Sinai Peninsula, returning it to Egypt. The intifada, a Palestinian rebellion that began in 1987, took hundreds of lives before peace negotiations resulted in a 1993 accord that granted Palestinian self-rule in the Gaza Strip and the West Bank city of Jericho. The Israeli military withdrew from all West Bank cities by 1997—and also left southern Lebanon in 2000. However, peace talks stalled; a second intifada started in September 2000, and most of the West Bank was reoccupied by 2002. Israel annexed the Golan Heights in 1981 after capturing it in 1967—Syria still claims this territory.

A "final status" agreement, leading to a Palestinian state, has yet to be reached between Israel and the Palestinians. Stumbling blocks include:

Jerusalem – Palestinians want their capital in Jerusalem. Israel claims that Jerusalem is its capital and that its status is not negotiable.

Gaza Strip – In 2004 Israel offered to withdraw its forces and Jewish settlements. Palestinians suspect that Israel will keep land in the West Bank after the Gaza pullout.

West Bank – Responding to suicide bombers, Israel started building a West Bank barrier in 2002. Palestinians complain that Israel is using the wall to grab land inside the Green Line—the boundary between Israel and the West Bank based on the 1949 armistice line.

Palestinian Areas
AREAS OF SPECIAL STATUS

Gaza Strip

CONTINENT Asia
AREA 365 sq km (141 sq mi)
POPULATION 1,299,000
RELIGION Muslim (mostly Sunni), Jewish
LANGUAGE Arabic, Hebrew, English

West Bank

CONTINENT Asia
AREA 5,655 sq km (2,183 sq mi)
POPULATION 2,260,000
RELIGION Muslim (mostly Sunni), Jewish, Christian
LANGUAGE Arabic, Hebrew, English

Some 3.3 million Palestinians in the West Bank and Gaza Strip endure the seemingly endless frustration of not having their own country. Peace with Israel seemed close in 2000, but it faded as violence and reprisals prevailed.

Italy
ITALIAN REPUBLIC

CONTINENT Europe
AREA 301,333 sq km (116,345 sq mi)
POPULATION 57,166,000
CAPITAL Rome 2,628,000
RELIGION Roman Catholic
LANGUAGE Italian, German, French, Slovene
LITERACY 99%
LIFE EXPECTANCY 80
CURRENCY euro
GDP PER CAPITA $25,100
ECONOMY IND: tourism, machinery, iron and steel, chemicals. AGR: fruits, vegetables, grapes, potatoes, beef, fish. EXP: engineering products, textiles and clothing, production machinery, motor vehicles, transport equipment.

Italy consists of a mountainous peninsula in southern Europe extending into the Mediterranean Sea and includes the islands of Sicily, Sardinia, and about 70 other smaller islands. The Alps form Italy's border with France,

Switzerland, Austria, and Slovenia. Most of Italy has warm, dry summers and mild winters, with northern Italy experiencing colder, wetter winters. There are some notable active volcanoes: Vesuvius (near Naples), Etna (on Sicily), and Stromboli (north of Sicily).

Although decades of struggle unified Italy in 1871, two Italys exist today: the prosperous, industrialized north and the less developed agricultural south, known as the Mezzogiorno (land of the midday sun). Their differences reach back to the Renaissance, when northern city-states flourished while the Kingdom of Naples and Sicily languished under French and Spanish rule. The government confronts corruption, which is traceable to organized crime and an unemployment rate in the south more than twice that of the north. To address regional inequalities, a constitutional referendum was held in 2001—the results favored giving greater autonomy to the country's 20 regions in tax, education, and environmental policies.

Milan reigns as Italy's first city of commerce, and the Po River plain is both Italy's agricultural heartland and southern Europe's most advanced industrial region. Turin, the capital of heavy industry, is home to Fiat—one of the world's largest car producers. A major attraction for pilgrims and tourists is the "Holy Shroud" in Turin's cathedral—tradition holds that this was Christ's burial cloth. Florence was the birthplace of the Renaissance and is home to great works of civic and religious architecture, sculpture, and paintings. Rome, Italy's capital, exhibits the architectural and artistic grandeur of ancient civilizations.

Italy has to import almost all its raw materials and energy. Italy's economic strength is in the processing and manufacturing of goods, primarily in small and medium size family-owned firms. Its major industries include precision machinery, motor vehicles, fashion, clothing, and footwear. A founding member of both NATO and the European Union, Italy's superb transportation system, from airports to high-speed trains, connects it with the rest of Europe.

Jamaica
JAMAICA

CONTINENT	North America
AREA	10,991 sq km
	(4,244 sq mi)
POPULATION	2,646,000
CAPITAL	Kingston 575,000
RELIGION	Protestant, Roman Catholic, other spiritual beliefs
LANGUAGE	English, patois English
LITERACY	88%
LIFE EXPECTANCY	75
CURRENCY	Jamaican dollar
GDP PER CAPITA	$3,800
ECONOMY	IND: tourism, bauxite, textiles, food processing. AGR: sugarcane, bananas, coffee, citrus, poultry. EXP: alumina, bauxite; sugar, bananas, rum.

Jamaica is a mountainous Caribbean island just south of Cuba. Columbus landed here in 1494, and the Spanish soon brought in slaves as the native Arawak Indians died out—today more than 90 percent of the population is of African descent. The British seized the island in 1655, granting independence in 1962. Tourism is a steady earner, but reliance on unpredictably priced commodities, such as bauxite, causes uneven growth. The island is a major transit point for South American cocaine en route to the U.S. and Europe. Other problems include illicit cultivation of marijuana and heavy deforestation.

Japan
JAPAN

CONTINENT	Asia
AREA	377,887 sq km
	(145,902 sq mi)
POPULATION	127,508,000
CAPITAL	Tokyo 35,327,000
RELIGION	Shinto, Buddhist
LANGUAGE	Japanese
LITERACY	99%
LIFE EXPECTANCY	81
CURRENCY	yen
GDP PER CAPITA	$28,700
ECONOMY	IND: motor vehicles, electronic equipment, machine tools, steel and nonferrous metals. AGR: rice, sugar beets, vegetables, fruit; pork, fish. EXP: motor vehicles, semiconductors, office machinery, chemicals.

Japan, a country of islands, extends along the Pacific coast of Asia. The main island is Honshu, and the country has three other large islands—Hokkaido to the north and Shikoku and Kyushu to the south. More than 4,000 smaller islands surround the four largest. A modern transportation system connects the main islands, including the Seikan Tunnel linking Honshu to Hokkaido—the world's longest railroad tunnel at 54 kilometers (33 miles). Japan's high-speed trains (known as *shinkansen,* or bullet trains) connect major urban areas.

About 73 percent of Japan is mountainous, and all its major cities, except the ancient capital of Kyoto, cling to narrow coastal plains. Only an estimated 18 percent of Japan's territory is suitable for settlement—so Japan's cities are large and densely populated. Tokyo, the capital, is the planet's largest urbanized area at 35 million people. However, Tokyo has a worrisome environmental history of destructive earthquakes and tsunamis (seismic sea waves). A major earthquake in 1923 killed an estimated 143,000 people.

One of the most traditional and isolated societies on Earth when Commodore Matthew C. Perry sailed an American fleet into Tokyo Bay in 1853, Japan is democratic and outward-looking today. Among the top three exporters of manufactured goods, the nation has the second largest economy after that of the U.S.

Aggressive expansion across the Pacific led to war with the U.S. in 1941. Defeat ended Japan's dream of ruling Asia, and the U.S. occupation imposed a parliamentary constitution, free labor unions, and stringent land reform. Despite a lack of raw materials, the economy was revived with the help of U.S. grants, high rates of labor productivity, personal savings, and capital investment.

Emperor Hirohito's death in 1989 marked the start of an era in which Japan faces the challenges of an aging population, rising inequality of wealth, the changing role of women in society, and growing concern about security and the environment. Current problems include unemployment—the highest since the end of World War II—and low economic growth. Relations with North Korea are tense because of that country's nuclear weapons program and its abduction of Japanese citizens in the 1970s and 1980s. Japan's ties with Russia are hampered because of some small islands east of Hokkaido known as the Northern Territories—the Habomai Islands, Shikotan, Kunishiri, and Etorofu (called Iturup by Russia). Japan still claims these Russian-held islands that were taken at the end of World War II.

Jordan
HASHEMITE KINGDOM OF JORDAN

CONTINENT	Asia
AREA	89,342 sq km
	(34,495 sq mi)
POPULATION	5,480,000
CAPITAL	Amman 1,237,000
RELIGION	Sunni Muslim, Christian
LANGUAGE	Arabic, English
LITERACY	91%
LIFE EXPECTANCY	69
CURRENCY	Jordanian dinar
GDP PER CAPITA	$4,300
ECONOMY	IND: phosphate mining, pharmaceuticals, petroleum refining, cement, potash. AGR: wheat, barley, citrus, tomatoes, sheep. EXP: phosphates, fertilizers, potash, agricultural products, manufactures.

Located on desert plateaus in southwest Asia, Jordan is almost landlocked but for a short coast on the Gulf of Aqaba. In 1923, after the dissolution of the Ottoman Empire, Transjordan was designated a British mandate. Independence came in 1946. Following the Arab-Israeli conflict in 1948–49, the country annexed the West Bank—but lost it to Israel in the 1967 war. The Arab-Israeli wars have brought this small, poor country some 1.5 million Palestinian refugees. Jordan has a constitutional monarchy, with an economy based on agriculture and phosphates.

Kazakhstan
REPUBLIC OF KAZAKHSTAN

CONTINENT Asia

AREA 2,717,300 sq km
(1,049,155 sq mi)

POPULATION 14,787,000

CAPITAL Astana 332,000

RELIGION Muslim, Russian
Orthodox

LANGUAGE Kazakh (Qazaq),
Russian

LITERACY 98%

LIFE EXPECTANCY 66

CURRENCY tenge

GDP PER CAPITA $7,200

ECONOMY IND: oil, coal, iron ore,
manganese, chromite,
lead, zinc, copper, tita-
nium. AGR: grain
(mostly spring wheat),
cotton, livestock. EXP:
oil and oil products, ferrous metals, chemicals,
machinery, grain.

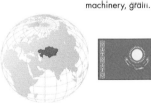

Stretching across Central Asia, Kazakhstan is a landlocked and mostly
dry land. Flat in the west, it rises to high
mountains in the east. More than a hundred ethnic groups live in Kazakhstan;
28 percent of the population is Russian—most live in the north near the
Russian border. Second in size only
to Russia among the former Soviet
Republics, Kazakhstan contained the
main Soviet test area for nuclear
weapons. From 1949 to 1989 there were
456 nuclear blasts at the Semipalatinsk
site, 116 in the air—this highly radioactive range was closed by the Kazakh government in 1991. Russia still uses the
Baykonur Cosmodrome in Kazakhstan,
the principal site for Soviet space
launches and the world's oldest and
largest spaceport.

In the 15th century the Kazakhs
emerged as nomadic stock herders
of the steppe, speaking a Turkic language and practicing Islam. Imperial
Russia colonized the region in the 19th
century. An estimated one million
Kazakhs died during Soviet campaigns
in the 1930s to forcibly settle the
nomads. The nation confronts a legacy
of environmental abuse left behind by
the Soviets, who dictated industrial
development in this mineral-rich republic. Kazakhstan faces ecological disaster in the Aral Sea area and is trying
to preserve the northern part of the sea
in order to prevent desertification. The
country is enjoying strong economic
growth because of its large oil, gas, and
mineral reserves.

Kenya
REPUBLIC OF KENYA

CONTINENT Africa

AREA 580,367 sq km
(224,081 sq mi)

POPULATION 31,639,000

CAPITAL Nairobi 2,818,000

RELIGION Protestant, Roman
Catholic, indigenous
beliefs, Muslim

LANGUAGE English, Kiswahili,
numerous indigenous
languages

LITERACY 85%

LIFE EXPECTANCY 46

CURRENCY Kenyan shilling

GDP PER CAPITA $1,100

ECONOMY IND: small-scale consumer goods (plastic,
furniture), agricultural
products processing;
oil refining. AGR: tea,
coffee, corn, wheat,
dairy products. EXP:
tea, horticultural prod-
ucts, coffee, petroleum
products, fish.

The East African country of Kenya
rises from a low coastal plain on the
Indian Ocean to mountains and
plateaus at its center. Most Kenyans
live in the highlands, and Nairobi, the
capital, is here at an altitude of 1,700
meters (5,500 feet). Even though
Nairobi is near the Equator, its high
elevation brings cooler air. To the west
of Nairobi the land descends to the
north-south running Great Rift Valley—the valley floor is at its lowest near
Lake Turkana in the deserts of northern Kenya. Around Lake Turkana, scientists have discovered some of
humankind's earliest ancestors—a fossil known as Kenya Man was dated at
3.5 to 3.2 million years old.

Both free enterprise and a measure of political debate helped make
Kenya one of Africa's most stable
nations after it achieved independence
from Britain in 1963. But, more
recently, corruption has been an under-
mining force, and the government—
pressured for reform—moved to a

multiparty system in the late 1990s. Barriers to progress are high population growth, electricity shortages, and inefficiency in key sectors.

Forty ethnic groups, including Kikuyu farmers and Masai cattle herders, crowd the countryside, still home to three-quarters of Kenya's people. Intense competition for arable land drives thousands to cities, where unemployment is high. In Nairobi, East Africa's commercial hub, skyscrapers abruptly give way to slums.

The government has stepped up efforts to stem poaching, particularly of the elephant and black rhino. Tourism is essential to the economy, and Kenya is one of Africa's major safari destinations.

Kiribati
REPUBLIC OF KIRIBATI

CONTINENT	Australia/Oceania
AREA	811 sq km (313 sq mi)
POPULATION	98,000
CAPITAL	Tarawa 42,000
RELIGION	Roman Catholic, Protestant
LANGUAGE	English, I-Kiribati
LITERACY	98%
LIFE EXPECTANCY	62
CURRENCY	Australian dollar
GDP PER CAPITA	$800
ECONOMY	IND: fishing, handicrafts. AGR: copra, taro, breadfruit, sweet potatoes, fish. EXP: copra, coconuts, seaweed, fish.

Scattered over five million square kilometers, the 33 islands were formerly the Gilberts of the British Gilbert and Ellice Islands Colony. These mostly low-lying coral islands in the Pacific became the Republic of Kiribati in 1979. In addition to fishing and copra, the country relies on foreign financial aid, particularly from the U.K. and Japan. As its phosphate reserves diminish, the island has been forced to rely on a reserve trust. Income from citizens now working abroad augments the economy.

Korea, North
DEMOCRATIC PEOPLE'S REPUBLIC OF KOREA

CONTINENT	Asia
AREA	120,538 sq km (46,540 sq mi)
POPULATION	22,661,000
CAPITAL	Pyongyang 3,228,000
RELIGION	Buddhist, Confucianist
LANGUAGE	Korean
LITERACY	99%
LIFE EXPECTANCY	63
CURRENCY	North Korean won
GDP PER CAPITA	$1,000
ECONOMY	IND: military products, machine building, electric power, chemicals, mining. AGR: rice, corn, potatoes, soybeans, cattle. EXP: minerals, metallurgical

products, manufactures (including armaments), textiles.

The Democratic People's Republic of Korea, or North Korea, occupies the northern part of the Korean peninsula in East Asia, with mountains covering more than 80 percent of the land. A Japanese possession from 1910 to 1945, Korea was then divided, with Soviet troops occupying the north and the U.S. the south. In 1950, two years after they had been made separate states, North Korea invaded South Korea. This began the devastating Korean War (1950–53), with the North, receiving Soviet and Chinese help, fighting a U.S.-led coalition of UN forces. The war caused an estimated three million casualties. After an armistice in 1953, a UN-monitored demilitarized zone was set up along the cease-fire line, dividing the two nations.

One of the few remaining communist states, reclusive North Korea has been one of the world's most secretive societies. The country lost subsidized trade relationships with the fall of the Soviet Union in 1991. After the death of dictator President Kim Il Sung in 1994, his son, Kim Jong Il, took control. North Korea, lagging far behind South Korea in economic development, devotes large amounts of money to the military, while its people suffer from chronic food shortages. Some experts estimate that at least 2.5

million people have died of starvation or hunger-related diseases since 1994. An estimated 300,000 North Koreans have fled to China as of 2004, where they hide from Chinese authorities who do not recognize them as refugees —and would force them to return. Despite severe shortages of food and electricity, North Korea maintains the world's fourth largest army, a nuclear weapons program, and missiles that threaten South Korea and Japan.

Korea, South
REPUBLIC OF KOREA

CONTINENT	Asia
AREA	99,250 sq km (38,321 sq mi)
POPULATION	47,939,000
CAPITAL	Seoul 9,592,000
RELIGION	Christian, Buddhist
LANGUAGE	Korean, English
LITERACY	98%
LIFE EXPECTANCY	76
CURRENCY	South Korean won
GDP PER CAPITA	$19,600
ECONOMY	IND: electronics, automobile production, chemicals, shipbuilding, steel, textiles. AGR: rice, root crops, barley, vegetables, cattle, fish. EXP: electronic products, machinery and equipment, motor vehicles, steel, ships, textiles.

The Republic of Korea, or South Korea, consists of the southern half of the Korean peninsula in East Asia and many islands lying off the western and southern coasts. The largest island, Jeju, has the highest mountain in South Korea at 1,950 meters (6,398 feet). The terrain is mountainous, though less rugged than that of North Korea.

Major population and industrial centers are located in the northwest (Seoul-Incheon area) and southeast. To overcome distance and traffic congestion, South Korea launched a high-speed rail service between Seoul and Busan in 2004. English is taught as a second language in most schools, and more than 1.5 million ethnic Koreans reside in the U.S. In 2000 the government introduced a new phonetic system for transcribing Korean into English, changing names like Cheju to Jeju and Pusan to Busan.

Starting in the late 1970s this capitalist prodigy matured into the world's 12th largest trading nation and a major exporter of cars, consumer electronics, and computer components—due in part to huge export-oriented conglomerates called *jaebols*. From 1980 to 1990 economic growth averaged 10 percent a year—more than three times that of the U.S. and twice that of Japan. However, the Asian financial crisis caused a deep recession. Economic growth rebounded in 1999, and the economy continues to perform well thanks to vibrant exports.

After the Korean War, South Korean society has shifted from being 75 percent rural to being 82 percent urban. Since 1987 it has grown as a multiparty democracy, and the government has pursued peace initiatives and trade with the unpredictable North Korean regime. Road and railway projects are under way to link the two Koreas, and in 2003 more than half a million South Koreans visited the North — and 1,023 North Koreans traveled to South Korea, mainly for joint sporting events. The potential military threat posed by North Korea keeps some 37,000 U.S. troops here.

Kosovo
See listing under Serbia and Montenegro.

Kuwait
STATE OF KUWAIT

CONTINENT	Asia
AREA	17,818 sq km (6,880 sq mi)
POPULATION	2,384,000
CAPITAL	Kuwait City 1,222,000
RELIGION	Sunni and Shiite Muslim, Christian, Hindu, Parsi
LANGUAGE	Arabic, English
LITERACY	84%
LIFE EXPECTANCY	78
CURRENCY	Kuwaiti dinar
GDP PER CAPITA	$17,500
ECONOMY	IND: petroleum, petrochemicals, desalination, food processing. AGR: practically no crops, fish. EXP: oil and refined products, fertilizers.

A small, oil-rich country on the Persian Gulf, Kuwait is flat and arid, but oil wealth makes this an attractive place for immigrants. Kuwaiti Arabs make up a third of the population, with other Arabs (Egyptians, Palestinians) 22 percent, and non-Arabs (mostly South Asians) 38 percent. Founded in the 18th century, the ruling al-Sabah dynasty was in place in 1899 when Kuwait came under British protection. Full independence was achieved in 1961. Iraq invaded Kuwait in 1990, but a U.S.-led coalition routed Iraqi forces. Kuwait was the principal platform for U.S. military operations against Saddam Hussein in 2003.

Kyrgyzstan
KYRGYZ REPUBLIC

CONTINENT	Asia
AREA	199,900 sq km
	(77,182 sq mi)
POPULATION	5,033,000
CAPITAL	Bishkek 806,000
RELIGION	Muslim, Russian Orthodox
LANGUAGE	Kyrgyz, Russian
LITERACY	97%
LIFE EXPECTANCY	69
CURRENCY	Kyrgyzstani som
GDP PER CAPITA	$2,900
ECONOMY	IND: small machinery, textiles, food processing, cement, shoes, logs. AGR: tobacco, cotton, potatoes, vegetables, sheep. EXP: cotton, wool, meat, tobacco, gold, mercury, uranium, natural gas.

A rugged nation in Central Asia, Kyrgyzstan shares the snowcapped Tian Shan with China. Some 75 percent of the land is mountainous. In their mountain fastness, the nomadic Kyrgyz, a Turkic-speaking people with loose ties to Islam, bred horses, cattle, and yaks for centuries. The Kyrgyz came under tsarist Russian rule during the 19th century, and thousands of Slavic farmers migrated into the region. Kyrgyzstan gained independence in 1991. Kyrgyz make up two thirds of the population, and there are large Uzbek and Russian minorities. Raising livestock still remains the main agricultural activity today.

Laos
LAO PEOPLE'S DEMOCRATIC REPUBLIC

CONTINENT	Asia
AREA	236,800 sq km
	(91,429 sq mi)
POPULATION	5,593,000
CAPITAL	Vientiane 716,000
RELIGION	Buddhist, animist, other
LANGUAGE	Lao, French, English, various ethnic languages
LITERACY	53%
LIFE EXPECTANCY	54
CURRENCY	kip
GDP PER CAPITA	$1,800
ECONOMY	IND: tin and gypsum mining, timber, electric power, agricultural processing. AGR: sweet potatoes, vegetables, corn, coffee, water buffalo. EXP: wood products, garments, electricity, coffee, tin.

Laos is a poor, landlocked, and mountainous country in Southeast Asia. Agriculture, mostly subsistence farming, dominates the economy. Most people live in the valleys of the Mekong River and its tributaries, where rice can be grown on fertile floodplains. Soon after independence from France in 1953, the country fell into turmoil; in 1975 the communist Pathet Lao seized power with help from North Vietnam. Many fled the regime, and the U.S. resettled some 250,000 Lao refugees. One of the few remaining communist states, the economy is hampered by poor roads, no railroad, and limited access to electricity.

Latvia
REPUBLIC OF LATVIA

CONTINENT	Europe
AREA	64,589 sq km
	(24,938 sq mi)
POPULATION	2,320,000
CAPITAL	Riga 733,000

RELIGION Lutheran, Roman
Catholic, Russian
Orthodox

LANGUAGE Latvian, Lithuanian,
Russian

LITERACY 100%

LIFE EXPECTANCY 71

CURRENCY Latvian lat

GDP PER CAPITA $8,900

ECONOMY IND: buses, vans,
street and railroad cars,
synthetic fibers. AGR:
grain, sugar beets, pota-
toes, vegetables, beef,
fish. EXP: wood and
wood products, machin-
ery and equipment,
metals, textiles.

Flat and forested, Latvia lies on the Baltic Sea in northern Europe. Few former Soviet republics experienced a more profound shift in character during their 50 years of domination than this Baltic country. From 1939–1989 the proportion of ethnic Latvians in Latvia dropped from 73 to 52 percent—due to heavy Russian immigration and Latvian emigration. Since independence in 1991, Latvian ethnicity has started to rebound and now constitutes 59 percent of the population—Russians are 29 percent. An industrial country with trade ties to the West, Latvia joined NATO and the EU in 2004.

Lebanon
LEBANESE REPUBLIC

CONTINENT Asia

AREA 10,452 sq km
(4,036 sq mi)

POPULATION 4,198,000

CAPITAL Beirut 1,792,000

RELIGION Muslim, Christian

LANGUAGE Arabic, French, English,
Armenian

LITERACY 87%

LIFE EXPECTANCY 73

CURRENCY Lebanese pound

GDP PER CAPITA $4,800

ECONOMY IND: banking, food
processing, jewelry,
cement, textiles, mineral
and chemical products.
AGR: citrus, grapes,
tomatoes, apples, sheep.
EXP: foodstuffs and
tobacco, textiles, chemi-
cals, precious stones,
metal products.

Lebanon is a small, mountainous country in the Middle East. After independence in 1943, Lebanon prospered as a banking, resort, and university center. It is estimated that two-thirds of the resident population are Muslim, with the rest being Christian. No census has been taken since 1932 due to political sensitivity over religious affiliation. Fighting between Christian and Muslim militias escalated into civil war from 1975 to 1991. Democracy was restored in 1992—allocating

government positions based on religion. During the civil war both Israel and Syria sent troops into Lebanon. Israel withdrew its army in 2000; some 16,000 Syrian soldiers remain.

Lesotho
KINGDOM OF LESOTHO

CONTINENT Africa

AREA 30,355 sq km
(11,720 sq mi)

POPULATION 1,800,000

CAPITAL Maseru 170,000

RELIGION Christian, indigenous
beliefs

LANGUAGE English, Sesotho, Zulu,
Xhosa

LITERACY 85%

LIFE EXPECTANCY 37

CURRENCY loti, South African rand

GDP PER CAPITA $2,700

ECONOMY IND: food, beverages,
textiles, apparel assem-
bly, handicrafts. AGR:
corn, wheat, pulses,
sorghum, livestock. EXP:
manufactures (clothing,
footwear, road vehicles),
wool, mohair.

In the early 19th century Basuto Chief Moshoeshoe united tribes in this mountainous land surrounded by South Africa. To establish a buffer against Boer expansion, he asked the British to administer the kingdom. Lesotho became independent in 1966.

In 1993 it became a constitutional monarchy. Water is Lesotho's major natural resource. Completion of a large hydropower plant in 1998 helps the economy expand through the sale of water to South Africa.

Liberia
REPUBLIC OF LIBERIA

CONTINENT	Africa
AREA	111,370 sq km (43,000 sq mi)
POPULATION	3,317,000
CAPITAL	Monrovia 572,000
RELIGION	Indigenous beliefs, Christian, Muslim
LANGUAGE	English, 20 ethnic languages
LITERACY	58%
LIFE EXPECTANCY	49
CURRENCY	Liberian dollar
GDP PER CAPITA	$1,000
ECONOMY	IND: rubber processing, palm oil processing, timber, diamonds. AGR: rubber, coffee, cacao, rice, sheep, timber. EXP: rubber, timber, iron, diamonds, cacao.

Freed American slaves began settling on the West African coast in 1820. In 1847 Liberia was declared an independent republic—Africa's first under a constitution modeled on that of the U.S. In 1989 civil war erupted, ending seven years later with the Abuja Peace Accords. In 1999 the government of Charles Taylor was accused of supporting rebels in Sierra Leone, and it fought a border war with Guinea in 2000. Taylor was forced into exile in 2003, and the new government works to rebuild the nation.

Libya
GREAT SOCIALIST PEOPLE'S LIBYAN ARAB JAMAHIRIYA

CONTINENT	Africa
AREA	1,759,540 sq km (679,362 sq mi)
POPULATION	5,499,000
CAPITAL	Tripoli 2,006,000
RELIGION	Sunni Muslim
LANGUAGE	Arabic, Italian, English
LITERACY	83%
LIFE EXPECTANCY	76
CURRENCY	Libyan dinar
GDP PER CAPITA	$6,200
ECONOMY	IND: petroleum, food processing, textiles, handicrafts, cement. AGR: wheat, barley, olives, dates, cattle. EXP: crude oil, refined petroleum products.

Water-poor, oil-rich Libya has the highest per capita income of continental Africa. Most Libyans live on the Mediterranean coast, many in Tripoli and Banghazi. The largest water development project ever devised, the Great Man-Made River Project, brings water from aquifers under the Sahara to the coastal cities. Since 1969 this former Italian colony, independent since 1951, has been an authoritarian socialist state under Muammar Qaddafi—whose backing of terrorism led to a U.S. bombing in 1986 and UN sanctions in 1992. In 2003 Libya ended its international isolation and abandoned its weapons programs.

Liechtenstein
PRINCIPALITY OF LIECHTENSTEIN

CONTINENT	Europe
AREA	160 sq km (62 sq mi)
POPULATION	35,000
CAPITAL	Vaduz 5,000
RELIGION	Roman Catholic, Protestant
LANGUAGE	German, Alemannic dialect
LITERACY	100%
LIFE EXPECTANCY	79
CURRENCY	Swiss franc
GDP PER CAPITA	$25,000
ECONOMY	IND: electronics, metal manufacturing, dental products, ceramics, pharmaceuticals. AGR: wheat, barley, corn, potatoes, livestock. EXP: small specialty machinery, connectors for audio and video, parts for motor vehicles, dental products.

Liechtenstein is a tiny independent state between Switzerland and Austria. In 1719 the princely House of Liechtenstein, which still rules this constitutional monarchy, purchased a strip of Rhine floodplain and adjacent mountains. Because of liberal tax policies and banking laws, it counts more companies than citizens.

Lithuania
REPUBLIC OF LITHUANIA

CONTINENT	Europe
AREA	65,300 sq km
	(25,212 sq mi)
POPULATION	3,458,000
CAPITAL	Vilnius 549,000
RELIGION	Roman Catholic, Lutheran, Russian Orthodox, Protestant
LANGUAGE	Lithuanian, Polish, Russian
LITERACY	100%
LIFE EXPECTANCY	72
CURRENCY	litas
GDP PER CAPITA	$8,400
ECONOMY	IND: metal-cutting machine tools, electric motors, television sets, refrigerators and freezers. AGR: grain, potatoes, sugar beets, flax, beef, fish. EXP: mineral products, textiles and clothing, machinery and equipment, chemicals.

Lithuania is in northern Europe, on the eastern shores of the Baltic Sea. The landscape consists of gently rolling plains and extensive forests. Beginning at about the same time as movements in the other Baltic republics of Estonia and Latvia, Lithuania quickly surged ahead. In March 1990 democratically elected representatives voted for independence, lost in 1940 with annexation by the Soviet Union. Lithuania, embracing market reform since independence, joined both the European Union and NATO in 2004.

Luxembourg
GRAND DUCHY OF LUXEMBOURG

CONTINENT	Europe
AREA	2,586 sq km
	(998 sq mi)
POPULATION	452,000
CAPITAL	Luxembourg 77,000
RELIGION	Roman Catholic
LANGUAGE	Luxembourgish, German, French
LITERACY	100%
LIFE EXPECTANCY	78
CURRENCY	euro
GDP PER CAPITA	$48,900
ECONOMY	IND: banking, iron and steel, food processing, chemicals. AGR: barley, oats, potatoes, wheat, livestock products. EXP: machinery and equipment, steel products, chemicals, rubber products, glass.

Luxembourg, a landlocked Western European country, has heavily forested hills in the north and open, rolling countryside in the south. It became a member of a customs union in 1948 that evolved into today's European Union. Although small in size, Luxembourg's central location, political stability, and multilingual population, along with tax incentives, have proved advantageous to it as a financial center. Foreign investment in light manufacturing and services has offset the decline in steel, once the nation's major industry.

Macau
See listing under China.

Macedonia
FORMER YUGOSLAV REPUBLIC OF MACEDONIA

CONTINENT	Europe
AREA	25,713 sq km
	(9,928 sq mi)
POPULATION	2,059,000
CAPITAL	Skopje 447,000
RELIGION	Macedonian Orthodox, Muslim
LANGUAGE	Macedonian, Albanian, Turkish
LITERACY	NA
LIFE EXPECTANCY	73
CURRENCY	Macedonian denar
GDP PER CAPITA	$5,100

ECONOMY IND: coal, metallic
chromium, lead, zinc,
ferronickel, textiles.
AGR: rice, tobacco,
wheat, corn, beef.
EXP: food, beverages,
tobacco, miscellaneous
manufactures, iron
and steel.

The landlocked and mostly mountainous country of Macedonia, in southeastern Europe, proclaimed independence from Yugoslavia in September 1991. The UN officially calls the country "The Former Yugoslav Republic of Macedonia"—due to Greece's fear that use of "Macedonia" might imply territorial ambitions toward the Greek region of Macedonia. The democratic government faced a 2001 rebellion launched by ethnic Albanians, who make up 25 percent of population. Negotiations led to laws making Albanian an official language and providing other minority rights.

Madagascar
REPUBLIC OF MADAGASCAR

CONTINENT Africa
AREA 587,041 sq km
(226,658 sq mi)
POPULATION 16,980,000
CAPITAL Antananarivo 1,678,000
RELIGION Indigenous beliefs,
Christian, Muslim
LANGUAGE French, Malagasy

LITERACY 69%
LIFE EXPECTANCY 55
CURRENCY Malagasy franc
GDP PER CAPITA $800
ECONOMY IND: meat processing,
soap, breweries, tanneries. AGR: coffee,
vanilla, sugarcane,
cloves, livestock products. EXP: coffee, vanilla,
shellfish, sugar, cotton
cloth, chromite, petroleum products.

Off Africa's southeast coast in the Indian Ocean, Madagascar is the world's fourth largest island after Greenland, New Guinea, and Borneo. A stunning diversity of plant and animal species found nowhere else evolved after the island broke away from the African continent 165 million years ago. It has a mountainous central plateau and coastal plains. The first settlers were of African and Asian origin, and 18 separate ethnic groups emerged, derived from an African and Malayo-Indonesian mixture. Asian features are most predominant in the central highlands people, and coastal people tend to show features of African origin. Most of the population depend on subsistence farming, based on rice and cattle, with coffee, vanilla, and seafood being important exports.

French colonial rule began in 1896; independence came in 1960. In 1990, after almost 20 years of Marxism,

Madagascar lifted a ban on opposition parties, and a new president was elected in 1993. Elections in 2001 resulted in a period of civil unrest, lasting for several months, until Marc Ravalomanana was declared winner of the presidential election. Environmental degradation is a major concern as damaging agricultural practices cause deforestation, soil erosion, and desertification. The island is heavily exposed to tropical cyclones, which brought destructive floods in 2004.

Malawi
REPUBLIC OF MALAWI

CONTINENT Africa
AREA 118,484 sq km
(45,747 sq mi)
POPULATION 11,651,000
CAPITAL Lilongwe 587,000
RELIGION Protestant, Roman
Catholic, Muslim
LANGUAGE English, Chichewa
LITERACY 63%
LIFE EXPECTANCY 39
CURRENCY Malawian kwacha
GDP PER CAPITA $600
ECONOMY IND: tobacco, tea,
sugar, sawmill products.
AGR: tobacco, sugarcane, cotton, tea,
groundnuts, cattle.
EXP: tobacco, tea,
sugar, cotton, coffee.

PLACES COUNTRY BY COUNTRY

Malawi lies landlocked in southeast Africa, with Lake Malawi taking up about a fifth of the landscape. Independent from Britain since 1964, it endured the one-party rule of President for Life Hastings Kamuzu Banda for more than 25 years. Democratic elections in 1994 ushered in new leadership of this country, nearly self-sufficient in food. Transportation costs for exports skyrocketed as a result of civil war in next-door Mozambique, which disrupted rail links to the sea. Since the 1992 peace accord in Mozambique, rail links have been reestablished, and Malawi is slowly recovering.

Malaysia
MALAYSIA

CONTINENT Asia
AREA 329,847 sq km
(127,355 sq mi)
POPULATION 25,061,000
CAPITAL Kuala Lumpur 1,352,000
RELIGION Muslim, Buddhist, Daoist,
Hindu, Christian, Sikh,
Shamanist
LANGUAGE Bahasa Melayu, English,
Chinese dialects, other
regional dialects and
indigenous languages
LITERACY 89%
LIFE EXPECTANCY 73
CURRENCY ringgit
GDP PER CAPITA $8,800
ECONOMY IND: rubber and oil
palm processing and
manufacturing, light
manufacturing industry,
logging, petroleum pro-
duction and refining.
AGR: rubber, palm oil,

cacao, rice. EXP: electronic equipment, petroleum and liquefied natural gas, wood and wood products, palm oil.

Comprising the territories of Malaya, Sarawak, and Sabah, Malaysia stretches from peninsular Malaysia to northeastern Borneo in Southeast Asia. Central mountains divide peninsular Malaysia (Malaya), separating the narrow eastern coast from the fertile western plains, with its sheltered beaches and bays. Sarawak and Sabah share the island of Borneo with Indonesia and Brunei, where swamps rise to jungle-covered mountains. Malays make up half the population, and almost all Malays are Muslims. Ethnic Chinese constitute a quarter of Malaysia's people, and Indians some 7 percent—both groups are concentrated on the peninsula's west coast.

In the mid-19th century the United Kingdom began importing Chinese to work the tin mines of Muslim sultanates on the Malay Peninsula; by the turn of the century new rubber plantations employed transported Indian laborers. In 1957 the Federation of Malaya gained independence from Britain. Six years later the colonies of Sarawak and Sabah, on the island of Borneo, and Singapore joined Malaya to form the Federation of Malaysia; Singapore withdrew in 1965.

Malaysia is one of the world's largest exporters of semiconductors, electrical goods, and appliances. After

a long period of economic growth, Malaysia—like many countries—was hit hard by the Asian financial crisis in the late 1990s. Kuala Lumpur, the nation's capital, anchors the new Multimedia Super Corridor, Asia's equivalent of the U.S.'s Silicon Valley. The government, a federal democracy with a ceremonial king, has ambitious plans to make Malaysia a leading producer and developer of high-tech products, including software.

Maldives
REPUBLIC OF MALDIVES

CONTINENT Asia
AREA 298 sq km (115 sq mi)
POPULATION 285,000
CAPITAL Male 83,000
RELIGION Sunni Muslim
LANGUAGE Maldivian Dhivehi (dialect
of Sinhala), English
LITERACY 97%
LIFE EXPECTANCY 67
CURRENCY rufiyaa
GDP PER CAPITA $3,900
ECONOMY IND: fish processing,
tourism, shipping,
boat building, coconut
processing. AGR:
coconuts, corn, sweet
potatoes, fish. EXP:
fish, clothing.

The island nation of Maldives is south of India in the Indian Ocean. The islands are small, and none rise more

418

than 1.8 meters (6 feet) above sea level. Like necklaces draped along an undersea plateau, 1,200 coral islands—about 200 inhabited—form the Maldives. In 1968, three years after independence from Britain, the sultanate gave way to an Islamic republic. Tourism and fishing sustain the economy. In 2001 more than 460,000 tourists visited this tiny nation.

Mali
REPUBLIC OF MALI

CONTINENT	Africa
AREA	1,240,192 sq km
	(478,841 sq mi)
POPULATION	11,626,000
CAPITAL	Bamako 1,264,000
RELIGION	Muslim, indigenous beliefs
LANGUAGE	French, Bambara, numerous African languages
LITERACY	46%
LIFE EXPECTANCY	45
CURRENCY	CFA franc
GDP PER CAPITA	$900
ECONOMY	IND: food processing; construction, phosphate and gold mining. AGR: cotton, millet, rice, corn, cattle. EXP: cotton, gold, livestock.

The landlocked West African country of Mali is mostly desert or semidesert. Trans-Saharan caravans once enriched Timbuktu and Gao, trading hubs on the Niger River. Descendants of the empires of Ghana, Malinke, and Songhai came under French rule in the late 19th century and gained independence in 1960. In the 1980s economic woes worsened by drought and famine led to deregulation and privatization. Desertification forces nomadic herders south to the subsistence-farming belt. With increased gold mining operations, Mali is becoming a major gold exporter.

Malta
REPUBLIC OF MALTA

CONTINENT	Europe
AREA	316 sq km (122 sq mi)
POPULATION	396,000
CAPITAL	Valletta 83,000
RELIGION	Roman Catholic
LANGUAGE	Maltese, English
LITERACY	93%
LIFE EXPECTANCY	77
CURRENCY	Maltese lira
GDP PER CAPITA	$17,200
ECONOMY	IND: tourism, electronics, ship building and repair. AGR: potatoes, cauliflower, grapes, wheat, pork. EXP: machinery and transport equipment, manufactures.

Malta's position in the Mediterranean, midway between Europe and Africa, has made it a strategic prize. Here, in the 16th century, the Knights of St. John repelled 30,000 soldiers of Süleyman the Magnificent's Ottoman Empire. It withstood Axis bombs during World War II. In 1964, after almost 150 years as a British colony, the Maltese islands won independence. Tourism is the cornerstone of the nation's economy, and it joined the EU in 2004.

Marshall Islands
REPUBLIC OF THE MARSHALL ISLANDS

CONTINENT	Australia/Oceania
AREA	181 sq km (70 sq mi)
POPULATION	55,000
CAPITAL	Majuro 25,000
RELIGION	Christian (mostly Protestant)
LANGUAGE	English, Marshallese, Japanese
LITERACY	94%
LIFE EXPECTANCY	68
CURRENCY	US dollar
GDP PER CAPITA	$1,600
ECONOMY	IND: copra, fish, tourism, craft items from shell, wood, and pearls. AGR: coconuts, tomatoes, melons, taro, pigs. EXP: copra cake, coconut oil, handicrafts, fish.

Tropical islands in the western Pacific, the Marshall Islands form two parallel island groups—the Ratak (sunrise)

Chain and Ralik (sunset) Chain. These atolls, reefs, and islets include Kwajalein, test range for U.S. missiles and home to the world's largest lagoon, and Enewetak, where the United States exploded the first hydrogen bomb in 1952. Bikini Atoll is still uninhabitable because of past nuclear tests. In 1986 the former trust territory became self-governing in free association with the United States, which is responsible for its defense and foreign affairs.

Martinique (France)
FRENCH OVERSEAS DEPARTMENT

CONTINENT	North America
AREA	1,100 sq km (425 sq mi)
POPULATION	392,000
CAPITAL	Fort-de-France 93,000
RELIGION	Roman Catholic, Protestant
LANGUAGE	French, Creole patois

Martinique, in the eastern Caribbean, has volcanic peaks in the north, with hills and beaches in the south. Major exports consist of bananas, pineapples, and rum, mostly going to France.

Mauritania
ISLAMIC REPUBLIC OF MAURITANIA

CONTINENT	Africa
AREA	1,030,700 sq km (397,955 sq mi)
POPULATION	2,914,000
CAPITAL	Nouakchott 600,000
RELIGION	Muslim
LANGUAGE	Hassaniya Arabic, Wolof, Pulaar, Soninke, French

LITERACY	42%
LIFE EXPECTANCY	54
CURRENCY	ouguiya
GDP PER CAPITA	$1,700
ECONOMY	IND: fish processing, mining of iron ore and gypsum. AGR: dates, millet, sorghum, rice, cattle. EXP: iron ore, fish and fish products, gold.

Part of French West Africa until independence in 1960, Mauritania is influenced by Arab as well as African cultures. Crop growing is largely confined to the floodplain of the Sénégal River, straining relations with the country of Senegal over use of the river. Some of the world's richest fishing grounds lie off the coast. The population still largely depends on agriculture and livestock for their livelihood, even though recurring droughts forced most nomads and many subsistence farmers into the cities. The country has been further strained by internal racial divisions between blacks and Arabs.

Mauritius
REPUBLIC OF MAURITIUS

CONTINENT	Africa
AREA	2,040 sq km (788 sq mi)
POPULATION	1,221,000
CAPITAL	Port Louis 143,000
RELIGION	Hindu, Christian, Muslim, Protestant

LANGUAGE	English, French, Creole, Hindi, Urdu, Hakka, Bhojpuri
LITERACY	86%
LIFE EXPECTANCY	72
CURRENCY	Mauritian rupee
GDP PER CAPITA	$10,100
ECONOMY	IND: food processing (largely sugar milling), textiles, clothing, chemicals, metal products. AGR: sugarcane, tea, corn, potatoes, bananas, cattle, fish. EXP: clothing and textiles, sugar, cut flowers, molasses.

The island country of Mauritius lies in the Indian Ocean east of Madagascar. The volcanic main island of Mauritius is ringed by coral reefs; there are some 20 smaller islands. The islands were ruled in turn by the Dutch, French, and British, whose legacy includes a parliamentary form of government. Independence came in 1968. Diversifying sugarcane farming with manufacturing and tourism strengthens the nation, a blend of Africans, Indians, Europeans, and Chinese.

Mayotte (France)
FRENCH OVERSEAS TERRITORIAL COLLECTIVITY

CONTINENT	Africa
AREA	374 sq km (144 sq mi)
POPULATION	167,000
CAPITAL	Mamoudzou 1,000
RELIGION	Muslim, Christian
LANGUAGE	Mahorian, French

Lying in the Indian Ocean off the coast of Africa, Mayotte was the only one of the four Comoros islands that did not opt for independence in 1976. The economy depends on agriculture and on aid from France.

Mexico
UNITED MEXICAN STATES

CONTINENT	North America
AREA	1,964,375 sq km (758,449 sq mi)
POPULATION	104,878,000
CAPITAL	Mexico City 19,013,000
RELIGION	Roman Catholic, Protestant
LANGUAGE	Spanish, various Mayan, Nahuatl, and other indigenous languages
LITERACY	92%
LIFE EXPECTANCY	75
CURRENCY	Mexican peso
GDP PER CAPITA	$8,900
ECONOMY	IND: food and beverages, tobacco, chemicals, iron and steel. AGR: corn, wheat, soybeans, rice, beef, wood products. EXP: manufactured goods, oil and oil products, silver, fruits, vegetables.

Mexico straddles the southern part of North America, with coastal plains along the Pacific and Atlantic coasts rising to a central plateau. Northern Mexico is desert-like, while the south is a mountainous jungle containing Maya and Aztec ruins. Most people live in the densely populated "waist" of the country, including the cities of Veracruz, Mexico City, and Guadalajara. Most Mexicans are of mixed Spanish and Indian descent, but about 30 percent are Indian—and millions still speak Indian languages in the southeast.

A 3,115-kilometer (1,936-mile) common border, commerce, and tourism link the world's largest Spanish-speaking country to its northern neighbor. Mexico is one of the world's largest oil producers—oil and gas provide a third of the government's revenue. Mexico exports oil to the U.S., which returns manufactured goods and foodstuffs. Agriculture remains an important employer. Mexico's system of communal farms, or *ejidos*, was reformed in the 1990s to promote private investment and large-scale agriculture. The North American Free Trade Agreement (NAFTA) makes Mexico highly dependent on exports to the U.S., and the downturn in U.S. business in 2001 resulted in little or no growth in the Mexican economy.

The nation is blessed with abundant minerals—notably silver, copper, sulfur, lead, and zinc—advanced technology, and a huge workforce. It profits from its maquiladora border industry: products are assembled at mostly U.S.-owned plants, then sent to the U.S. and elsewhere. The foreign plant owners gain from the lower cost of doing business in Mexico, and Mexicans gain jobs. However, many poor Mexicans try to cross the border for jobs in the U.S.—an estimated five million Mexican immigrants are in the U.S. illegally.

Mexico's declining birthrate promises some relief from the crushing pressure of its population. In 2000 Mexico became the 11th country in the world to have 100 million people—more than double its 1970 population of 48 million. With more than 19 million people, many living in barrio slums, Greater Mexico City is one of the world's largest urbanized areas. Tough environmental restrictions have been enacted to cope with increasingly dangerous levels of air and water pollution.

Tax reform, privatization of state-run industries, and more open trade policies have improved competitiveness and boosted exports. Education funding is increasing, and authority is being transferred from the federal to state governments to improve accountability. New four-lane highways provide a network helping business and tourism.

Micronesia
FEDERATED STATES OF
MICRONESIA

CONTINENT	Australia/Oceania
AREA	702 sq km (271 sq mi)
POPULATION	115,000
CAPITAL	Palikir 7,000
RELIGION	Roman Catholic, Protestant
LANGUAGE	English, Trukese, Pohnpeian, Yapese, Kosraean, Ulithian
LITERACY	89%
LIFE EXPECTANCY	68
CURRENCY	US dollar
GDP PER CAPITA	$2,000
ECONOMY	IND: tourism, construction, fish processing, specialized aquaculture, craft items from shell, wood, and pearls. AGR: black pepper, tropical fruits and vegetables, coconuts, cassava (tapioca), pigs. EXP: fish, garments, bananas, black pepper.

Micronesia consists of the Caroline Islands Archipelago in the western Pacific Ocean. In 1899 Spain sold the islands to Germany. Japan later occupied the region and fortified the islands just before World War II. In 1986 these 600 islands and atolls, formerly part of the U.S.-administered Trust Territory of the Pacific Islands, became self-governing in free association with the United States. American aid is crucial to the islands' economy.

Moldova
REPUBLIC OF MOLDOVA

CONTINENT	Europe
AREA	33,800 sq km (13,050 sq mi)
POPULATION	4,253,000
CAPITAL	Chişinău 662,000
RELIGION	Eastern Orthodox
LANGUAGE	Moldovan, Russian, Gagauz
LITERACY	99%
LIFE EXPECTANCY	68
CURRENCY	Moldovan leu
GDP PER CAPITA	$2,600
ECONOMY	IND: food processing, agricultural machinery, foundry equipment, refrigerators and freezers. AGR: vegetables, fruits, wine, grain, beef. EXP: foodstuffs, textiles, machinery.

Landlocked Moldova lies in eastern Europe between Romania and Ukraine. It consists of hilly grassland drained by the Prut and Dniester Rivers, and the economy is mainly agricultural. Most of Moldova was part of Romania before World War II, and two-thirds of Moldovans speak Romanian. Soviets annexed Moldova in 1940, and Russians and Ukrainians settled in the industrial region east of the Dniester (known as Transdniestria). After Moldova gained independence in 1991, Transdniestria seceded, making Tiraspol its capital.

Moldova does not recognize Transdniestria's independence and works to resolve the conflict.

Monaco
PRINCIPALITY OF MONACO

CONTINENT	Europe
AREA	2 sq km (1 sq mi)
POPULATION	34,000
CAPITAL	Monaco 34,000
RELIGION	Roman Catholic
LANGUAGE	French, English, Italian, Monegasque
LITERACY	99%
LIFE EXPECTANCY	79
CURRENCY	euro
GDP PER CAPITA	$27,000
ECONOMY	IND: tourism, banking, construction, small-scale industrial and consumer products. AGR: NA EXP: NA

Monaco occupies a mostly rocky strip of land on France's Mediterranean coast. An unparalleled luxury resort since the mid-19th century, Monaco has a reputation that belies its size. Millions come to Monaco each year for the beachfront hotels, the yacht harbor, the Opera House, and the famous Monte Carlo Casino. Wealthy residents benefit from no income tax. The House of Grimaldi has ruled since 1297, except between 1793 and 1814. Tourism and gambling drive the economy, and it is also a major banking center.

Mongolia
MONGOLIA

CONTINENT Asia
AREA 1,564,116 sq km (603,909 sq mi)
POPULATION 2,500,000
CAPITAL Ulaanbaatar 812,000
RELIGION Tibetan Buddhist, Lamaism
LANGUAGE Khalkha Mongol, Turkic, Russian
LITERACY 99%
LIFE EXPECTANCY 65
CURRENCY togrog/tugrik
GDP PER CAPITA $1,900
ECONOMY IND: construction materials, mining, oil, food and beverages, processing of animal products. AGR: wheat, barley, potatoes, forage crops; sheep. EXP: copper, gold, livestock, animal products, cashmere, wool, hides.

Mongolia is a large landlocked country between two larger countries—Russia and China. Located on mountains and plateaus, it is one of the world's highest countries, with an average elevation of 1,580 meters (5,180 feet). Mongolia suffers temperature extremes, and southern Mongolia is dominated by the Gobi desert. Genghis Khan's Mongol horsemen conquered much of Asia and Europe during the 13th century. Mongolia became a communist country in 1924, but in 1990 multiparty elections were held. Poverty is a major concern, but copper, cashmere, and gold exports help the economy.

Montserrat (U.K.)
BRITISH OVERSEAS TERRITORY

CONTINENT North America
AREA 102 sq km (39 sq mi)
POPULATION 5,000
CAPITAL Plymouth (abandoned)
RELIGION Anglican, Methodist, Roman Catholic
LANGUAGE English

In 1995 the Soufriere Hills volcano became active in southern Montserrat. The eruptions forced the evacuation of the southern half of the island (including the capital, Plymouth), and some 6,000 Montserratians have left the island.

Morocco
KINGDOM OF MOROCCO

CONTINENT Africa
AREA 710,850 sq km (274,461 sq mi)
POPULATION 30,366,000
CAPITAL Rabat 1,759,000
RELIGION Muslim
LANGUAGE Arabic, Berber dialects, French
LITERACY 52%
LIFE EXPECTANCY 70
CURRENCY Moroccan dirham
GDP PER CAPITA $3,900
ECONOMY IND: phosphate rock mining and processing, food processing, leather goods, textiles. AGR: barley, wheat, citrus, wine, livestock. EXP: clothing, fish, inorganic chemicals, transistors, crude minerals, fertilizers.

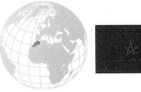

Lying in the northwest corner of Africa, Morocco is dominated by the Atlas Mountains, which separate the fertile coastal regions from the harsh Sahara. The high mountains helped protect Morocco from European colonialism until 1912. From 1912 to 1956 the country was divided into French and Spanish zones—two small Spanish enclaves remain, Ceuta and Melilla. Mosques, minarets, and bazaars typify Morocco, 99 percent of whose inhabitants are Muslims. King Mohammed VI, who has ruled since 1999, claims descent from the Prophet Muhammad. Morocco today is one of only three kingdoms left on the continent of Africa—the others, Lesotho and Swaziland, are small, southern African countries.

Most Moroccans live in cities such as Fez, Casablanca, and Marrakech, on the coastal plain. Although rural people are crowding into cities, Morocco remains primarily a nation of farmers. Many Moroccans emigrate to Spain and other European Union countries for better economic opportunities. Drought, unemployment, and dispute over control of phosphate-rich Western Sahara (formerly Spanish Sahara) have taxed the country. In spite of a 1991 UN-supervised cease-fire, sporadic warfare continues between the Moroccan Army and Algerian-backed Polisario (the Western Sahara independence movement based in Tindouf, Algeria). Moroccan forces built a 2,500 kilometer (1,500 mile) sand wall to keep Polisario fighters out. A UN-sponsored referendum of Western Sahara residents is now planned to determine the status of the area, but disputes regarding the referendum remain unresolved.

Mozambique
REPUBLIC OF MOZAMBIQUE

CONTINENT	Africa
AREA	799,380 sq km
	(308,642 sq mi)
POPULATION	17,479,000
CAPITAL	Maputo 1,221,000
RELIGION	Indigenous beliefs,
	Christian, Muslim
LANGUAGE	Portuguese, indigenous
	dialects
LITERACY	48%
LIFE EXPECTANCY	34
CURRENCY	metical
GDP PER CAPITA	$1,100
ECONOMY	IND: food, beverages,
	chemicals, aluminum,
	petroleum products,
	textiles. AGR: cotton,
	cashew nuts, sugar-
	cane, tea, beef. EXP:
	aluminum, prawns,
	cashews, cotton, sugar.

Mozambique, on the east coast of southern Africa, is mainly a savanna plateau drained by the mighty Limpopo and Zambezi Rivers, with highlands to the north. It has a tropical climate that can produce heavy flooding along the rivers. In 2001 flooding along the Zambezi River valley forced 70,000 people to flee their homes, and the World Bank estimated that a total of 491,000 were displaced by floods throughout the country. Most people live along the coasts or in the river valleys.

Infusions of aid are essential to a country devastated by decades of war, drought, and floods. After nearly five centuries of Portuguese presence, Mozambique won independence in 1975. An exodus of skilled Portuguese workers followed, and the country became a one-party state allied to the Soviet bloc. Mozambique was drawn into a long struggle against white rule in Rhodesia and South Africa. In 1989 the government renounced Marxism, and a democratic constitution was written in 1990. Fighting between the government and right-wing guerrillas, which claimed a hundred thousand lives and displaced more than four million, ended in 1992. In 1994 multiparty elections ushered in a new government, which has focused on diversifying the country's economy away from small-scale agriculture. Production of food and manufactured goods is steadily increasing, and a

large-scale aluminum smelter started in 2000. Solid economic success bodes well for the future.

Myanmar
UNION OF MYANMAR

CONTINENT	Asia
AREA	676,552 sq km
	(261,218 sq mi)
POPULATION	49,481,000
CAPITAL	Yangon 3,874,000
RELIGION	Buddhist, Christian,
	Muslim
LANGUAGE	Burmese, minor
	languages
LITERACY	83%
LIFE EXPECTANCY	57
CURRENCY	kyat
GDP PER CAPITA	$1,700
ECONOMY	IND: agricultural
	processing; knit and
	woven apparel, wood
	and wood products;
	copper, tin, tungsten,
	iron, gems, jade. AGR:
	rice, pulses, beans,
	sesame, hardwood
	(teak), fish. EXP: gas,
	wood products, pulses,
	beans, fish, rice.

In 1989 the largest nation of mainland Southeast Asia changed its name from Burma to Myanmar and that of its capital from Rangoon to Yangon. The name changes were made by an unelected military regime, and many continue to use the old names. Geographically, the country's Irrawaddy basin is surrounded on three sides by densely forested mountains and plateaus. Most people live in the fertile valley and delta of the Irrawaddy River.

The majority of Myanmar's people are ethnic Burmans, and other ethnic groups (including Shans, Karens, and Kachins) add up to some 30 percent of the population. Ethnic minorities are dominant in border and mountainous areas: Shan in the north and northeast (Indian and Thai borders), Karen in the southeast (Thai border), and Kachin in the far north (Chinese border). The military regime has brutally suppressed ethnic groups wanting rights and autonomy, and many ethnic insurgencies operate against it.

Independence from Britain in 1948 was followed by isolationism and socialism. Military governments have ruled Myanmar since 1962 and have been accused of corruption, heroin trafficking, and human rights violations—including forcible relocation of civilians and use of forced labor. In 1988 military forces killed more than a thousand pro-democracy demonstrators. In 1990 national elections were held for parliament, but the military refused to recognize the results. Myanmar is a resource-rich country with a strong agricultural base, and is a leading producer of gems, jade, and teak. However, military rule prevents the economy from developing, and the Burmese people remain poor and are getting poorer.

Namibia
REPUBLIC OF NAMIBIA

CONTINENT	Africa
AREA	824,292 sq km (318,261 sq mi)
POPULATION	1,927,000
CAPITAL	Windhoek 237,000
RELIGION	Christian, indigenous beliefs
LANGUAGE	English, Afrikaans, German, indigenous languages
LITERACY	84%
LIFE EXPECTANCY	49
CURRENCY	Namibian dollar, South African rand
GDP PER CAPITA	$6,900
ECONOMY	IND: meatpacking, fish processing, dairy products, mining (diamonds, lead, zinc). AGR: millet, sorghum, peanuts, livestock, fish. EXP: diamonds, copper, gold, zinc, lead.

Namibia is a large and sparsely populated country on Africa's southwest coast. The low population can be attributed to the country's harsh geography—the coastal Namib Desert, central semiarid mountains, and Kalahari Desert east of the mountains. About 87 percent of the residents are black, and 6 percent are white. During World War I, South Africa captured the area from Germany; ethnic Germans are still a sizable minority.

Independence from South Africa was achieved in 1990. The multiparty, multiracial democracy inherits an economy based on mining (mostly diamonds), sheep and cattle ranching, and fishing.

Nauru
REPUBLIC OF NAURU

CONTINENT	Australia/Oceania
AREA	21 sq km (8 sq mi)
POPULATION	12,000
CAPITAL	Yaren 670
RELIGION	Protestant, Roman Catholic
LANGUAGE	Nauruan, English
LITERACY	NA
LIFE EXPECTANCY	61
CURRENCY	Australian dollar
GDP PER CAPITA	$5,000
ECONOMY	IND: phosphate mining, offshore banking, coconut products. AGR: coconuts. EXP: phosphates.

Nauru is a small oval-shaped island in the western Pacific. The interior phosphate plateau, comprising 60 percent of the land area, has been extensively mined, leaving a jagged and pitted landscape. Germany annexed Nauru in 1888, and Australia took it over in 1914. After World War II it was a joint trust territory of Australia, Britain, and New Zealand until it became independent in 1968. Phosphate exports earned economic stability for the country, but deposits could run out by 2005.

Nepal
KINGDOM OF NEPAL

CONTINENT	Asia
AREA	147,181 sq km
	(56,827 sq mi)
POPULATION	25,164,000
CAPITAL	Kathmandu 741,000
RELIGION	Hindu, Buddhist,
	Muslim
LANGUAGE	Nepali, English, many
	other languages and
	dialects
LITERACY	45%
LIFE EXPECTANCY	59
CURRENCY	Nepalese rupee
GDP PER CAPITA	$1,400
ECONOMY	IND: tourism, carpets,
	textiles; small rice,
	jute, sugar, and oilseed
	mills, cigarettes, cement
	and brick production.
	AGR: rice, corn, wheat,
	sugarcane, milk. EXP:
	carpets, clothing,
	leather goods, jute
	goods, grain.

Nepal lies between China and India in South Asia. The king of this constitutional monarchy traces his lineage to the ruler of Gorkha, who unified the area in the late 18th century. The present monarch, King Gyanendra, came to the throne in 2001 after the tragic murder of the previous king. Violent political protest in early 1990 opened the way to multiparty government. Since then, no single party has been able to form a majority, resulting in a number of incompatible coalitions. The king postponed elections in 2002 because Maoist insurgents controlled nearly half of Nepal—mostly the poorer western region. This political instability has not fostered economic reforms, and Nepal remains one of the world's poorest countries.

Most Nepalese live in the central, hilly region, which embraces the Kathmandu Valley, and in the southern plain known as the Terai. The cutting of trees for fuel—increased by demands of a booming tourist industry— causes erosion. Rivers that spring from the Himalaya generate electricity for local use and potentially for export. Nepal possesses the greatest altitude variation on the Earth, from the lowlands near sea level to Mount Everest at 8,850 meters (29,035 feet). Everest, named after British surveyor Sir George Everest, is known as Chomolungma by the local Sherpas, meaning "Goddess Mother of the World"; related to this is the Chinese name Qomolangma. The Nepali word for Everest, Sagarmatha, translates as "Forehead of the Sky." Sherpas benefit from the mountaineering boom and tourism in the Everest region, owning much of the lodging and transportation. They teach visitors about Sherpa culture and Buddhism's love of the land.

Netherlands
KINGDOM OF THE NETHERLANDS

CONTINENT	Europe
AREA	41,528 sq km
	(16,034 sq mi)
POPULATION	16,237,000
CAPITAL	Amsterdam 1,145,000
RELIGION	Roman Catholic,
	Protestant, Muslim
LANGUAGE	Dutch, Frisian
LITERACY	99%
LIFE EXPECTANCY	78
CURRENCY	euro
GDP PER CAPITA	$27,200
ECONOMY	IND: agroindustries,
	metal and engineering
	products, electrical
	machinery and equip-
	ment, chemicals. AGR:
	grains, potatoes, sugar
	beets, fruits, livestock.
	EXP: machinery and
	equipment, chemicals,
	fuels, foodstuffs.

The Netherlands faces the North Sea in western Europe. The Dutch have a saying: "God made the Earth, but the Dutch made Holland." The first defenses against the sea went up some 800 years ago. Today more than 2,400 kilometers (1,491 miles) of dikes shield the low, flat land—almost half of which lies below sea level—from invasion by the North Sea. Without the existing dikes 65 percent of the country would be flooded daily.

Reclamation of the Zuider Zee has created 165,000 hectares (407,700 acres) of arable land—a precious commodity in this densely populated nation. About 60 percent of the country is farmed, with super-efficiency, by just 2 percent of the workforce. Only the U.S. and France export more agricultural goods. Located at the mouth of the Rhine River, the Netherlands is a gateway to northwestern Europe and participates in the European common currency, the euro. Rotterdam, the world's largest and busiest general-cargo port, includes Europoort, a petroleum-refining center.

For several decades natural gas production has subsidized a welfare system. Funds are needed for continued flood-control efforts, for cleaning up the Rhine and the North Sea, and for combating damage to forests by acid rain. The government seeks to cut back on all forms of pollution by up to 90 percent. Tourism is important to the country, and many come to see Dutch art, architecture—and the flowers. Tulips are a major industry, and the Dutch produce billions of bulbs a year—more than any other country.

The Netherlands was a major colonial power, but its largest colonies, Indonesia and Suriname, gained independence decades ago. The islands of Aruba and the Netherlands Antilles still form part of the Kingdom of the Netherlands.

OVERSEAS REGIONS

Aruba

Netherlands Antilles

Netherlands Antilles
OVERSEAS REGION OF THE NETHERLANDS

CONTINENT	North America
AREA	800 sq km (309 sq mi)
POPULATION	179,000
CAPITAL	Willemstad 134,000
RELIGION	Roman Catholic, Protestant
LANGUAGE	Dutch, Papiamento, English, Spanish

The Netherlands Antilles consists of two main island groups: Curaçao and Bonaire in the southern Caribbean, and Saba, St. Eustatius (Statia), and the southern part of St. Martin (called Sint Maarten) in the northern Caribbean.

New Caledonia (France)
FRENCH OVERSEAS TERRITORY

CONTINENT	Australia/Oceania
AREA	19,060 sq km (7,359 sq mi)
POPULATION	222,000
CAPITAL	Nouméa 140,000
RELIGION	Roman Catholic, Protestant
LANGUAGE	French, 33 Melanesian-Polynesian dialects

New Caledonia consists of a large island and some smaller islands located in the South Pacific east of Australia. An emerald green lagoon surrounds the main island; the economy is based on tourism, nickel mining, and French aid.

New Zealand
NEW ZEALAND

CONTINENT	Australia/Oceania
AREA	270,534 sq km (104,454 sq mi)
POPULATION	4,008,000
CAPITAL	Wellington 343,000
RELIGION	Protestant, Roman Catholic
LANGUAGE	English, Maori
LITERACY	99%
LIFE EXPECTANCY	78
CURRENCY	New Zealand dollar
GDP PER CAPITA	$20,100
ECONOMY	IND: food processing, wood and paper products, textiles, machinery. AGR: wheat, barley, potatoes, pulses, wool, fish. EXP: dairy products, meat, wood and wood products, fish, chemicals, wool, mutton.

New Zealand is a fertile and mountainous group of islands in the southwestern Pacific Ocean. "It is a land uplifted high," wrote Abel Tasman, a Dutch navigator who was the first European to sight New Zealand, in 1642. Snowy peaks, fjord-scarred shores, and pastures dotted with sheep define this country.

New Zealand, a parliamentary democracy modeled on that of the United Kingdom, has been a self-governing British dominion since 1907. It became a founding member of the British Commonwealth in 1926.

One in three citizens—Kiwis—lives in or around the city of Auckland. Rugby clubs with names such as Canterbury and Wellington reveal a nation peopled mostly by descendants of British settlers. The indigenous Maori constitute about 14 percent of New Zealanders; recent immigrants—primarily from Samoa and Fiji—give Auckland one of the world's largest Polynesian populations.

The export-driven country, whose chief trading partner used to be the United Kingdom, faltered in 1973 when Britain joined the European Union. The loss of preferential treatment prompted a search for new markets. Japan, Australia, and the U.S. now buy half of all exports, which include wool, mutton, lamb, beef, cheese, fish, and chemicals.

New Zealand plays an active role in helping democratic nations and emerging Pacific island economies. It sent troops to East Timor when violence broke out in 1999, and it provided millions of dollars to the South Pacific island of Niue after it was devastated by a tropical cyclone in 2004. Niue and the Cook Islands enjoy a status of self-government in free association with New Zealand.

ASSOCIATED ISLANDS

Cook Islands

Niue

Tokelau

Nicaragua
REPUBLIC OF NICARAGUA

CONTINENT	North America
AREA	130,000 sq km (50,193 sq mi)
POPULATION	5,482,000
CAPITAL	Managua 1,098,000
RELIGION	Roman Catholic, Protestant
LANGUAGE	Spanish, English, indigenous languages
LITERACY	68%
LIFE EXPECTANCY	69
CURRENCY	gold cordoba
GDP PER CAPITA	$2,200
ECONOMY	IND: food processing, chemicals, machinery and metal products, textiles. AGR: coffee, bananas, sugarcane, cotton, beef, veal. EXP: coffee, shrimp and lobster, cotton, tobacco, bananas.

Natural disasters and the consequences of civil war have beset this largest Central American country. Volcanoes and earthquakes along the Pacific coast are a constant threat, and hurricanes hit the low-lying Caribbean coast. With the Sandinista's overthrow of Anastasio Somoza in 1979, ending his family's 42-year dictatorship, Nicaragua came under the control of a junta. Eight years of civil war between the Sandinista regime and the U.S.-funded rebels (contras)

ended in 1988. Peace brought democracy, but poverty and corruption are major problems.

Niger
REPUBLIC OF NIGER

CONTINENT	Africa
AREA	1,267,000 sq km (489,191 sq mi)
POPULATION	12,073,000
CAPITAL	Niamey 890,000
RELIGION	Muslim, indigenous beliefs, Christian
LANGUAGE	French, Hausa, Djerma
LITERACY	18%
LIFE EXPECTANCY	45
CURRENCY	CFA franc
GDP PER CAPITA	$800
ECONOMY	IND: uranium mining, cement, brick, textiles. AGR: cowpeas, cotton, peanuts, millet, cattle. EXP: uranium ore, livestock, cowpeas, onions.

Landlocked in the west of Africa, Niger uses the Niger River as a link to the distant sea. Most people live in the southern savanna region; the north is consumed by the Sahara. This crossroads of ancient trading empires became a French colony in 1922 and gained independence in 1960. Military coups and governments occurred between 1974 and 1999; free elections in 1999 restored democracy. Economic problems have been exacerbated by falling world demand for uranium, the country's most valuable resource. At the mercy of drought and desertification, Niger relies on foreign aid.

Nigeria
FEDERAL REPUBLIC OF NIGERIA

CONTINENT	Africa
AREA	923,768 sq km (356,669 sq mi)
POPULATION	133,882,000
CAPITAL	Abuja 452,000
RELIGION	Muslim, Christian, indigenous beliefs
LANGUAGE	English, Hausa, Yoruba, Igbo, Fulani
LITERACY	68%
LIFE EXPECTANCY	52
CURRENCY	naira
GDP PER CAPITA	$900
ECONOMY	IND: crude oil, coal, tin, columbite, palm oil, peanuts, cotton, rubber, wood, hides and skins, textiles. AGR: cacao, peanuts, palm oil, corn, cattle, timber, fish. EXP: petroleum and petroleum products, cacao, rubber.

Located in West Africa, Nigeria is the population giant of Africa, with more than 130 million people. The terrain changes from the oil-rich Niger Delta in the south to a belt of rain forests inland and to high savanna-covered plateaus in the north.

The population is as diverse as it is large, with some 250 ethnic groups. Nigeria's three largest ethnic groups are: Hausa-Fulani (29 percent of the population), Yoruba (21 percent) and Igbo, or Ibo (18 percent). Northern Nigeria is mostly Islamic and dominated by the Hausa-Fulani ethnic group. Southern Nigeria is more westernized and urbanized than the north, with the Yoruba in the southwest and the Igbo in the southeast. It is estimated that about half the Yorubas are Christian and half Muslim, though many maintain traditional beliefs. The Igbo in the southwest tend to be Christian; many are Roman Catholic.

A century of British rule ended in 1960. After independence ethnic tensions increased, deepened by the rift between the poor north and the more prosperous south. Civil war raged from 1967 to 1970, when the Igbo fought unsuccessfully for autonomy as the Republic of Biafra. The end of the civil war did not mark a return to political stability. After decades of military coups and military rule, free elections were held in 1999 that brought Nigeria back on the road to democracy. The system of government is based on the United States model with a federal government and 36 states, with a Federal Capital Territory at Abuja. The nation's capital moved from Lagos to Abuja in 1991. The federal form of government and location of the capital seek to balance the three major ethnic groups—and subdue ethnic and regional conflict. However, introduction of sharia (criminal code based in Islamic law) in 12 northern states in 2000 provoked violence between Christians and Muslims, leading to thousands of deaths.

Since the oil boom of the 1970s Nigeria has had an unhealthy dependence on crude oil. In 2002 oil and gas exports accounted for 98 percent of export earnings—providing 83 percent of the federal government's revenue. Agriculture suffers from years of mismanagement and corruption. The country's poor transportation infrastructure hinders economic development.

Niue (New Zealand)
SELF-GOVERNING IN FREE ASSOCIATION WITH NEW ZEALAND

CONTINENT	Australia/Oceania
AREA	263 sq km (102 sq mi)
POPULATION	1,600
CAPITAL	Alofi 1,000
RELIGION	Ekalesia Niue (a Protestant Church)
LANGUAGE	Niuean, English

Norfolk Island (Australia)
AUSTRALIAN EXTERNAL TERRITORY

CONTINENT	Australia/Oceania
AREA	35 sq km (14 sq mi)
POPULATION	2,500
CAPITAL	Kingston 1,000
RELIGION	Protestant, Roman Catholic
LANGUAGE	English, Norfolk

Northern Mariana Islands (U.S.)
COMMONWEALTH IN
POLITICAL UNION WITH
THE UNITED STATES

CONTINENT	Australia/Oceania
AREA	464 sq km (179 sq mi)
POPULATION	78,000
CAPITAL	Saipan 62,400
RELIGION	Christian, traditional beliefs
LANGUAGE	English, Chamorro, Carolinian

Located in the northwest Pacific, the Northern Mariana Islands consists of 14 islands, though 90 percent of the people live on Saipan. Many World War II battles occurred here. Tourism is a major employer in this self-governing territory.

Norway
KINGDOM OF NORWAY

CONTINENT	Europe
AREA	323,758 sq km (125,004 sq mi)
POPULATION	4,568,000
CAPITAL	Oslo 795,000
RELIGION	Evangelical Lutheran
LANGUAGE	Norwegian
LITERACY	100%
LIFE EXPECTANCY	79
CURRENCY	Norwegian krone
GDP PER CAPITA	$33,000
ECONOMY	IND: petroleum and gas, food processing, shipbuilding, pulp and paper products. AGR: barley, wheat, potatoes, pork, fish.

EXP: petroleum and petroleum products, machinery and equipment, metals, chemicals, ships, fish.

In northern Europe, the thinly populated Kingdom of Norway, whose dominion includes the Arctic islands of Svalbard and Jan Mayen, is partitioned by mountains and has a fjord-gashed shoreline that exceeds 21,000 kilometers (13,050 miles). Ever since Vikings left home waters in the ninth century, Norway has drawn strength from the sea. Today its merchant and oil-tanker fleets are among the world's largest, and its fishing flotilla lands Western Europe's biggest catch.

Wealth from oil and gas in the North Sea, first tapped in the early 1970s, subsidizes public health and welfare programs. Recession required austerity in the 1980s, but since then Norway has enjoyed a higher economic growth rate than many other European countries. In 2002 Norway was the world's third largest oil exporter. A member of NATO, the UN, and the European Free Trade Association, Norway voted against joining the European Union in 1994. The country is home to some 30,000 Sami, or Lapps—the largest population of the Arctic reindeer herders.

Norway opened the world's longest road tunnel in 2000, with a length of 24.5 kilometers (15.3 miles). The Laerdal Tunnel is a third longer than the St. Gotthard Tunnel in Switzerland—previously the longest. Norwegians hope the tunnel, on the main Oslo-Bergen highway, will boost tourism to the spectacular fjords. The tunnel itself is something of a tourist attraction, featuring immense caverns that simulate sunrise— to help refresh drivers, or give them a chance to pull over and rest.

DEPENDENCIES

Bouvet Island

Jan Mayen

Svalbard

Oman
SULTANATE OF OMAN

CONTINENT	Asia
AREA	309,500 sq km (119,500 sq mi)
POPULATION	2,637,000
CAPITAL	Muscat 638,000
RELIGION	Ibadhi Muslim, Sunni Muslim, Shiite Muslim, Hindu
LANGUAGE	Arabic, English, Baluchi, Urdu, Indian dialects
LITERACY	76%
LIFE EXPECTANCY	73
CURRENCY	Omani rial
GDP PER CAPITA	$8,300
ECONOMY	IND: crude oil production and refining, natural gas production, construction. AGR: dates, limes, bananas, alfalfa; camels, fish. EXP: petroleum, reexports, fish, metals, textiles.

At the mouth of the Persian Gulf and in the path of trade routes to East Africa and the Orient, Oman built a commercial empire centuries ago. After the mid-19th century, power struggles weakened the sultanate, strengthening bonds to the British Empire. In 1970 British-educated Qaboos bin Said deposed his father and, as sultan, began modernizing. Oman allows the United States to use port and air base facilities. Oil, exported since 1967, has financed roads, schools, and hospitals. The majority of Omanis still farm or fish, and protection of fisheries and coastal zones is promoted.

Pakistan
ISLAMIC REPUBLIC OF PAKISTAN

CONTINENT	Asia
AREA	796,095 sq km
	(307,374 sq mi)
POPULATION	149,147,000
CAPITAL	Islamabad 698,000
RELIGION	Sunni and Shiite Muslim, Christian, Hindu
LANGUAGE	Punjabi, Sindhi, Siraiki, Pashtu, Urdu, English
LITERACY	46%
LIFE EXPECTANCY	60
CURRENCY	Pakistani rupee
GDP PER CAPITA	$2,000
ECONOMY	IND: textiles and apparel, food processing, beverages,

construction materials. AGR: cotton, wheat, rice, sugarcane, milk. EXP: textiles, rice, leather, sports goods, carpets and rugs.

Pakistan is in the northwest part of South Asia. The eastern and southern parts of the country are dominated by the Indus River and its tributaries. Most of Pakistan's population lives along the Indus. West of the Indus the land becomes increasingly arid and mountainous. To the north the land rises to the great mountains of the Hindu Kush and Karakoram — including K2, the world's second highest mountain after Everest, at 8,611 meters (28,250 feet).

The military has loomed large in Pakistan, the western portion of a bifurcated country created for Muslims when the British relinquished predominantly Hindu India in 1947. Relations with New Delhi, embittered by claims to Kashmir, worsened as a result of India's role in East Pakistan's rebirth as Bangladesh in 1971. Military rule followed Gen. Zia ul-Haq's 1977 coup that toppled Prime Minister Ali Bhutto. After Zia's death in a plane crash in 1988, Benazir Bhutto, daughter of the former prime minister, became the first woman elected to lead a Muslim nation. She restored civil rights, but was plagued by problems: continuing tension and a presumed nuclear arms rivalry with India, 3.5 million refugees from the war in neighboring Afghanistan, and a growing trade in heroin. Ousted in 1990, Bhutto was reelected in 1993; her government was dismissed in 1996 on charges of corruption. Political unrest and a general failure of government followed the election of a new prime minister in 1997. Two years later a military coup led to General Pervez Musharraf becoming president. National and provincial elections were held in 2002, with these assemblies giving President Musharraf a vote of confidence in 2004.

Tensions with India came to a head in 1998 when both that country as well as Pakistan conducted nuclear tests. Kashmir is the key issue for India and Pakistan. Pakistan's interest in Kashmir focuses on protecting the Muslim population in that region and in securing the headwaters of the Indus River, the country's lifeline. Agriculture is concentrated in the extensively irrigated Indus Basin. Despite an increase in cotton, wheat, and rice production, feeding the growing population is a constant challenge. Favorable relations with China have been a pillar of Pakistan's foreign policy—helping to offset the power of India. Since 2001 Pakistan has been a key U.S. ally in the war on terrorism and in bringing democracy to neighboring Afghanistan.

Palau
REPUBLIC OF PALAU

CONTINENT	Australia/Oceania
AREA	489 sq km (189 sq mi)
POPULATION	20,000
CAPITAL	Koror 14,000
RELIGION	Roman Catholic, Protestant, Modekngei (indigenous)
LANGUAGE	English, Palaun, Japanese, 3 additional local languages
LITERACY	92%
LIFE EXPECTANCY	69
CURRENCY	US dollar
GDP PER CAPITA	$9,000
ECONOMY	IND: tourism, craft items, construction, garment making. AGR: coconuts, copra, cassava (tapioca), sweet potatoes. EXP: shellfish, tuna, copra, garments.

Located in the western Pacific, the more than 250 islands that constitute Palau—a Japanese stronghold during World War II—were assigned to U.S. administration by the United Nations in 1947. Economically tied to the U.S., the territory became an independent nation in October 1994. About 70 percent of Palauans live in the capital city of Koror on the island of Koror. Tourism is the country's main industry, with the rich marine environment inviting snorkeling and scuba diving.

Palestinian Areas
See listing under Israel.

Panama
REPUBLIC OF PANAMA

CONTINENT	North America
AREA	75,517 sq km (29,157 sq mi)
POPULATION	2,981,000
CAPITAL	Panama City 930,000
RELIGION	Roman Catholic, Protestant
LANGUAGE	Spanish, English
LITERACY	93%
LIFE EXPECTANCY	74
CURRENCY	balboa, US dollar
GDP PER CAPITA	$6,200
ECONOMY	IND: construction, petroleum refining, brewing, cement and other construction materials. AGR: bananas, rice, corn, coffee, livestock, shrimp. EXP: bananas, shrimp, sugar, coffee, clothing.

Panama is a narrow land bridge connecting North and South America. The Panama Canal, built by the United States after Panama's independence from Colombia in 1903, joins the Atlantic and Pacific Oceans. In 1989 U.S. troops overthrew Gen. Manuel Noriega, following his indictment for complicity in drug trafficking. Panama's first woman president, Mireya Moscoso, was elected in 1999—the same year that her country assumed full control of the Panama Canal.

Papua New Guinea
INDEPENDENT STATE OF PAPUA NEW GUINEA

CONTINENT	Australia/Oceania
AREA	462,840 sq km (178,703 sq mi)
POPULATION	5,525,000
CAPITAL	Port Moresby 275,000
RELIGION	Protestant, indigenous beliefs, Roman Catholic
LANGUAGE	715 indigenous languages
LITERACY	66%
LIFE EXPECTANCY	57
CURRENCY	kina
GDP PER CAPITA	$2,100
ECONOMY	IND: copra crushing, palm oil processing, plywood production, wood chip production. AGR: coffee, cacao, coconuts, palm kernels, poultry. EXP: oil, gold, copper ore, logs.

Papua New Guinea, an island country in the western Pacific, gained independence from Australia in 1975. An abundance of minerals and petroleum brightens the outlook for this tropical nation, comprising eastern New Guinea and many small islands — including Bougainville and the Bismarck

Archipelago. A patchwork of mountains, jungles, and swamplands, the country is home to some 700 Papuan and Melanesian tribes, each with its own language. Most of the inhabitants are subsistence farmers, although some grow cash crops.

Paraguay
REPUBLIC OF PARAGUAY

CONTINENT South America
AREA 406,752 sq km
(157,048 sq mi)
POPULATION 6,188,000
CAPITAL Asunción 1,639,000
RELIGION Roman Catholic
LANGUAGE Spanish, Guarani
LITERACY 94%
LIFE EXPECTANCY 71
CURRENCY guarani
GDP PER CAPITA $4,300
ECONOMY IND: sugar, cement, textiles, beverages. AGR: cotton, sugarcane, soybeans, corn, beef, timber. EXP: soybeans, feed, cotton, meat, edible oils.

Landlocked in central South America, the Paraguay River divides the country into a hilly, forested east and a flat plain (known as the Chaco) in the west. The Chaco, marshy near the river and turning semidesert farther west, takes up 60 percent of the country but contains only 2 percent of the people. Paraguayans, mostly a mixture of Spanish and Guaraní Indian, are the most racially homogeneous in South America. Most understand Spanish but prefer to speak Guaraní.

The War of the Triple Alliance, 1865 to 1870, looms large in the minds of Paraguayans. The war claimed the lives of nine-tenths of Paraguay's adult male population as it battled Argentina, Brazil, and Uruguay over access to the sea. Paraguay won the Chaco from Bolivia during a war in the 1930s that claimed 36,000 Paraguayan soldiers. This history helps explain the country's low population density. However, in recent decades high population growth has led to a dramatic increase in landless Paraguayan families. The situation is aggravated by the influx of an estimated 400,000 Brazilians, who have crossed into Paraguay to flee high land prices in their own country.

Paraguay possesses plenty of electric power thanks to hydroelectric dams such as Itaipú, the world's largest, built and operated jointly with Brazil. Democracy replaced dictatorship by 1993, but the government faces problems with a poor agricultural population and rapid deforestation. The government works with the U.S. to reduce illegal weapons and terrorism-financing activity in Ciudad del Este—a city on the border with Argentina and Brazil.

Peru
REPUBLIC OF PERU

CONTINENT South America
AREA 1,285,216 sq km
(496,224 sq mi)
POPULATION 27,126,000
CAPITAL Lima 8,180,000
RELIGION Roman Catholic
LANGUAGE Spanish, Quechua, Aymara
LITERACY 91%
LIFE EXPECTANCY 69
CURRENCY nuevo sol
GDP PER CAPITA $5,000
ECONOMY IND: mining of metals, petroleum, fishing, textiles, clothing, food processing. AGR: coffee, cotton, sugarcane, rice, poultry, fish. EXP: fish and fish products, gold, copper, zinc, crude petroleum and byproducts.

Peru lies on the Pacific coast of South America just south of the Equator. To the Quechua Indians Peru means "land of abundance." Sites such as Machu Picchu and Cusco recall the wealth of the Inca civilization, destroyed in the early 16th century by Spaniards, who built an empire on Peru's gold and silver. Today Peru ranks among the world's top producers of silver, copper, lead, and zinc. Its petroleum industry is one of the world's oldest, and its fisheries are among the world's richest.

The Inca capital was Cusco, but the Spanish founded Lima in 1535 along the coast and made it their capital. The Spanish preferred the lowland coast because of the climate and for trade links to Spain. The western seaboard is desert, where rain seldom falls. Lima is an oasis containing more than a quarter of Peru's population—most of European descent or mestizo. The Andean highlands occupy about a third of the country and contain mostly Quechua-speaking Indians. Quechua was the language of the Inca Empire. East of the Andes lies a sparsely populated jungle; the major city of this region is Iquitos. Iquitos can be reached by ocean-going vessels coming 3,700 kilometers (2,300 miles) up the Amazon River; recent oil discoveries have brought more people.

Peru's recent history has seen it switch between periods of democracy and dictatorship. The desperate poverty of the Indian population gave rise to the ruthless Maoist guerrilla organization Sendero Luminoso (Shining Path). The guerrillas were largely defeated but problems with poverty and illegal coca production persist.

Philippines
REPUBLIC OF THE PHILIPPINES

CONTINENT Asia

AREA 300,000 sq km (115,831 sq mi)

POPULATION 81,578,000

CAPITAL Manila 10,677,000

RELIGION Roman Catholic, Protestant, Muslim, Buddhist

LANGUAGE Filipino (based on Tagalog), English, and 8 major dialects

LITERACY 96%

LIFE EXPECTANCY 70

CURRENCY Philippine peso

GDP PER CAPITA $4,600

ECONOMY IND: textiles, pharmaceuticals, chemicals, wood products. AGR: rice, coconuts, corn, sugarcane, pork, fish. EXP: electronic equipment, machinery and transport equipment, garments, coconut products.

The Philippines, in southeastern Asia, consists of 7,107 islands lying between the South China Sea and the Pacific Ocean. The islands of Luzon and Mindanao account for two-thirds of the land area. Even though the Philippines lies just north of the world's largest Muslim state, Indonesia, it is about 94 percent Christian—mostly Roman Catholic. About five percent of Filipinos are Muslim, mostly living on the islands of Mindanao and Palawan—islands closest to the Muslim countries of Malaysia and Indonesia.

In 1521 Ferdinand Magellan claimed the Philippines for Spain, which ceded the islands to the U.S. in 1898. Independence came in 1946, after Japanese occupation ended. Widespread poverty and political corruption sparked social unrest starting in the 1970s. In 1986 President Ferdinand Marcos was compelled to hold an election. Despite his fraudulent claim to victory, Marcos was forced into exile, and Corazon Aquino, widow of a murdered opposition leader, became president.

The government continues to make progress in negotiations with Muslim rebels with a cease-fire in 2003, and it works to provide political representation and economic development to the Autonomous Region of Muslim Mindanao.

Pitcairn Islands (U.K.)
BRITISH OVERSEAS TERRITORY

CONTINENT Australia/Oceania

AREA 47 sq km (18 sq mi)

POPULATION 45

CAPITAL Adamstown 45

RELIGION Seventh-Day Adventist

LANGUAGE English, Pitcairnese

Poland
REPUBLIC OF POLAND

CONTINENT Europe

AREA 312,685 sq km (120,728 sq mi)

POPULATION 38,599,000

CAPITAL Warsaw 2,200,000

RELIGION Roman Catholic

LANGUAGE Polish

LITERACY 100%

LIFE EXPECTANCY 74

CURRENCY zloty

GDP PER CAPITA $9,700

ECONOMY IND: machine building, iron and steel, coal mining, chemicals, shipbuilding. AGR: potatoes, fruits, vegetables, wheat, poultry. EXP: machinery and transport equipment, intermediate manufactured goods.

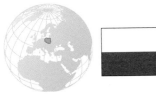

The largest country in central Europe, most of Poland is low-lying, with woods and lakes. Unlike many of its neighbors, Poland has only a minuscule minority population. Poles as a nation are unified by the Polish language and a common religion—Roman Catholicism.

Buffered by the Baltic Sea in the north and the Carpathian Mountains in the south, Poland enjoys no such natural protection to the east and west. Nazi Germany invaded in 1939 and built the Auschwitz concentration camp, where 1.35 million Jews and more than 100,000 others were murdered. After World War II, Joseph Stalin seized a chunk of eastern Poland for the Soviet Union.

Communists took power in 1947 but did not win Poles away from Roman Catholicism. In 1980 soaring prices and tumbling wages spawned Solidarity, the Eastern bloc's first free-trade union. In 1989 Solidarity swept Poland's first free elections in more than 40 years and began moving the U.S.S.R.'s largest, most populous satellite toward democracy and free enterprise. It was the first Eastern European country to overthrow communist rule.

Faced with triple-digit inflation, Poland in 1990 introduced a bold economic reform plan. It developed a market-oriented economy and joined the European Union in 2004. Poland joined NATO in 1999, and it increased its profile on the international stage by joining the U.S.-led military campaign in Iraq. A Polish-led international force, including 2,400 Polish troops, took over responsibility for south-central Iraq in September 2003.

Portugal
PORTUGUESE REPUBLIC

CONTINENT	Europe
AREA	92,345 sq km
	(35,655 sq mi)
POPULATION	10,446,000
CAPITAL	Lisbon 1,962,000
RELIGION	Roman Catholic
LANGUAGE	Portuguese, Mirandese
LITERACY	93%
LIFE EXPECTANCY	77
CURRENCY	euro
GDP PER CAPITA	$19,400
ECONOMY	IND: textiles and footwear; wood pulp, paper, and cork, metalworking. AGR: grain, potatoes, olives, grapes, sheep. EXP: clothing and footwear, machinery, chemicals, cork and paper products, hides.

Portugal, with its long Atlantic coast, lies on the western coast of the Iberian Peninsula in southwestern Europe—the most westerly country on the European mainland. The land consists of highland forests in the north and rolling lowland in the south. It tends to be wetter and cooler in the north. The south can be hot and parched, and it is dotted with reservoirs to conserve water. Most people live along the coast, with a third of the population living in the urban areas of Lisbon and Porto.

Established in the 12th century, Portugal came to preside over a vast realm that had its roots in the seafaring expeditions of the 1400s. In 1487–88 Bartolomeu Dias was the first European to round Africa's Cape of Good Hope. Breakup of the last great overseas empire came in the 1970s, when Portugal relinquished Angola, Mozambique, and other colonies; the influx of some 700,000 returning settlers, *retornados,* strained an already weak economy. Portugal includes the Azores and the Madeira Islands. Macau, the nation's last possession, reverted to China at midnight on December 19, 1999.

A coup in 1974 ended 42 years of dictatorship. Portugal joined the EU in 1986. EU loans funded infrastructure improvements but added to the burden of debt. To reverse the depopulation and desertification of its southeast region, the government built the Alqueva Dam on the Guadiana River. The hydroelectric dam was completed in 2002—the filling reservoir is creating Europe's largest manmade lake.

Puerto Rico (U.S.)
SELF-GOVERNING COMMON-
WEALTH IN ASSOCIATION WITH
THE UNITED STATES

CONTINENT North America
AREA 9,084 sq km
(3,507 sq mi)
POPULATION 3,879,000
CAPITAL San Juan 433,000
RELIGION Roman Catholic, Protestant
LANGUAGE Spanish, English

Puerto Rico, in the northern Caribbean, became a U.S. possession in 1898, and Puerto Ricans have been U.S. citizens since 1917. Enjoying a dynamic economy, voters in 1998 turned down options for statehood or independence.

Qatar
STATE OF QATAR

CONTINENT Asia
AREA 11,521 sq km
(4,448 sq mi)
POPULATION 629,000
CAPITAL Doha 286,000
RELIGION Muslim
LANGUAGE Arabic, English
LITERACY 83%
LIFE EXPECTANCY 72
CURRENCY Qatari rial
GDP PER CAPITA $20,100
ECONOMY IND: crude oil production
and refining, fertilizers,
petrochemicals, steel rein-
forcing bars, cement.
AGR: fruits, vegetables;
poultry, dairy products,
beef; fish. EXP: petroleum
products, fertilizers, steel.

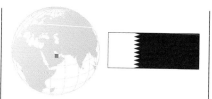

Qatar occupies a peninsula that extends into the Persian Gulf. This oil-rich nation, under British protection until 1971, chose not to join the United Arab Emirates. Qatar has exported oil since 1949, and as reserves decline, the nation has turned to its natural gas. The North Field (half the size of the entire country) is the largest single reservoir of natural gas in the world. The current emir has instituted political reforms, including allowing women to vote and hold office.

Réunion (France)
FRENCH OVERSEAS DEPARTMENT

CONTINENT Africa
AREA 2,507 sq km
(968 sq mi)
POPULATION 754,000
CAPITAL St.-Denis 178,000
RELIGION Roman Catholic, Hindu,
Muslim, Buddhist
LANGUAGE French, Creole

East of Madagascar in the Indian Ocean, Réunion is a mostly mountainous island with fertile coastal plains. The forested mountains are more than 3,000 meters (10,000 feet) high, and the plains feature sugarcane plantations.

Romania
ROMANIA

CONTINENT Europe
AREA 238,391 sq km
(92,043 sq mi)
POPULATION 21,622,000
CAPITAL Bucharest 1,853,000
RELIGION Eastern Orthodox,
Protestant, Catholic
LANGUAGE Romanian, Hungarian,
German
LITERACY 98%
LIFE EXPECTANCY 71
CURRENCY leu
GDP PER CAPITA $7,600
ECONOMY IND: textiles and foot-
wear, light machinery
and auto assembly,
mining, timber. AGR:
wheat, corn, barley,
sugar beets, eggs.
EXP: textiles and
footwear, metals and
metal products, machin-
ery and equipment, min-
erals and fuels.

Romania lies on the Black Sea coast of southeastern Europe. The Carpathian Mountains and the Transylvanian Alps divide the country into three physical and historical regions: Wallachia in the south, Moldavia in the northeast, and Transylvania in the country's center. The majority of the people are Romanian (89 percent), but the Hungarian minority, living in the Transylvanian basin, numbers some 1.7 million.

Communists took power in 1947 and installed a Soviet-style government. Under President Nicolae Ceausescu, however, Romania steered its own course, refusing to participate in Warsaw Pact maneuvers and conducting half its trade with the West. Police arrested dissidents and monitored contacts with foreigners.

A producer of grain and oil, Romania, so named because it was a colony of imperial Rome, is also a favored Black Sea vacation spot. But Romanian citizens enjoyed little of the bounty under communism. To help repay bank loans, petroleum and agricultural produce were exported during the 1980s, while imports were restricted, electricity was rationed, and shop shelves lay bare. With decline in production, basic commodities remained scarce and exports slowed.

In 1989 government security police killed demonstrators in Timisoara and Bucharest, igniting a revolution. The ensuing execution of Ceausescu and his wife ended their reign of repression, deprivation, and ethnic discrimination. The governments that followed have been laboring under massive foreign debt. Significant levels of public and private corruption impede economic growth and undercut public trust in new democratic institutions. Romania joined NATO in 2004, and is planning to join the EU as early as 2007.

Russia
RUSSIAN FEDERATION

CONTINENT	Europe/Asia
AREA	17,075,400 sq km
	(6,592,850 sq mi)
POPULATION	145,546,000
CAPITAL	Moscow 10,672,000
RELIGION	Russian Orthodox, Muslim, other
LANGUAGE	Russian
LITERACY	100%
LIFE EXPECTANCY	65
CURRENCY	Russian ruble
GDP PER CAPITA	$9,700
ECONOMY	IND: mining and extractive industries, machine building, shipbuilding, road and rail transportation equipment, communications equipment. AGR: grain, sugar beets, sunflower seed, vegetables, beef. EXP: petroleum and petroleum products, natural gas, wood and wood products, metals, fur.

Stretching from Europe to Asia, Russia spans 11 time zones. Heartland of the former Soviet Union, Russia is today a democratic federation, with ethnic groups such as the Tatars and Bashkirs politically represented in their own autonomous republics.

The country has rich mineral and energy resources. The mighty Volga, Europe's longest river, flows from northern Russia into the Caspian Sea. A bleak behemoth, Siberia encompasses more than half the territory but is home to less than 20 percent of the population. Siberian workers toil at prying natural gas, oil, coal, gold, and diamonds from the frozen earth. Commodities such as fur and timber also earn coveted foreign currency.

Invading Mongols controlled Russia from 1240 to 1380. In 1547 Ivan IV, a Muscovite prince, adopted the ancient title of caesar (tsar in Russian). He and his successors unified fragmented lands and began taking the region that is today Siberia.

Russia looked westward after 1698, when Peter the Great returned from his travels in Europe. Conquering territory along the Baltic Sea, he built his mostly landlocked realm a port capital, St. Petersburg (known from 1924 until 1991 as Leningrad), and established Russia's first navy. Russia entered the 20th century as enormous and imperial.

The forced abdication of Nicholas II in March 1917 ended tsarist rule. In November Vladimir Ilyich Lenin, a Marxist, gained power and moved the capital to Moscow, deep in the Russian interior. The new communist state would look inward, expanding and confronting the West. Eventually the Soviet Union came to consist of 15 republics. Soviet planners relocated entire peoples, to reward or punish. Relocation often moved minority peoples eastward (often to Siberia) and replaced them with Russians—who came to teach the Russian language, to organize (and often dominate) the local Communist Party, and to implement Moscow's decisions. Military power and Soviet security forces held the empire together, extending Soviet control into Eastern Europe after World War II.

Mikhail Gorbachev took office in 1985 and unveiled sweeping plans for economic restructuring (perestroika), soon followed by unprecedented political openness (glasnost). The Soviet Union dissolved after a failed coup in 1991, producing Russia and 14 independent republics—with Russian minorities totaling some 20 million. Russia seeks to protect these minorities, maintain its economic influence on resources (like oil), and confront separatism at home (as in Chechnya).

Rwanda
REPUBLIC OF RWANDA

CONTINENT	Africa
AREA	26,338 sq km (10,169 sq mi)
POPULATION	8,306,000
CAPITAL	Kigali 656,000
RELIGION	Roman Catholic, Protestant, Adventist, Muslim
LANGUAGE	Kinyarwanda, French, English, Kiswahili
LITERACY	70%
LIFE EXPECTANCY	40
CURRENCY	Rwandan franc
GDP PER CAPITA	$1,200
ECONOMY	IND: cement, agricultural products, small-scale beverages, soap. AGR: coffee, tea, pyrethrum, bananas, livestock. EXP: coffee, tea, hides, tin ore.

Rwanda, just south of the Equator in central Africa, is a mountainous land. This tiny, landlocked country, the continent's most densely populated, gained independence from Belgium in 1962. Conflict and civil war between ethnic Hutus and Tutsis have marked the country's history. In 1994 the genocide of some 800,000 Tutsis by Hutus occurred before Tutsi forces could gain control of Rwanda. Hutu militias fled Rwanda and continued to attack Tutsis from Zaire until Rwandan forces invaded Zaire in 1997 —where they remained until 2002, when the Democratic Republic of the Congo (formerly Zaire) agreed to help disarm Hutu gunmen.

Saint Helena (U.K.)
BRITISH OVERSEAS TERRITORY

CONTINENT	Africa
AREA	411 sq km (159 sq mi)
POPULATION	7,000
CAPITAL	Jamestown 2,000
RELIGION	Anglican (majority)
LANGUAGE	English

The South Atlantic island of St. Helena also administers remote Tristan da Cunha (population 275) to the south and Ascension Island (population 1,000) to the north. The economy depends on fishing and Ascension's military bases.

Saint Kitts and Nevis
FEDERATION OF SAINT KITTS AND NEVIS

CONTINENT	North America
AREA	269 sq km (104 sq mi)
POPULATION	46,000
CAPITAL	Basseterre 13,000
RELIGION	Anglican, other Protestant
LANGUAGE	English
LITERACY	97%
LIFE EXPECTANCY	71
CURRENCY	East Caribbean dollar
GDP PER CAPITA	$8,800
ECONOMY	IND: sugar processing, tourism, cotton, salt. AGR: sugarcane, rice, yams, vegetables, fish. EXP: machinery, food, electronics, beverages, tobacco.

Once known as the Gibraltar of the West Indies, the massive 17th-century fortress atop Brimstone Hill on St. Kitts recalls colonial occupation. Independent of Britain since 1983, the nation is diversifying the economy away from sugar towards tourism, banking, and light manufacturing. The twin islands are both volcanic with sandy beaches and a warm, wet climate. The island of Nevis (population 7,000) has been pursuing the constitutional process of secession since 1996.

Saint Lucia
SAINT LUCIA

CONTINENT	North America
AREA	616 sq km (238 sq mi)
POPULATION	162,000
CAPITAL	Castries 14,000
RELIGION	Roman Catholic, Protestant
LANGUAGE	English, French patois
LITERACY	67%
LIFE EXPECTANCY	72
CURRENCY	East Caribbean dollar
GDP PER CAPITA	$5,400
ECONOMY	IND: clothing, assembly of electronic components, beverages, corrugated cardboard boxes. AGR: bananas, coconuts, vegetables, citrus. EXP: bananas, clothing, cacao, vegetables.

Saint Lucia, a tropical island in the eastern Caribbean Sea, gained independence from Britain in 1979. Tropical forests cloak a mountainous interior flanked by twin volcanic peaks, known as pitons. Oil transshipment via a U.S.-built terminal and the export of manufactured goods supplement agriculture and tourism. The government is challenged by unemployment and drug-related crime.

Saint-Pierre and Miquelon (France)
FR. OVERSEAS TERRITORIAL COLLECTIVITY

CONTINENT	North America
AREA	242 sq km (93 sq mi)
POPULATION	7,000
CAPITAL	St.-Pierre 6,000
RELIGION	Roman Catholic
LANGUAGE	French

Saint Vincent and the Grenadines
SAINT VINCENT AND THE GRENADINES

CONTINENT	North America
AREA	389 sq km (150 sq mi)
POPULATION	110,000
CAPITAL	Kingstown 29,000
RELIGION	Anglican, Methodist, Roman Catholic, other Protestant
LANGUAGE	English, French patois
LITERACY	96%
LIFE EXPECTANCY	72
CURRENCY	East Caribbean dollar
GDP PER CAPITA	$2,900
ECONOMY	IND: food processing, cement, furniture, clothing, starch. AGR: bananas, coconuts, sweet potatoes, spices, cattle, fish. EXP: bananas, dasheen (taro), arrowroot starch, tennis racquets.

This eastern Caribbean country consists of volcanic St. Vincent island and the Grenadines, 32 smaller islands and cays. St. Vincent is hilly with rich volcanic soils, and its volcano, Soufrière, last erupted in 1979—the year of independence from Britain. Two hydroelectric plants help power St. Vincent's diversifying economy, dependent in part on exports of bananas and arrowroot, valuable as a starch in carbonless copy paper. Tourism is of growing importance.

Samoa
INDEPENDENT STATE OF SAMOA

CONTINENT	Australia/Oceania
AREA	2,831 sq km (1,093 sq mi)
POPULATION	172,000
CAPITAL	Apia 40,000
RELIGION	Christian
LANGUAGE	Samoan, English
LITERACY	100%
LIFE EXPECTANCY	69
CURRENCY	tala
GDP PER CAPITA	$5,600
ECONOMY	IND: food processing, building materials, auto parts. AGR: coconuts, bananas, taro, yams. EXP: fish, coconut oil and cream, copra, taro.

Western political institutions and Polynesian social structure combine in Samoa, whose location at the crossroads of South Pacific shipping lanes attracted European powers. Germany took over the western part of the Samoan archipelago in 1900. After World War I, New Zealand administered the islands until they became independent in 1962. Membership in the Commonwealth came in 1970. An enlarged airport brings growing numbers of tourists.

San Marino
REPUBLIC OF SAN MARINO

CONTINENT	Europe
AREA	61 sq km (24 sq mi)
POPULATION	30,000
CAPITAL	San Marino 5,000
RELIGION	Roman Catholic
LANGUAGE	Italian
LITERACY	96%
LIFE EXPECTANCY	81
CURRENCY	euro
GDP PER CAPITA	$34,600
ECONOMY	IND: tourism, banking, textiles, electronics. AGR: wheat, grapes, corn, olives, cattle. EXP: building stone, lime, wood, chestnuts, wheat, wine, baked goods, hides, ceramics.

Originally a medieval city-state, the world's oldest republic perches atop a mountain in north-central Italy. San Marino takes pride in its finely minted coins, ceremonial guard, and postage stamps. Well-preserved castles and sweeping vistas of the Adriatic coast enchant 3.5 million visitors a year.

Sao Tomé and Principe
DEMOCRATIC REPUBLIC OF SAO TOMÉ AND PRINCIPE

CONTINENT	Africa
AREA	1,001 sq km (386 sq mi)
POPULATION	176,000
CAPITAL	São Tomé 54,000
RELIGION	Roman Catholic, Evangelical, Protestant, Seventh-Day Adventist
LANGUAGE	Portuguese
LITERACY	79%
LIFE EXPECTANCY	65
CURRENCY	dobra
GDP PER CAPITA	$1,200
ECONOMY	IND: light construction, textiles, soap, beer, fish processing, timber. AGR: cacao, coconuts, palm kernels, copra, poultry, fish. EXP: cacao, copra, coffee, palm oil.

In the 15th century, Portuguese navigators discovered these two volcanic islands off the coast of West Africa in the Gulf of Guinea. São Tomé is the larger island with about 90 percent of the population. Africa's smallest nation gained independence in 1975. In 1991 the formerly Marxist government made a complete transition to democracy. The new leaders have moved to liberalize the economy and reduce dependence on plantation crops. New offshore oil discoveries could translate into substantial oil revenue in the near future.

Saudi Arabia
KINGDOM OF SAUDI ARABIA

CONTINENT	Asia
AREA	1,960,582 sq km (756,985 sq mi)
POPULATION	24,070,000
CAPITAL	Riyadh 5,126,000
RELIGION	Muslim
LANGUAGE	Arabic
LITERACY	79%
LIFE EXPECTANCY	72
CURRENCY	Saudi riyal
GDP PER CAPITA	$11,400
ECONOMY	IND: crude oil production, petroleum refining, basic petrochemicals, cement. AGR: wheat, barley, tomatoes, melons, mutton. EXP: petroleum and petroleum products.

Saudi Arabia occupies most of the Arabian Peninsula and is the largest country in area in the Middle East, but 95 percent of the land is desert. Mountains running parallel to the Red Sea slope down to plains along the Persian Gulf (called Arabian Gulf by Arab states). Below the arid landscape, oil has made this desert kingdom one of the wealthiest nations in the world.

The oil-enriched economy has brought some 5.6 million resident foreigners, mostly from Arab states or South Asia. The mismatch between the job skills of Saudi graduates and the needs of the job market, as well as constraints on employment for Saudi-women, are reasons for the large number of foreign workers. Unemployment for Saudi males is high.

Pumping the lifeblood of industrial economies, Saudi Arabia exports more oil than any other nation and holds 25 percent of the world's proven reserves. To increase non-oil exports, economic diversification emphasizes more dependence on manufacturing and on irrigated farming, which draw on limited underground water supplies.

King Abd al-Aziz Al Saud merged warring Bedouin tribes to form Saudi Arabia in 1932. Succession has fallen in turn to his sons, governing through consultation with others in the royal family, religious leaders, and technocrats. A major supporter of the 1990–91 Persian Gulf War, the country served as operations base for coalition forces. Relations with the U.S. were strained after the September 11, 2001, terrorist attacks, carried out mainly by Saudi citizens.

In this conservative society, underpinned by Islamic law, women live in veiled segregation. Saudi Arabia is keeper of Islam's most sacred cities: Mecca, where the Prophet Muhammad received the word of Allah, and Medina, where Muhammad died in A.D. 632.

Senegal
REPUBLIC OF SENEGAL

CONTINENT Africa
AREA 196,722 sq km
(75,955 sq mi)
POPULATION 10,580,000
CAPITAL Dakar 2,167,000
RELIGION Muslim, Christian
LANGUAGE French, Wolof, Pulaar, Diola, Jola, Mandinka
LITERACY 40%
LIFE EXPECTANCY 53
CURRENCY CFA franc
GDP PER CAPITA $1,500
ECONOMY IND: agricultural and fish processing, phosphate mining, fertilizer production, petroleum refining. AGR: peanuts, millet, corn, sorghum, cattle, fish. EXP: fish, groundnuts (peanuts), petroleum products, phosphates, cotton.

Senegal lies on West Africa's Atlantic coast. One of the first multiparty democracies in today's Africa, this river-laced land of marshes and plains was once ruled by Wolof chieftains. Three centuries of French administration ended in 1960. The moderate socialist government has initiated economic reforms. A deep natural harbor at Dakar makes the cosmopolitan city a major West African port. Peanuts from the drought-prone interior are an important export.

Serbia and Montenegro
SERBIA AND MONTENEGRO

CONTINENT Europe
AREA 102,173 sq km
(39,450 sq mi)
POPULATION 10,684,000
CAPITAL Belgrade (administrative capital) 1,118,000; Podgorica (judicial capital) 163,493
RELIGION Orthodox, Muslim, Roman Catholic
LANGUAGE Serbian, Albanian
LITERACY 93%
LIFE EXPECTANCY 73
CURRENCY new Yugoslav dinar, euro
GDP PER CAPITA $2,200
ECONOMY IND: machine building, metallurgy, mining, consumer goods. AGR: cereals, fruits,

vegetables, tobacco, cattle. EXP: manufactured goods, food and live animals, raw materials.

Located in southwestern Europe, Serbia and Montenegro features a fertile Danube plain in the north, rising to mountains in the south. The name "Yugoslavia" passed into history on February 4, 2003, replaced by the union of "Serbia and Montenegro." On that day the Yugoslav parliament ratified a constitutional charter creating a new government and country name. Belgrade is the administrative capital, and Podgorica is the judicial capital.

Yugoslavia, meaning "land of the southern Slavs," was born in 1929 in an attempt to unify the Kingdom of Serbs, Croats, and Slovenes, founded in 1918. From the beginning, Serbia, the largest Yugoslav republic in area and population, sought to dominate Yugoslavia. The country was held together by force, first under kings then under a communist government, until 1991–92 when Slovenia, Croatia, Macedonia, and Bosnia and Herzegovina declared independence. The Serb-dominated Yugoslav army invaded Slovenia and Croatia, and civil war erupted in Croatia and Bosnia. All four countries eventually won their independence.

By 1992, all that was left of Yugoslavia was Serbia and Montenegro. Serbia's brutal war in Kosovo, starting in 1998, caused Montenegro to distance itself from Slobodan Milosevic and his Yugoslav government. Kosovo, a province of Serbia, has an ethnic Albanian majority that wanted, and still wants, independence. The war in Kosovo ended in 1999 only after NATO bombed Serbia and the UN made Kosovo an international protectorate.

Today Serbia and Montenegro is a democratic union of two republics with a small central government. Both republics agreed to the union because of their desire to join the European Union. However, the constitutional charter allows either Serbia or Montenegro to hold an independence referendum in 2006.

Seychelles
REPUBLIC OF SEYCHELLES

CONTINENT	Africa
AREA	455 sq km (176 sq mi)
POPULATION	87,000
CAPITAL	Victoria 25,000
RELIGION	Roman Catholic, Anglican
LANGUAGE	English, French, Creole
LITERACY	58%
LIFE EXPECTANCY	70
CURRENCY	Seychelles rupee
GDP PER CAPITA	$7,800
ECONOMY	IND: fishing, tourism, processing of coconuts and vanilla, coir (coconut fiber) rope. AGR: coconuts, cinnamon, vanilla, sweet potatoes, broiler chickens, tuna fish. EXP: canned tuna, frozen

fish, cinnamon bark, copra, petroleum products (reexports).

Some 115 tropical islands form the Republic of Seychelles in the Indian Ocean—independent from Britain since 1976. An international airport opened in 1971 on the largest island, Mahé, increasing tourism, the economic mainstay. An estimated 88 per cent of the population live on Mahé. After 15 years of one-party rule, the first elections in the country's history were held in 1993.

Sierra Leone
REPUBLIC OF SIERRA LEONE

CONTINENT	Africa
AREA	71,740 sq km (27,699 sq mi)
POPULATION	5,733,000
CAPITAL	Freetown 921,000
RELIGION	Muslim, indigenous beliefs, Christian
LANGUAGE	English, Mende, Temne, Krio
LITERACY	31%
LIFE EXPECTANCY	43
CURRENCY	leone
GDP PER CAPITA	$500
ECONOMY	IND: mining (diamonds); small-scale manufacturing (beverages, textiles); petroleum refining. AGR: rice, coffee, cacao, palm kernels,

poultry, fish. EXP: diamonds, rutile, cacao, coffee.

Sierra Leone is on the Atlantic coast of West Africa, with coastal swamps rising to interior plateaus and mountains. Named "lion mountain" by a 15th-century Portuguese explorer, Sierra Leone was a British colony from the early 19th century until 1961. In the 1990s democratically elected leaders were overthrown but subsequently regained power, and major hostilities have demoralized the population and destabilized the economy. In 2002 Sierra Leone emerged from a decade of civil war, with the help of some 17,000 UN peacekeepers.

Singapore
REPUBLIC OF SINGAPORE

CONTINENT	Asia
AREA	660 sq km (255 sq mi)
POPULATION	4,372,000
CAPITAL	Singapore 4,372,000
RELIGION	Buddhist, Muslim, Christian, Hindu, Sikh, Taoist, Confucianist
LANGUAGE	Chinese, Malay, Tamil, English
LITERACY	93%
LIFE EXPECTANCY	79
CURRENCY	Singapore dollar
GDP PER CAPITA	$25,200
ECONOMY	IND: electronics, chemicals, financial services,

oil drilling equipment, petroleum refining. AGR: rubber, copra, fruit, orchids, poultry. EXP: machinery and equipment (including electronics), consumer goods, chemicals, mineral fuels.

The country of Singapore, consisting of Singapore island and some 50 smaller islands, is located in Southeast Asia at the tip of the Malay Peninsula. More than 3,000 multinational companies have offices on this tropical island at the entrance to the Strait of Malacca, the shortest sea route between the Indian Ocean and the South China Sea. As a trade center of the British Empire, Singapore attracted thousands of Chinese settlers, now 77 percent of the population. Independent since 1965, Singapore is Southeast Asia's financial hub and the world's busiest container port.

Slovakia
SLOVAK REPUBLIC

CONTINENT	Europe
AREA	49,035 sq km (18,932 sq mi)
POPULATION	5,355,000
CAPITAL	Bratislava 425,000
RELIGION	Roman Catholic, Atheist, Protestant
LANGUAGE	Slovak, Hungarian
LITERACY	99%

LIFE EXPECTANCY	74
CURRENCY	Slovak koruna
GDP PER CAPITA	$12,400
ECONOMY	IND: metal and metal products, food and beverages, electricity, gas. AGR: grains, potatoes, sugar beets, hops, pigs, forest products. EXP: machinery and transport equipment, miscellaneous manufactured goods.

A landlocked country in central Europe, Slovakia is mostly mountainous except for southern lowlands along the Danube, where the capital, Bratislava, is found. This country's split from the more affluent, industrialized Czech Republic in 1993 was prompted by Slovak nationalism and grievances over rapid economic reforms instituted by the Czechoslovak government in Prague—reforms that left many Slovaks without jobs. Slovakia's industrial economy is market oriented. It joined NATO and the European Union in 2004.

Slovenia
REPUBLIC OF SLOVENIA

CONTINENT	Europe
AREA	20,273 sq km (7,827 sq mi)
POPULATION	1,999,000
CAPITAL	Ljubljana 256,000

RELIGION	Roman Catholic, other
LANGUAGE	Slovenian, Serbo-Croatian
LITERACY	100%
LIFE EXPECTANCY	76
CURRENCY	tolar
GDP PER CAPITA	$19,200
ECONOMY	IND: ferrous metallurgy and aluminum products, lead and zinc smelting, electronics, trucks. AGR: potatoes, hops, wheat, sugar beets, cattle. EXP: manufactured goods, machinery and transport equipment, chemicals, food.

Slovenia is an Alpine-mountain state in southern Europe consisting mainly of Roman Catholic Slovenes. In 1918 Slovenia joined the Kingdom of Serbs, Croats, and Slovenes, subsequently named Yugoslavia. Slovenia proclaimed its independence in June 1991, prompting a ten-day conflict that brought defeat to the Serb-dominated Yugoslav Army. It is the most prosperous of the former Yugoslav republics, with the region's highest standard of living. Its Western outlook and economic stability won Slovenia membership in both NATO and the EU in 2004.

Solomon Islands
SOLOMON ISLANDS

CONTINENT	Australia/Oceania
AREA	28,370 sq km (10,954 sq mi)
POPULATION	491,000
CAPITAL	Honiara 56,000
RELIGION	Protestant, Roman Catholic, indigenous beliefs
LANGUAGE	Melanesian pidgin, 120 indigenous languages, English
LITERACY	62%
LIFE EXPECTANCY	71
CURRENCY	Solomon Islands dollar
GDP PER CAPITA	$1,700
ECONOMY	IND: fish (tuna), mining, timber. AGR: cacao, coconuts, palm kernels, rice; cattle, timber, fish. EXP: timber, fish, copra, palm oil.

Northeast of Australia in the South Pacific, the Solomon Islands chain consists of six main islands that are volcanic, mountainous, and forested. Guadalcanal is the most populous island. A strategic battleground during World War II, the islands gained independence from Britain in 1978. About 85 percent of the islanders are Melanesians, who speak some 92 indigenous languages. Ethnic tension between natives of Guadalcanal and settlers from nearby Malaita island escalated into armed conflict from 1998 until 2003, when an Australian-led peacekeeping force restored order.

Somalia
SOMALIA

CONTINENT	Africa
AREA	637,657 sq km (246,201 sq mi)
POPULATION	8,025,000
CAPITAL	Mogadishu 1,175,000
RELIGION	Sunni Muslim
LANGUAGE	Somali, Arabic, Italian, English
LITERACY	38%
LIFE EXPECTANCY	46
CURRENCY	Somali shilling
GDP PER CAPITA	$600
ECONOMY	IND: a few light industries, including sugar refining, textiles, petroleum refining (mostly shut down). AGR: bananas, sorghum, corn, coconuts, cattle, fish. EXP: livestock, bananas, hides, fish, charcoal.

Somalia is a semiarid land in the Horn of Africa, and it is flat in the south, with mountains in the north reaching more than 2,000 meters (6,500 feet). In 1960 northern British Somaliland voted to join southern Italian Somaliland to create Somalia. The Somalis are one of the most homogeneous peoples in Africa, but unity is thwarted by clan-based rivalries. Civil war ended a 21-year dictatorship in 1991, and Somalia has been without a national government since that time. UN efforts from 1992 to 1995 to stop clan fighting failed, and UN and U.S. forces left after suffering high casualties. In southern Somalia the absence of a government means declining health and increasing poverty.

More prosperous is the "Republic of Somaliland," which seceded from the rest of Somalia in 1991, within the old borders of British Somaliland. Somaliland's capital is Hargeysa, and its port, Berbera, provides goods to landlocked Ethiopia. The independence of Somaliland, while widely acknowledged, is not officially recognized by the international community, as efforts continue to reunify Somalia. Somaliland's peace, stability, and democracy have been recently threatened by drought and by regional warlords.

South Africa
REPUBLIC OF SOUTH AFRICA

CONTINENT Africa
AREA 1,219,090 sq km (470,693 sq mi)
POPULATION 44,024,000
CAPITAL Pretoria (administrative capital) 1,209,000; Bloemfontein (judicial capital) 381,000; Cape Town (legislative capital) 3,103,000
RELIGION Christian, indigenous beliefs, Muslim, Hindu
LANGUAGE Afrikaans, English, Ndebele, Pedi, Sotho, Swazi, Tsonga, Tswana, Venda, Xhosa, Zulu
LITERACY 86%
LIFE EXPECTANCY 53
CURRENCY rand
GDP PER CAPITA $10,000
ECONOMY IND: mining (platinum, gold, chromium), automobile assembly, metalworking, machinery. AGR: corn, wheat, sugarcane, fruits, beef. EXP: gold, diamonds, platinum, other metals and minerals, machinery and equipment.

South Africa, Africa's southernmost nation, is also Africa's largest and most developed economy. Diamond and gold strikes in the late 19th century began transforming this land of African tribespeople, Boer farmers, and British traders into an industrial colossus. Today South Africa produces high-tech equipment and is a world leader in the output of gold and diamonds. On South Africa's high grassland plateau, or veld, lies its premier city, Johannesburg (usually shortened to Jo'burg). Johannesburg and its satellite cities are home to more than 8 million people—generating 9 percent of all economic activity in Africa.

From 1948 to 1991 South Africa's political system was dominated by apartheid, a policy of segregation that isolated blacks in so-called homelands and overcrowded townships. Blacks, or Africans, compose 78 percent of the population, with whites at 10 percent, coloreds (mixed race) 8.7 percent, and Asians (Indians) 2.5 percent. In 1989 a reform-minded government, spurred by international economic sanctions as well as domestic protests, began the process of dismantling apartheid. A year later Nelson Mandela, the long-jailed leader of a black nationalist group, was released.

By the middle of 1991 all remaining apartheid legislation was revoked, including the Population Registration Act, which classified South Africans by race and was widely considered the cornerstone of apartheid. In March 1992 white South Africans voted in favor of negotiations aimed at replacing the constitution with a nonracial one, accomplished in 1993. The first multiracial parliament was elected in 1994. Nelson Mandela, winner of the Nobel Peace Prize in 1993 in conjunction with former President Frederik W. de Klerk, became the new president, and the black homelands were abolished.

In the 21st century, South Africa is a democratic country representing all its diverse people, often called the rainbow nation. The government recognizes 11 official languages (including English). Today South Africa is making up for decades of social disruption and lost education, but high unemployment and the AIDS epidemic threaten economic progress. Without effective prevention and treatment, AIDS deaths could total 5 to 7 million by 2010.

Spain
KINGDOM OF SPAIN

CONTINENT Europe
AREA 505,988 sq km
(195,363 sq mi)
POPULATION 41,334,000
CAPITAL Madrid 5,145,000
RELIGION Roman Catholic
LANGUAGE Castilian Spanish, Cata-
lan, Galician, Basque
LITERACY 98%
LIFE EXPECTANCY 79
CURRENCY euro
GDP PER CAPITA $21,200
ECONOMY IND: textiles and
apparel, food and bev-
erages, metals and
metal manufactures,
chemicals. AGR: grain,
vegetables, olives, wine
grapes, beef, fish. EXP:
machinery, motor vehi-
cles, foodstuffs, other
consumer goods.

Spain occupies most of the Iberian Peninsula in southwest Europe, and its territory includes the Balearic Islands in the Mediterranean and the Canary Islands in the Atlantic. Much of the mainland is high plateau, with mountain ranges, including the Pyrenees, in the north. The plateau experiences hot summers and cold winters; it is cooler and wetter to the north.

About 200 B.C. the Romans occupied this crossroads between Europe and Africa. Moors invaded in A.D. 711, ruling for almost 800 years before Christian armies routed them. Enriched by its New World empire, Spain dominated Europe during the 16th and 17th centuries; today it rules only the North African territories of Ceuta and Melilla.

Gen. Francisco Franco wielded power from 1936 until his death in 1975, when Juan Carlos became king. Three years later a new constitution confirmed Spain as a parliamentary monarchy. After 1986, when the Socialist Party under Felipe González Márquez led Spain into the European Union, the economy grew faster than any other member nation's. Yet the government's pro-business policies in the 1990s were blamed for widening the gap between rich and poor and for the bankruptcy of noncompetitive industries—all contributing to high unemployment. Separatist agitation born of historical regional differences, most pronounced in the Basque country and in Catalonia, still challenges national unity, but a strong national peace movement has developed to counteract terrorist activities.

Unemployment continues to be a problem, but recent economic growth makes the country's future outlook more positive. Spain is one of the European Union nations participating in the euro currency.

Sri Lanka
DEMOCRATIC SOCIALIST
REPUBLIC OF SRI LANKA

CONTINENT Asia
AREA 65,525 sq km
(25,299 sq mi)
POPULATION 19,273,000
CAPITAL Colombo 648,000
RELIGION Buddhist, Hindu,
Christian, Muslim
LANGUAGE Sinhala, Tamil,
English
LITERACY 92%
LIFE EXPECTANCY 72
CURRENCY Sri Lankan rupee
GDP PER CAPITA $3,700
ECONOMY IND: rubber processing,
tea, coconuts, other agri-
cultural commodities,
clothing, cement. AGR:
rice, sugarcane, grains,
pulses; milk. EXP: textiles
and apparel, tea, dia-
monds, coconut prod-
ucts, petroleum products.

Sri Lanka (formerly known as Ceylon) is a tropical island lying close to the southern tip of India and near the Equator. From the coast, the land rises to a central plateau, where tea plantations are found. Sinhalese form the country's major ethnic group (74 percent) and Tamils are the largest minority, at 18 percent. Population density is highest in the island's southwest corner—where Colombo, the capital, is located. The Tamil minority tends to be geographically concentrated along the eastern and northern coastal areas.

Under European control for some 450 years, Ceylon won independence from the United Kingdom in 1948 and changed its name in 1972. Since then a segment of the Tamil Hindu minority has pressured the Sinhalese

Buddhist majority for a separate state. Conflict broke out in 1983 and escalated to civil war. Violence continued after the assassination of President Premadasa in 1993 by Tamil separatists. Thousands of lives have been lost. Hundreds of thousands of Tamil civilians have fled Sri Lanka. The civil war has had a negative impact on economic growth, which is based largely on tea and garment manufacture. A cease-fire and political agreement between the government and Tamil rebels in 2002, raised hopes for a lasting peace.

Sudan
REPUBLIC OF THE SUDAN

CONTINENT	Africa
AREA	2,505,813 sq km
	(967,500 sq mi)
POPULATION	38,114,000
CAPITAL	Khartoum 4,286,000
RELIGION	Sunni Muslim, indigenous
	beliefs, Christian
LANGUAGE	Arabic, Nubian, Ta
	Bedawie, many local
	dialects
LITERACY	61%
LIFE EXPECTANCY	57
CURRENCY	Sudanese dinar
GDP PER CAPITA	$1,400
ECONOMY	IND: oil, cotton ginning,
	textiles, cement, edible
	oils, sugar, soap dis-
	tilling, shoes, petroleum
	refining. AGR: cotton,
	groundnuts (peanuts),
	sorghum, millet, wheat,
	sheep. EXP: oil and
	petroleum products, cot-
	ton, sesame, livestock,
	groundnuts.

Africa's largest country in land area, Sudan is dominated by the Nile and its tributaries, with mountains rising along its Red Sea coast and along the western border with Chad. Sudan's name in Arabic means, "land of the blacks."

Since independence from Britain in 1956, a north-south war has dominated Sudan's history, pitting Arab Muslims in the northern desert against black Christians and animists in the southern wetlands. Muslim Arabs control the government in Khartoum, but are only about 39 percent of the population. Blacks, or Africans, make up 52 percent of Sudanese, and are most numerous in southern and western Sudan. The country is further divided with hundreds of black, Arab, and non-Arab ethnicities, tribes, and languages.

Sudan's political history has been unstable. Gen. Muhammad Nimeiri, who seized control in the 1970s, was deposed in 1985. In 1989 another military coup, led by then-Col. Omar al-Bashir, toppled the elected government. The military dictatorship, so far, has been incapable of stopping the civil war. Indeed its intensity rose with the discovery and exploitation of oil fields in the south. In 2004 a rebel uprising by blacks in western Sudan's Dafur region brought army reprisals, creating 100,000 refuges. Pro-government Arab militias carried out systematic killings of Darfur's blacks, who are mostly Muslim.

Suriname
REPUBLIC OF SURINAME

CONTINENT	South America
AREA	163,265 sq km
	(63,037 sq mi)
POPULATION	444,000
CAPITAL	Paramaribo 253,000
RELIGION	Hindu, Protestant
	(Moravian), Roman
	Catholic, Muslim
LANGUAGE	Dutch, English, Sranang
	Tongo (Taki-Taki), Hindus-
	tani, Javanese
LITERACY	93%
LIFE EXPECTANCY	70
CURRENCY	Surinamese guilder
GDP PER CAPITA	$3,400
ECONOMY	IND: bauxite and gold
	mining, alumina pro-
	duction, oil, lumbering,
	food processing. AGR:
	paddy rice, bananas,
	palm kernels, beef,
	forest products; shrimp.
	EXP: alumina, crude
	oil, lumber, shrimp and
	fish, rice.

Along the north coast of South America, Suriname is a small, but ethnically diverse, country. Most people are descendants of African slaves and Indian or Indonesian servants brought over by the Dutch to work in agriculture.

Suriname, formerly known as Dutch Guiana, gained independence in 1975. Most Surinamers live in the narrow, northern coastal plain. Access to the interior rain forest and forest people is limited. Bauxite mining and alumina exports dominate trade; inexpensive power from the hydroelectric plant at Afobaka helps the economy. Boundary disputes with Guyana and French Guiana persist.

Svalbard (Norway)
NORWEGIAN DEPENDENCY

CONTINENT Europe
AREA 61,020 sq km
(23,560 sq mi)
POPULATION 2,800
CAPITAL Longyearbyen 1,500
RELIGION Evangelical Lutheran
LANGUAGE Russian, Norwegian

The Arctic islands of Norway's Svalbard territory are rugged and often icebound. Coal mining is the main economic activity, with most settlements being company towns run by Norwegians or Russians.

Swaziland
KINGDOM OF SWAZILAND

CONTINENT Africa
AREA 17,363 sq km
(6,704 sq mi)
POPULATION 1,161,000
CAPITAL Mbabane (administrative capital) 70,000; Lobamba (legislative and royal capital) 4,400
RELIGION Indigenous beliefs, Roman Catholic, Muslim
LANGUAGE English, Swati
LITERACY 82%

LIFE EXPECTANCY 45
CURRENCY lilangeni
GDP PER CAPITA $4,800
ECONOMY IND: mining (coal), wood pulp, sugar, soft drink concentrates. AGR: sugarcane, cotton, corn, tobacco, cattle. EXP: soft drink concentrates, sugar, wood pulp, cotton yarn.

Swaziland, consisting mostly of high plateaus and mountains, is in southern Africa. In 1949 the British government rejected a South African request for control of this small, landlocked nation. Independence was granted in 1968. The death of King Sobhuza in 1982 led to the coronation of 18-year-old King Mswati III in 1986. The king is an absolute monarch with supreme executive, legislative, and judicial powers. Nearly 60 percent of Swazi territory is held by the crown.

Sweden
KINGDOM OF SWEDEN

CONTINENT Europe
AREA 449,964 sq km
(173,732 sq mi)
POPULATION 8,960,000
CAPITAL Stockholm 1,697,000
RELIGION Lutheran, Roman Catholic
LANGUAGE Swedish
LITERACY 99%
LIFE EXPECTANCY 80

CURRENCY Swedish krona
GDP PER CAPITA $26,000
ECONOMY IND: iron and steel, precision equipment, wood pulp and paper products, processed foods. AGR: barley, wheat, sugar beets, meat. EXP: machinery, motor vehicles, paper products, pulp and wood, iron and steel products, chemicals.

Armed neutrality has kept Sweden out of war for nearly two centuries. Low unemployment, a low birthrate, and one of the world's highest life expectancies have characterized modern Sweden. Success has been credited to a blending of socialism and capitalism, including cooperation between the government and labor unions, which represent 90 percent of workers. High taxes finance advanced social programs, from education to health and child care and paid paternal leave.

In the 1980s a flood of immigrants from Asia, Africa, and Latin America sought the Swedish utopia but further taxed expensive social programs. Mounting economic problems led to cutbacks in 1991, when Sweden reassessed its social policies and elected a conservative government. The Social Democrats returned to power in 1994 with a commitment to stringent economic controls. By 1998 they were operating from a weakened power base—the lowest vote share in 78 years. Sweden joined the EU in 1995. Inflation is low and unemployment is down.

Radioactive fallout from Chornobyl underscored Sweden's resolve to dismantle its nuclear power plants, a process which was begun in 1997.

Switzerland
SWISS CONFEDERATION

CONTINENT	Europe
AREA	41,284 sq km
	(15,940 sq mi)
POPULATION	7,341,000
CAPITAL	Bern 320,000
RELIGION	Roman Catholic,
	Protestant
LANGUAGE	German, French, Italian,
	Romanisch
LITERACY	99%
LIFE EXPECTANCY	80
CURRENCY	Swiss franc
GDP PER CAPITA	$32,000
ECONOMY	IND: machinery, chemicals, watches, textiles, precision instruments. AGR: grains, fruits and vegetables, meat. EXP: machinery, chemicals, metals, watches, agricultural products.

A history of political stability and expertise in technology and commerce help explain the Swiss phenomenon: a post-industrial economy that reported one of the highest per capita incomes in the world.

Founded in 1291 as a union of three cantons chafing against Hapsburg rule, Switzerland has been independent since 1815; its borders now encompass 26 cantons embracing three official languages, German, French, and Italian. Foreigners make up 25 percent of the workforce. Switzerland competes in global markets with exports that make up almost half of the nation's economy; however the Swiss in a 2001 referendum voted against joining the European Union.

The UBS, Switzerland's largest bank, suffered big losses as the secrecy practices of Swiss banks continued to come under relentless attack.

Elaborate civil defense measures and a strong militia back up the Swiss policy of permanent neutrality. Switzerland is firmly committed to world peace, and in 2002 became a member of the United Nations.

Syria
SYRIAN ARAB REPUBLIC

CONTINENT	Asia
AREA	185,180 sq km
	(71,498 sq mi)
POPULATION	17,537,000
CAPITAL	Damascus 2,228,000
RELIGION	Sunni, Alawite, Druze and other Muslim sects, Christian
LANGUAGE	Arabic, Kurdish, Armenian, Aramaic, Circassian, French, English
LITERACY	77%
LIFE EXPECTANCY	70
CURRENCY	Syrian pound
GDP PER CAPITA	$3,700
ECONOMY	IND: petroleum, textiles, food processing, beverages. AGR: wheat, barley, cotton, lentils, beef. EXP: crude oil, petroleum products, fruits and vegetables, cotton fiber, clothing.

Syria is in southwest Asia in the heart of the Middle East. The Mediterranean coastal plain is backed by a low range of hills, followed by a vast interior desert plateau. Most people live near the coast or near the Euphrates River, which brings life to the desert plateau. Damascus, capital of this desert country, was built on an oasis and is said to be the world's oldest continuously inhabited settlement.

Syrians are mostly Arab, although about 9 percent are Kurds, living mainly in the northeast corner of Syria. Syria's population is about 90 percent Muslim, mostly Sunni, but the Alawite minority (12 percent of Syrians) is politically dominant. The Alawite-controlled Baath (Renaissance) Party has ruled Syria since 1963.

Part of the Ottoman Empire for four centuries, Syria came under French mandate in 1920 and gained independence in 1946. Dreams of a "Greater Syria" were dashed when the smaller states of Lebanon, Palestine, and Jordan were created by Britain and France in the 1920s. Together with Egypt, it formed the United Arab

Republic between 1958 and 1961. Syria has fought four wars with Israel, losing the Golan Heights in 1967. Recovering the Golan has been a matter of fierce national pride for Syrians.

The 30-year rule of Hafez al-Assad was marked by authoritarian government, an anti-Israeli policy, and military intervention in Lebanon. Some fear that Syria's 16,000 troops in Lebanon are being used to create Greater Syria. Terrorist groups, such as Hezbollah, get Syrian backing to attack Israel from bases in Lebanon. Bashar al-Assad succeeded his father as president in 2000 and continues his father's harsh policies.

Taiwan
See listing under China.

Tajikistan
REPUBLIC OF TAJIKISTAN

CONTINENT	Asia
AREA	143,100 sq km
	(55,251 sq mi)
POPULATION	6,574,000
CAPITAL	Dushanbe 554,000
RELIGION	Sunni and Shiite Muslim
LANGUAGE	Tajik, Russian
LITERACY	99%
LIFE EXPECTANCY	68
CURRENCY	somoni
GDP PER CAPITA	$1,300
ECONOMY	IND: aluminum, zinc, lead, chemicals and fertilizers. AGR: cotton, grain, fruits, grapes, vegetables, cattle, sheep, goats. EXP: aluminum, electricity, cotton, fruits, vegetable oil, textiles.

Mountains cover more than 90 percent of this Central Asian republic, whose river valleys are home to a majority of the people. About two-thirds of the people are ethnic Tajiks, but about a quarter are Uzbeks. Shortly after independence in 1991, Tajikistan endured a five-year civil war between the Moscow-backed government and the Islamist-led opposition. A peace agreement was signed in 1997, but the political turmoil has depressed the economy. Tajikistan relies heavily on Russian assistance, and there are some 23,000 Russian troops guarding Tajikistan's borders—against weapons, drugs, and Islamic extremists.

Tanzania
UNITED REPUBLIC OF TANZANIA

CONTINENT	Africa
AREA	945,087 sq km
	(364,900 sq mi)
POPULATION	35,363,000
CAPITAL	Dar es Salaam (administrative capital) 2,683,000; Dodoma (legislative capital) 155,000
RELIGION	Christian, Muslim, indigenous beliefs
LANGUAGE	Kiswahili, Kiungujo, English, Arabic, many local languages
LITERACY	78%
LIFE EXPECTANCY	45
CURRENCY	Tanzanian shilling
GDP PER CAPITA	$600
ECONOMY	IND: agricultural processing (sugar, beer, cigarettes, sisal twine), diamond and gold mining, oil refining. AGR: coffee, sisal, tea, cotton, cattle. EXP: gold, coffee, cashew nuts, manufactures, cotton.

Tanzania, the largest country in East Africa, includes the spice islands of Zanzibar, Pemba, and Mafia and contains Africa's highest point, Kilimanjaro, at 5,895 meters (19,340 feet). Kilimanjaro, a dormant volcano, is snowcapped even though it is near the Equator. The African population consists of more than 120 ethnic groups.

Tanganyika, a British-controlled UN trust territory, gained independence in 1961; and Zanzibar, a British protectorate with an Arab population, became independent in 1963. Tanganyika and Zanzibar united to form Tanzania in 1964. Until resigning as president in 1985, independence leader Julius K. Nyerere guided two decades of socialism, adapted to the *ujamaa* policy of village farming. A multiparty system was established in 1992 after a constitutional amendment.

Some 80 percent of Tanzanians farm or fish at subsistence levels; in many areas tse-tse fly infestation hampers successful animal husbandry. Deteriorating roads and railways and high energy costs are major problems. The Ngorongoro Crater and Serengeti National Park are rich in wildlife, although poaching endangers some species. Tourism remains important. Dar es Salaam is the administrative capital, but Dodoma is the designated future capital and current home to Tanzania's legislature.

Thailand
KINGDOM OF THAILAND

CONTINENT	Asia
AREA	513,115 sq km
	(198,115 sq mi)
POPULATION	63,063,000
CAPITAL	Bangkok 6,604,000
RELIGION	Buddhist, Muslim
LANGUAGE	Thai, English, ethnic and
	regional dialects
LITERACY	96%
LIFE EXPECTANCY	71
CURRENCY	baht
GDP PER CAPITA	$7,000
ECONOMY	IND: tourism, textiles
	and garments, agri-
	cultural processing,
	beverages, tobacco,
	cement. AGR: rice, cas-
	sava (tapioca), rubber,
	corn. EXP: computers,
	transistors, seafood,
	clothing, rice.

Thailand, in Southeast Asia, is dominated by the Chao Phraya River basin, which contains Bangkok—the capital and largest city, with some 6.6 million people. Bangkok presents a distinctive Buddhist landscape, with gold-layered spires, graceful pagodas, and giant Buddha statues. To the east rises the Khorat Plateau, a sandstone plateau with poor soils supporting grasses and woodlands. The long southern region, connecting with Malaysia, is hilly and forested. The highest mountains are in northern Thailand, and the rich soils in the remote mountain valleys produce opium poppies.

The population is largely homogeneous, with most being ethnic Thai and professing Buddhism. Some three million Muslims live in the south near the border with Malaysia.

Two 19th-century kings of Siam, Mongkut and his son Chulalongkorn, introduced Western education and technology but preserved the character of a devout Buddhist society. The only nation in Southeast Asia to escape colonial rule, Siam changed its name in 1939 to Thailand, meaning "land of the free." However, Thailand has not escaped military coups—more than a dozen since 1932, when a revolution transformed the government from an absolute to a constitutional monarchy. Resentment against leaders of the 1991 coup sparked demonstrations by a pro-democracy movement. Reforms did take place, and a new constitution went into effect in 1997. The 2001 elections confirmed Thailand's democracy credentials as the people voted in the new Thai Rak Thai ("Thais Love Thais") Party.

The government enjoys one of the world's fastest-growing economies, but faces the challenge of spreading the wealth to poorer regions—the infertile eastern plateau is the poorest. Opium production has been reduced, but heroin trafficking is still a problem. The long, mountainous border with Myanmar (Burma) brings refugees, illegal immigrants, and drugs into the country. Some 140,000 Burmese refugees live in Thailand. A Muslim separatist struggle flared up in southern Thailand in 2004.

Timor-Leste
DEMOCRATIC REPUBLIC OF TIMOR-LESTE

CONTINENT	Asia
AREA	14,609 sq km
	(5,640 sq mi)
POPULATION	778,000
CAPITAL	Dili 49,000
RELIGION	Christian (mostly Roman
	Catholic)
LANGUAGE	Tetum, Portuguese, Bahasa
	Indonesian, English
LITERACY	48%
LIFE EXPECTANCY	49
CURRENCY	US dollar
GDP PER CAPITA	$500
ECONOMY	IND: printing, soap man-
	ufacturing, handicrafts,
	woven cloth. AGR: cof-
	fee, rice, maize, cas-
	sava. EXP: coffee,
	sandalwood, marble.

The first new nation of the 21st century is located in Southeast Asia, just north of Australia. A Portuguese colony from the 17th century until 1975, Timor-Leste (the Portuguese name for East Timor) shares the island of Timor with Indonesia. When the Portuguese left in 1975, Indonesia invaded and annexed East Timor. The United Nations condemned Indonesia's occupation, and in 1999 a UN-organized referendum showed that most East Timorese wanted independence. Militias caused damage after the vote, but the UN guided East Timor to independence on May 20, 2002. The new country is the poorest in Asia, but oil in the Timor Sea promises future revenue.

Togo
TOGOLESE REPUBLIC

CONTINENT	Africa
AREA	56,785 sq km
	(21,925 sq mi)
POPULATION	5,429,000
CAPITAL	Lomé 799,000
RELIGION	Indigenous beliefs, Christian, Muslim
LANGUAGE	French, Ewe, Mina, Kabye, Dagomba
LITERACY	61%
LIFE EXPECTANCY	54
CURRENCY	CFA franc
GDP PER CAPITA	$1,400
ECONOMY	IND: phosphate mining, agricultural processing, cement, handicrafts. AGR: coffee, cacao, cotton, yams, livestock, fish. EXP: reexports, cotton, phosphates, coffee, cacao.

Togo is a long, narrow country in West Africa, with an interior plateau rising to mountains in the north. From the late 17th to the mid-19th century, slave traders prowled Togo's forests and savannas. In 1922 the eastern part of the German protectorate of Togoland passed into French hands, becoming independent in 1960. Military rule finally yielded to some democratic reforms amid civil unrest in the 1990s. However, massive electoral fraud has marred recent elections. Togo is a poor agricultural country with a dismal human-rights record.

Tokelau (New Zealand)
TERRITORY OF NEW ZEALAND

CONTINENT	Australia/Oceania
AREA	12 sq km (5 sq mi)
POPULATION	1,500
CAPITAL	Administered from Wellington, N.Z.
RELIGION	Congregational Christian Church
LANGUAGE	Tokelauan, English

Tonga
KINGDOM OF TONGA

CONTINENT	Australia/Oceania
AREA	748 sq km (289 sq mi)
POPULATION	107,000
CAPITAL	Nuku'alofa 35,000
RELIGION	Christian

LANGUAGE	Tongan, English
LITERACY	99%
LIFE EXPECTANCY	68
CURRENCY	pa'anga
GDP PER CAPITA	$2,200
ECONOMY	IND: tourism, fishing. AGR: squash, coconuts, copra, bananas, fish. EXP: squash, fish, vanilla beans, root crops.

The ruler of this last remaining Polynesian kingdom in the South Pacific traces his lineage back a thousand years. The mutiny on the British ship *Bounty* occurred in Tongan waters in 1789. The 170 islands and islets, 36 of them inhabited, were under the protection of Great Britain for 70 years, until independence came in 1970. The economy relies heavily on agriculture, with tourism and light industry becoming increasingly important.

Trinidad and Tobago
REPUBLIC OF TRINIDAD AND TOBAGO

CONTINENT	North America
AREA	5,128 sq km
	(1,980 sq mi)
POPULATION	1,309,000
CAPITAL	Port-of-Spain 55,000
RELIGION	Roman Catholic, Hindu, Anglican, Muslim, Presbyterian
LANGUAGE	English, Hindi, French, Spanish

LITERACY 99%

LIFE EXPECTANCY 71

CURRENCY Trinidad and Tobago dollar

GDP PER CAPITA $10,000

ECONOMY IND: petroleum, chemicals, tourism, food processing. AGR: cacao, sugarcane, rice, citrus, poultry. EXP: petroleum and petroleum products, chemicals, steel products, fertilizer.

The Caribbean islands of Trinidad and Tobago, while geographically close, are far apart in their tempo of life: Steel-band music and a multiethnic population, including many of African and East Indian descent, give flamboyant Trinidad a fast beat; small farms and quiet resorts give scenic Tobago a slower rhythm. In addition to oil and natural gas, Trinidad contains Pitch Lake, a huge asphalt deposit. High priorities for the economy are increased gas production, aggressive promotion of foreign investment, and industrial and agricultural diversification.

Tunisia
TUNISIAN REPUBLIC

CONTINENT Africa

AREA 163,610 sq km (63,170 sq mi)

POPULATION 9,898,000

CAPITAL Tunis 1,996,000

RELIGION Muslim

LANGUAGE Arabic, French

LITERACY 74%

LIFE EXPECTANCY 73

CURRENCY Tunisian dinar

GDP PER CAPITA $6,800

ECONOMY IND: petroleum, mining, tourism, textiles, footwear. AGR: olives, olive oil, grain, dairy products. EXP: textiles, mechanical goods, phosphates and chemicals, agricultural products.

Gaining its independence in 1956 after 75 years under French control, this North African nation was ruled by President for Life Habib Bourguiba until his ouster in 1987. Political and economic reforms have since pulled Tunisia from the brink of collapse. The fluctuating economy is based on agriculture, particularly market gardening of vegetables, as well as phosphates and petroleum. Tunisia's sunny Mediterranean coast and the nation's ancient history, spectacularly preserved at Carthage, make for a robust tourist industry.

Turkey
REPUBLIC OF TURKEY

CONTINENT Europe/Asia

AREA 779,452 sq km (300,948 sq mi)

POPULATION 71,224,000

CAPITAL Ankara 3,428,000

RELIGION Muslim (mostly Sunni)

LANGUAGE Turkish, Kurdish, Arabic, Armenian, Greek

LITERACY 87%

LIFE EXPECTANCY 69

CURRENCY Turkish lira

GDP PER CAPITA $7,300

ECONOMY IND: textiles, food processing, autos, mining, steel, petroleum. AGR: tobacco, cotton, grain, olives, livestock. EXP: apparel, foodstuffs, textiles, metal manufactures, transport equipment.

Straddling the continents of Europe and Asia, Turkey tries to be a bridge between West and East. The portion of Turkey's land in Europe may be small (about 5 percent), but the country's largest city, Istanbul, is there. With 9.7 million people, Istanbul is the third most populous European urban area, after Moscow and Paris. The Asian part of Turkey is dominated by the dry plateau of Anatolia; the coastal areas of Anatolia consist of fertile lowlands. The country, especially northern Turkey, suffers from severe earthquakes. Mount Ararat, the highest point in Turkey at 5,137 meters (16,854 feet), is the biblical resting-place of Noah's ark.

The Ottoman Empire, which was centered here for 600 years, commanded vast stretches of northern Africa, southeastern Europe, and western Asia until it fell to the Allied armies during World War I. In 1923 Mustafa Kemal, known as Atatürk, Father of the Turks, founded the republic and sought to transform a conservative Islamic society into a secular, westernized state. Atatürk insisted that Turkish be written with the Latin alphabet instead of Arabic characters. He wanted women liberated from the Islamic veil, and he led the fight to win them the vote in 1934.

Turkey joined the UN in 1945, and NATO in 1952. Although Turkey and Greece both belong to NATO, disputes over the Aegean Sea and Cyprus strain relations between the two countries. Turkish forces invaded Cyprus in 1974 to protect the Turkish-Cypriot community during a military coup, it still maintains some 30,000 troops in northern Cyprus. Southeastern Turkey saw years of civil war in the 1980s and 1990s between Turkish forces and Kurds from the Kurdistan Workers' Party (PKK), who wanted to form an independent Kurdish state. Relations improved when the Turkish parliament passed laws giving more rights to Kurds.

In 1990 Turkey supported the West against Iraq following Iraq's invasion of Kuwait and in 2003 allowed U.S. forces to use Turkish air space in the Iraq war. In 1999 Turkey gained approval as a candidate country for membership in the European Union. There are more than three million Turks working and living in EU countries, most in Germany. Most trade is with Europe, and many European vacationers come to Turkey for the climate, fine beaches, resorts, Roman ruins, and Crusader castles.

Turkmenistan
TURKMENISTAN

CONTINENT	Asia
AREA	488,100 sq km (188,456 sq mi)
POPULATION	5,703,000
CAPITAL	Ashgabat 574,000
RELIGION	Muslim, Eastern Orthodox
LANGUAGE	Turkmen, Russian, Uzbek
LITERACY	98%
LIFE EXPECTANCY	67
CURRENCY	Turkmen manat
GDP PER CAPITA	$6,700
ECONOMY	IND: natural gas, oil, petroleum products, textiles. AGR: cotton, grain, livestock. EXP: gas, oil, cotton fiber, textiles.

Turkmenistan, a desert nation, has the second lowest population density (after Kazakhstan) in former Soviet Central Asia. Nomadic herdsmen for centuries, Turkmen were subdued by Russia during the late 19th century, gaining independence in 1991. Begun in the 1950s, the Garagum Canal, one of the world's longest, drained water away from the Amu Darya River to southern Turkmenistan, but the old canal leaks and creates salt deserts. Also, by diverting water from the Amu Darya, the canal contributed greatly to the drying up of the Aral Sea.

Turkmenistan's hope lies in its sector of the Caspian Sea, where oil and natural gas fields are concentrated. The country's natural gas reserves rank fifth in the world, but development of gas exports is hampered by a lack of gas-pipeline routes out of landlocked Turkmenistan. Russia controls most of the pipelines and has refused to export Turkmen natural gas to hard-currency markets. A gas pipeline through Afghanistan to Pakistan gained approval in 2002, but the security situation in Afghanistan remains an obstacle. Disputes between Azerbaijan, Iran, Kazakhstan, Russia, and Turkmenistan over Caspian Sea seabed and maritime boundaries limits international investment in new gas fields and pipelines. Revenue from oil and gas production benefits few because of an authoritarian and corrupt government.

Turks and Caicos Islands (U.K.)
BRITISH OVERSEAS TERRITORY

CONTINENT	North America
AREA	430 sq km (166 sq mi)
POPULATION	19,000
CAPITAL	Cockburn Town 6,000
RELIGION	Baptist, Anglican, Methodist
LANGUAGE	English

Some 40 islands and cays form the Turks and Caicos Islands; only six of them are inhabited. These Caribbean islands are mostly barren and dry, making the surrounding waters clear — bringing in tourists for snorkeling and scuba diving.

Tuvalu
TUVALU

CONTINENT	Australia/Oceania
AREA	26 sq km (10 sq mi)
POPULATION	10,000
CAPITAL	Funafuti 6,000
RELIGION	Church of Tuvalu (Congregationalist), Seventh-Day Adventist, Baha'l
LANGUAGE	Tuvaluan, English, Samoan
LITERACY	NA
LIFE EXPECTANCY	66
CURRENCY	Australian dollar, Tuvaluan dollar
GDP PER CAPITA	$1,100
ECONOMY	IND: fishing, tourism, copra. AGR: coconuts, fish. EXP: copra, fish.

This cluster of South Pacific atolls, once part of the Gilbert and Ellice Islands Colony, achieved independence from Britain and joined the Commonwealth in 1978. Farming is limited due to poor soils and lack of fresh water. Mainstays are fishing, sales of copra and postage stamps, and remittances from Tuvaluans working abroad. Revenue also comes from a trust fund created by the U.K. and Pacific-area sponsors.

Uganda
REPUBLIC OF UGANDA

CONTINENT	Africa
AREA	241,139 sq km (93,104 sq mi)
POPULATION	25,262,000
CAPITAL	Kampala 1,246,000
RELIGION	Roman Catholic, Protestant, indigenous beliefs, Muslim
LANGUAGE	English, Ganda or Luganda, many local languages
LITERACY	70%
LIFE EXPECTANCY	44
CURRENCY	Ugandan shilling
GDP PER CAPITA	$1,200
ECONOMY	IND: sugar, brewing, tobacco, cotton textiles, cement. AGR: coffee, tea, cotton, tobacco, beef. EXP: coffee, fish and fish products, tea, gold, cotton.

Uganda, a landlocked country in East Africa, consists of savanna plateau with mountains and lakes. "The pearl of Africa," wrote Winston Churchill of this former British protectorate that embraces Lake Victoria, source of the White Nile, and the misty Ruwenzori Mountains—a home of the endangered mountain gorilla.

Prosperous at independence in 1962, the country was brutalized under the chaotic regimes of Milton Obote and Idi Amin, when as many as 800,000 Ugandans were murdered. In 1986 Yoweri Museveni, leader of the National Resistance Army, came to power and, despite continued military and economic challenges, restored stability to a large extent. Museveni was elected president in May 1996 in the first popular election for president since independence, and reelected in 2001.

In 1998 Uganda sent troops into the Democratic Republic of the Congo's civil war. It withdrew the last of its forces in 2003, and thousands of Congolese crossed into Uganda to seek asylum. An insurgent militia, known as the Lord's Resistance Army, continues to terrorize northern Uganda, abducting some 20,000 children over the years and making them soldiers. Fertile soil keeps farms and coffee plantations flourishing, but AIDS, epidemic in some areas, may now be the country's greatest enemy.

Ukraine
UKRAINE

CONTINENT	Europe
AREA	603,700 sq km (233,090 sq mi)
POPULATION	47,793,000
CAPITAL	Kiev 2,618,000
RELIGION	Ukrainian Orthodox, Ukrainian Catholic (Uniate), Protestant, Jewish

LANGUAGE Ukrainian, Russian,
Romanian, Polish,
Hungarian
LITERACY 100%
LIFE EXPECTANCY 68
CURRENCY hryvnia
GDP PER CAPITA $4,500
ECONOMY IND: coal, electric
power, ferrous and
nonferrous metals,
machinery and transport
equipment. AGR: grain,
sugar beets, sunflower
seed, beef. EXP: ferrous
and nonferrous metals,
fuel and petroleum
products, chemicals,
machinery and transport
equipment.

The Carpathian Mountains rise in the west and the Crimean Mountains in the south, but the heartland of Ukraine, slightly larger than France, is the rich flat earth that stretches for 1,600 kilometers (1,000 miles), known as the steppe. Once called the bread-basket of the Soviet Union, Ukraine also has huge deposits of coal and iron that feed heavy industry, particularly in the Donbas (Donets Basin) and Kryvyy Rih regions.

Such natural wealth tempted conquerors. In 988 Vladimir the Great adopted Christianity, which evolved into Russian Orthodoxy, to unify the Kievan Rus, a confederation of Slavic peoples. The Mongols overran the land in the 13th century, followed by the Lithuanians in the 14th century. Poland asserted dominion in 1569. Defying their Polish masters, rebel-minded peasants, the Cossacks, gathered under warlike leaders called *hetmans* in the vastness of the steppe. After a revolt led by Bohdan Khmel-nytsky, the Cossacks formed their own state in 1649. But in 1654, still fighting the Poles, they entered a pact with Russia, which soon exerted control.

At its greatest extent, about 1880, the Russian Empire encompassed 85 percent of present-day Ukraine; the remainder was under the influence of Austria-Hungary. After the Russian Revolution, Ukraine enjoyed brief independence. Despite Lenin's promises, however, the Red Army invaded, and by 1920 most of Ukraine was Bolshevik ruled. Joseph Stalin, fearing Ukrainian nationalism, killed the intelligentsia, and, through his policy of collectivization, engineered a famine in 1932 and 1933 that took at least five million lives. Nazi occupation scourged the country during the "Great Patriotic War." The republic lost 7.5 million people, 4 million of them civilians and 2.2 million deported to Germany as laborers. After World War II, Soviet rule prevailed.

Ukraine suffered the world's worst recorded nuclear accident. On the morning of April 26, 1986, reactor No. 4 at the Chornobyl Nuclear Power Plant, 80 miles north of Kiev, exploded, sending radioactive contaminants three miles up into the atmosphere and out over parts of Europe, Asia, and North America.

A political meltdown occurred in December 1991, when 90 percent of Ukrainians voted for independence, in effect dissolving the Soviet Union. Now Ukraine faces ongoing border disputes with Russia. The new millennium has brought economic growth, with rising industrial output and falling inflation. In 2001 the country destroyed its last Soviet-era nuclear missile silo, and in 2002 it announced plans to join NATO.

United Arab Emirates
UNITED ARAB EMIRATES

CONTINENT Asia
AREA 77,700 sq km
(30,000 sq mi)
POPULATION 3,888,000
CAPITAL Abu Dhabi 475,000
RELIGION Sunni and Shiite Muslim,
Christian, Hindu
LANGUAGE Arabic, Persian, English,
Hindi, Urdu
LITERACY 78%
LIFE EXPECTANCY 74
CURRENCY Emirati dirham
GDP PER CAPITA $22,100
ECONOMY IND: petroleum, fishing,
petrochemicals, con-
struction materials.
AGR: dates, vegetables,
watermelons, poultry,
fish. EXP: crude oil,
natural gas, reexports,
dried fish, dates.

Seven sheikhdoms on the Arabian Peninsula combined to form a federation after Britain pulled out of this barren coastal region in 1971. The United Arab Emirates comprises Abu Dhabi, seat of the federal government and the oil capital; Dubayy (Dubai), the main port and commercial-industrial hub; Ajman; Umm al Qaywayn; Ras al Khaymah; Al Fujayrah; and Sharjah. Oil, discovered in 1958, is the major income earner. Oil wealth brought foreign workers, who now make up about three-quarters of the population. A favorite destination for tourists, the country has a liberal attitude toward other cultures and beliefs.

United Kingdom
UNITED KINGDOM OF GREAT BRITAIN AND NORTHERN IRELAND

CONTINENT	Europe
AREA	242,910 sq km (93,788 sq mi)
POPULATION	59,200,000
CAPITAL	London 7,615,000
RELIGION	Anglican, Roman Catholic, other Protestant, Muslim
LANGUAGE	English, Welsh, Scottish form of Gaelic
LITERACY	99%
LIFE EXPECTANCY	78
CURRENCY	British pound
GDP PER CAPITA	$25,500
ECONOMY	IND: machine tools, electric power equipment, automation equipment, railroad equipment, shipbuilding. AGR: cereals, oilseed, potatoes, vegetables, cattle, fish. EXP: manufactured goods, fuels, chemicals, food, beverages, tobacco.

Separated from the European continent by the North Sea and English Channel, the United Kingdom (informally referred to as Britain) includes England, Scotland, Wales, and Northern Ireland. England and Wales were united in 1536. The addition of Scotland in 1707 created Great Britain, renamed the United Kingdom in 1801 when Ireland was added. The Republic of Ireland fought itself free of British rule in 1922, leaving volatile Northern Ireland as a province of the United Kingdom. About 55 percent of Northern Ireland's 1.6 million people trace their ancestry to Scotland or England, are Protestants, and favor continued union with Britain; however, many of the Roman Catholic population (44 percent) want to join the Republic of Ireland.

England is the most populous part of the U.K., with 49 million inhabitants. Almost one third of England's people live in the prosperous southeastern part of the country centered on London, one of the largest cities in Europe. Scotland, with one third of Britain's area, is a mountainous land with 5 million people, most of them (75 percent) concentrated in the lowland area where Glasgow and Edinburgh (Scotland's capital) are located. The Scottish nation can be traced to the Scoti, a Gaelic-speaking Celtic tribe. Wales, with 2.9 million people,

is also mountainous with a Celtic culture, the country is called Cymru (pronounced CUM-ree) in the Welsh language, and its capital, Cardiff, features castles and museums highlighting Welsh culture. Since 1997 the government has been pursuing a policy of devolution, leading in 1999 to an elected Scottish parliament and Welsh assembly. In 2000 Londoners elected their first mayor and assembly.

The industrial revolution was born in Britain in the 18th century, making it the world's first industrialized nation. The British Empire, a worldwide system of dependencies, fed raw materials to British industry and spread British culture. Most dependencies gained independence in the 20th century. Part of the legacy of empire is that Britain is home to a growing multicultural population. The 2001 census counted more than 2.5 million Asians (mostly Indians and Pakistanis) and 1.1 million blacks (from Africa and the Caribbean). Most of the remaining dependencies consist of small islands in the Atlantic and Caribbean.

DEPENDENCIES AND TERRITORIES

Anguilla

Bermuda

British Indian Ocean Territory

British Virgin Islands

Cayman Islands

Channel Islands

Falkland Islands

Gibraltar

Isle of Man

Montserrat

Pitcairn Islands

Saint Helena

South Georgia and the
South Sandwich Islands

Turks and Caicos Islands

United States
UNITED STATES OF AMERICA

CONTINENT	North America
AREA	9,826,630 sq km
	(3,794,083 sq mi)
POPULATION	291,512,000
CAPITAL	Washington, D.C.
	4,190,000
RELIGION	Protestant, Roman
	Catholic, Jewish
LANGUAGE	English, Spanish
LITERACY	97%
LIFE EXPECTANCY	77
CURRENCY	US dollar
GDP PER CAPITA	$36,300
ECONOMY	IND: petroleum, steel,
	motor vehicles, aero-
	space, telecommunica-
	tions, chemicals,
	electronics. AGR:
	wheat, corn, other
	grains, fruits, beef,
	forest products, fish.
	EXP: capital goods,
	automobiles, industrial

supplies and raw materials, consumer goods, agricultural products.

In 1776, after 169 years of distant and undemocratic British rule, the 13 Colonies declared independence. With the Constitution of 1787, the United States of America's four million people embarked on a political experiment: a democratic republic with representation at the local, state, and federal levels, with a built-in balance among executive, legislative, and judicial branches. From 1861 to 1865, the union of the United States was tested by the Civil War; it survived after a great loss of life.

On September 11, 2001, America witnessed death not seen on its soil since the Civil War, when 3,025 people died as four passenger jets were hijacked and turned into terrorist bombs, crashing in New York, Pennsylvania, and Virginia. The images of a destroyed World Trade Center and damaged Pentagon had a momentous impact on the country, and the government launched a "war on terror." In October 2001 the U.S. led a military campaign against terrorists in Afghanistan, driving the Taliban regime from power. In November 2002 the Department of Homeland Security was created to protect the nation against terrorist attacks. In March 2003 a U.S.-led coalition invaded Iraq, toppling Saddam Hussein's hostile dictatorship.

COMMONWEALTHS AND TERRITORIES

American Samoa

Baker Island

Guam

Howland Island

Jarvis Island

Johnston Atoll

Kingman Reef

Midway Islands

Navassa Island

Northern Mariana Islands

Palmyra Atoll

Puerto Rico

Virgin Islands, U.S.

Wake Island

Uruguay
ORIENTAL REPUBLIC OF URUGUAY

CONTINENT	South America
AREA	176,215 sq km
	(68,037 sq mi)
POPULATION	3,380,000
CAPITAL	Montevideo 1,341,000
RELIGION	Roman Catholic
LANGUAGE	Spanish, Portunol,
	Brazilero

LITERACY	98%
LIFE EXPECTANCY	75
CURRENCY	Uruguayan peso
GDP PER CAPITA	$7,900
ECONOMY	IND: food processing, electrical machinery, transportation equipment, petroleum products. AGR: rice, wheat, corn, barley, livestock, fish. EXP: meat, rice, leather products, wool, vehicles.

Situated in southeastern South America, Uruguay consists mostly of low, rolling grasslands. Ranchers raise cattle and sheep on the well-watered pastures. It has one of the highest urbanization and literacy rates in South America as well as the lowest poverty and population growth rates. About 93 percent of Uruguayans, most of Spanish or Italian descent, live in cities, with Montevideo home to one-third. Education is compulsory and free, but many Uruguayans emigrate to places like Spain for better job opportunities.

Uruguay's economy remains dependent on agriculture. Economic diversification, including development of hydroelectric power, has spread optimism; tourists flock to Atlantic beach resorts like Punta del Este.

Uzbekistan
REPUBLIC OF UZBEKISTAN

CONTINENT	Asia
AREA	447,400 sq km (172,742 sq mi)
POPULATION	25,672,000
CAPITAL	Tashkent 2,155,000
RELIGION	Muslim, Eastern Orthodox
LANGUAGE	Uzbek, Russian, Tajik
LITERACY	99%
LIFE EXPECTANCY	70
CURRENCY	Uzbekistani sum
GDP PER CAPITA	$2,600
ECONOMY	IND: textiles, food processing, machine building, metallurgy, natural gas. AGR: cotton, vegetables, fruits, grain, livestock. EXP: cotton, gold, energy products, mineral fertilizers, ferrous metals.

Uzbekistan, a landlocked country dominated by the Qizilqum desert, is Central Asia's most populous country. About 80 percent of the country is flat desert, with mountain ranges rising in the far southeast and northeast. The Fergana Valley in the northeast is the country's most fertile region, containing many cities and industries.

The ancient oasis cities of Tashkent, Samarqand, and Bukhara all evoke the old Silk Road to China. Uzbeks, third largest ethnic group of the former Soviet Union (after Russians and Ukrainians), descend from Turkic people and are rooted in the Sunni Muslim faith.

Most of the population lives in rural areas, where cotton crops, imposed by Soviet planners at horrendous cost to the environment, made Uzbekistan one of the world's top five producers. The Aral Sea, fed by rivers extensively tapped for irrigation, has shrunk to a fraction of its 1960s extent, and may become a vast desert by the year 2018.

Economic growth and living standards are among the lowest in the former Soviet Union. Uzbekistan is still one of the largest exporters of cotton, and the world's largest open-pit gold mine is at Muruntau in the Qizilqum desert; some geologists claim it is Earth's largest gold deposit. However, the economic climate is poor because of smothering state control. The government is authoritarian and is becoming more rigid as it is threatened by Islamist groups. The country faced a wave of violence from suicide bombings in 2004.

Vanuatu
REPUBLIC OF VANUATU

CONTINENT	Australia/Oceania
AREA	12,190 sq km (4,707 sq mi)
POPULATION	214,000
CAPITAL	Port-Vila 34,000
RELIGION	Protestant, Catholic, indigenous beliefs
LANGUAGE	English, French, more than 100 local languages
LITERACY	53%
LIFE EXPECTANCY	67

CURRENCY vatu

GDP PER CAPITA $2,900

ECONOMY IND: food and fish freezing, wood processing, meat canning. AGR: copra, coconuts, cacao, coffee, fish. EXP: copra, beef, cacao, timber, kava.

France and the U.K. jointly administered some 80 South Pacific islands known as the New Hebrides for 74 years until independence in 1980. During World War II the U.S. launched attacks from here against Japanese troops in the Solomon Islands and New Guinea, inspiring James Michener's *Tales of the South Pacific.* Tourism augments income from copra, tuna processing, meat canning, and timber sales. South American hardwoods have been introduced to expand forests.

Vatican City
STATE OF THE VATICAN CITY (THE HOLY SEE)

CONTINENT Europe

AREA 0.4 sq km (0.2 sq mi)

POPULATION 1,000

RELIGION Roman Catholic

LANGUAGE Italian, Latin, French

LITERACY 100%

LIFE EXPECTANCY 78

CURRENCY euro

GDP PER CAPITA $25,500

ECONOMY IND: printing, production of coins, medals, postage stamps, worldwide banking and financial activities. AGR: NA EXP: NA

From this state within the city of Rome, the Pope ministers to a flock of more than a billion Roman Catholics. The Lateran Treaty between Italy and the Holy See created an independent Vatican City in 1929. The Holy See maintains diplomatic relations with more than 150 countries; issues passports, coins, and stamps; has a radio station and a publishing house; and billets a force of Swiss Guards. John Paul II, the first non-Italian Pope in 455 years, was elected in 1978.

Venezuela
BOLIVARIAN REPUBLIC OF VENEZUELA

CONTINENT South America

AREA 912,050 sq km (352,144 sq mi)

POPULATION 25,698,000

CAPITAL Caracas 3,226,000

RELIGION Roman Catholic

LANGUAGE Spanish, indigenous dialects

LITERACY 93%

LIFE EXPECTANCY 73

CURRENCY bolivar

GDP PER CAPITA $5,400

ECONOMY IND: petroleum, iron ore mining, construction materials, food processing. AGR: corn, sorghum, sugarcane, rice, beef, fish. EXP: petroleum, bauxite and aluminum, steel, chemicals, agricultural products.

Venezuela, in northern South America, was named for Italy's Venice by 15th-century European explorers who found native houses on stilts above Lake Maracaibo.

The Lake Maracaibo basin splits the Andes into two mountain ranges. Mild temperatures exist on the mountains while the Maracaibo basin swelters in tropical heat. Most people live in cities on the range near the Caribbean coast, from Caracas to Barquisimeto. South of the mountains is the Orinoco River basin, a vast plain of savanna grasses known as the *Llanos* (YAH-nohs). South of the Orinoco are the Guiana Highlands, with the world's highest waterfall, Angel Falls. Almost half of Venezuela's land is south of the Orinoco, but this region contains only 5 percent of the population.

Venezuela is one of the oldest democracies in South America (elections since 1958). A founding member of the Organization of Petroleum Exporting Countries (OPEC), the nation has the largest proven oil reserves in the Western Hemisphere— and the second largest natural gas reserves (after the U.S.). The petroleum industry accounts for half the government's revenue; however, few Venezuelans benefit from its wealth and most live in poverty, which contributes to political instability.

Vietnam
SOCIALIST REPUBLIC OF VIETNAM

CONTINENT	Asia
AREA	331,114 sq km (127,844 sq mi)
POPULATION	80,786,000
CAPITAL	Hanoi 3,977,000
RELIGION	Buddhist, Hoa Hao, Cao Dai, Christian, indigenous beliefs, Muslim
LANGUAGE	Vietnamese, English, French, Chinese, Khmer, local languages
LITERACY	94%
LIFE EXPECTANCY	72
CURRENCY	dong
GDP PER CAPITA	$2,300
ECONOMY	IND: food processing, garments, shoes, machine-building. AGR: paddy rice, corn, potatoes, rubber, poultry, fish. EXP: crude oil, marine products, rice, coffee, rubber.

Vietnam, in Southeast Asia, stretches 1,600 kilometers (1,000 miles) north to south, but is only about 40 kilometers (25 miles) wide at its narrowest point near the country's center. The Red River delta lowlands in the north are separated from the huge Mekong Delta in the south by long, narrow coastal plains backed by the forested Annam highlands. Hanoi, the capital, is the main city on the Red River and Ho Chi Minh City, or Saigon, is the main city on the Mekong.

Independent for almost a thousand years, Vietnam fell prey to French colonialism in the mid-19th century. During Japanese occupation in World War II, communist leader Ho Chi Minh formed the Vietminh, an alliance of communist and noncommunist nationalist groups. Armed struggle won independence in 1954 and led to the partition of Vietnam.

For two decades noncommunist South Vietnam, aided by the U.S., fought North Vietnam, backed by China and the Soviet Union. American troops withdrew in 1973, and two years later South Vietnam fell. In 1976 the country was reunified under a communist regime.

To replace support lost when the U.S.S.R. dissolved, economic policy encouraged a free-market system as well as trade with the West. Vietnam saw dramatic economic progress throughout most of the 1990s. In 1995 the U.S. resumed diplomatic relations. Economic growth stalled, however, with the Asian financial crisis. A stock exchange was launched in 2000, and Vietnam has seen increasing levels of foreign investment.

Virgin Islands (U.S.)
TERRITORY OF THE UNITED STATES

CONTINENT	North America
AREA	386 sq km (149 sq mi)
POPULATION	110,000
CAPITAL	Charlotte Amalie 11,000
RELIGION	Protestant, Roman Catholic
LANGUAGE	English, Spanish, Creole

The U.S. Virgin Islands, in the northern Caribbean, consist of three main islands: St. Thomas, St. John, and St. Croix. St. Thomas is a lively tourist destination, most of St. John is a national park, and St. Croix has industry.

Wallis and Futuna (France)
FRENCH OVERSEAS TERRITORY

CONTINENT	Australia/Oceania
AREA	161 sq km (62 sq mi)
POPULATION	15,000
CAPITAL	Matâ'utu 1,000
RELIGION	Roman Catholic
LANGUAGE	French, Wallisian (indigenous Polynesian language)

West Bank
See listing under Israel.

Western Sahara (Morocco)
AREA OF SPECIAL STATUS

CONTINENT Africa
AREA 252,120 sq km
(97,344 sq mi)
POPULATION 263,000
RELIGION Muslim
LANGUAGE Hassaniya Arabic,
Moroccan Arabic

Yemen
REPUBLIC OF YEMEN

CONTINENT Asia
AREA 536,869 sq km
(207,286 sq mi)
POPULATION 19,350,000
CAPITAL Sanaa 1,469,000
RELIGION Sunni and Shiite
Muslim
LANGUAGE Arabic
LITERACY 50%
LIFE EXPECTANCY 60
CURRENCY Yemeni rial
GDP PER CAPITA $800
ECONOMY IND: crude oil produc-
tion and petroleum
refining, small-scale pro-
duction of cotton textiles
and leather goods.
AGR: grain, fruits,
vegetables, pulses,
dairy products, fish.
EXP: crude oil, coffee,
dried and salted fish.

Ancient kingdoms flourished in south-western Arabia (now Yemen), a crossroads of trade from the Orient and Africa to the Mediterranean. At the time of Christ, camel caravans carried as much as 3,000 tons of frankincense each year to Greece and Rome. Marib, capital of Saba (biblical Sheba), was the queen city of incense; nearby a huge dam irrigated thousands of hectares of farmland. Today a new dam and oil pump life into Marib. In Yemen's highlands volcanic soils yield cereal crops. Most coffee groves (Yemen invented the drink in the 11th century, and mocha is named for the Red Sea port, Al Mukha) have been replaced by fields of kat, chewed as a stimulant.

Beginning in the 1500s the Turks periodically dominated the region's interior. After 1839 Britain controlled the port of Aden and surrounding coastal area; Aden boomed after the Suez Canal opened in 1869. In 1904 the Turks and the British established a boundary between their territories, known as North Yemen and South Yemen (Aden).

Following the 1918 collapse of the Ottoman Empire, tribal imams closed the doors of North Yemen. It re-emerged in 1962, when army officers proclaimed the Yemen Arab Republic, sparking an eight-year civil war. South Yemen won independence from Britain in 1967 after two years of Marxist-guerrilla warfare; it became the People's Democratic Republic of Yemen in 1970. Ideological differences provoked conflicts between pro-Soviet South Yemen and pro-Western North Yemen in 1972 and 1979.

In May 1990 the two nations, spurred by reforms in Eastern Europe and the U.S.S.R. and drawn together by ancient cultural bonds, merged in an uneasy alliance that erupted into several weeks of civil war in 1994. Yemen's modest oil reserves provide most of the revenue, but it is the poorest country in the Middle East.

Zambia
REPUBLIC OF ZAMBIA

CONTINENT Africa
AREA 752,614 sq km
(290,586 sq mi)
POPULATION 10,896,000
CAPITAL Lusaka 1,394,000
RELIGION Christian, Muslim,
Hindu
LANGUAGE English, indigenous
languages
LITERACY 81%
LIFE EXPECTANCY 41
CURRENCY Zambian kwacha
GDP PER CAPITA $800
ECONOMY IND: copper mining
and processing, con-
struction, foodstuffs,
beverages. AGR: corn,
sorghum, rice, peanuts,
cattle. EXP: copper,
cobalt, electricity,
tobacco, flowers.

A landlocked country in central Africa, Zambia occupies an elevated plateau, flanked in the south by the Zambezi River—and Victoria Falls. There are more than 70 ethnic groups, and most of them live in Lusaka, the capital, or in the cities of the Copperbelt; the two largest, Ndola and Kitwe, have more than 400,000 people each. It is one of sub-Saharan Africa's most highly urbanized countries.

Endowed with huge copper reserves and fertile farmland, Zambia looked to the future with optimism after independence from Britain in 1964. But copper prices plummeted in the mid-1970s, and transport costs soared. The economy has been in decline ever since, and copper, vital to Zambia's economy, suffers from declining prices.

Farming will become increasingly important; only a fifth of the arable land is cultivated. Thundering Victoria Falls and other power sources bring self-sufficiency in hydroelectricity.

Zambia's first multiparty elections in 19 years were held in 1991, in which President Frederick Chiluba was elected. He won reelection in 1996, but international observers cited harassment of opposition parties. A coup was attempted in 1997 but suppressed, and there were alleged voting irregularities in the 2001 elections.

More than 70 percent of Zambians live in poverty, and unemployment is a serious problem. Zambia depends on copper for most of its foreign earnings so the economy suffers when copper prices decline. AIDS is blamed for decimating the cream of Zambian professionals, including engineers and political leaders. It kills around 100,000 people each year.

Zimbabwe
REPUBLIC OF ZIMBABWE

CONTINENT	Africa
AREA	390,757 sq km
	(150,872 sq mi)
POPULATION	12,577,000
CAPITAL	Harare 1,469,000
RELIGION	Syncretic (part Christian, part indigenous beliefs), Christian, indigenous beliefs
LANGUAGE	English, Shona, Sindebele
LITERACY	91%
LIFE EXPECTANCY	41
CURRENCY	Zimbabwean dollar
GDP PER CAPITA	$2,100
ECONOMY	IND: mining (coal, gold), steel, wood products, cement, chemicals. AGR: corn, cotton, tobacco, wheat, cattle. EXP: tobacco, gold, ferro-alloys, textiles, clothing.

This landlocked plateau country in southern Africa takes its name from Great Zimbabwe, a fortified trading hub built in medieval times and used by the majority Shona tribe (about 70 percent of today's population). In 1890 the first British settlers arrived; colony status came in 1923 under the name Southern Rhodesia. In 1965 the white minority government unilaterally declared independence, as Rhodesia. International sanctions and guerrilla warfare against the illegal regime led to legitimately independent Zimbabwe in 1980. Although nominally a multiparty state, in practice the party of President Robert Mugabe dominates the political system.

The economy centers on farming, mining (Zimbabwe holds a tenth of the world's chromite), and manufacturing. Until drought struck in the early 1990s, the nation fed itself. Whites still own choice tracts, and land redistribution is a charged issue. Mugabe's government suddenly started seizing all white-owned commercial agricultural land in 2000. African settlers were being dumped on the land without required government support (including seed, water, and fertilizer). This chaotic land reform is causing massive declines in food production, and millions of Zimbabweans are at risk of famine. The economy is in crisis, with high inflation and unemployment rates.

Canadian Provinces and Territories

Alberta

AREA 661,848 sq km
(255,541 sq mi)
POPULATION 3,153,700
CAPITAL Edmonton 666,104
GDP PER CAPITA $48,320
ENTERED Sept. 1, 1905
ECONOMY IND: service industries, petroleum and natural gas production, chemical manufacturing, food processing. AGR: cattle, wheat, canola, barley, timber.

Wheat and cattle are the traditional mainstays of this prairie province. Farmers and ranchers were joined by petroleum geologists and roughnecks during the mid-20th century, when Alberta was found to hold the bulk of Canada's energy resources. The province is the world's second-largest natural gas exporter. And Alberta has the largest oil sands resources in the world, with more than 300 billion barrels to be recovered.

Tax incentives stimulate manufacturing and service industries. Personal income tax is the lowest in Canada and Albertans pay no retail sales tax.

At more than three million people, Alberta's population derives from British, German, Ukrainian, Chinese, and Dutch stock. More than half of the people live in Edmonton, seat of the provincial government, and in Calgary, Alberta's commercial capital. For sheer wildness little can surpass backcountry adventures in

Banff, Jasper, and Waterton Lakes National Parks in the Rocky Mountains.

British Columbia

AREA 944,735 sq km
(364,764 sq mi)
POPULATION 4,146,600
CAPITAL Victoria 74,125
GDP PER CAPITA $32,447
ENTERED July 20, 1871
ECONOMY IND: service industries, wood and paper products, food processing, refined-fuel products, primary metals, tourism. AGR: nursery stock and ornamental flowers, dairy products, greenhouse vegetables, berries, salmon and other seafood.

A fertile plateau west of the Rocky Mountains gives way to rain forests along the fjord-indented Pacific coast. Here, the yearly precipitation of 380 centimeters (150 inches) sustains enormous tracts of Sitka spruce, hemlock, and red cedar, a tenth of North America's standing timber.

Explorers and fur trappers arrived in the late 1700s; gold seekers, lumberjacks, and fishermen followed, depleting resources that had served native peoples for centuries. Clearcutting in old-growth forests prompts demands for a stricter national logging policy, with emphasis on sustainable-yield harvesting. Known for their diversity of plant and animal life, the Queen Charlotte Islands, home to the Haida people, are threatened by logging.

In 1971 British Columbia established ecological reserves; more than 150 now preserve sedge meadows, peat bogs, and dunes for study and limited recreation.

Salmon canning began along the Fraser River in 1870. More than a century later overfishing, pollution, and damming of spawning grounds for hydropower, have diminished Canada's most productive salmon fisheries. The federal and provincial governments have invested 565 million dollars to restore fish stocks.

Much of British Columbia's coal fuels Japanese and South Korean steel mills. Vancouver, Canada's gateway to Asia, will host the 2010 Olympic Winter Games.

Manitoba

AREA 647,797 sq km
(250,116 sq mi)
POPULATION 1,162,800
CAPITAL Winnipeg 619,544
GDP PER CAPITA $31,735
ENTERED July 15, 1870
ECONOMY IND: service industries, food processing, transportation equipment manufacturing, nickel mining. AGR: canola, wheat, potatoes, oats, flax-seed, hogs, beef cattle, dairy products.

Once headquarters of the Hudson's Bay Company, which grew fat on the fur trade in the early 1800s, Winnipeg is now the distribution center for Canada's grain industry. Manitoba farms grow more sunflower seed and beans than any other province and produce the second-largest flaxseed crop. But agricultural wealth comes at a price: Almost all the indigenous tallgrass prairie has disappeared, and efforts are under way to preserve remaining flora and fauna.

Lumbering and mining of nickel, copper, and zinc buttress northern Manitoba, along with production of hydroelectric power—a third of which is exported to Ontario, Saskatchewan, and the United States. In the southwest, one potash reserve offers 120 million minable metric tons. Sparkling glacial lakes—100,000 of them—are alive with 80 fish species, but the yearly catch of summer tourists earns more cash than the take of fish.

New Brunswick

AREA 72,908 sq km
(28,150 sq mi)
POPULATION 750,600
CAPITAL Fredericton 47,560
GDP PER CAPITA $27,593
ENTERED July 1, 1867
ECONOMY IND: service industries, food and beverage processing, wood and paper products, mining, tourism. AGR: lobsters, crabs, and other seafood, potatoes, dairy products, poultry, nursery stock, livestock.

Arrow-straight pines logged in the mid-19th century from vast forests in New Brunswick made masts for ocean-crossing square-riggers, built in Moncton and St. John. Today maple, birch, spruce, and fir supply pulp and paper mills—gearing toward production of high-quality stock. New Brunswick ranks first in Canada's output of bismuth, lead, and zinc; antimony, peat, potash, copper, and silver are also important.

In the Bay of Fundy tides that rise 15 meters (49 feet) keep the port of St. John, site of North America's first mainland oil supertanker terminal, ice free year-round. The city, chartered in 1785, is New Brunswick's industrial engine.

Newfoundland and Labrador

AREA 405,212 sq km
(156,453 sq mi)
POPULATION 563,000
CAPITAL St. John's 99,182
GDP PER CAPITA $19,360
ENTERED March 31, 1949
ECONOMY IND: service industries, oil production, iron ore mining, paper products manufacturing, fish food products. AGR: dairy products, poultry, eggs, nursery stock.

Nicknamed the Rock, independent-spirited Newfoundland resisted federation until 1949, when it joined mainland Labrador to become the newest province. By the early 1990s overharvesting had depleted the cod around Grand Banks, one of the world's richest fisheries for 500 years. In 1992 the government imposed a moratorium on cod fishing, and hundreds of fishermen were put out of work. Energy and mining projects along Canada's Atlantic coast have recently brought modest economic growth, including pumping oil from the Hibernia oil field. Population declines have stabilized and the unemployment rate has continued to slowly drop.

Northwest Territories

AREA	1,346,106 sq km
	(519,734 sq mi)
POPULATION	40,000
CAPITAL	Yellowknife 16,541
GDP PER CAPITA	$43,510
ENTERED	July 15, 1870
ECONOMY	IND: diamond mining, gold mining, oil and natural gas production. AGR: NA.

Boreal forests in the south fade to barren tundra in the north of this back-of-beyond spanning two time zones. The short growing season of only about 70 frost-free days makes commercial agriculture impossible.

Native peoples make up more than half the sparse population and continue to support themselves by trapping muskrat, lynx, mink, beaver, and marten. Inuit, who live in the northern reaches of the territories, have suffered economically since the collapse of the sealskin market. Great Slave Lake, rich in whitefish and trout, is the major center for fishing.

In 1999 the region divided into two territories. The smaller, western area continues to be known as Northwest

Territories; the eastern portion, which incorporates 60 percent of the original area, became Nunavut.

Nova Scotia

AREA	55,284 sq km
	(21,345 sq mi)
POPULATION	936,000
CAPITAL	Halifax 359,111
GDP PER CAPITA	$27,717
ENTERED	July 1, 1867
ECONOMY	IND: service industries, food processing, paper products, mining. AGR: blueberries, apples, livestock, dairy products, lobsters and other seafood.

In 1497 John Cabot anchored at Cape Breton Island, just west of the teeming Atlantic fishing banks. By the time Britain defeated France in North America in 1763, thousands of French-speaking Acadians had been expelled. New Englanders replaced them, many settling in the fertile Annapolis Valley. Immigration of Scottish Highlanders began in 1773, and annual events such as the Nova Scotia International Tattoo preserve their heritage.

The sea still largely defines life on this 550-kilometer-long peninsula with 7,500 kilometers of coastline. Lobster, scallops, and haddock are major exports. Lumbering and industry, from textiles to aircraft engines, diversify the economy. Ice-free Halifax, Atlantic Canada's most populous city, also contains the country's largest naval and coast guard bases.

The water in Northumberland Strait is the warmest for ocean swimming in the province. The Bay of Fundy, with the world's highest tides, is one of Nova Scotia's main tourist attractions.

Nunavut

AREA	2,093,190 sq km
	(808,185 sq mi)
POPULATION	29,400
CAPITAL	Iqaluit 5,236
GDP PER CAPITA	$33,607
ENTERED	April 1, 1999
ECONOMY	IND: government and education services, mining of zinc, gold, and lead, fishing, trapping. AGR: NA

Carved out of the eastern half of the Northwest Territories, Nunavut ("our land" in the Inuktitut language) became Canada's third territory on April 1, 1999. Nunavut is roughly the size of Alaska and California combined. Its population is nearly 30,000; 85 percent of the people are Inuit. Their once-nomadic lifestyle has yielded to a community-based existence, though traditional heritage is strong. Nunavut winters last for nine months, and the average temperature in January hovers around minus 6°C. The Polaris lead-zinc mine on Little Cornwallis Island is the world's northernmost metal mine.

Ontario

AREA	1,076,395 sq km
	(415,598 sq mi)
POPULATION	12,238,300
CAPITAL	Toronto 2,481,494
GDP PER CAPITA	$38,993
ENTERED FEDERATION	July 1, 1867
ECONOMY	IND: service industries, automobile manufacturing, telecommunications and computer equipment, food processing, mining. AGR: dairy products, cattle, hogs, feed crops, nursery stock, fruits, and vegetables.

Bounded north and south by water, Ontario is Canada's most populous and most industrialized province. Second in size, it covers an area larger than France and Spain combined. Ninety percent of the population lives on less than 15 percent of the land—in the fertile St. Lawrence lowlands and in the "golden horseshoe" stretching from Oshawa to the Niagara peninsula. The St. Lawrence Seaway, one of the world's busiest shipping lanes, gives oceangoing vessels access to the interior of North America.

As Canada's heartland, Ontario contributes more than 40 percent of the gross domestic product. Automobiles are manufactured in cities such as Oshawa, Windsor, and St. Thomas. Newsprint is the principal wood product; about a fifth of Canada's pulp and paper issues from the seemingly endless expanse of conifers that clothe the northern areas of the province. Toronto, the main port of entry for Canada's immigrants, ranks also as the nation's chief publishing, commercial, and banking center.

Hydroelectricity is a power train between the resource-rich north and the industrially energetic south. The Ottawa, Niagara, and St. Lawrence Rivers are major sources of hydroelectric power, but Ontario Hydro's network of nuclear power stations now produces close to half the province's electricity. The provincial government owns some nine-tenths of the forest lands and grants logging licenses to private companies. Public debate raises questions about the future of the boreal forests.

Prince Edward Island

AREA	5,660 sq km
	(2,185 sq mi)
POPULATION	137,800
CAPITAL	Charlottetown 32,245
GDP PER CAPITA	$26,907
ENTERED	July 1, 1873
ECONOMY	IND: service industries, farm and fish products processing, aircraft parts manufacturing. AGR: potatoes, barley, blueberries, vegetables, dairy products, cattle, hogs, lobsters and other seafood.

When Jacques Cartier came ashore in 1534, Indians had lived on Abegweit, meaning "cradled on the waves," for 10,000 years. Surprisingly warm waves wash its beaches, and the island's population increases five-fold in summer. Visitors enjoy Victorian villages, tidy farmsteads, and lobster suppers in steepled churches. Manicured Charlottetown, site of the 1864 meeting that led to Canada's union, draws thousands to its annual arts festival.

In May 1997, the Confederation Bridge, an eight-mile span between Prince Edward Island and New Brunswick, was completed. This longest bridge over ice-covered waters in the world brings a new age of transportation to Atlantic Canada. Bounty from the sea includes shellfish and Irish moss, a source of carrageenan, a food stabilizer.

Quebec

AREA	1,542,056 sq km
	(595,391 sq mi)
POPULATION	7,487,200
CAPITAL	Quebec 508,000
GDP PER CAPITA	$32,584
ENTERED	July 1, 1867
ECONOMY	IND: service industries, manufacturing of newsprint and other paper products, transportation equipment, timber production, mining. AGR: dairy products, hogs, poultry, cattle, vegetables, apples, corn, nursery stock, maple products.

An island of francophones in the English-speaking sea of North America, La Belle Province de Quebec has long struggled to retain its mother tongue and clarify its relationship with the rest of Canada. More than four-fifths of the population in this oldest and largest province speaks French. In 1980, 60 percent of Quebec voters rejected a referendum advocating separation, but the province still seeks special status within Canada to preserve its cultural heritage. Limitations on possible future secession were further addressed in the Clarity Act of 2000.

Quebec's economy has enriched Canada. The province produces 21 percent of the country's manufactured goods and a major portion of its iron, copper, asbestos, paper, and lumber. Profitable sales of hydroelectricity to New York and New England spur further exploitation of the great rivers that course through the subarctic. Distant James Bay, home to some 10,000 Indians and Inuit and one of the world's largest caribou herds, is the site of an enormous hydropower project. The first phase alone, completed in 1985, rerouted four rivers, flooded an area the size of Connecticut, and built 203 dikes, 1,500 kilometers of roads, and 8 dams. Phase two was put on hold, but a recent agreement cleared the way for its completion.

Most of Quebec's people live in a fertile region along the St. Lawrence River, where a wealth of fruits and vegetables is grown. With more than 3.4 million residents, Greater Montréal makes up close to half the province's population. Montrealers seek refuge from long, bitter winters in the Underground City, a series of boutiques and restaurants connected by pedestrian walkways and subway tunnels.

Saskatchewan

AREA 651,036 sq km
(251,366 sq mi)

POPULATION 994,800
CAPITAL Regina 178,225
GDP PER CAPITA $34,117
ENTERED Sept. 1, 1905
ECONOMY IND: service industries, petroleum and natural gas production, mining of potash and uranium. AGR: wheat, canola, barley, cattle.

Granary of Canada, this prairie province grows a sizable portion of the nation's wheat and much of its oats, barley, flaxseed, and canola. Three centuries ago wealth was counted in the furs of animals.

Regina, Saskatchewan's capital, has Canada's only training academy for the Royal Canadian Mounted Police. Saskatoon, the province's largest city, is home to many high-tech firms, as well as the University of Saskatchewan.

The name "Saskatchewan" comes from the Cree word for "river that flows swiftly." Immigrants soon swamped the Indians and Métis — offspring of French fur traders and Indian women, who number about 44,000 today. Descendants of British pioneers exceed those of German, Polish, Scandinavian, French, Dutch, and Ukrainian settlers. Onion-domed Ukrainian churches dot the landscape.

The rocky expanse of the Canadian Shield to the north, covered by lakes and stunted conifers, contains gold and uranium deposits. The uranium mine at Key Lake is the world's largest, and potash mines in the south supply 35 percent of world demand. A large coal-fired power plant in the southeast uses advanced

technology that ensures clean burning.

Yukon Territory

AREA 482,443 sq km
(186,272 sq mi)
POPULATION 31,100
CAPITAL Whitehorse 19,058
GDP PER CAPITA $39,767
ENTERED June 13, 1898
ECONOMY IND: mining of metal ores, lumber, tourism. AGR: greenhouse vegetables, hunting, fur trade, fishing.

Ever since the cry "Klondike!" drew prospectors in the 1890s, the Yukon's economy has boomed and slumped with the price of gold and other minerals. During the late 1980s annual earnings topped 500 million dollars. Yet today white-water rafters and fishermen outnumber gold panners. More than 313,000 tourists make summer pilgrimages to the Arctic frontier. They find glaciated peaks, untouched wilderness, scenic splendors, and abundant wildflowers and wildlife. Athapaskan Indians still hunt caribou and moose. Resolution of native land claims could spur mineral exploitation, but disputes between miners and conservationists augur intensifying debate on levels of development. The sparseness of the year-round population is no accident: Winter temperatures fall to minus 50°C, permafrost makes construction costly, and fresh produce is limited, even in Whitehorse, where the majority of the population lives.

The Tennessee-Tombigbee Waterway, a transportation corridor for industry and agriculture, connects 16,000 miles of U.S. inland waterways to the Gulf of Mexico via Mobile's port.

United States

Alabama
HEART OF DIXIE

AREA 135,765 sq km
(52,419 sq mi)

POPULATION 4,500,752

CAPITAL Montgomery 201,425

PCI $25,096

STATEHOOD December 14, 1819;
22nd state

ECONOMY IND: retail and wholesale trade, services, government, finance, insurance, real estate, transportation, construction, electrical equipment. AGR: poultry, forest products, cattle, nursery stock, cotton, eggs, peanuts, soybeans.

Home to five Native American tribes (Creek, Cherokee, Chickasaw, Choctaw, and Seminole), this area of fertile soils was the heart of the "cotton kingdom" before the Civil War.

Today Alabama lies at the center of a revitalized Deep South. Birmingham, the state's largest city, has become a focus for medical research, and is a major manufacturer of steel, iron, and coal. Leading industries include textiles and apparel, rubber and plastics, paper, chemicals, primary metals, and automobile manufacturing. The Alabama Research and Education Network, first state-funded computer network in the U.S., links universities and school systems across the state.

Alaska
GREAT LAND

AREA 1,717,854 sq km
(663,267 sq mi)

POPULATION 648,818

CAPITAL Juneau 30,751

PCI $31,792

STATEHOOD January 3, 1959;
49th state

ECONOMY IND: petroleum products, state and local government, services, trade, federal government. AGR: shellfish, seafood, nursery stock, vegetables, dairy products, feed crops.

In 1867 Secretary of State William H. Seward paid Russia 7.2 million dollars for a huge region derided as "Seward's Icebox." Today this land of overwhelming beauty, abundant resources, and few people is a battleground between conservationists and energy and mining interests. More than a third of the mineral-rich state is forested; a quarter is set aside as parks, refuges, and wilderness. Fisheries teem with salmon, halibut, and shellfish. Alaska natives, who number some 100,000, administer 13 regional corporations established under the 1971 Alaska Native Claims Settlement Act.

Arizona
GRAND CANYON STATE

AREA 295,254 sq km
(113,998 sq mi)
POPULATION 5,580,811
CAPITAL Phoenix 1,371,960
PCI $26,157
STATEHOOD February 14, 1912;
48th state
ECONOMY IND: real estate,
manufactured goods,
retail, state and local
government, trans-
portation and public
utilities, wholesale
trade, health services.
AGR: vegetables,
cattle, dairy products,
cotton, fruit, nursery
stock, nuts.

Ghostly cliff dwellings, such as Canyon de Chelly, contrast sharply with Phoenix and Tucson, where one out of three Arizonans live. The state's population has doubled in the past 20 years, and today smog often shrouds Phoenix, once favored by asthmatics for its clean, dry air. Service industries and high-tech and aerospace manufacturing have eclipsed the "three C's" — copper, cattle, and cotton; irrigated agribusiness remains important. The multibillion-dollar Central Arizona Project river diversion helps slake Arizona's thirst.

Arkansas
NATURAL STATE

AREA 137,732 sq km
(53,179 sq mi)
POPULATION 2,725,714
CAPITAL Little Rock 184,055
PCI $23,417
STATEHOOD June 15, 1836
25th state
ECONOMY IND: services, food processing, paper products,
transportation, metal
products, machinery,
electronics. AGR: poultry
and eggs, rice, soy-
beans, cotton, wheat.

Migrants from the southern Appalachians settled the forested Ozark Plateau and Ouachita Mountains in the early 19th century. Meanwhile, rich black soils along the Mississippi River attracted cotton planters to the east and south. Arkansas now leads the nation in rice, ranks third in turkeys, and places high in soybeans and sorghum. No other place in the world produces more bromine. Hot Springs National Park and the scenic Ozarks attract a growing number of visitors.

California
GOLDEN STATE

AREA 423,970 sq km
(163,696 sq mi)
POPULATION 35,484,453
CAPITAL Sacramento 435,245
PCI $32,898
STATEHOOD September 9, 1850
31st state
ECONOMY IND: electronic components and equipment,
aerospace, film produc-
tion, food processing,
petroleum, computers
and computer software,
tourism. AGR: vegeta-
bles, fruits and nuts,
dairy products, cattle,
nursery stock, grapes.

"Eureka, I have found it" is the apt motto for the nation's most populous state, home to one in eight Americans. The gold rush of 1849 created California's image as a promised land. By 1900 almost half the population was clustered around San Francisco and Los Angeles, each the focus of intense competition for water.

In 1913 the Los Angeles Aqueduct began tapping water from Owens Valley to feed the city's continued unchecked growth. In 1934 San Francisco satisfied its thirst with water from Hetch Hetchy Reservoir, a project that spawned the modern conservation movement. The Colorado River Aqueduct eased the water cravings of burgeoning southern California in the 1940s. The state's largest water transfer, the Central Valley Project, greens some 128,000 square kilometers in the San Joaquin and Sacramento Valleys, where migrant farm laborers find seasonal work.

More immigrants reside here than in any other state, and nearly half of all Californians are immigrants or the children of immigrants. Between 1990 and 2000, California's population increased by 4.1 million. More than 90 percent of the state is urban. Among leading industries is the manufacture of high-tech equipment, centered in the areas of biotechnology, aerospace-defense, and computers.

Colorado
CENTENNIAL STATE

AREA	269,601 sq km
	(104,094 sq mi)
POPULATION	4,550,688
CAPITAL	Denver 560,415
PCI	$33,170
STATEHOOD	August 1, 1876
	38th state
ECONOMY	IND: real estate, state and local government, durable goods, communications, health and other services, nondurable goods, transportation. AGR: cattle, corn, wheat, dairy products, hay.

"What a splendid field it is for new expeditions," wrote 19th-century mountaineer Frederick Chapin. Gold and silver drew mining expeditions in the early days. Today some 26 million visitors a year make excursions to the highest state. As ski resorts and tourist centers expand, wildlife habitat and the "splendid field" suffer. In Aspen and Telluride particulate pollution, mainly from wood-burning stoves, can be worse than in Denver; regulations limit fireplaces in new homes.

Pueblo, once dependent on steel, has broadened its industrial base; the city is also the gateway to rafting and camping in the southern mountains. Colorado's economy, long reliant on energy and minerals, would be in the doldrums without tourism, a $7 billion industry. Near Denver, energy capital of the mountain West, the Rocky Flats weapons plant, which once pumped 300 million dollars a year into state coffers, no longer manufactures nuclear arms. An environmental cleanup of the site, with a $7 billion price tag, is nearing completion.

Connecticut
CONSTITUTION STATE

AREA	14,357 sq km
	(5,543 sq mi)
POPULATION	3,483,372
CAPITAL	Hartford 124,558
PCI	$42,829
STATEHOOD	January 9, 1788
	5th state
ECONOMY	IND: transportation equipment, metal products, machinery, electrical equipment, printing and publishing, scientific instruments. AGR: nursery stock, dairy products, poultry, eggs, shellfish.

The industrious colonial Yankee would be at home today in this business-minded state with the nation's second highest per capita income. Many firms whose products include robotic and

fiber-optic equipment have their corporate headquarters in Stamford. Hartford, the nation's insurance capital, rises above the cleaned-up Connecticut River, again a spawning ground for the Atlantic salmon. Hartford is also home to the *Hartford Courant,* the oldest (1764) newspaper in the U.S. that is still being published.

Delaware
FIRST STATE

AREA	6,447 sq km
	(2,489 sq mi)
POPULATION	817,491
CAPITAL	Dover 32,581
PCI	$32,307
STATEHOOD	December 7, 1787
	1st state
ECONOMY	IND: food processing, chemicals, rubber and plastic products, scientific instruments, printing and publishing. AGR: poultry, soybeans, nursery stock, corn, vegetables, dairy products.

E. I. du Pont de Nemours & Co., the chemicals and plastics giant, established a gunpowder plant near Wilmington in 1802. Today more than 300,000 companies are incorporated in Delaware, which offers a business-friendly community. The favorable business climate has particularly spurred growth in banking and financial services. In 1971, in an effort to protect beaches and wetlands, the state legislature passed the nation's first coastal-zone act, banning waterfront industries that pollute.

District of Columbia
JUSTITIA OMNIBUS (JUSTICE TO ALL)

AREA 177 sq km (68 sq mi)
POPULATION 570,898
PCI $43,371
ECONOMY IND: government, service, tourism.

In 1791 President George Washington commissioned Pierre Charles L'Enfant to design a capital fit for "this vast empire." Two centuries later the center of the nation's political universe is also a world of monuments and museums, visited by close to 20 million people each year.

The 570,898 residents, 60 percent of whom are African Americans, have the nation's highest per capita income. The District of Columbia is administered by an elected mayor and city council. Its two Congressional delegates are permitted to vote in committees, but not on the House floor. Efforts toward statehood for the District have, so far, been unsuccessful.

Florida
SUNSHINE STATE

AREA 170,304 sq km
(65,755 sq mi)
POPULATION 17,019,068

CAPITAL Tallahassee 155,171
PCI $29,559
STATEHOOD March 3, 1845
27th state
ECONOMY IND: health services, business services, communications, banking, electronic equipment, insurance, tourism. AGR: citrus, vegetables, field crops, nursery stock, cattle, dairy products.

This subtropical playground, long favored by retirees, is one of the fastest growing states. But nature suffers as life-giving wetlands are drained, overbuilding accelerates coastal erosion, and the drawing down of the water table causes saltwater intrusion and sinkholes. Miami, the bilingual gateway for Latin America, is more than half Hispanic, mostly Cuban. In central Florida, Sea World, Universal Studios, and Walt Disney World make big money. High-tech industry supplements tourism and agriculture, mainly citrus, tomatoes, landscaping plants, and sugarcane.

Georgia
EMPIRE STATE OF THE SOUTH

AREA 153,909 sq km
(59,425 sq mi)
POPULATION 8,684,715
CAPITAL Atlanta 424,868
PCI $28,703
STATEHOOD January 2, 1788;
4th state

ECONOMY IND: textiles and clothing, transportation equipment, food processing, paper products, chemicals, electrical equipment, tourism. AGR: poultry and eggs, cotton, peanuts, vegetables, sweet corn, melons, cattle.

Atlanta began as a railhead in 1837, was burned to the ground in the Civil War, and rose again to become the transportation hub of the new South. Savannah, an industrial port and resort, was the nation's first planned city, laid out in 1733. In the northeastern uplands, tourism now supersedes textile manufacturing and farming, while pine forests in the southeast make Georgia a leading supplier of wood pulp. Farmers on the fertile Coastal Plain grow almost half the nation's peanuts, which generate some $500 million in annual revenue.

Hawai'i
ALOHA STATE

AREA 28,311 sq km
(10,931 sq mi)
POPULATION 1,257,608
CAPITAL Honolulu 378,155
PCI $30,040
STATEHOOD August 21, 1959
50th state
ECONOMY IND: tourism, trade, finance, food processing, petroleum refining, stone, clay, and glass products.

AGR: sugar cane, pineapples, nursery stock, tropical fruit, livestock, macadamia nuts.

More than six million vacationers, most of them from the continental U.S. or Japan, spend close to 11 billion dollars a year in this tropical archipelago. Defense, centered on U.S. military bases at Pearl Harbor, is the second largest moneymaker. Descendants of Asians, who immigrated in the 19th and early 20th century to work on sugar plantations, add to the mix of people in this only state with no ethnic majority: Caucasians constitute 24 percent; Japanese, 18 percent; Filipino, 12 percent. The remainder includes ethnic Chinese and those of Hawaiian ancestry. Ecologists estimate that 89 percent of Hawai'i's flowering plants and 97 percent of its land animals, among them the world's only predatory caterpillars, exist nowhere else on Earth.

Idaho
GEM STATE

AREA	216,446 sq km (83,570 sq mi)
POPULATION	1,366,332
CAPITAL	Boise 189,847
PCI	$25,042
STATEHOOD	July 3, 1890 43rd state
ECONOMY	IND: electronics and computer equipment, tourism, food processing, forest products, mining,

chemicals. AGR: potatoes, dairy products, cattle, wheat, alfalfa hay, sugar beets, barley, trout.

Mountains dominate Idaho, whose irrigated valleys yield more than one-third of the nation's potatoes. The Snake River alone waters a million hectares of farmland. Cascading through Hells Canyon, at 2,408 meters (7,900 feet) the deepest river gorge in the lower 48, the Snake also powers hydroelectric turbines. Second in silver mining, Idaho taps more antimony than any other state. Visitors pour into Sun Valley resort and to Craters of the Moon National Monument, a scene of cindery desolation.

Illinois
LAND OF LINCOLN

AREA	149,998 sq km (57,914 sq mi)
POPULATION	12,653,544
CAPITAL	Springfield 111,834
PCI	$33,320
STATEHOOD	December 3, 1818; 21st state
ECONOMY	IND: industrial machinery, electronic equipment, food processing, chemicals, metals, printing and publishing, rubber and plastics, motor vehicles. AGR: corn, soybeans, hogs, cattle, dairy products, nursery stock.

Farmland covers nearly 80 percent of the state, and Illinois is a major exporter of farm products. Chicago is home to nearly one-quarter of all Illinoisans. No other U.S. city moves more freight by train and truck. Linked to the Atlantic by the St. Lawrence Seaway, Chicago sends cargo into the U.S. interior; canals and the Illinois and Mississippi Rivers extend its reach to the Gulf of Mexico.

Indiana
HOOSIER STATE

AREA	94,321 sq km (36,418 sq mi)
POPULATION	6,195,643
CAPITAL	Indianapolis 783,612
PCI	$28,233
STATEHOOD	December 11, 1816; 19th state
ECONOMY	IND: transportation equipment, steel, pharmaceutical and chemical products, machinery, petroleum, and coal industries, mining. AGR: corn, soybeans, hogs, poultry and eggs, cattle, dairy products.

Like spokes in a wheel, interstates converge on Indiana's capital, where the first Indy 500 auto race was held in 1911. Total farm acreage in Indiana has decreased by 35 percent since 1900 due to factors such as bankruptcy or consolidation. Meanwhile, development has consumed thousands of acres of arable land. Manufacturing employs one out of four workers. Gary and Hammond anchor one of the world's great industrial regions.

Iowa
HAWKEYE STATE

AREA	145,743 sq km
	(56,272 sq mi)
POPULATION	2,944,062
CAPITAL	Des Moines 198,076
PCI	$28,141
STATEHOOD	December 28, 1846;
	29th state
ECONOMY	IND: real estate, health
	services, industrial
	machinery, food pro-
	cessing, construction.
	AGR: hogs, corn, soy-
	beans, oats, cattle,
	dairy products.

Bracketed by the Mississippi and Missouri Rivers, the territory that became Iowa was part of the Louisiana Purchase of 1803. Today 92 percent of this heartland state, blessed with fine prairie soil, is under cultivation. Farm income suffered during most of the 1980s, forcing many Iowa farmers out of business. Yet in 2002 the state still led the nation in corn and soybean production and hogs.

Kansas
SUNFLOWER STATE

AREA	213,096 sq km
	(82,277 sq mi)
POPULATION	2,723,507
CAPITAL	Topeka 122,103
PCI	$28,838
STATEHOOD	January 29, 1861;
	34th state
ECONOMY	IND: aircraft manu-
	facturing, transportation
	equipment, construction,
	food processing, print-
	ing and publishing,
	health care. AGR: cat-
	tle, wheat, sorghum,
	soybeans, hogs, corn.

Cattle towns such as Abilene have long since given way to manufacturing centers. Wichita turns out 70 percent of the general-aviation aircraft produced in the U.S.; Kansas City makes automobiles. Among the top states in crude-oil production, Kansas also banks on one of the nation's largest natural gas fields. Salt deposits near Hutchinson are the remnant of a shallow sea that once submerged the Great Plains. Although no other state grows more wheat — Mennonites from Europe introduced a hardy winter variety in the 1870s — livestock earns more for Kansas.

Kentucky
BLUEGRASS STATE

AREA	104,659 sq km
	(40,409 sq mi)
POPULATION	4,117,827
CAPITAL	Frankfort 27,660
PCI	$25,657
STATEHOOD	June 1, 1792
	15th state
ECONOMY	IND: manufacturing, serv-
	ices, government, finance,
	insurance, real estate,
	retail trade, transporta-
	tion, wholesale trade, con-
	struction, mining. AGR:
	tobacco, horses, cattle,
	corn, dairy products.

Bluegrass and the music named for it are rooted here. In 1769 Daniel Boone entered the region through the Cumberland Gap. Today Kentucky ranks with Wyoming and West Virginia as a leading coal producer. More than 20 percent of its farm receipts come from tobacco. The state has the largest concentration of whiskey distilleries in the U.S., although 70 of the 120 counties are dry. Derby Day has drawn horse racing enthusiasts to Churchill Downs every year since 1875.

Louisiana
PELICAN STATE

AREA	134,264 sq km
	(51,840 sq mi)
POPULATION	4,496,334
CAPITAL	Baton Rouge 225,702

PCI $25,370
STATEHOOD April 30, 1812;
18th state
ECONOMY IND: chemicals, petroleum products, food processing, health services, tourism, oil and natural gas extraction, paper products. AGR: forest products, poultry, marine fisheries, sugarcane, rice, dairy products, cotton, cattle, aquaculture.

Louisiana ranks fourth in the U.S. in crude-oil production and second in natural gas. Louisiana also ranks second nationally in crude-oil refining capacity. Petrochemical plants line the Mississippi River from New Orleans—the country's busiest port—to Baton Rouge, and the production of chemicals is the state's leading manufacturing activity. Tourism is the second most important industry. Vibrant New Orleans preserves its Creole heritage in the architecture and foods of the French Quarter. Lafayette, the heart of Cajun country, is home to descendants of the Acadians who were expelled from Canada in the mid-18th century.

Maine
PINE TREE STATE

AREA 91,646 sq km
(35,385 sq mi)
POPULATION 1,305,728
CAPITAL Augusta 18,551
PCI $27,804

STATEHOOD March 15, 1820;
23rd state
ECONOMY IND: health services, tourism, forest products, leather products, electrical equipment, food processing, textiles. AGR: seafood, potatoes, dairy products, poultry and eggs, livestock, apples, blueberries, vegetables.

Timber-products companies own nearly half of Maine's forests. Potatoes, apples, and blueberries are other offerings of the land. From the sea comes as much as 50 million pounds of lobster a year, a delicacy enjoyed by many of the almost nine million annual visitors.

As development edges north, the state and conservation groups are buying up pristine lands. French speakers predominate in many towns along the state's border with Canada's Quebec Province.

Maryland
OLD LINE STATE

AREA 32,133 sq km
(12,407 sq mi)
POPULATION 5,508,909
CAPITAL Annapolis 36,196
PCI $36,121
STATEHOOD April 28, 1788
7th state
ECONOMY IND: real estate, federal government, health services, business services, engineering services,

electrical and gas services, communications, banking, insurance. AGR: poultry and eggs, dairy products, nursery stock, soybeans, corn, seafood, cattle, vegetables.

"An immense protein factory," wrote Baltimorean H. L. Mencken of the Chesapeake Bay. One of the world's greatest estuaries, the Chesapeake contains thousands of species, many threatened by agricultural runoff, sewage, and urban wastes. Baltimore, with its revitalized Inner Harbor, is a major seaport, ranking as the country's second port for foreign tonnage. The state capital of Annapolis, an East Coast sailing hub, preserves the nation's largest concentration of 18th-century buildings and is home to the U.S. Naval Academy. The Delmarva Peninsula draws pleasure seekers to its Ocean City boardwalk as well as to undeveloped beaches. Hilly western Maryland is still largely rural. Elsewhere commuter suburbs, interstate highways, and shopping malls share space with the Potomac River and the C & O Canal.

Massachusetts
BAY STATE

AREA 27,336 sq km
(10,555 sq mi)
POPULATION 6,433,422
CAPITAL Boston 589,281
PCI $39,044

STATEHOOD   February 6, 1788;
6th state

ECONOMY   IND: electrical equipment, machinery, metal
products, scientific
instruments, printing and
publishing, tourism.
AGR: fruits, nuts and
berries, nursery stock,
dairy products.

From Cape Cod in the east and the Berkshires in the west, all roads lead to Boston. Described as "the thinking center of the continent" by Oliver Wendell Holmes in 1859, the Boston area is home to some 50 degree-granting institutions. A historic park in Lowell recalls the 19th-century heyday of the textile industry. Medical researchers in Worcester, the state's second largest city, developed the birth-control pill during the 1950s. Recession hit the state hard in the late 1980s after a biomedical and high-tech boom earlier in the decade. Once-polluted Boston Harbor, where Puritans landed in 1630, has seen a dramatic turnaround following a $4 billion cleanup project.

Michigan
GREAT LAKES STATE

AREA   250,494 sq km
(96,716 sq mi)

POPULATION   10,079,985

CAPITAL   Lansing 118,588

PCI $30,222

STATEHOOD   January 26, 1837;
26th state

ECONOMY   IND: motor vehicles and
parts, machinery, metal
products, office furniture,
tourism, chemicals. AGR:
dairy products, cattle,
vegetables, hogs, corn,
nursery stock, soybeans.

Michigan's economy has diversified and has gained some independence from the automobile industry. Auto manufacturing is still central, but it has managed to become more efficient, more diversified, and more high tech. More than half a million jobs in the state are connected with motor vehicles and transportation services. Between 1997 and 2003, Michigan attracted 10,229 business projects, ranking number one in the nation for new plants and expansions. Grand Rapids has been a furniture-making center since the late 19th century, while Battle Creek remains America's breakfast-cereal capital.

The Upper Peninsula, with its lakes and forests, continues to attract nature lovers. The recreation industry depends on a healthy environment, but even around Lake Superior's remote Isle Royale National Park fish contain toxic chemicals.

Minnesota
GOPHER STATE

AREA   225,171 sq km
(86,939 sq mi)

POPULATION   5,059,375

CAPITAL   St. Paul 284,037

PCI $33,895

STATEHOOD   May 11, 1858;
32nd state

ECONOMY   IND: health services,
tourism, real estate,
banking and insurance,
industrial machinery,
printing and publishing,
food processing, scientific equipment. AGR:
corn, soybeans, dairy
products, hogs, cattle,
turkeys, wheat.

Northernmost of the lower 48 states, Minnesota holds a pivotal position. The Great Lakes waterway system gives access to the Atlantic Ocean; the Mississippi River, which rises here, provides a link to the Gulf of Mexico. Duluth is one of the world's largest inland ports, transshipping low-grade iron ore, much of it from the nearby Mesabi Range, source of more than three-quarters of U.S. output. Agriculture dominates the state's south; agribusiness clusters in metropolitan Minneapolis-St. Paul, also a magnet for high-tech growth and the cultural center of the upper Midwest. With 15,000 lakes, Minnesota attracts many summer visitors.

Mississippi
MAGNOLIA STATE

AREA   125,434 sq km
(48,430 sq mi)

POPULATION   2,881,281

CAPITAL   Jackson 180,881

PCI $22,370

STATEHOOD December 10, 1817;
20th state

ECONOMY IND: petroleum products,
health services, electronic
equipment, transportation,
banking, forest products,
communications. AGR:
poultry, eggs, cotton, catfish, soybeans, cattle, rice,
dairy products.

Mississippi ranks first in world production of catfish and third among
states in cotton. It leads in the manufacturing of upholstered furniture.

Between 1950 and 1980 soaring
demand for soybeans prompted reclamation of more than four-fifths of the
almost five million acres of wetlands.
Federal laws have now made it less
profitable to convert waterfowl habitat to cropland, but agriculture is still
the state's leading industry. A heritage
corridor along the Mississippi River
includes antebellum houses and Civil
War battlefields.

Missouri
SHOW ME STATE

AREA 180,533 sq km
(69,704 sq mi)
POPULATION 5,704,484
CAPITAL Jefferson City 39,079
PCI $28,841
STATEHOOD August 10, 1821;
24th state
ECONOMY IND: transportation equip-
ment, food processing,
chemicals, electrical

equipment, metal prod-
ucts. AGR: cattle, soy-
beans, hogs, corn, poultry
and eggs, dairy products.

Frenchmen began mining lead here
in the early 1700s; today southeastern
Missouri is the country's foremost
lead producer. In 1764 French traders
founded St. Louis near the confluence
of the Missouri and Mississippi Rivers.
In 1803, the U.S. acquired Missouri as
part of the Louisiana Purchase. Subsequently, the influx of white settlers drove
out the original inhabitants, members
of numerous Native American tribes.
Most were gone by 1836. The fur trade
and the Santa Fe Trail brought increased
prosperity, and St. Louis became the
gateway for pioneers headed west.
Today, both St. Louis and Kansas City
thrive as transport hubs.

Agribusiness centers on Kansas City.
St. Louis, the locus of aerospace and
automobile manufacturing, is headquarters of the world's largest brewing
company. More than seven million
tourists a year enjoy live country music
in Branson's celebrity theaters.

Montana
TREASURE STATE

AREA 380,838 sq km
(147,042 sq mi)
POPULATION 917,621
CAPITAL Helena 26,353
PCI $24,906
STATEHOOD November 8, 1889;
41st state

ECONOMY IND: forest products,
food processing, min-
ing, construction,
tourism. AGR: wheat,
cattle, barley, hay,
sugar beets, dairy
products.

Only Texas has more land devoted
to agriculture than Montana, which
celebrated its statehood centennial in
1989. In the Rocky Mountains, scene
of a gold rush before statehood, mining of copper, gold, and silver is still
profitable. Montana has 50 billion
tons of stripable coal, one-third of the
nation's reserve, but strip mining
arouses opposition from ranchers and
environmentalists. Some 16 million
acres of national forest land bristle
with fir and pine trees in Big Sky Country. Conservationists press for stricter
logging limits.

Nebraska
CORNHUSKER STATE

AREA 200,345 sq km
(77,354 sq mi)
POPULATION 1,739,291
CAPITAL Lincoln 232,362
PCI $29,544
STATEHOOD March 1, 1867;
37th state
ECONOMY IND: food processing,
machinery, electrical
equipment, printing and
publishing. AGR: cattle,
corn, hogs, soybeans,
wheat, sorghum.

During the mid-1800s one visitor called Nebraska's prairies "fat indeed compared to your New England pine plains." Today nearly 95 percent of the land is in farms and ranches. In the Sand Hills many spreads are so large that herds are tracked from the air. Corn and soybeans cover rolling eastern prairies; wheat grows on the drier central and western plains. The Ogallala aquifer irrigates corn, sugar beets, and alfalfa. More than a third of Nebraskans live in Omaha and Lincoln.

Nevada
SILVER STATE

AREA	286,351 sq km
	(110,561 sq mi)
POPULATION	2,241,154
CAPITAL	Carson City 54,311
PCI	$30,169
STATEHOOD	October 31, 1864;
	36th state
ECONOMY	IND: tourism and gaming, mining, printing and publishing, food processing, electrical equipment. AGR: cattle, hay, dairy products.

No rivers flow to the sea from the Great Basin, whose high ridges and alkali sinks dominate this driest state. In 1859 prospectors struck the Comstock Lode in western Nevada. Today gold and silver mines are reopening, but casino gambling, legalized in 1931, added more than nine billion dollars of revenues to the economy in 2002 — more than agriculture, manufacturing, and mining combined. Nevada is 87 percent federally owned. It is the fastest growing state; its population has more than doubled since 1980.

New Hampshire
GRANITE STATE

AREA	24,216 sq km
	(9,350 sq mi)
POPULATION	1,287,687
CAPITAL	Concord 41,404
PCI	$34,276
STATEHOOD	June 21, 1788
	9th state
ECONOMY	IND: machinery, electronics, metal products. AGR: nursery stock, poultry and eggs, fruits and nuts, vegetables.

"Representation without taxation" could be this state's motto, as one of only two states with neither general sales nor personal income taxes. Revenues come mainly from corporate income, real estate, tourism, and liquor sales. Property taxes, set at town meetings, fund schools and community services. After a recession in the early 1990s, economic conditions improved. But high-tech companies in southern areas were impacted by a slowdown early in the new century. The White Mountains, the lakes, and the seacoast regions attract vacationers in all seasons. New Hampshire boasts more forests and wildlife today than a century ago.

New Jersey
GARDEN STATE

AREA	22,588 sq km
	(8,721 sq mi)
POPULATION	8,638,396
CAPITAL	Trenton 85,650
PCI	$39,567
STATEHOOD	December 18, 1787;
	3rd state
ECONOMY	IND: chemicals, printing and publishing, food processing, machinery, electronics. AGR: nursery stock, vegetables, grain and hay, fruits and berries, dairy products.

Wedged between Philadelphia and New York City, New Jersey has long been a distributor of goods and a giant of industry. The state ranks first in production of pharmaceuticals, which account for 12 percent of all jobs. Along a narrow strip, part of the megalopolis stretching from Boston to Washington, D.C., population density far exceeds that of India, yet 57 percent of the state remains forested or in farms. In million-acre Pinelands National Reserve, a model of pragmatic environmental management, development is diverted from the most fragile areas. Ocean resorts from Cape May to Asbury Park were fashionable by the late 19th century. Today some $300 million in taxes collected from Atlantic City casinos are used to benefit the state's elderly and handicapped.

New Mexico
LAND OF ENCHANTMENT

AREA 314,915 sq km
(121,590 sq mi)
POPULATION 1,874,614
CAPITAL Santa Fe 65,127
PCI $23,908
STATEHOOD January 6, 1912;
47th state
ECONOMY IND: electronic equipment, state and local government, real estate, business services, federal government, oil and gas extraction, health services. AGR: cattle, dairy products, hay, chilies, onions.

Diverse cultures — Indian, Spanish, and Mexican — have shaped New Mexico, which ranks fifth in area but 36th in population. Most New Mexicans, nearly half of whom are Hispanic, live along the Rio Grande, many in the Albuquerque area. Income from copper and potash supplements earnings from crude oil and natural gas. The boom in weapons and energy-related research brought scientists to Los Alamos and Sandia National Laboratories. Santa Fe, seat of government since 1610 when it was a dusty Spanish outpost, beckons artists, writers, and tourists.

New York
EMPIRE STATE

AREA 141,299 sq km
(54,556 sq mi)
POPULATION 19,190,115
CAPITAL Albany 93,779
PCI $35,708
STATEHOOD July 26, 1788
11th state
ECONOMY IND: printing and publishing, machinery, computer products, finance, tourism. AGR: dairy products, cattle and other livestock, vegetables, nursery stock, apples.

Historic gateway to the New World, New York City welcomed millions of immigrants ashore at Ellis Island from the 1890s to the 1920s. The nation's most populous metropolis, still a major entry point, is a world center of finance, communications, fashion, and culture. Commerce has long flourished along the Hudson River, and the completion of the Erie Canal in 1825 opened up a key trade route to the interior. The St. Lawrence Seaway and power projects in the north have led to industrial expansion in that part of the state.

North Carolina
TAR HEEL STATE

AREA 139,389 sq km
(53,819 sq mi)
POPULATION 8,407,248
CAPITAL Raleigh 306,944

PCI $27,566
STATEHOOD November 21, 1789;
12th state
ECONOMY IND: real estate, health services, chemicals, tobacco products, finance, textiles. AGR: poultry, hogs, tobacco, nursery stock, turkeys, cotton, soybeans.

Manufacturing and agriculture are the major industries in North Carolina, which leads the U.S. in the production of bricks, furniture, tobacco, and textiles. With four major universities and world-renowned Research Triangle Park all within miles of one another, North Carolina is rapidly becoming a global leader in high technology. Vast natural areas attract more than 44 million tourists annually, bringing in some $12 billion in revenues yearly.

North Dakota
FLICKERTAIL STATE

AREA 183,112 sq km
(70,700 sq mi)
POPULATION 633,837
CAPITAL Bismarck 56,234
PCI $26,567
STATEHOOD November 2, 1889;
39th state
ECONOMY IND: services, government, finance, construction, transportation, oil and gas. AGR: wheat, cattle, sunflowers, barley, soybeans.

Vast Dakota Territory, at the geographic center of the continent, was divided into North and South Dakota in 1889. Rich loess soils favor agriculture, and farms cover 90 percent of the state's land area. But the practice of draining glacier-formed prairie potholes to increase cropland destroys critical habitat for migrating birds. Extremes in weather and in world markets subject wheat farmers to cycles of boom and bust, but petroleum and lignite production assists the economy. Garrison Dam, on the Missouri River, produces 400,000 kilowatts of electricity, providing extensive irrigation for the surrounding area.

Ohio
BUCKEYE STATE

AREA	116,096 sq km
	(44,825 sq mi)
POPULATION	11,435,798
CAPITAL	Columbus 725,228
PCI	$29,317
STATEHOOD	March 1, 1803;
	17th state
ECONOMY	IND: transportation, equipment, metal products, machinery, food processing, electrical equipment. AGR: soybeans, dairy products, corn, hogs, cattle, poultry and eggs.

Blessed with the navigable waters of Lake Erie and the Ohio River, a thousand kilometers open to barge traffic, Ohio enjoyed an early boom in manufacturing and commerce. The industrial cities of Toledo, Akron, and Cleveland turn out rubber, automobiles, glass, and steel; today Ohio ranks third in U.S. manufacturing employment. As ports, Cleveland and Toledo benefit from foreign trade-zone status. Farms cover 56 percent of Ohio, which lies within the bountiful midwestern grain belt.

Oklahoma
SOONER STATE

AREA	181,036 sq km
	(69,898 sq mi)
POPULATION	3,511,532
CAPITAL	Oklahoma City 519,034
PCI	$25,136
STATEHOOD	November 16, 1907;
	46th state
ECONOMY	IND: natural gas production, manufacturing, services, food processing. AGR: cattle, wheat, hogs, poultry, nursery stock.

Descendants of more than 60 tribes make Oklahoma second only to California in Indian population. In the early 1800s the region was established as Indian Territory, a vast reserve for displaced Native Americans. The eastern Indian Territory remained separate until 1906. Eventually land-hungry farmers swallowed up Indian Territory,

and Oklahoma was born in 1907. This state of "soil, oil, and toil" has weathered hard times. The droughts and dust storms of the Depression years forced Okies westward in desperate search of work. The state's economy has been broadly diversified from its traditional oil and agriculture base. Natural gas production, machinery manufacturing, and food processing are among the leading industries.

Oregon
BEAVER STATE

AREA	254,805 sq km
	(98,381 sq mi)
POPULATION	3,559,596
CAPITAL	Salem 140,977
PCI	$28,533
STATEHOOD	February 14, 1859;
	33rd state
ECONOMY	IND: real estate, retail and wholesale trade, electronic equipment, health services, construction, forest products, business services. AGR: nursery stock, hay, cattle, grass seed, wheat, dairy products, potatoes.

The end of the road for wagon trains bound westward over the Oregon Trail, the Pacific Northwest became more accessible after completion of a transcontinental railroad line to Portland in 1883. Today location on the Pacific Rim is one of Oregon's greatest assets: Most of its international trade is with Asia, and Portland serves as a large distribution center for Japanese autos.

Hydroelectricity generated on the Columbia River has powered industry since World War II. Forests cover half the state, and the lumber and wood-products industries bring in some $3.3 billion annually. Fruits and vegetables grow in the Willamette Valley, and wheat grows east of the Cascades. Natural resources are managed carefully, and Oregon's land-use and recycling laws set standards for the nation.

Pennsylvania
KEYSTONE STATE

AREA	119,283 sq km
	(46,055 sq mi)
POPULATION	12,365,455
CAPITAL	Harrisburg 48,540
PCI	$31,663
STATEHOOD	December 12, 1787;
	2nd state
ECONOMY	IND: machinery, printing
	and publishing, forest
	products, metal products.
	AGR: dairy products,
	poultry and eggs,
	mushrooms, cattle,
	hogs, grains.

In 1681 William Penn, an English Quaker, received a royal proprietorship to what became Pennsylvania. Almost a century later, his capital, Philadelphia, witnessed the signing of the Declaration of Independence and the framing of the U.S. Constitution. European immigrants, many of them iron- and steelworkers, founded trade unions that evolved into the American Federation of Labor (AFL) and Congress of Industrial Organizations (CIO). German, Slav, and Italian neighborhoods enliven Pittsburgh today.

Still heavily industrialized, Pennsylvania produces much of the nation's steel, but retail, manufacturing, and other services employ more workers.

Rhode Island
OCEAN STATE

AREA	4,002 sq km
	(1,545 sq mi)
POPULATION	1,076,164
CAPITAL	Providence 175,901
PCI	$31,107
STATEHOOD	May 29, 1790
	13th state
ECONOMY	IND: health services,
	business services,
	silver and jewelry, metal
	products. AGR: nursery
	stock, vegetables, dairy
	products, eggs.

This smallest state is highly industrialized and the second most densely populated, after New Jersey. Entrepreneurs in Pawtucket launched New England's textile industry soon after the American Revolution. Today Rhode Island is known for making silverware and fine jewelry. Providence, a wholesale distribution center and seaport, lies at the head of Narragansett Bay and is home to one in six Rhode Islanders. Newport is famous for its yacht races, jazz festivals, and oceanside mansions.

South Carolina
PALMETTO STATE

AREA	82,932 sq km
	(32,020 sq mi)
POPULATION	4,147,152
CAPITAL	Columbia 117,394
PCI	$25,395
STATEHOOD	May 23, 1788
	8th state
ECONOMY	IND: service industries,
	tourism, chemicals, tex-
	tiles, machinery, forest
	products. AGR: poultry,
	tobacco, nursery stock,
	dairy products, cotton.

Before 1860, rice, indigo, and cotton cultivated by a large slave population enriched tidewater planters. Shattered by the Civil War, which raged from 1861–65, South Carolina came to depend on textile goods and tobacco. Historic Charleston and many nearby antebellum plantations evoke the Old South. Today, retirement communities and resorts such as Myrtle Beach and Hilton Head line the fast-developing seacoast.

South Dakota
MOUNT RUSHMORE STATE

AREA	199,731 sq km
	(77,117 sq mi)
POPULATION	764,309
CAPITAL	Pierre 14,012
PCI	$26,694
STATEHOOD	November 2, 1889;
	40th state
ECONOMY	IND: finance, services,
	manufacturing, gov-
	ernment, retail trade,
	transportation and utili-
	ties, wholesale trade,
	construction, mining.
	AGR: cattle, corn, soy-
	beans, wheat, hogs,
	hay, dairy products.

Ranchers raise cattle and sheep on the Great Plains west of the Missouri. East of the river farmers grow corn and other grains. Sioux Falls, with 130,491 people, is the state's largest city. The Black Hills, spiritual homeland of the Sioux, contain one of North America's largest gold mines, the Homestake. Rapid City, once the gateway to mining camps, now channels tourists to Mount Rushmore and Custer State Park, which protects a publicly owned herd of 1,500 buffalo.

Tennessee
VOLUNTEER STATE

AREA	109,151 sq km
	(42,143 sq mi)
POPULATION	5,841,748
CAPITAL	Nashville 545,915
PCI	$27,378
STATEHOOD	June 1, 1796
	16th state
ECONOMY	IND: service industries,
	chemicals, transportation
	equipment, processed
	foods, machinery. AGR:
	cattle, cotton, dairy
	products, hogs, poultry,
	nursery stock.

Foreign companies drawn by non-unionized labor and access to U.S. markets are invigorating Tennessee, and the population is now predominantly urban. The Tennessee River, whose dams generate abundant electricity, trisects the state. In the West, biomedical, telecommunications, and transportation industries lead an economic resurgence centered on Memphis that has brought billions to the local economy. Soybean and cotton growers in this region struggle to conserve easily eroded soils. Nashville, in the middle of the state, is America's country-music capital.

Texas
LONE STAR STATE

AREA	695,621 sq km
	(268,581 sq mi)
POPULATION	22,118,509
CAPITAL	Austin 671,873
PCI	$28,401
STATEHOOD	December 29, 1845;
	28th state
ECONOMY	IND: chemicals, machin-
	ery, electronics and

computers, food products, petroleum and natural gas, transportation equipment. AGR: cattle, sheep, poultry, cotton, sorghum, wheat, rice, hay, peanuts, pecans.

Texas began as a vast frontier, where 19th century settlers carved out cotton farms in the humid east and cattle ranches in the arid west. A century later, Texas fueled the automobile revolution with a booming oil industry. At the start of the 21st century, the economy has been transformed again. Predominantly urban, with some 22 million residents, the Lone Star State is an industrial giant. One in four Texans is Hispanic, and Texans generally embrace the cultural influences of Mexico. National parks in Big Bend country and the Guadalupe Mountains of West Texas preserve the state's natural heritage.

Utah
BEEHIVE STATE

AREA	219,887 sq km
	(84,899 sq mi)
POPULATION	2,351,467
CAPITAL	Salt Lake City 181,266
PCI	$24,157
STATEHOOD	January 4, 1896;
	45th state
ECONOMY	IND: government,
	manufacturing, real
	estate, construction,
	health services, business
	services, banking.

AGR: cattle, dairy products, hay, poultry and eggs, wheat.

Mormon religious refugees settled along Great Salt Lake in 1847. Today Mormons make up nearly three-quarters of Utah's population. Family values are strong, and the birthrate far exceeds that of the U.S. as a whole. Although biomedical technology and the manufacturing of aerospace equipment and computer software buoy the economy, growth has slowed. Attracted by five national parks and Salt Lake City — financial, retail, and transportation hub of the western Rockies and site of the 2002 Winter Olympics — 17 million visitors a year enjoy Utah's scenic diversity. The U.S. government owns 66 percent of the land; conservationists and developers disagree over its use.

Vermont
GREEN MOUNTAIN STATE

AREA 24,901 sq km
(9,614 sq mi)
POPULATION 619,107
CAPITAL Montpelier 8,026
PCI $29,464
STATEHOOD March 4, 1791;
14th state
ECONOMY IND: health services,
tourism, finance, real
estate, computer
components, electrical
parts, printing and
publishing, machine

tools. AGR: dairy products, maple products, apples.

Strict environmental guidelines, including a billboard ban and a 1988 law requiring towns to join in regional planning and statewide forest and farmland conservation efforts, aim to protect the rustic beauty of this least populous state in the East. The forests of the Green Mountains, Vermont's backbone, turn crimson and gold in the autumn; sugar maples, whose syrup sweetens the economy, highlight the glorious display of color. In winter, when snow blankets hillsides, skiers descend on more than 50 resorts. Vermont is the leading producer of monument granite and marble.

Virginia
OLD DOMINION

AREA 110,785 sq km
(42,774 sq mi)
POPULATION 7,386,330
CAPITAL Richmond 197,456
PCI $32,676
STATEHOOD June 25, 1788
10th state
ECONOMY IND: food processing,
communication and
electronic equipment,
transportation equip-
ment, printing, ship-
building, textiles. AGR:
tobacco, poultry, dairy
products, beef cattle,
soybeans, hogs.

Home to both the first permanent English settlement and one of the largest concentrations of Internet companies in America, Virginia offers a balance of modern and historical riches. The entrepreneurial spirit of the pioneers at Jamestown nearly 400 years ago lives on in Virginia's highly diversified modern economy. One-third of all jobs are in service areas, particularly business and medicine. Virginia leads the nation in coal and tobacco exports, hosts a shipbuilding industry, and is home base for the U.S. Navy's Atlantic Fleet. From the Atlantic to the Blue Ridge, the state offers abundant recreational and cultural attractions.

Washington
EVERGREEN STATE

AREA 184,665 sq km
(71,300 sq mi)
POPULATION 6,131,445
CAPITAL Olympia 43,519
PCI $32,661
STATEHOOD November 11, 1889;
42nd state
ECONOMY IND: aerospace, tourism,
food processing, forest
products, paper prod-
ucts, industrial machin-
ery, printing and
publishing, metals.
AGR: seafood, apples,
dairy products, wheat,
cattle, potatoes, hay.

Western hemlock and Douglas fir support Washington's timber industry. The state is second in the nation for timber production, but the cutting of old-growth forests has caused much litigation between environmentalists and loggers. Snowmelt from the Cascade Range waters the Yakima Valley, where bumper apple crops grow. The harnessed Columbia River irrigates the Columbia Basin, an oasis of wheat, fruits, and vegetables.

West Virginia
MOUNTAIN STATE

AREA	62,755 sq km
	(24,230 sq mi)
POPULATION	1,810,354
CAPITAL	Charleston 51,702
PCI	$23,628
STATEHOOD	June 20, 1863
	35th state
ECONOMY	IND: coal mining, chemicals, metal manufacturing, forest products, stone, clay, oil, glass products. AGR: poultry and eggs, cattle, dairy products, apples.

Bituminous coal, oil, natural gas, and silica underlie this rugged Appalachian state. Only Wyoming produces more coal. Declines in steel and chemical manufacturing prompted corporate tax incentives to diversify the economy; as a result the service sector is burgeoning. Hardwoods cover three-fourths of the state and feed lumber mills, but forests also shelter wildlife, including an endangered species of flying squirrel.

Wisconsin
BADGER STATE

AREA	169,639 sq km
	(65,498 sq mi)
POPULATION	5,472,299
CAPITAL	Madison 215,211
PCI	$29,996
STATEHOOD	May 29, 1848
	30th state
ECONOMY	IND: industrial machinery, paper products, food processing, metal products, electronic equipment, transportation. AGR: dairy products, cattle, corn, poultry and eggs, soybeans.

Once home to more than a dozen Indian tribes, Wisconsin received an influx of Scandinavians, Germans, and other northern Europeans in the late 1800s. Today Wisconsin farmers rank second in the nation in the production of milk and butter, and first in cheese. Most than 15,000 lakes, 47 state parks, and 13 forests stimulate tourism. In 1854 the Republican Party was born here.

Wyoming
EQUALITY STATE

AREA	253,336 sq km
	(97,814 sq mi)
POPULATION	501,242
CAPITAL	Cheyenne 53,658
PCI	$30,494
STATEHOOD	July 10, 1890
	44th state
ECONOMY	IND: mining, generation of electricity, chemicals, tourism. AGR: cattle, sugar beets, sheep, hay, wheat.

Ninth largest state in area, Wyoming ranks 50th in population. Oil, natural gas, coal, and uranium are plentiful. In 1869 this frontier territory pioneered women's suffrage in the U.S. In 1924 Wyoming became the first state to elect a woman governor. Yellowstone, the first national park in the U.S., was established in 1872. Today neighboring Grand Teton National Park's alpine peaks also attract more than four million visitors annually.

PART IV

MAPS & GLOSSARY

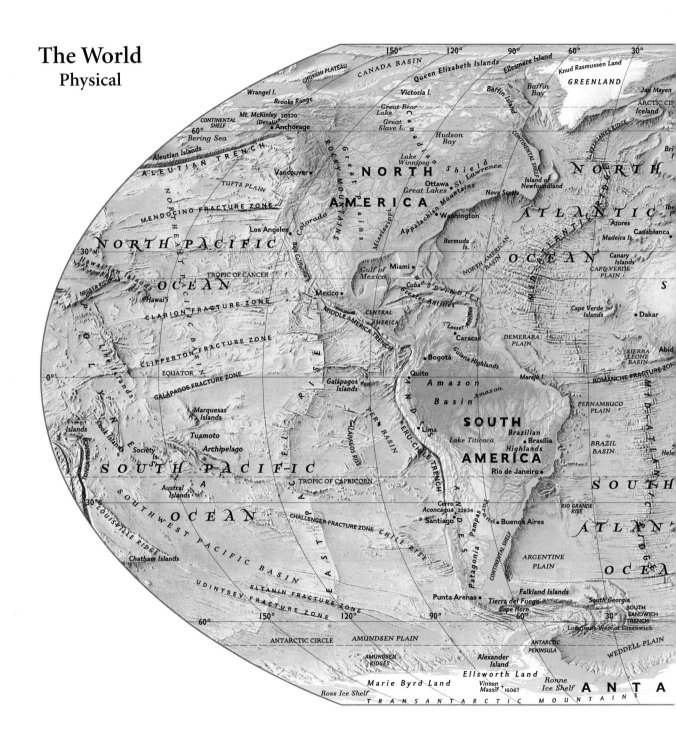

The World
Physical

CHUKCHI PLATEAU
CANADA BASIN
Queen Elizabeth Islands
Ellesmere Island
Knud Rasmussen Land
GREENLAND

Wrangel I.
Brooks Range
Victoria I.
Baffin Island
Baffin Bay
Jan Mayen
ARCTIC CIR
Iceland

Great Bear Lake
CONTINENTAL SHELF
Mt. McKinley 20320 (Denali)
Anchorage
Great Slave L.
Hudson Bay

60°
Bering Sea
Aleutian Islands
ALEUTIAN TRENCH
TUFTS PLAIN
Vancouver
Lake Winnipeg
Shield
St. Lawrence
Nova Scotia
Island of Newfoundland

NORTH AMERICA
Rocky Mountains
Great Plains
Ottawa
Great Lakes
Washington
Appalachian Mountains
Mississippi

NORTH
REYKJANES RIDGE
Bri

MENDOCINO FRACTURE ZONE
NORTHEAST PACIFIC BASIN
NORTH PACIFIC
Los Angeles
Colorado
ATLANTIC
Azores
Casablanca
Madeira Is.

30°N
TROPIC OF CANCER
OCEAN
Bermuda Is.
NORTH AMERICAN BASIN
OCEAN
Canary Islands
CAPE VERDE PLAIN

Hawaiian Islands
Hawai'i
Gulf of Mexico
Miami
Cuba
West Indies
Greater Antilles
Cape Verde Islands
Dakar

MECKER RIDGE
CLARION FRACTURE ZONE
Mexico
CENTRAL AMERICA
MIDDLE AMERICA TRENCH
Lesser Antilles
Caracas
DEMERARA PLAIN
SIERRA LEONE BASIN
Abid

CLIPPERTON FRACTURE ZONE
Bogotá
Guiana Highlands
ROMANCHE FRACTURE ZON

0°
EQUATOR
GALAPAGOS FRACTURE ZONE
Galápagos Islands
Quito
Amazon Basin
Amazon
Marajó I.
PERNAMBUCO PLAIN

Line Islands
Marquesas Islands
PERU BASIN
Lima
SOUTH
Brazilian Highlands
Brasília
BRAZIL BASIN
Hele

Cook Islands
Islands
Tuamoto Archipelago
EAST PACIFIC RISE
Lake Titicaca
AMERICA
Rio de Janeiro
SOUTH

TONGA TRENCH
Society Is.
Austral Islands
SOUTH PACIFIC
TROPIC OF CAPRICORN
Cerro Aconcagua 22834
Santiago
Andes
Pampas
Buenos Aires
RIO GRANDE RISE
ATLANT

30°S
SOUTHWEST PACIFIC
OCEAN
CHALLENGER FRACTURE ZONE
CHILE RISE
Patagonia
CONTINENTAL SHELF
ARGENTINE PLAIN
OCEA

LOUISVILLE RIDGE
Chatham Islands
SOUTHWEST PACIFIC BASIN
Falkland Islands
South Georgia
SOUTH SANDWICH TRENCH

ELTANIN FRACTURE ZONE
Punta Arenas
Tierra del Fuego
Cape Horn
60°
Longitude West of Greenwich

UDINTSEV FRACTURE ZONE
60°
150°
120°
90°
30°

ANTARCTIC CIRCLE
AMUNDSEN PLAIN
ANTARCTIC PENINSULA
WEDDELL PLAIN

AMUNDSEN RIDGES
Alexander Island
Ellsworth Land
Vinson Massif 16067
Ronne Ice Shelf
ANTA

Ross Ice Shelf
Marie Byrd Land
TRANSANTARCTIC MOUNTAINS

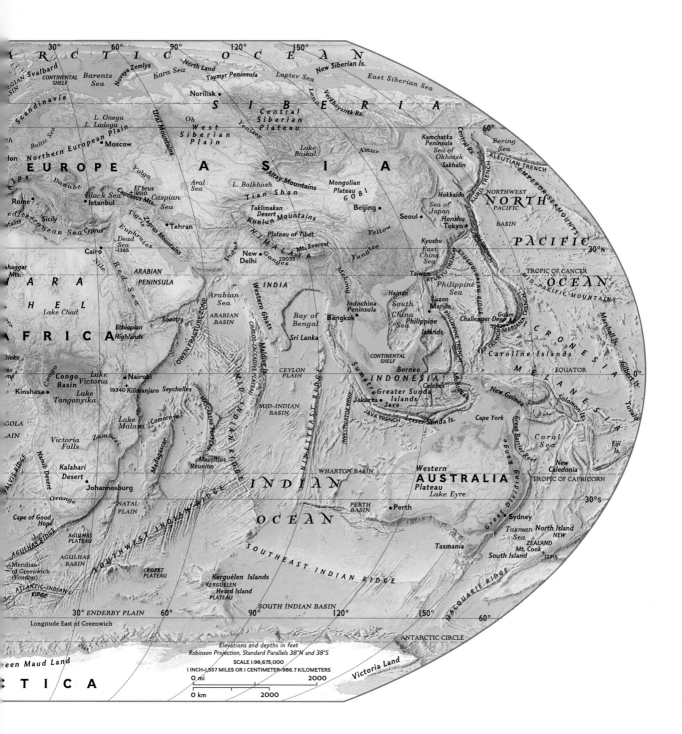

ARCTIC OCEAN

Svalbard · Continental Shelf · Barents Sea · Novaya Zemlya · Kara Sea · North Land · Taymyr Peninsula · Laptev Sea · New Siberian Is. · East Siberian Sea

Norilsk

S I B E R I A

Scandinavia · L. Onega · L. Ladoga · Ob · West Siberian Plain · Yenisey · Central Siberian Plateau · Lena · Verkhoyansk Ra.

Baltic Sea · Northern European Plain · Moscow · Ural Mountains · Lake Baikal · Amur · Kamchatka Peninsula · Sea of Okhotsk · Sakhalin · Central Ra. · Bering Sea

60°

E U R O P E · A S I A

Alps · Rome · Danube · Volga · El'brus 18510 · Caspian Sea · Aral Sea · Altay Mountains · Mongolian Plateau · GOBI · Beijing · Hokkaido · Sea of Japan · Honshu · Tokyo · NORTHWEST PACIFIC · NORTH BASIN · PACIFIC

Sicily · Black Sea · Istanbul · Caucasus Mts. · Tigris · Zagros Mountains · Tehran · L. Balkhash · Tian Shan · Taklimakan Desert · Kunlun Mountains · Seoul · Kyushu · East China Sea · 30°N

Mediterranean Sea · Cyprus · Euphrates · Dead Sea -1365 · Plateau of Tibet · H I M A L A Y A S · Mt. Everest 29035 · Yellow · Yangtze · RYUKYU TRENCH · Philippine Sea · TROPIC OF CANCER · MID-PACIFIC MOUNTAINS · OCEAN

Cairo · Nile · Red Sea · ARABIAN PENINSULA · Indus · New Delhi · Ganges · I N D I A · Mekong · Taiwan · RYUKYU-PALAU RIDGE · Guam · Challenger Deep · Marshalls · MARIANA TRENCH

SAHARA · Mts. · Lake Chad · Arabian Sea · ARABIAN BASIN · Western Ghats · Bay of Bengal · Bangkok · Indochina Peninsula · Hainan · South China Sea · Manila · Luzon · Philippine Islands · Challenger Deep · M I C R O N E S I A · Caroline Islands · Gilbert Is.

SAHEL · A F R I C A · Socotra · OWEN FRACTURE ZONE · CHAGOS-LACCADIVE PLATEAU · Maldive Is. · CEYLON PLAIN · Sri Lanka · CONTINENTAL SHELF · Borneo · Celebes · Philippine Trench

Bioko · Congo Basin · Lake Victoria · Nairobi · Kilimanjaro 19340 · Seychelles · MASCARENE PLATEAU · MID-INDIAN BASIN · Sumatra · I N D O N E S I A · Moluccas · New Guinea · Solomon Is. · Tuvalu · EQUATOR · 0°

Kinshasa · Lake Tanganyika · Comoro Is. · NINETYEAST RIDGE · Greater Sunda Islands · Jakarta · Java · JAVA TRENCH · Lesser Sunda Is. · Cape York

ANGOLA · Victoria Falls · Zambezi · Lake Malawi · Madagascar · INVESTIGATOR RIDGE · Mauritius · Réunion · WHARTON BASIN · Western AUSTRALIA Plateau · Lake Eyre · Coral Sea · New Caledonia · Fiji Is. · TROPIC OF CAPRICORN

Namib Desert · Kalahari Desert · Johannesburg · I N D I A N · O C E A N · Great Dividing Range · 30°S

Orange · NATAL PLAIN · SOUTHWEST INDIAN RIDGE · PERTH BASIN · Perth · Sydney

Cape of Good Hope · Aguhlas Ridge · AGULHAS PLATEAU · CROZET PLATEAU · Mauritius · Réunion · SOUTHEAST INDIAN RIDGE · Tasmania · Tasman Sea · North Island · NEW ZEALAND · Mt. Cook 12316 · South Island

WALVIS RIDGE · Meridian of Greenwich (London) · ATLANTIC-INDIAN RIDGE · AGULHAS BASIN · ENDERBY PLAIN · Kerguélen Islands · KERGUELEN Heard Island PLATEAU · SOUTH INDIAN BASIN · 60°

30° · 60° · 90° · 120° · 150° · 60°

Longitude East of Greenwich

ANTARCTIC CIRCLE

Green Maud Land

ANTARCTICA

Victoria Land

Elevations and depths in feet
Robinson Projection, Standard Parallels 38°N and 38°S

SCALE 1:98,675,000
1 INCH=1,557 MILES OR 1 CENTIMETER=986.7 KILOMETERS

0 mi ———— 2000

0 km ———— 2000

ARCTIC OCEAN

Wandel Sea · Nord

GREENLAND SEA

LINCOLN SEA

2006 North ✳ Magnetic Pole

QUEEN ELIZABETH ISLANDS

Peary Land
Daneborg

Oodaaq I.
Alert
Knud Rasmussen Land

Ittoqqortoormiit

ARCTIC CIRCLE

Borden Island
Axel Heiberg I.
Ellesmere Island
SVERDRUP ISLANDS
Mackenzie King I.
Prince Patrick I.

GREENLAND (KALAALLIT NUNAAT) Denmark

Attu I.

ALEUTIAN ISLANDS

St. Lawrence I.

BERING SEA

Point Hope

Point Barrow

BEAUFORT SEA

PARRY ISLANDS
Melville Island

Devon I.
Baffin Bay

Qaanaaq

Kangersuatsiaq

Seward Peninsula
Hooper Bay
Togiak
Nunivak Island
Bethel

Kaktovik

Sachs Harbour
Banks Island

Somerset I.
Brodeur Pen.
Arctic Bay
Borden Pen.

Clyde River

Qeqertarsuaq

Umnak I.
Unalaska I.
Unimak I.

ALASKA U.S.

Mt. McKinley (Denali) 6194 m 20320 ft
Fairbanks
Anchorage
Valdez

Inuvik
Holman

Prince of Wales I.
Victoria Island
Boothia Peninsula
Melville Peninsula

Baffin Island

Pangnirtung

Nuuk (Godthåb)

Qaqortoq
Nunap Isua (Cape Farewell)

Alaska Peninsula
Kodiak I.

YUKON TERRITORY

Whitehorse

NORTHWEST TERRITORIES

Lupin

Baker Lake

Southampton Island

Iqaluit
Kimmirut

LABRADOR SEA

GULF OF ALASKA

Glacier Bay
Skagway
Sitka
Juneau

Mayo

Hay River

Yellowknife
Great Slave L.

Chesterfield Inlet

Ivujivik

HUDSON STRAIT

NEWFOUNDLAND & LABRADOR

Cartwright

Alexander Archipelago

Queen Charlotte Islands

Kitimat
Prince George

BRITISH COLUMBIA

CANADA

Great Bear L.

Uranium City

Fort McMurray

Churchill
Gillam

HUDSON BAY

Arviat

Belcher Islands

Inukjuak

Scheffervile

LABRADOR

St. Anthony
Island of Newfoundland

Nain

Kuujjuarapik

Vancouver Island

ALBERTA

SASK.

MANITOBA

Fort Albany

QUEBEC

Chisasibi

Sept-Îles

St. John's
Avalon Peninsula

Vancouver
Seattle
WA.
Portland
OR.
Eugene

Edmonton
Red Deer
Calgary

Saskatoon
Regina
Medicine Hat

Winnipeg

Thunder Bay

ONTARIO

Waskaganish

Baie-Comeau

Chicoutimi
Québec

N.B.

Sydney
Cape Breton Island
NOVA SCOTIA

Île d'Anticosti
Gulf of St. Lawrence

P.E.I.

PACIFIC OCEAN

Cape Mendocino

Butte
Spokane
ID.
Boise
Idaho Falls

MT.

Bismarck
N.D.
Billings

Fargo

Minneapolis
MN.
St. Paul
WI.
Sioux Falls
S.D.

Kouyn-Noranda

Ottawa
Toronto
Detroit
N.Y.

Montréal

VT.
Concord, N.H.
Boston, MA.
Providence, R.I.
Hartford, CT.
N.J.

ME.
Fredericton
Bangor

ATLANTIC OCEAN

Sacramento
San Francisco
San Jose
CA.
Fresno

Reno
NV.

Cheyenne
WY.
Salt Lake City
UT.

-86 m -282 ft
Great Salt Lake
Grand Canyon

Des Moines
Omaha
NE.

Denver
CO.

UNITED

Indianapolis

St. Louis
IL. IN.
Louisville
KY.

Chicago
Columbus
OH. W.V.

New York
Philadelphia, PA.
DE.
Washington, D.C.
MD.
Richmond
VA.

Bermuda Islands
U.K.

Los Angeles
San Diego
Las Vegas
Tijuana

Sierra Nevada
AZ.
Phoenix

Santa Fe
N.M.

Oklahoma City
OK.
Tulsa

STATES

KS.
Wichita
MO.

AR.

TN.
Nashville
Memphis
MS.
Jackson

Charlotte
N.C.
Atlanta
GA.
Birmingham
AL.

Charleston, S.C.

Tucson
El Paso

Fort Worth
TX.
Dallas
Austin
San Antonio
LA.

New Orleans

Jacksonville
Tallahassee
FL.

TROPIC OF CANCER

Ciudad Juárez
Chihuahua

Baja California

Sierra Madre Occidental

Monterrey

San Luis Potosí

GULF OF MEXICO

Tampa
Miami
Key West

BAHAMAS
Nassau

La Paz
Cabo San Lucas
Mazatlán

Sierra Madre Oriental

Mérida

Havana
Santiago de Cuba

CUBA
HAITI

DOMINICAN REPUBLIC

Santo Domingo

San Juan
PUERTO RICO U.S.

BARBADOS
DOMINICA
ANTIGUA AND BARBUDA
ST. KITTS & NEVIS

Islas Revillagigedo
Mexico

Guadalajara
Mexico
Acapulco

MEXICO

Veracruz
Belmopan
BELIZE
San Pedro Sula

Belize City

Yucatan Pen.

U.K. Cayman Is.
JAMAICA
Kingston

Port-au-Prince

ST. LUCIA
ST. VINCENT AND THE GRENADINES
GRENADA

Punta Eugenia

Gulf of Tehuantepec

GUATEMALA
Guatemala
San Salvador
EL SALVADOR

Tegucigalpa
HONDURAS
NICARAGUA
Managua
COSTA RICA
San José
PANAMA
Panama

CARIBBEAN SEA

Port-of-Spain
TRINIDAD AND TOBAGO

PANAMA CANAL

Isla del Coco
Costa Rica

I. de Coiba
Gulf of Panama

EQUATOR

MAP SYMBOLS
⊛ ★ ⊙ Capitals
∴ Ruin
Below sea level
Dry salt lake
Glacier
Sand
Swamp

North America
Azimuthal Equidistant Projection

0 mi   600
0 km   600

488

South America

Azimuthal Equidistant Projection

0 mi — 600
0 km — 600

CARIBBEAN SEA

ATLANTIC OCEAN

Santa Marta
Barranquilla
Cartagena
Maracaibo
Barquisimeto
Valencia **Caracas**
Maracay
Lago de Maracaibo
Ciudad Guayana
Orinoco
Cúcuta
Bucaramanga
San Cristóbal
VENEZUELA
Georgetown
Paramaribo
Medellín
LLANOS
GUYANA
SURINAME
Cayenne
FRENCH GUIANA France
Manizales
Ibagué
Bogotá
COLOMBIA
Angel Falls
GUIANA HIGHLANDS
Amapá
Malpelo I.
Colombia
Boundary claimed by Suriname
Esmeraldas
Pasto
Marajó I.
Belém
EQUATOR
Quito
ECUADOR
A
Negro
Amazon
Santarém
São Luís
Parnaíba
Fortaleza
GALÁPAGOS ISLANDS
(ARCHIPIÉLAGO DE COLÓN)
Ecuador
Guayaquil
Cuenca
Iquitos
Marañón
N
Manaus
Amazon
Teresina
Natal
João Pessoa
Piura
Purus
AMAZON
BASIN
Porto Velho
Madeira
Tapajós
Xingu
Campina Grande
Recife
Chiclayo
Trujillo
Chimbote
Rio Branco
D
Teles Pires
B R A Z I L
Maceió
Aracaju
Feira de Santana
Callao
Lima
Ayacucho
Machu Picchu
Cusco
Lago Titicaca
Trinidad
São Francisco
Salvador
(Bahia)
E
BOLIVIA
Cuiabá
BRAZILIAN
Ilhéus
Arequipa
La Paz
Cochabamba
Oruro
Santa Cruz
Sucre
Brasília
Goiânia
HIGHLANDS
Arica
S
Altiplano
Uberlândia
Uberaba
Governador Valadares
Iquique
Salar de Uyuni
Tarija
Campo Grande
São José
do Rio Preto
Ribeirão Preto
Belo Horizonte
Londrina
Campinas
Nova Iguaçu
Antofagasta
Salta
Asunción
PARAGUAY
São Paulo
Rio de Janeiro
Santos
GRAN CHACO
Iguazú Falls
Curitiba
San Miguel de Tucumán
Resistencia
Corrientes
Passo Fundo
Florianópolis
San Félix I.
Chile
San Ambrosio I.
Uruguaiana
Porto Alegre
Santa Maria
PACIFIC
La Serena
Coquimbo
Córdoba
PAMPAS
Santa Fe
Rosario
Cerro Aconcagua
6960 m
Mendoza
URUGUAY
Juan Fernández Is.
Chile
Valparaíso
Santiago
Buenos Aires
La Plata
Montevideo
Río de la Plata
OCEAN
Talca
Concepción
Mar del Plata
ATLANTIC
Temuco
Bahía Blanca
Negro
Viedma
Puerto Montt
-40 m
-131 ft
Valdés Peninsula
OCEAN
Isla Grande
de Chiloé
A
R
G
E
N
T
I
N
A
Comodoro Rivadavia
Golfo
San Jorge
Taitao
Peninsula
PATAGONIA
FALKLAND ISLANDS
(ISLAS MALVINAS)
U.K.
Wellington I.
Río Gallegos
Stanley
Administered by United Kingdom
(claimed by Argentina)
Punta Arenas
Strait of Magellan
TIERRA DEL FUEGO
Ushuaia
Cape Horn
South Georgia I.
U.K.

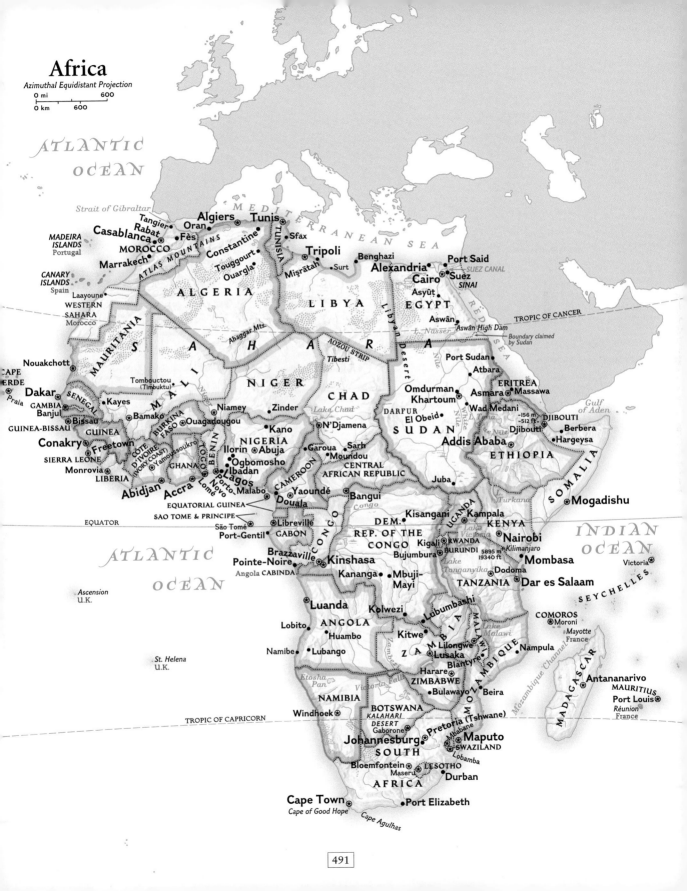

Africa

Azimuthal Equidistant Projection

0 mi 600
0 km 600

ATLANTIC OCEAN

MEDITERRANEAN SEA

Strait of Gibraltar

MADEIRA ISLANDS
Portugal

Tangier
Rabat
Casablanca
Fès
Oran
Algiers
Tunis
Sfax

MOROCCO
Constantine
Marrakech
Touggourt
Ouargla
Tripoli
Benghazi
Surt
Miṣrātah

CANARY ISLANDS
Spain
Laayoune
WESTERN SAHARA
Morocco

ALGERIA

LIBYA

ATLAS MOUNTAINS

TUNISIA

Port Said
Alexandria
Cairo
Suez
SINAI
Asyût

EGYPT

Aswân
Aswân High Dam
L. Nasser
TROPIC OF CANCER
Boundary claimed by Sudan

Libyan Desert

Nile

RED SEA

CAPE VERDE
Nouakchott

MAURITANIA

Dakar
Praia
GAMBIA
Banjul
Bissau
GUINEA-BISSAU

SENEGAL
Kayes

MALI
Tombouctou (Timbuktu)

Ahaggar Mts.

SAHARA

AOZOU STRIP
Tibesti

NIGER

CHAD

DARFUR
El Obeid
Omdurman
Khartoum
Wad Medani

Port Sudan
Atbara

ERITREA
Asmara
Massawa

GUINEA
Conakry
Freetown
SIERRA LEONE
Monrovia
LIBERIA

CÔTE D'IVOIRE (IVORY COAST)

Niamey
Zinder
Kano

BURKINA FASO
Ouagadougou
Yamoussoukro
GHANA
Abidjan
Accra
Lomé
TOGO
BENIN
Porto Novo
Ilorin
Ogbomosho
Ibadan
Lagos
Malabo

NIGERIA
Abuja

Garoua
Moundou
Sarh

CENTRAL AFRICAN REPUBLIC

N'Djamena
Lake Chad

SUDAN

-156 m
L. Tana
DJIBOUTI
Djibouti
Berbera
Hargeysa

Addis Ababa

ETHIOPIA

White Nile
Blue Nile

Juba

EQUATOR

SAO TOME & PRINCIPE
São Tomé

EQUATORIAL GUINEA
Bamako
Bamako
Bissau

CAMEROON
Yaoundé
Douala
Libreville
GABON
Port-Gentil

Bangui

DEM. REP. OF THE CONGO
Kisangani
Congo

UGANDA
Kampala
Kigali
RWANDA
Bujumbura
BURUNDI
Lake Victoria

KENYA
Nairobi
Kilimanjaro
5895 m
19340 ft
Mombasa

L. Turkana

SOMALIA
Mogadishu

INDIAN OCEAN

Victoria
SEYCHELLES

ATLANTIC OCEAN

Ascension
U.K.

Brazzaville
Pointe-Noire
Angola CABINDA
Kinshasa
Kananga
Mbuji-Mayi

Dodoma
Lake Tanganyika

TANZANIA
Dar es Salaam

Luanda
Lobito
ANGOLA
Huambo
Namibe
Lubango

Kolwezi
Lubumbashi
Kitwe

Z A M B I A
Lusaka
Lilongwe
MALAWI
Blantyre
Lake Malawi
Nampula
COMOROS
Moroni
Mayotte
France

St. Helena
U.K.

Kunene

Etosha Pan

Victoria Falls

Harare
ZIMBABWE
Bulawayo
Beira

MOZAMBIQUE

Mozambique Channel

MADAGASCAR

Antananarivo
MAURITIUS
Port Louis
Réunion
France

TROPIC OF CAPRICORN

NAMIBIA
Windhoek

BOTSWANA
KALAHARI DESERT
Gaborone

Pretoria (Tshwane)
Mbabane
Maputo
Johannesburg
SWAZILAND
Lobamba

SOUTH AFRICA

Bloemfontein
Maseru
LESOTHO
Durban

Cape Town
Cape of Good Hope
Port Elizabeth
Cape Agulhas

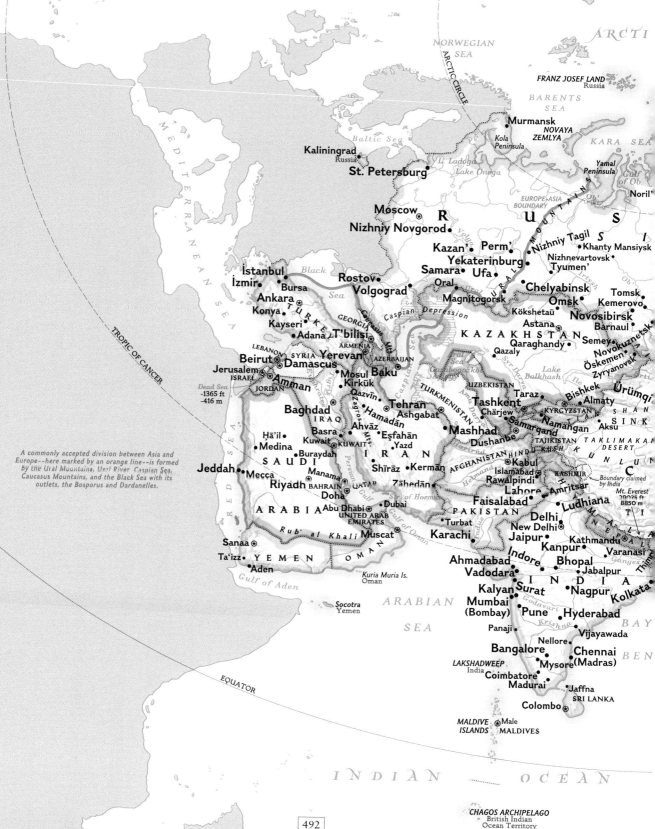

ATLANTIC OCEAN

ARCTI

NORWEGIAN
SEA

ARCTIC CIRCLE

FRANZ JOSEF LAND
Russia

BARENTS
SEA

Murmansk
Kola
Peninsula

NOVAYA
ZEMLYA

KARA SEA

Baltic Sea

L. Ladoga
Lake Onega

Yamal
Peninsula

Gulf
of Ob

Noril'

Kaliningrad
Russia

St. Petersburg

EUROPE-ASIA
BOUNDARY

Moscow

R

Nizhniy Novgorod

Kazan'   Perm'   Nizhniy Tagil
Khanty Mansiysk

Yekaterinburg   Nizhnevartovsk

Samara   Ufa   Tyumen'

Oral   Magnitogorsk   Chelyabinsk   Omsk   Tomsk
Kemerovo

İstanbul   Black   Rostov   Kökshetaū   Novosibirsk
İzmir   Sea   Volgograd   Astana   Barnaul

Bursa   Caspian   Depression   KAZAKHSTAN
Ankara   Qaraghandy   Semey   Novokuznetsk
Konya   Magnitogorsk   Qazaly   Öskemen   Zyryanovsk
Kayseri   Caspian   GEORGIA   Aral   Sea

Adana   T'bilisi   ARMENIA   Lake
Balkhash   Ürümqi

LEBANON   SYRIA   Yerevan   AZERBAIJAN   Baku   UZBEKISTAN   Taraz   Bishkek   Almaty   SINK
Beirut   Damascus   Mosul   TURKMENISTAN   Tashkent   KYRGYZSTAN   Namangan   Aksu
Jerusalem   Amman   Kirkūk   Chärjew   Samarqand   TAKLIMAKAN
ISRAEL   Qazvīn   Tehrān   Ashgabat   TAJIKISTAN   DESERT
Dead Sea   JORDAN   Hamadān   Mashhad   Dushanbe
-1365 ft   Baghdad   Ahvāz   Dushanbe   HINDU   KUNLUN   C
-416 m   IRAQ   Esfahān   IRAN   AFGHANISTAN   KUSH   Kabul
Ha'il   Basra   Yazd   Islamabad   KASHMIR   Mt. Everest
Medina   Kuwait   Shīrāz   Kerman   Rawalpindi   Boundary claimed   29028 ft
Buraydah   KUWAIT   Zāhedān   Lahore   by India   8850 m
Jeddah   SAUDI   Manama   Kandahar   Faisalabad   Amritsar
Mecca   BAHRAIN   Doha   Zāhedān   PAKISTAN   Ludhiana
Riyadh   QATAR   New Delhi   TI
ARABIA   Abu Dhabi   Dubai   Turbat   Delhi
Sanaa   UNITED ARAB   Muscat   Karachi   Jaipur   Kathmandu
EMIRATES   Kanpur   Varanasi
Rub' al Khali   OMAN   Indore   Bhopal
Ta'izz   YEMEN   Ahmadabad   Jabalpur
Aden   Vadodara   INDIA   Nagpur   Kolkata
Gulf of Aden   Kuria Muria Is.   Surat
Oman   Kalyan   Hyderabad
Socotra   ARABIAN   Mumbai   Pune
Yemen   (Bombay)   Godavari
SEA   Panaji   Krishna   Vijayawada   BAY
Nellore
LAKSHADWEEP   Bangalore   Chennai   BEN
India   Mysore   (Madras)
Coimbatore
EQUATOR   Madurai   Jaffna
Colombo   SRI LANKA

MALDIVE   Male
ISLANDS   MALDIVES

INDIAN   OCEAN

A commonly accepted division between Asia and
Europe--here marked by an orange line--is formed
by the Ural Mountains, Ural River, Caspian Sea,
Caucasus Mountains, and the Black Sea with its
outlets, the Bosporus and Dardanelles.

TROPIC OF CANCER

CHAGOS ARCHIPELAGO
British Indian
Ocean Territory

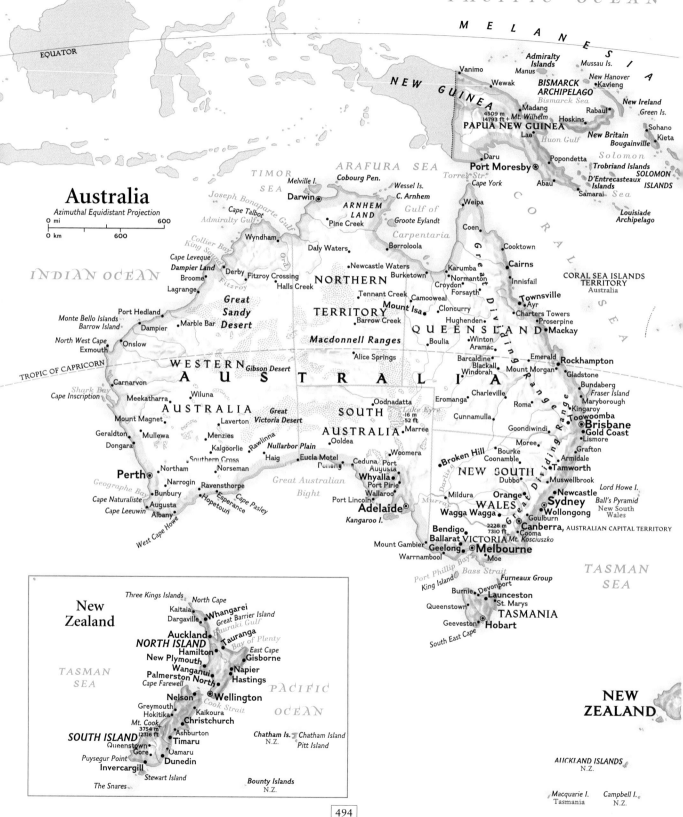

Australia
Azimuthal Equidistant Projection

0 mi | 600
0 km | 600

PACIFIC OCEAN

MELANESIA

EQUATOR

INDIAN OCEAN

TIMOR SEA

ARAFURA SEA

NEW GUINEA

Vanimo
Wewak
Admiralty Islands
Manus
Mussau Is.
New Hanover
Kavieng
New Ireland
Green Is.
BISMARCK ARCHIPELAGO
Bismarck Sea
Madang
Hoskins
Rabaul
4509 m
14793 ft + Mt. Wilhelm
PAPUA NEW GUINEA
Lae
New Britain
Bougainville
Sohano
Kieta
Huon Gulf
Daru
Popondetta
Port Moresby ⊛
Abau
Samarai
Trobriand Islands
D'Entrecasteaux Islands
SOLOMON ISLANDS
Solomon Sea
Louisiade Archipelago

Melville I.
Cobourg Pen.
Darwin
C. Arnhem
Wessel Is.
Cape York
Weipa
Torres Str.
Cape Talbot
Joseph Bonaparte Gulf
Admiralty Gulf
Collier Bay
King Sound
Cape Leveque
Dampier Land
Broome
Derby
Fitzroy Crossing
Halls Creek
Lagrange
Wyndham
Pine Creek
ARNHEM LAND
Groote Eylandt
Gulf of Carpentaria
Coen
Cooktown
Cairns
Innisfail
CORAL SEA ISLANDS TERRITORY
Australia
Daly Waters
Borroloola
Newcastle Waters
Burketown
Karumba
Normanton
Croydon
Forsayth
Camooweal
Tennant Creek
Mount Isa
Cloncurry
Hughenden
Charters Towers
Proserpine
Mackay
Townsville
Ayr
Coral Sea
Great Dividing Range

NORTHERN TERRITORY

Barrow Creek

Macdonnell Ranges

Great Sandy Desert

Marble Bar
Port Hedland
Dampier
Monte Bello Islands
Barrow Island
North West Cape
Exmouth
Onslow

TROPIC OF CAPRICORN

WESTERN AUSTRALIA

Gibson Desert

Alice Springs

Boulia
Winton
Aramac
Barcaldine
Blackall
Windorah
Mount Morgan
Emerald
Rockhampton
Gladstone
Bundaberg
Fraser Island
Maryborough
Kingaroy
Toowoomba
Brisbane
Gold Coast
Lismore
Grafton
Armidale
Tamworth
Muswellbrook

QUEENSLAND

SOUTH AUSTRALIA

Carnarvon
Shark Bay
Cape Inscription
Meekatharra
Wiluna
Great Victoria Desert
Oodnadatta
Lake Eyre
-16 m
-52 ft
Marree
Eromanga
Charleville
Cunnamulla
Roma
Goondiwindi
Moree
Bourke
Coonamble
Dubbo
Orange
Newcastle
Sydney
Wollongong

Mount Magnet
Geraldton
Mullewa
Dongara
AUSTRALIA
Laverton
Menzies
Kalgoorlie
Rawlinna
Southern Cross
Haig
Nullarbor Plain
Eucla Motel
Penong
Ceduna
Port Augusta
Woomera
Whyalla
Port Pirie
Wallaroo
Broken Hill
NEW SOUTH WALES
Mildura
Wagga Wagga
Canberra, AUSTRALIAN CAPITAL TERRITORY
Cooma
Goulburn

Lord Howe I.
Ball's Pyramid
New South Wales

Perth
Northam
Narrogin
Bunbury
Ravensthorpe
Esperance
Cape Pasley
Great Australian Bight
Port Lincoln
Adelaide
Kangaroo I.
Geographe Bay
Cape Naturaliste
Augusta
Cape Leeuwin
Albany
West Cape Howe
Hopetoun

Bendigo
Ballarat
Geelong
VICTORIA
Melbourne
Moe
2228 m
7310 ft
Mt. Kosciuszko
Mount Gambier
Warrnambool
Port Phillip Bay
King Island
Bass Strait
TASMAN SEA
Burnie
Devonport
Launceston
St. Marys
Queenstown
TASMANIA
Geeveston
Hobart
South East Cape
Furneaux Group

NEW ZEALAND

New Zealand

TASMAN SEA

Three Kings Islands
North Cape
Kaitaia
Dargaville
Whangarei
Great Barrier Island
Hauraki Gulf
Auckland
Tauranga
Bay of Plenty
NORTH ISLAND
Hamilton
East Cape
Gisborne
New Plymouth
Wanganui
Napier
Palmerston North
Hastings
Cape Farewell
PACIFIC OCEAN
Nelson
Wellington
Greymouth
Hokitika
Kaikoura
Cook Strait
Mt. Cook
3754 m
12316 ft
Christchurch
Ashburton
Chatham Is.
N.Z.
Chatham Island
Pitt Island
SOUTH ISLAND
Timaru
Queenstown
Gore
Oamaru
Puysegur Point
Dunedin
Invercargill
Stewart Island
The Snares
Bounty Islands
N.Z.

AUCKLAND ISLANDS
N.Z.

Macquarie I.
Tasmania
Campbell I.
N.Z.

494

Antarctica

Azimuthal Equidistant Projection

0 mi   600

0 km   600

ATLANTIC OCEAN

SCOTIA SEA

South Sandwich Is.
U.K.

South Orkney Is.

INDIAN OCEAN

ANTARCTIC CIRCLE

FIMBUL ICE SHELF

Cape Norvegia

Riiser-Larsen Peninsula

Lützow-Holm Bay

RIISER–LARSEN ICE SHELF

QUEEN MAUD LAND

ENDERBY LAND

South Shetland Islands

Joinville I.

WEDDELL SEA

WEDDELL SEA

COATS LAND

Cape Darnley

AMERY ICE SHELF
Prydz Bay

LARSEN ICE SHELF

3184m Mt. Jackson

Alexander I.

ANTARCTIC PENINSULA

RONNE ICE SHELF

Berkner Island

PENSACOLA MTS.

AMERICAN HIGHLAND

BELLINGSHAUSEN SEA

WEST ICE SHELF

ELLSWORTH LAND

T R A N S A N T A R C T I C   M O U N T A I N S

POLAR PLATEAU

+ South Pole

EAST ANTARCTICA

4897m Vinson Massif
16067ft

ELLSWORTH MTS.

WEST ANTARCTICA

Thurston I.

SHACKLETON ICE SHELF

AMUNDSEN SEA

MARIE BYRD LAND

4528m

ROSS ICE SHELF

Cape Poinsett

WILKES LAND

GETZ ICE SHELF

Roosevelt I.

Ross I.
Mt. Erebus
3794m

Porpoise Bay

ROSS SEA

McMurdo Sound

VICTORIA LAND

INDIAN OCEAN

4165m Mt. Minto

Cape Adare

* 2006 South Magnetic Pole

PACIFIC OCEAN

Balleny Islands

ANTARCTIC CIRCLE

GLOSSARY

A

albedo percentage of electromagnetic radiation reflected from a surface

alluvial fan feature found in deserts, the result of stream deposits accumulating where stream channels emerge from the base of mountains and channel gradients level out

anticline upfold of layered rocks in an archlike structure

anticyclone center of high atmospheric pressure; it spins in the opposite direction from a cyclone

aphelion point on the Earth's elliptical orbit at which the sun is farthest from the Earth

aquifer rock mass or layer with high porosity and high permeability; stores and transmits ground water

artesian spring groundwater that flows to the surface as a result of hydrostatic pressure that forces water upward

asthenosphere soft layer of the upper mantle of the Earth, beneath the lithosphere

atmosphere thin layer of gasses surrounding the Earth that is the medium for weather and climate

atoll circular coral reef enclosing a shallow lagoon

avalanche general term for extremely rapid slides and falls of snow, ice, rocks, and trees

B

barrier island landform created by long linear wave deposits that form parallel to shorelines

barrier reef coral reef separated from shoreline by a lagoon

biodiversity number of species present in an ecosystem

biogeography study of the distribution patterns of plants and animals and the processes that produce those patterns

biomass dry weight of living organic matter in a particular ecosystem; units are grams of organic matter per square meter

biotic communities local communities of interdependent plants and animals that are often found together

birth rate number of live births per 1,000 population in a given year

braided stream small waterway with shallow channels that carry multiple flows

breakwaters piles of rock built parallel to the shore to prevent damage to watercraft or construction

C

calving breaking off of blocks of glacial ice into the ocean, forming icebergs

capitalism form of economic organization characterized by resource allocation through markets. The means of production are privately owned, and production organized around profit maximization.

carbon cycle material cycle in which carbon flows through an ecosystem

cartogram simplified map designed to present a single idea in a diagrammatic way, usually not to scale

cash crops crops grown for sale on the market, that is, for profit

cataract waterfall that forms a single long drop

central business district (CBD) central nucleus of commercial land uses in a city.

central place theory interpretation of city systems set forth by German geographer Walter Christaller in 1933 that centers on consumer demand, including the maximum distance consumers will travel for a given good and the minimum market size necessary to sustain them

centrality functional domininace of cities, in terms of econmoic, political, and cultural activity, witin an urban system.

chinook very dry wind that occurs when air that is blown up the windward side of the Rocky Mountains descends

cirque bowl-shaped basin that holds the collecting ground and firn of an Alpine glacier

City Beautiful movement early 20th century attempt to renovate cities to reflect the higher values of society, using neoclassical architecture, grandiose street plans, parks, and inspirational monuments and statues

colloids extremely small mineral particles that can remain in suspension in water indefinitely

colonialism economic and political system by which some nations dominate others

comparative advantage principle that some regions can produce some goods and services more efficiently or profitably than others

condensation change of water from vapor to a liquid or a solid.

continental drift theory that today's continents were formed by the breakup of prehistoric supercontinents and slowly drifted to their present positions

continental shelf offshore extensions of continents.

continental suture mountain range formed by the convergence of continental plates—for example, the Appalachians, the Alps, and the Himalayas.

core area nation's or culture's historic homeland

Coriolis effect result of the Earth's rotation that causes any freely moving object or fluid to turn toward the right in the northern hemisphere and to the left in the southern hemisphere

creole any pidgin language used widely enough to become a population's primary tongue

cultural divergence theory that explains the formation of early cultures as the result of groups dividing, migrating, and slowly changing in response to new ideas and environmental stress

cultural revivalism rediscovery of a former cultural identity in which members of a culture that has been overwhelmed by a dominant culture seek to regain their former culture

culture way of life of a group that is transmitted between generations and includes a shared system of meanings, beliefs, values, and social relations; it includes such things as language, religion, clothing, music, laws, and entertainment.

culture hearth specific place where a distinctive culture originated

cumulative causation spiral buildup of mutually reinforcing advantages that occurs in specific geographic settings as a result of the development of external economies and localization economies

cyberspace world of electronic computerized spaces encompassed by the Internet and related technologies such as the World Wide Web

D

deindustrialization loss of comparative advantage in manufacturing, brought about by technological displacement and the rise of foreign competitors

delta flat, low-lying area formed by sediment deposited by a stream entering a body of standing water

demographic equation calculation of population change expressed as change equals births minus deaths plus net migration

demographic transition historical shift of birth and death rates from high to low levels in a population; the decline of mortality usually precedes the decline in fertility, thus resulting in rapid population growth during the transition period.

deposition laying down of sediment

desert region that has little or no vegetation and averages less that 10 inches of precipitation a year

desertification invasion of desert into nondesert regions.

devolution violent or nonviolent process in which power from a state government is returned to a political region or homeland, in forms ranging from limited authority (e.g., over local issues) to statehood

dialects regional variations of one language with differences in vocabulary, accent, pronunciation, and syntax

diminishing returns principle first set forth by Thomas Malthus: increased investments in production capacity yield steadily smaller marginal increases in output.

drumlin low, linear hill shaped by glaciers whose long axes parallel the direction of the glacier's movement

E

ecosystem group of organisms and the environment with which they interact

edge cities nodal concentrations of retail and office space that are situated on the outer fringes of metropolitan areas, typically near major highway intersections

El Niño periodic reversal of current flow and water temperatures in the mid-Pacific Ocean

elasticities of demand measure of the degree to which the aggregate demand for a commodity rises or falls with changes in income and price

energy ability to move solids, liquids, and gases

environment sum of the conditions that surround and influence an organism

environmental degradation damage to the environment resulting from human development and use of natural resources; human involvement that changes the environment or interferes with biological and environmental processes

environmental determinism often misapplied belief that the environment causes human events and activities

epicenter ground surface point directly above the earthquake focus, the point in the earth where the energy of the earthquake is first released

equinox an astronomical event occuring twice a year, when the subsolar point falls on the Earth's Equator and the circle of the sun's illumination passes through both poles. The vernal equinox occurs on March 20 or 21; the autumnal equinox occurs on September 22 or 23.

erosion general term for the removal of sediment by the force of water, air, or ice, or by the impact of solid particles carried by a fluid

escarpment cliff or steep rock face that separates two comparatively level land surfaces

esker long ridge formed by the stress of water that flowed underneath a glacier

estuary broadened seaward end or extension of a river

ethnicity minority group with a collective self-identity within a larger host population

eutrophication process that occurs when large amounts of nutrients from fertilizers or animal wastes enter a water body and bacteria break down the nutrients. The bacterial action depletes the water of dissolved oxygen.

export base industry or group of firms that export most of their output from a region, capitalizing on its comparative advantage and earning nonlocal revenues

external economies cost savings and other benefits that result from circumstances beyond a firm's own organization and methods of production—in particular, savings and benefits that accrue to producers in geographic settings that encompass the specialized business services that they need

F

fault offset fractures or breaks in rock where the sides of the break are displaced in any direction relative to each other

federal state governmental organization in which political power is shared and derived from both the national government and a number of subnational governments

fertility actual reproductive performance of an individual, a couple, or a population.

fertility rate number of live births per 1,000 women ages 15–44 in a given year

fjord narrow, deep ocean valley that reaches far inland and partially fills a glacial trough

firn old snow that has become granular and forms a surface layer in a glacier

First World theoretical grouping that includes the economically developed nations of Europe, Japan, North America, and Australia and New Zealand, in contrast to the developing or underdeveloped Third World; also called the global North

flood basalts huge lava flows that produce thick accumulations of basalt layers over a large area

folds rock layers lifted up or pushed down relative to the surrounding area

food chain organization of an ecosystem through which energy flows as organisms at each level consume energy stored in the bodies of organisms of the next lower level

Fordism form of economic organization put in place by Henry Ford, characterized by the vertically integrated production of homogeneous goods, mass markets, and large economies of scale

foreign direct investment (FDI) tangible investments in productive capacity made by firms from one nation in another.

fossil fuels remains of ancient plants and animals trapped in sediment that are used for fuel in the form of coal, petroleum, and natural gas

front low-pressure boundary between two unlike high-pressure air masses

G

gateway city urban area that serves as a link between one country or region and others because of its physical situation.

geographic information system (GIS) computer system used to store, revise, analyze, manipulate, model, and display geographic data.

geomorphology study of the Earth's surface features

gerrymandering redistricting of the boundaries of an electoral district to provide an unfair advantage to one political faction over others; named for a salamander-shaped district formed to favor Elbridge Gerry, a 19th-century Massachusetts politician

glacier large natural accumulation of ice on land that moves

global positioning system (GPS) space-based system of 24 satellites that provide three-dimensional positional, velocity, and time information to suitably equipped users anywhere at or near the Earth's surface

great circle largest circle that can be drawn around a sphere such as a globe; a great circle route is the shortest route between two points

green revolution sustained effort to introduce high-yield, high-protein crops in Mexico, India, the Philippines, Indonesia, Bangladesh, and elsewhere in the 1960s

greenhouse effect property of the atmosphere that allows the short-wave radiation of sunlight to pass easily to the Earth's surface but makes it difficult for heat in the form of long-range radiation to escape back toward space

greenhouse gases group of gases including water vapor, carbon dioxide, ozone, nitrous oxide, and methane

groin a pile of rocks extending out from a beach.

gross national product (GNP) sum total of the value of goods and services produced by a nation-state in one year

growth rate rate at which a population is increasing or decreasing in a given year due to natural increase and net migration, expressed as a percentage of the base population

H

habitat natural environment of a plant or animal having a certain combination of controlling physical factors

heartland theory assertion by Halford Mackinder (1861–1947) that the relative location and environmental challenges of Russia made it a nearly unconquerable area

hinterlands broadly, the area of influence surrounding a city.

hot spot intensely hot region deep within the Earth

hot springs thermal springs that discharge heated ground water at temperatures of more than 98.6 degrees Fahrenheit (average human body temperature)

humidity general term for the amount of water vapor present in the atmosphere

humus dark organic matter on or in the soil that is made up of decomposed vegetation

hurricane rotating tropical storm with winds of at least 74 mph (119km/h); called a typhoon when it forms in the Western Pacific Ocean, and called a cyclone when it forms over the Bay of Bengal and the northern Indian Ocean.

I

imperialism policy of dominating colonies or other states and maintaining those relationships to increase state power

inexhaustible resources natural resources such as solar and tidal energy that are generated continuously and production is not reduced through mismanagement

informal sector economic activities that take place beyond official record, not subject to formalized systems of regulation or remuneration

infrastructure transportation and communications networks that allow goods, people, and information to flow across space.

insolation sun's radiation emitted toward Earth

Intertropical convergence zone (ITCZ) low-pressure zone created by intense solar radiation and heating

isobar line on a map connecting all points of equal atmospheric pressure

isotherm line on a map connecting all points of equal air temperature

J–K

jet stream high-speed winds flowing in narrow zones within the upper air westerlies

karst area of land underlain by limestone formations such as sink-holes, underground streams, and caves

Kondratieff waves long-term economic periodicities (50 to 75 years) in capitalist production, including changes in innovations, output, prices, and employment, linked to the emergence of critical new industries

L

La Niña meteorological event that occurs when the equatorial waters in the Pacific Ocean become colder than normal

labor theory of value classic theorization of how value is created; it holds that all value is ultimately produced only through human labor

lagoon shallow, narrow body of water lying between a barrier island and the mainland

lahar volcanic mudflow

land reclamation process in which mined land is recontoured, resoiled, and revegetated, thus alleviating much of the degradation caused by mining

language divergence evolution of different languages over the last 100,000 years as members of cultural groups separated

language families groups of languages that diverged from a single ancestral language, for example, the Indo-European family

lingua franca established language used by a population as a common language

lithification process of formation of sedimentary rock

lithosphere general term for the entire solid Earth; in plate tectonics, the rigid outer layer of Earth shell, above the asthenosphere

localization economies cost savings that accrue to particular industries as a result of clustering together at a specific loca- tion

loess dust composed mostly of silt and clay

longshore current current in the breaker zone running parallel with the shoreline that transports and deposits sand to beaches and spits

M

mafic rocks igneous rocks rich in magnesium and iron.

magma molten, mobile rock that lies beneath the Earth's surface

map projection system of transferring information about a round object such as a globe to a flat piece of paper or other surface

mass wasting downward movement of rock, soil, and sediment in response to gravity

meanders winding bends of a river that form when it flows around a curve

merchant capitalism form of economic organization in which capital accumulation is based on trade in primary products (agricultural, fish, and forest products, minerals) and handicrafts

mesa broad, flat-topped landform surrounded by cliffs; they become buttes as they grow smaller

microclimate climate of a shallow layer of air near the ground

migration movement of people across a specified boundary for the purpose of establishing a new residence

mineral inorganic solid that has a characteristic chemical composition and specific crystal structure that affect its physical characteristics

modern movement intellectual movement with origins in the early 20th century that incorporated the idea that buildings and cities should be designed and run like machines

Modified Mercalli scale earthquake intensity scale that uses 12 intensity levels relating to the phenomena observed during an earthquake

monsoon seasonal change in the direction of the prevailing wind

mortality deaths as a component of population change

multinational corporations (MNCs) large firms with operations in more than one nation-state; also called transnational corporations

N

nation distinct society dedicated to its own region or homeland, whether or not it is a state; usually an ethnic group that shares the same language, religion, history, and icons, or symbols of their distinctiveness

nationalism concept that nations deserve the right to self-determination, as autonomous regions or as sovereign states

nation-state state in which the homeland of a nation coincides with the territory of a state

natural disaster a natural event that adversely affects humans; in the United States, the term is usually reserved for events in which more than 100 people die or more than $1,000,000 in damage occurs

natural increase surplus of births over deaths in a population in a given period

neocolonialism economic dependence of former colonies on their former colonizers

New World Order an optimistic view of the arrangement of power in the world. The New World Order ostensibly took effect in 1991, after the Union of Soviet Socialist Republics dissolved and Soviet communism was discredited. It holds that the bipolar aspects of the Cold War were over and could be replaced by increasing connections between states and nations, that supranationalist organizations would emerge to balance the superpowers, and that multinational actions would replace unilateral decisions and actions.

nonpoint source pollution pollutants that are emitted through runoff from streets and agricultural fields, for example

O

offshore banking financial activities in deregulated sites, often but not always located in small island states and nations

overurbanization condition in which cities grow more rapidly than the jobs and housing they can sustain

oxbow lake crescent-shaped lake or swamp (such as a bayou) that occupies an abandoned channel left by a meandering stream

oxidation chemical weathering process in which oxygen combines with minerals to produce mineral oxides

ozone layer stratospheric area containing ozone, which protects life on Earth by absorbing the sun's ultraviolet rays

P

Pangaea hypothetical supercontinent that is thought to be the parent of today's continents

pedology science of soil

perihelion point on the Earth's orbit at which the Earth is closest to the sun

permafrost condition of permanently frozen ground in subarctic and arctic regions

pH measure of the concentration of hydrogen ions in a solution on a scale in which 7 represents neutrality. Lower numbers indicate increasing acidity and higher numbers indicate increasing alkalinity.

photosynthesis process by which plants convert carbon dioxide to oxygen

pidgin language simplified version of language used among people who do not share a common language

plate tectonics theory that Earth's lithospheric plates slide or shift over the asthenosphere and their interactions cause geologic events such as volcanic explosions, movement of landmasses, and earthquakes

plateau extensive elevated area bounded by a steep cliff

playa temporary lake or lake bed in a desert

pluton a mass of rock formed by the cooling of magma and the crystallization of igneous rock beneath the Earth's surface

point source pollution the emission of pollutants from a specific and limited area, such as a sewage pipe or factory

pollutant substance, usually a result of human waste or human activity, that contaminates the surrounding environment. A substance can become a pollutant when its concentration increases to levels that threaten the health of living things or change atmospheric conditions.

polyglot state country with a mixture of cultural groups with different languages

population momentum tendency for population growth to continue beyond the time that replacement level fertility has been achieved because of a relatively high concentration of people in the childbearing years

primacy condition in which the population of the largest city in an urban system is disproportionately large in relation to the second- and third-largest cities in that system

primary sector cluster of economic activities that pertain to the extraction of raw materials from Earth's surface, including agriculture, forestry, fishing, and mining

prime meridian line of 0 degrees longitude, which runs through Greenwich, England

producer services service industries that cater primarily to corporate clients rather than households

production platform geographic setting involving a tightly knit cluster of towns and cities with specialized, interrelated manufacturing activities bound together by the creation and exploitation of external economies

protectionism government policy or web of policies designed to protect domestic producers from foreign competition by limiting imports through tariffs, quotas, and nontariff barriers

push-pull model theory of migration stating that circumstances at the place of origin repel, or push, people out of that place to other places that exert a positive attraction, or pull

Q–R

quotas form of protectionism in which governments limit the absolute volume of imports in an industry, effectively driving up the market price

race visible differences or phenotypes among people such as skin color, eye shape, and hair color

racism social practice of discrimination based on appearance

radiation balance condition of balance between energy coming from the sun and energy radiated and reflected from the Earth

rain forest moist, densely wooded area usually found in a warm, tropical climate; annual rainfall is about 80 inches (200 cm) a year or more

rain shadow dry region on the downwind (leeward) side of a mountain range

rank-size rule statistical regularity in the population-size distribution of cities and regions

regional trading blocs associations of countries designed to reduce protectionism and enhance economic intercourse among member states

relative humidity measure of the moisture content of the air, expressed as the amount of water vapor present relative to the maximum that can exist at the current temperature

remote sensing measurement of some property of an object by means other than direct contact, usually from aircraft or satellites

renewable resources natural resources that can be regenerated by either biologic reproduction or by environmental processes; mismanagement of these resources can cause their depletion

replacement level fertility average number of children needed to replace both parents in the population; in the United States today, total fertility rate of about 2.1 is considered to be the replacement level

residential mobility movement of households from one residential location to another within a city

resource management decisions made on what natural resources should (and should not) be developed, to what degree, in what manner, and for whom

resources substances, qualities, or organisms that have use and value to a society

Richter scale logarithmic scale of numbers that represent the relative amount of energy released by an earthquake

rift valley long valley formed by the depression of a block between two parallel faults

Ring of Fire nearly complete arc of volcanoes that circles much of the Pacific Ocean

S

salinization accumulation of soluble salts in the soil

savanna tropical grassland with widely spaced trees that experiences distinct wet and dry seasons

seamount submerged volcano

Second World now defunct theoretical grouping of the nations of the former Soviet bloc (USSR, eastern Europe, Mongolia, and Cuba) characterized by socialist economies

secondary sector industries that transform raw materials into finished goods; i.e., manufacturing and construction

securities markets one of a series of financial markets and institutions involved in buying and selling equities, foreign currency exchange, and investment management

seep small amount of water that flows slowly out of the ground and eventually into a stream

sensible heat heat felt or sensed as warmth, measurable by a thermometer

silt type of sediment transported and deposited by water, ice, or wind

sinkhole surface depression resulting from the ground collapsing into a cave

soil horizon horizontal layer of the soil that is set apart from other layers by differences in chemical and physical composition, organic content, structure, or a combination of those properties

solstice celestial event that occurs twice a year, when the sun appears directly overhead to observers at the Tropic of Cancer or the Tropic of Capricorn; summer solstice occurs on June 21 or 22, and winter solstice occurs on December 21 or 22.

spheres of influence theory popular in the 1960s that the United States, maritime Europe, the Soviet Union, and China were the Earth's four geostrategic regions

spit beach extension that forms along a shoreline with bays and other indentations

state independent political unit that claims jurisdiction over a defined territory and the people in it; used interchangeably with country.

steppe semiarid treeless regions that receive between 10 and 20 inches of precipitation yearly

strait narrow passage of water connecting two larger bodies of water

subduction process by which the downbent edge of a crustal plate is forced underneath another plate

sublimation process by which water vapor (gas) changes to ice (solid) or vice versa

sustained yield amount of food and material species that can be harvested annually without depletion

T

tariffs a surcharge on imports levied by a state, a form of protectionism designed to increase imports' market price and thus inhibit their consumption

tarn small lake that occupies a basin in a cirque or glacial trough

temperature inversion atmospheric condition, caused by rapid reradiation, in which air at lower altitudes is cooler than air higher up

territoriality societal action of defending a state's territory from outside threats

tertiary sector industries and activities that produce and transmit intangibles, i.e., services including finance, producer, and consumer services, transportation and communications, education, health care, nonprofit organizations, and the public sector

thermal inertia resistance of land and water to temperature change

Third World theoretical grouping roughly synonymous with the global South, essentially comprising former European colonies in Latin America, Africa, and Asia

tide regular rise and fall of the ocean level, caused by gravitational pull between the Earth and moon in combination with Earth rotation

till unsorted mixture of rock fragments deposited beneath moving glacier ice

time-space compression idea that distance can be measured in terms of the time required to cross it via transportation and communications

toponymy study of place names

tornado violently rotating column of air that descends to the ground during intense thunderstorm activity

total fertility rate average number of children that would be born alive to a woman during her lifetime if she were to pass through all her childbearing years conforming to the age-specific fertility rates of a given year

tsunami series of ocean waves caused by the vertical displacement of the seafloor during an earthquake or volcanic explosion

tundra cold region characterized by low vegetation

U–V

urban heat island dome of heat over a city resulting from urban activities and conditions

urban system group of cities in which a disproportionate part of the world's most important business is conducted

vertical integration the consolidation of different stages in the production or distribution process within the confines of an individual corporation

volcanic arc long, narrow chain of composite volcanoes that forms parallel to a convergent boundary

W

water table upper level of the saturated zone of soil and rock, where it meets unsaturated soil and rock

wave refraction angle at which waves meet shorelines, usually about 5 degrees from parallel

weathering geomorphic process that causes physical disintegration and chemical decomposition of rock and soil

wetland area of land covered by water or saturated by water sufficiently enough to support vegetation adapted to wet conditions

world system theory that the only meaningful unit of social analysis is the global system of states and markets, and that individual places can only be understood via their position within this worldwide system

X–Z

xerophytes plants that can survive in a dry environment

zone in transition area of mixed commercial and residential land uses surrounding the central business district of a city

zoning process of subdividing urban areas as a basis for land-use planning and policy

INDEX

A

ACKNOWLEDGEMENTS

Michael Bell
International Research Institute
for Climate Prediction

Carl Haub
Population Reference Bureau

Jason Kowal
TeleGeography Research

Marco Noordeloos
ReefBase Project

Christopher R. Scotese
PALEOMAP Project

Gregory Yetman
CIESIN (Center for International
Earth Science Information Network)

Jeroen Zandberg
UNPO (Unrepresented Nations
and Peoples Organization)

CREDITS

COVER: Printed from digital image © 1996 CORBIS

PART I: WHAT IS GEOGRAPHY?
6, NASA Goddard Space Flight Center, Image by Reto Stockli, Enhancements by Robert Simmon; 8, SeaWiFS; 10, By permission of the British Library: 18, David G. Long, Brigham Young University; 19, United States Census Bureau, Population by Census Tract, 2000; 20, Joseph H. Bailey; 21 (up), Library of Congress; 21 (low), The Bodleian Library, University of Oxford; 22, Courtesy Jim Siebold; 23 (reefs), World Resources Institute, 2002 (data), ReefBase, 2002 (maps); 23 (map), United States Department of the Interior, Geological Survey, 7.5 minute series (topographic). Eden Prairie, Minnesota, 1993; 24 (up), S.C. DeBrock/Lockheed Martin; 24 (low), WorldSat International/Photo Researchers, Inc.; 25, Stuart Armstrong/WorldSat International, Inc.

PART II: PHYSICAL GEOGRAPHY
32, George Ranalli/Photo Researchers, Inc.; 41, NASA image courtesy Greg Shirah, GSFC Scientific Visualization Studio, based on data from the TOMS science team; 52, National Oceanic and Atmospheric Administration, Surface Weather map and Station Weather at 7:00 AM EST, Tue., Nov. 5, 2002; 64, AVHRR Pathfinder SST data are courtesy of NASA's Physical Oceanography Distributed Active Archive Center (PO.DAAC), Jet Propulsion Laboratory/California Institute of Technology in collaboration with NOAA's National Oceanographic Data Center (NODC) and the University of Miami's Rosenstiel School of Marine and Atmospheric Sciences; 71, International Research Institute for Climate Prediction, CPC Merged Analysis of Precipitation (CMAP) datasets for January and July 1979-2000, November 2004; 72, NASA/NOAA; 77, Bruce Dale; 84, John Conrad/CORBIS; 89, James P. Blair; 91, Jim Brandenburg; 93, Priit

Vesilind/NGS Image Collection; 94, Raymond Gehman; 95, Richard T. Nowitz/NGS Image Collection; 106-107, PALEOMAP Project (www.scotese.com); 115, James L. Stanfield; 119, Bettmann/CORBIS; 122, Joseph H. Bailey; 124, Frans Lanting; 125, National Geographic Atlas of the Ocean, p. 86 and United States Geological Survey; 126, Steve Raymer; 127, Walter Meayers Edwards; 128, Craig Blacklock; 129, Howell Walker; 131, National Geographic Photographer Jodi Cobb; 138, David S. Boyer; 140, Printed from digital image © 1996 CORBIS; 146, Raymond Gehman; 147, Jacob Gayer (photo), National Speleological Survey, Distribution Map of Caves and Cave Animals in the United States, 1999 (data); 149, United States Geological Survey, National Atlas; 150, George F. Mobley; 151, Dick Durrance II; 152, Steve Raymer; 154, Richard C. Johnson/Visuals Unlimited; 160, U.S. Geological Survey; 161, United Nations Environment Programme-World Conservation Monitoring Centre, 2001 (data); ReefBase, (maps) 162, National Geographic Photographer Jodi Cobb; 163, Gregory Heisler; 164 & 166, George F. Mobley; 170, Marli Miller/Visuals Unlimited.

ART III: HUMAN GEOGRAPHY
184, National Geographic Photographer Jodi Cobb; 192-194, Population Reference Bureau; 196, Center for International Earth Science Information Network (CIESIN), Columbia University, Gridded Population of the World v.3, 2005; 198, Steve McCurry; 204, CIA World Factbook; 207, Sam Abell; 210, Population Reference Bureau; 219, Gordon Gahan; 223, Daniel R. Westergren, NGS; 227, Jim Richardson; 236, Thomas J. Abercrombie; 241 & 242, James L. Stanfield; 245, Ingrid Booz Morejohn/PictureWorks; 248, James L. Stanfield; 252, David Butow/CORBIS SABA; 254, United States Department of Agriculture, Economic Research Service, 2000; 258, Energy Information Administration (estimates from Oil and Gas Journal, 1/1/05); 262, National Geographic Photographer Jodi Cobb; 264, CIA World Factbook; 265, Bruce Dale; 270, Digital Vision/Getty Images; 273, TeleGeography Research, PriMetrica, Inc.; 275, Michael S. Yamashita/CORBIS; 281, Europa; 286, United Nations, Statistics Division, Millennium Development Indicators: World and Regional Groupings, 2004; 291, World Bank, World Development Indicators, 2005 (GNI), Energy Information Administration (consumption); 292, World Bank, World Development Indicators, 2005, Table 4.17: External debt management; 293, Michael Melford; 295, David Ball/CORBIS; 298, Bruce Dale; 300, James L. Stanfield; 308, Stuart Franklin; 312, David Keaton/CORBIS; 316, CORBIS; 323, UNPO (Unrepresented Nations and Peoples Organization); 328, Bob Krist/CORBIS; 333, Lester Lefkowitz/CORBIS; 337, World Trade Organization, International Trade Statistics 2004; 341, CHINA PHOTOS/Reuters/CORBIS; 344, United Nations, Statistics Division, Millennium Indicators, Water, percentage of population with access to improved drinking water sources, total (WHO-UNICEF), 2002; 345, Kevin Fleming; 346, Michelle Barnes; 348, Scorecard, the Pollution Information Site; 349, United States Department of Agriculture, Natural Resources Conservation Service, Global Desertification Vulnerability, 1998; 350, Bruce Dale; 351, Pete Saloutos/CORBIS; 353, James P. Blair.

PART IV: PLACES
360, Don Foley.

The text for the chapter in Part III, "Urban Geography" was derived from P.L. Knox and S.A Marston, *Human Geography: Places and Regions in Global Context*, Prentice Hall, 1988.

One of the world's largest nonprofit scientific and educational organizations, the National Geographic Society was founded in 1888 "for the increase and diffusion of geographic knowledge." Fulfilling this mission, the Society educates and inspires millions every day through its magazines, books, television programs, videos, maps and atlases, research grants, the National Geographic Bee, teacher workshops, and innovative classroom materials.

The Society is supported through membership dues, charitable gifts, and income from the sale of its educational products. This support is vital to National Geographic's mission to increase global understanding and promote conservation of our planet through exploration, research, and education.

For more information about the National Geographic Society and its educational programs and publications, please call 1-800-NGS-LINE (647-5463), or write to the following address:

National Geographic Society, 1145 17th Street N.W., Washington, D.C. 20036-4688 U.S.A.

Visit the Society's Web site at www.nationalgeographic.com.

PUBLISHED BY THE NATIONAL GEOGRAPHIC SOCIETY

John M. Fahey, Jr., *President and Chief Executive Officer*
Gilbert M. Grosvenor, *Chairman of the Board*
Nina D. Hoffman, *Executive Vice President*

Prepared by the Book Division

Kevin Mulroy, *Senior Vice President and Publisher*
Kristin Hanneman, *Illustrations Director*
Marianne R. Koszorus, *Design Director*
Carl Mehler, *Director of Maps*
Barbara Brownell Grogan, *Executive Editor*

Staff for this book

Scott Mahler, *Project Editor*
Martha Sharma, *Chief Consultant and Text Editor*
Cinda Rose, *Art Director*
Kristin Hanneman, *Illustrations Editor*
Barbara Seeber, Jane Sunderland, *Contributing Editors*
Gregory Ugiansky, *Map research and production*
Matt Chwastyk, XNR Productions, *Map edit, research, and production*

Britt Griswold, Cindy Min, *Graphic Artists*
Meredith Wilcox, *Illustrations Specialist*
Ric Wain, *Production Project Manager*
Sanaa Akkach, *Editorial Assistant*
Cataldo Perrone, *Design Intern*

Manufacturing and Quality Control

Christopher A. Liedel, *Chief Financial Officer*
Phillip L. Schlosser, *Managing Director*
John T. Dunn, *Technical Director*

Library of Congress
Cataloging-in-Publication Data
National Geographic almanac of geography
p.cm.
Includes bibliographical references
ISBN 0-7922-3877-X
1. Geography--Handbooks, manuals, etc. I. Title: Almanac of geography.
G123.N37 2005
910--dc22   2005049129